Advances in Intelliger

Volume 1330

The series "Advances in Intelligent Systems and Computing" contains publications on theory, applications, and design methods of Intelligent Systems and Intelligent Computing. Virtually all disciplines such as engineering, natural sciences, computer and information science, ICT, economics, business, e-commerce, environment, healthcare, life science are covered. The list of topics spans all the areas of modern intelligent systems and computing such as: computational intelligence, soft computing including neural networks, fuzzy systems, evolutionary computing and the fusion of these paradigms, social intelligence, ambient intelligence, computational neuroscience, artificial life, virtual worlds and society, cognitive science and systems, Perception and Vision, DNA and immune based systems, self-organizing and adaptive systems, e-Learning and teaching, human-centered and human-centric computing, recommender systems, intelligent control, robotics and mechatronics including human-machine teaming, knowledge-based paradigms, learning paradigms, machine ethics, intelligent data analysis, knowledge management, intelligent agents, intelligent decision making and support, intelligent network security, trust management, interactive entertainment, Web intelligence and multimedia.

The publications within "Advances in Intelligent Systems and Computing" are primarily proceedings of important conferences, symposia and congresses. They cover significant recent developments in the field, both of a foundational and applicable character. An important characteristic feature of the series is the short publication time and world-wide distribution. This permits a rapid and broad dissemination of research results.

Indexed by SCOPUS, DBLP, EI Compendex, INSPEC, WTI Frankfurt eG, zbMATH, Japanese Science and Technology Agency (JST), SCImago.

All books published in the series are submitted for consideration in Web of Science.

More information about this series at http://www.springer.com/series/11156

Álvaro Rocha · Carlos Ferrás ·
Paulo Carlos López-López ·
Teresa Guarda
Editors

Information Technology and Systems

ICITS 2021, Volume 1

Editors
Álvaro Rocha
ISEG
University of Lisboa
Lisboa, Portugal

Carlos Ferrás
Geografía
University of Santiago de Compostela
Santiago de Compostela, Spain

Paulo Carlos López-López
Ciencia Política y Sociología
Universidade de Santiago de Compostela
Santiago de Compostela, Spain

Teresa Guarda
Campus UPSE
Universidad Estatal Peninsula de Santa
Elena Sistemas
La Libertad, Ecuador

ISSN 2194-5357 ISSN 2194-5365 (electronic)
Advances in Intelligent Systems and Computing
ISBN 978-3-030-68284-2 ISBN 978-3-030-68285-9 (eBook)
https://doi.org/10.1007/978-3-030-68285-9

This Springer imprint is published by the registered company Springer Nature Switzerland AG
The registered company address is: Gewerbestrasse 11, 6330 Cham, Switzerland

Preface

This book is composed by the papers written in English and accepted for presentation and discussion at The 2021 International Conference on Information Technology & Systems (ICITS'21). This conference had the support of the Universidad Estatal Península de Santa Elena (UPSE), Information and Technology Management Association (ITMA), IEEE Systems, Man, and Cybernetics Society, and Iberian Association for Information Systems and Technologies (AISTI). It took place in Península de Santa Elena, Ecuador, February 10–12, 2021.

The 2021 International Conference on Information Technology & Systems (ICITS'21) is an international forum for researchers and practitioners to present and discuss the most recent innovations, trends, results, experiences, and concerns in the several perspectives of information technology and systems.

The Program Committee of ICITS'21 was composed of a multidisciplinary group of 168 experts and those who are intimately concerned with information systems and technologies. They have had the responsibility for evaluating, in a 'double-blind review' process, the papers received for each of the main themes proposed for the conference: A) Information and Knowledge Management; B) Organizational Models and Information Systems; C) Software and Systems Modeling; D) Software Systems, Architectures, Applications and Tools; E) Multimedia Systems and Applications; F) Computer Networks, Mobility and Pervasive Systems; G) Intelligent and Decision Support Systems; H) Big Data Analytics and Applications; I) Human–Computer Interaction; J) Ethics, Computers & Security; K) Health Informatics; and L) Information Technologies in Education.

ICITS'21 also included several workshop sessions taking place in parallel with the conference ones. They were sessions of WMETACOM 2021 – 4th Workshop on Media, Applied Technology and Communication.

ICITS'21 received more than 300 contributions from 28 countries around the world. The papers accepted for presentation and discussion at the conference are published by Springer (this book) and by AISTI, and will be submitted for indexing by ISI, Ei Compendex, Scopus, DBLP, and/or Google Scholar, among others.

We acknowledge all of those that contributed to the staging of ICITS'21 (authors, committees, workshop organizers, and sponsors). We deeply appreciate their involvement and support that was crucial for the success of ICITS'21.

February 2021

Álvaro Rocha
Carlos Ferras Sexto
Paulo Carlos López
Teresa Guarda

Contents

Intelligent and Decision Support Systems

Multimedia Systems and Applications

Software Systems, Architectures, Applications and Tools

Information and Knowledge Management

A Semantic Frame to Represent and Instantiate ASTM Laboratories

Ingrid-Durley Torres[1(✉)] and Jaime A. Guzmán-Luna[2]

[1] Universidad Católica Luis Amigó, Tv. 51a #67B-90, Medellín, Antioquia, Colombia
ingrid.torrespa@amigo.edu.co
[2] Universidad Nacional de Colombia, sede Medellín, Facultad de Minas, Kra 80 #65-223, Medellín, Antioquia, Colombia
jaguzman@unal.edu.co

Abstract. Modeling semantically lab test regarding civil engineering field so far have not been formalized, it demands to represent key aspects on every experimental activity which is defined from its documented side as well as the one from its workflow. The documented side represents all the static information, written in legible letters and comprehensible (so far) by humans; whereas the workflow models all the dynamic information, which means the procedure itself being organized in each activity, with its input data, operations, and output data. In this article, it is described the semantic representation modeling, being supported by a group of ontologies that allow to symbolize and reuse with enough level of detail all the documented information and workflow on experimental practice in civil engineering, based on the essays regarding ASTM regulations.

Keywords: Semantic · Civil engineering · ASTM

1 Introduction

The ASTM [2] policies have been developed by the *American Society for testing* and *Materials* (therefore, its denomination) as standards regarding the design of laboratory testing, establishing within them the most adequate conditions and procedures to obtain good results. The ASTM policies are classified as *test method*, *specification*, *classification*, *practice*, *guide* and *terminology*. The ASTM policies on testing mode, typically include a brief description and an organized procedure to determine a material's properties, material's assembly, or a product itself. The instructions to make the test must include all the essential details regarding the devices, sample, procedure, and the needed estimates to reach a satisfactory precision and bias. The ASTM test method must represent an accord, regarding the best possible testing procedure available now for provided usage and it must be supported by experience and adequate data obtained in the testing. In general, the ASTM policies on testing mode, constitute a guide that can propose a series of options or instructions without recommending a defined taking course; which means, it may as well constitute a guide but not necessarily establishing an only way of designing the laboratory practice, since in the majority of cases, this

Á. Rocha et al. (Eds.): ICITS 2021, AISC 1330, pp. 3–13, 2021.
https://doi.org/10.1007/978-3-030-68285-9_1

shall depend on the available device, or in the input data value, or in the executive command alternatives, or even all of them together as a combination of each.

Parallel, civil engineering [5] as the field that uses knowledge about calculations, mechanic, hydraulic and physics in order to be in charge of the designing, building and maintenance of the deployed infrastructure in the environment [12], it incorporates various branches in which different characteristics are analyzed, behaviors and determined elements properties; among them the most recognizable are [13]: ground analysis, structure analysis, fluid analysis and material analysis. Each one of these branches implement their own laboratories and experimental activities that specify observation, procedure and results of corresponding material analysis oriented by the ASTM policies.

The first intention is to demonstrate that it is possible to represent and instantiate the ASTM laboratory test, using the framework, and secondly, to improve the experimental information quality and those of the laboratory services that support this process. The implementation ends up being useful to evidence the automatization in management regarding the trial's instantiation, achieved by the interpretation of software agents.

Under the previous scenario, this article describes the semantic model which was adopted as a representation strategy of the ASTM policy in its most abstract level; as well as it extends such representation in order to model from each branch of the civil engineering field. The corresponding specific trial, extending the proposal includes achieving the already documented trial with the reached findings done in the laboratory.

This article is structured as follows: Sect. 2 describes the related works on experimental ontologies. Section 3 describes the framework of the ontology. Section 4 proposes a case study to validate the ontology. And finally, in Sect. 5 describe this work conclusions and future Works.

2 Related Works

Nowadays, there is an increasing number of ontological proposals of domain, specifically oriented towards the representation of experimental information, especially in the biology field With this, it is guaranteed that the common experimental metadata are restrained, trying in one way the ambiguity of natural languages by means of a standard taxonomy. Following, some of these efforts are cited.

OBI (Ontology for Biomedical Investigations) [9], is one of the most recent proposals to model the design on a biomedical research: the protocols, the instruments, and the used materials, besides the generated data in the biomedical field. This ontology allows the registration of all the structure of scientific investigations related with the how and why the research was executed, the reason why the field was chosen, the conclusions reached, the base of these conclusions, among others.

SMART (http://vocab.linkeddata.es/SMARTProtocols) [6], represents an ontology of experimental protocols in the field of plant experimentation, it divides the ontology in two fields: i) A protocol representation as a document, and ii) the procedure's representation (Workflow). To service Document: use the class *iao:InformationContentEntity* and its subclasses (*iao:Document*, *iao:DocumentPart*, *iao:TextualEntity* and *iao:DataSet*); those which were imported from Information Artifact Ontology (IAO https://code.google.com/p/information-artifact-ontology/). For the Workflow, the used representation proceeds from *p-plan:Plan* [7], *p-plan:Step* y *p-plan:Variable* des all of them belonging to P-Plan Ontology (P-Plan).

Expo (http://expo.sourceforge.net/) [14]); is an ontology which is based in the ideas regarding science philosophy: logic, probability, methodology and epistemology. EXPO [10] describes the general knowledge about the world, analyzes existent ontologies such as the bio-ontologies and theories as the experiments design; besides, it proposes knowledge in proper abstraction levels, making it feasible and desirable, as it is used in the experimental beginnings in order to have an efficient analysis, to reach the entry and interchange of results, contributing with the science objective. EXPO has used the methodology bottom-up on its design. The first step on this focus deals with anchoring EXPO to a standard superior ontology, one that describes the general knowledge about the world. Such standard superior ontology offers: template structures, terms and relations, along with the definitions and fundamental axioms in order to model the domain [3]; to achieve EXPO connecting to a concept of higher level, SUMO was chosen from which it was based in order to build what today is known as EXPO.

EXACT [11], is an ontology that that is used as the bases of a method to represent a laboratory experiment in the biological and bioinformatical field. EXACT provides a model to describe the experiment steps, it can be used in the representation of fully formalized protocols, or it can be combined with other formalities in order to describe biomedical researches and that way reproduce it in other environments.

MIBBI (http://www.dcc.ac.uk/resources/metadata-standards/mibbi-minimum-information-biological-and-biomedical-investigations) [14], to increase the clarity of projects associated to the domain of biological and biomedical science. Throughout biological and biomedical science, this representation is used as checklist that specify key information to be included in the experimental results report. This representation is available to experimentalists, analysts, editors and providers. It is about a type of (MI) "minimal" checklist, information regarding biological domains, modeled as semi-structured information presented in Excel sheets and XML Schema.

P-PLAN, is an extension PROV-O ontology [7], which "expresses the data model that provides a group of classes, properties and restrictions that can be used to represent and exchange precedence information, generated on different systems and different contexts" [3]. P-PLAN was developed as an ontology that describes the abstract scientific workflow, using the plans as a registry of affirmations that records the steps and precedence when executed. Augmenting PROV with Plans in P-PLAN.

As it is possible to appreciate, it already exists semantic representations that model experimental activities among others [3]; however, due to the fact that knowledge on each domain is different and given the adopted scope in this research, it focuses civil engineering, where there is no regulation regarding ontological representation. It is required to formulate an innovative semantic proposal.

3 Semantic Framework

With the objective of proposing a semantic representation made by a group of ontologies [1], it is submitted to one of the methodological guides more commonly accepted by the Ontology Development 101 [8]. The basic starting point of this methodology involves defining the domain of knowledge that it wishes to cover, in which from the beginning it specifies the ASTM Test type trials.

The second stage, consisted on verifying the reuse of existent ontologies, which is why they were considered on a first stage SMART- Protocol (prefix *sp*) and P-Plan (prefix *P-Plan*) ontologies, shown in Sect. 2. Both ontologies were compared with the needs on ASTM Test experimental scope, as it described on the third step of the methodology: Enumerate important terms in the ontology; the analysis made, verified that which was requested by the domain to represent, facing offers given by each ontology on its classes and relationships. Although the two of them ended up being required, they were not enough, which is why an import made from a new ontology was proposed one that references all the all elements being requested by the actual representation domain, such ontology was identified with the prefix *sc* that corresponds to e-Science Civil engineering, and over this one, it was established the achievement of step 4 of the methodology, defining its classes and the class hierarchy.

However, for such characterization, these were not the only ontologies that were reused. On Fig. 1, it is graphically described the ontology sc, on the immerse table it is cited each one of the prefixes that correspond to an ontology of precedence, imported, highlighting that in almost is totality, they were suggested by sp. Meanwhile, in Created, there exists a correspondence among the ontologies that were created in order to define better the aspects related to the ASTM domain. This description considered also, the definition of the properties of classes—slots. In summary, it was decided to import the Research Object ontology [1, 17] (prefix *ro*, identified in *sc* as OI by its translation to the Spanish language as Objeto de Investigación), to represent the element as a digital container of resources aggregation; to symbolize the aggregations of resources it was imported the Object Reuse ontology and Exchange (prefix ore), where specifically, ore: DataSet was used to define a data aggregation which group represents something common. To model the metadata, it was used various representations: the ontology friend to friend (prefix *foaf*), to relate the OI author as a creator agent, whereas with dublinCore (prefix *dc*), it was used the primitive datatime data, which made possible to represent the OI's date of birth.

In that scenario, the e-Science Civil engineering (prefix *sc*), was created as a result of the need on this domain, it is built on the following way: the experimental protocol, it is generated as a RO subclass, embodying within it *sp:Document* (*sp* is the precedence of Smart-Protocol), accompanied by all the textual information concepts, which are required by the experimental scope on the ASTM policy. In this case, it is also adopted elements such as Textual Entity, preceding from the Information Artifact Ontology, (*iao:Textual Entity*), which concretions are glyph patrons destinated to be

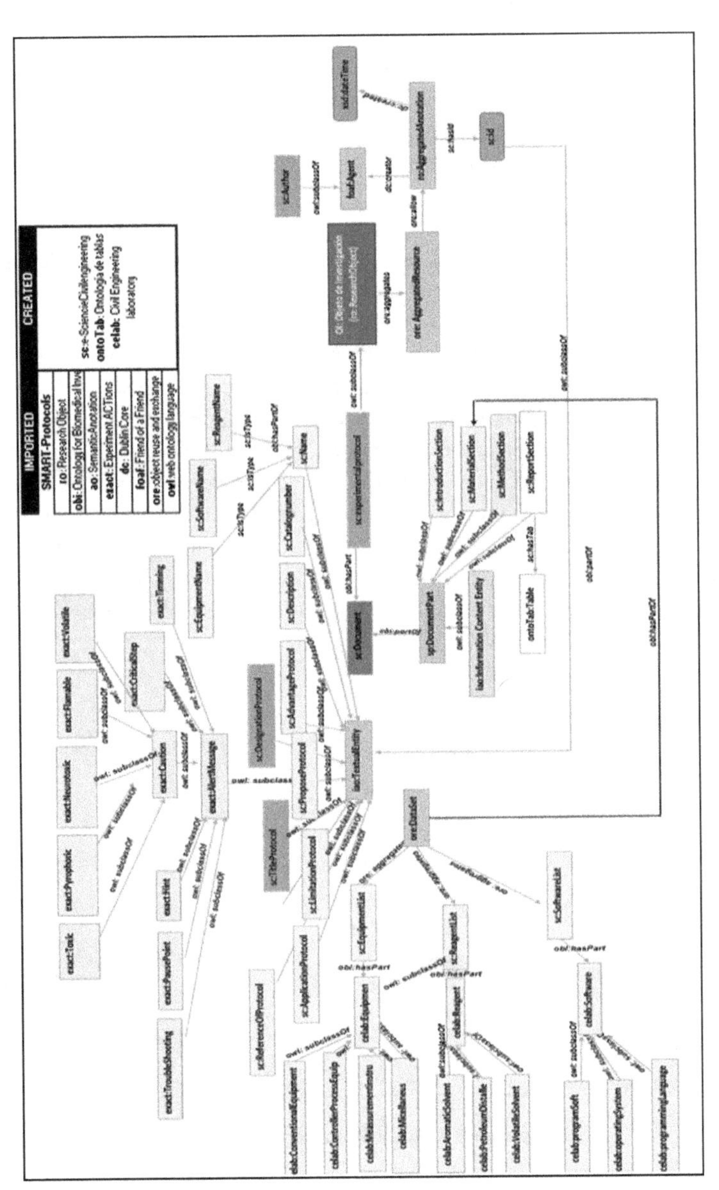

Fig. 1. Semantic representation map *sc:Document.*

interpreted as words, formulations, etc., whereby it was possible to represent all the textual entities, including the OI's identifier (*id)*; *sc*, as well as it was accompanied by ontologies that were exclusively developed for this own domain, such as the Ontology of Tables ontology (prefix *ontotab*), to represent the defined tables in the reports section and the ontology itself. The same way it was built the OI's documental aspect, described with *sc* from *sp*, it also built the workflow [7], from *p-plan*. This way, the workflow corresponds to a whole plan, which is described towards a composition of steps, which can have variables that act as input and output; these variables are classified by defined types from another unit ontologies (refix: *uo*), created specifically for ASTM. It even can be defined when a step is preceded by other, allowing with this, to order the step sequences. It can be seen in Fig. 2. Although, until today the execution part has not been properly described, it must be taken into account that the *OI* must have the necessary information in order to implement computational calculations to obtain in an automatic way the output value on each variable, being the result of a computational calculation. The implementation is shown in the Fig. 3, where, even though *P-plan* was imported, there are *sc:workflow* specifications being generated. This way, as an example, *sc:Formula* generated as a new concept of *sc:Workflow*, takes all the information that is needed to make the data conversion and the corresponding execution of some OI. The enactment specifies the details about how to accede the software, it means that it also keeps primarily the access protocol information and the message formats, the serialization, the transport, and the routing. In this case, the process is considered as mapping from an abstract specification to a concrete one of those elements that are described which are needed to interact with the software in particular, for our purposes the input and output variables of atomic processes are represented in this model as *sc:Step*.

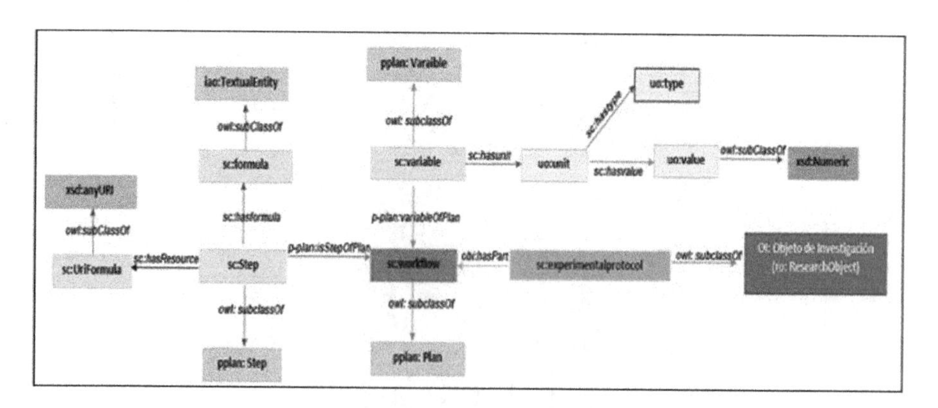

Fig. 2. Semantic representation map *sc:Workflow*.

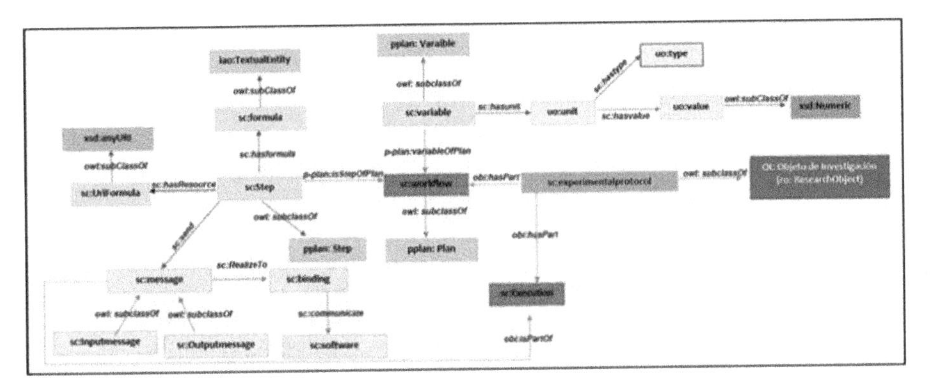

Fig. 3. Mapa de representación semántica *sc:Execution*

4 Use Case

One of the most used trials in ASTM experimental practices, is being represented in policy ASTM 4318 [4], titled Standard Test Methods for Liquid Limit, Plastic Limit, and Plasticity Index of Soils (seen in Fig. 4). These test methods cover the determination of the liquid limit, plastic limit, and the plasticity index of soils. This trial has three main steps, first it must be calculated the liquid's limit, then the plastic's limit to finally deliver the plastic index. However, to begin with step one, it is optional to use two possible methods that can be carried on, both of them with the same output variable, then the idea is to follow one of them in the practice. On Fig. 5, it is shown the options from their workflow. Both representations are built by means of a graphic interface, which does not require any technical or semantical knowledge from the laboratory technician or engineer who is in charged with the trial practice.

This way, the final user only fills in the fields with the corresponding information to its specific test practice. An important aspect to highlight is that the system has each policy sheet (as abstract ontology) already loaded, this way the laboratory technician will first choose the ASTM 4318 trial from those available and immediately he will proceed to document all the information that are required in the fields previously established by the abstract sheet. From those fields some are already designed and cannot be edited, these are, Tittle, description, reach, limitations, advantages, equipment (with no brand nor description being specified), among others.

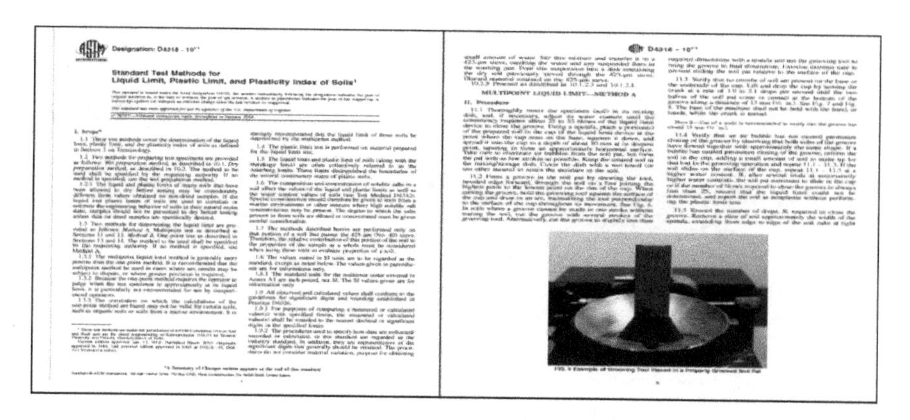
Standard Test Methods for
Liquid Limit, Plastic Limit, and Plasticity Index of Soils

Fig. 4. ASTM 4318 as document portion

The data that can be filled in are shown with numbers: 4, 5, 6, and 7 (just to mention some), in Fig. 6. A laboratory technician user is also in charge of using the previously designed workflow on an abstract way for each trial for this, it must be completed with the concrete data values for each variable (see Fig. 8). This information will allow to execute the software subsequently. In this case is tied to Wolfram Alpha's service, which will take care of solving the mathematical formula (Fig. 7, No. 17), associated to the values given to each one of the variables of such formula. Implementation or not of each one of the steps and variables. It is chosen only if the step has been enacted. (Figure 7, No. 13) in case it has not, for each variable (Fig. 7, No. 14) the value is entered (Fig. 7, No. 15) and the measurement unit (Fig. 7, No. 16). This will be done keeping the established precedence among the steps. In case of not existing such restriction, the laboratory technician would have to complete the sequence on a coherent way according to the order in which the experimental practice has been carried on in the laboratory. Finally, Fig. 7 shows a portion in protégé on the *sc:-workflow* ontology, being instantiated with the amounts on 4318 policy with the concrete values.

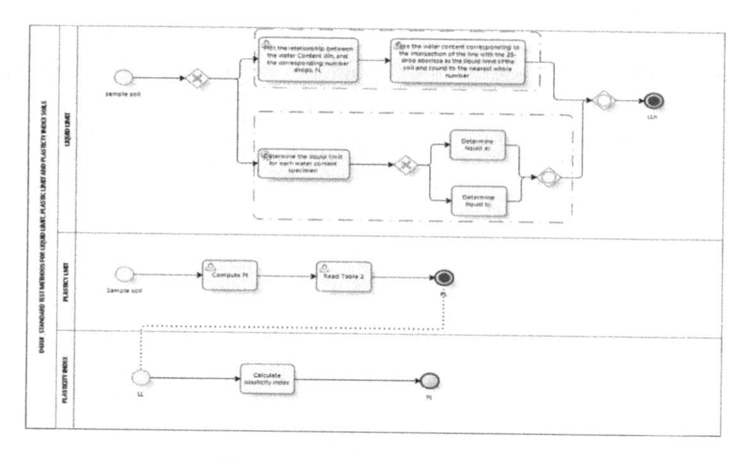

Fig. 5. ASTM 4318 as workflow

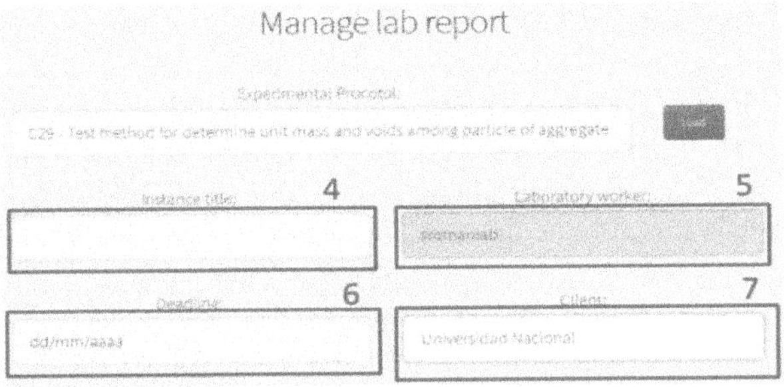

Fig. 6. Interface sc:Document ASTM 4318 for the laboratory technician

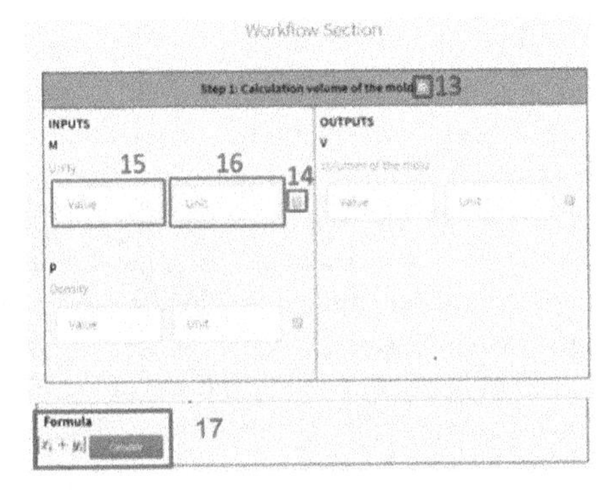

Fig. 7. Interface sc:workflow ASTM 4318 workflow for the laboratory technician

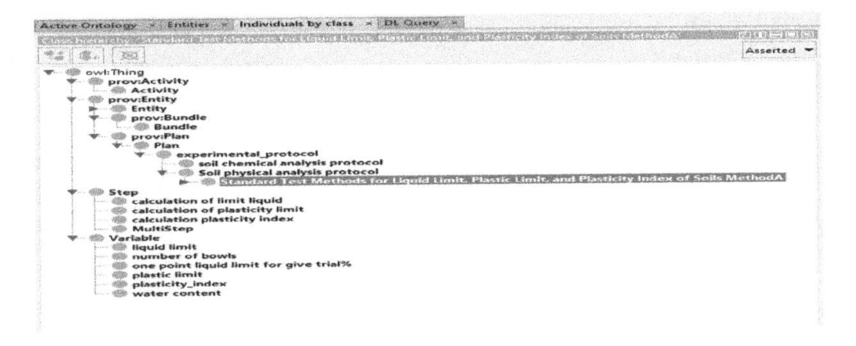

Fig. 8. Protégé sc:Workflow portion

5 Conclusions and Future Works

Given the fact that as far as the search reached and the bibliographical review done in this work, it was not found any ontologies or vocabulary with a semantic specific value to the meaning of concepts and relationships of civil engineering experimental field, it was included some ontologies with contributions to other fields, in order to use them as base for the construction of a new semantic representation that would answer to such requirements. As a result, this whole has been covered by means of the construction of an ontological infrastructure which control has been framed in OI, that acts as an information container, being reported on experimental practices of ASTM trials. To define the problem, it was considered the distribution and access to information from a "static" perspective as document and "dynamic" as workflow primarily inspired in SMART-Protocols. The reuse is represented and strategically split in three levels of abstraction: The abstract representation, which represents that of what will be understood semantically as ASTM trial; the semi-instantiated that properly represents an specific ASTM with a designation but without data, it means, the policy as it is published by ASTM and finally, the instantiated representation, which is the trial with all the concrete data as a result of the laboratory practice. In Fig. 1, the table titled imported/created, allows to cite all imported ontologies with their prefix and full name. In imported, all the ontologies that exist in the literature, but that were necessary to consider, are cited, highlighting that almost in their entirety, they are suggested by Smart-Protocols. Meanwhile, in Created, there is some correspondence with the ontologies that were created to better define the aspects of the civil engineering domain. With the above it is shown that although there are semantic models to reuse, they were not sufficient, given the conditions of the new domain, such is the case of the tables and the taxonomic description of the laboratory equipment, typical of an experimental practice.

In this work, only ASTM Test protocols have been represented, that is, laboratory practices. It is valuable for the domain to show that the model can be extended to NSR-10 and AASHTO, for example. Such a representation can also incorporate considerations of the WF description, specifically in the logical control mechanism, where until now complex workflows are not possible, so parallel, repetitive tasks or even human activities can be included. Besides, although the developed system addresses heterogeneous data with semantics, it does not incorporate the necessary mechanisms to deal with problems related to data integration. Integration can be defined as the process of combining data that is stored in different repositories.

References

1. Alper, P., Belhajjame, K., Goble, C.: Small is beautiful: summarizing scientific workflows using semantic annotations. In: IEEE International Congress on Big Data, Santa Clara, CA, pp. 318–325 (2013)
2. American Society for Testing and Materials: Form and Style for ASTM Standards (2015). www.astm.org

3. Arp, R., Barry, S.: Function, role, and disposition in basic formal ontology. Nature Precidings pre-publication research and preliminary findings, June 2008
4. ASTM 4318, E. ASTM 4318, 10E Standard Test Methods for Liquid Limit, Plastic Limit, and Plasticity Index of Soils (p. 16). ASTM, West Conshohocken United States (2014)
5. Beeby, W., Narayanan, R.S.: Introduction to Design for Civil Engineers, 1st Edición, Edición Kindle (2001)
6. Giraldo, O.: Smart protocols ontology. In: 4th Workshop on Linked Science 2014- Making Sense out of Data (LISC 2014), Italy, October (2014)
7. Garrijo, D., Gil, Y.: Augmenting PROV with plans in P-PLAN: scientific processes as linked data. In: Proceedings of the Second International Workshop on Linked Science Tackling Big Data, in conjunction with 11th International Semantic Web Conference, Boston, MA, pp. 25–33 (2012)
8. Nataly, F., McGuinness, L.: Ontology Development 101: A Guide to Creating Your First Ontology (2001). http://protege.stanford.edu/publications/ontology_devdevelopm/ontology101.pdf
9. OBI: The OWL of Biomedical Investigations. The ontology for biomedical investigations. PLoS One. **11**(4), 56–68 (2016). https://journals.plos.org/plosone/article?id=10.1371/journal.pone.0154556
10. Soldatova, L.N., King, R.D.: An ontology of scientific experiments. J. Royal Soc. Interface **3** (11), 795–803 (2013)
11. Soldatova, L., Aubrey, W., King, R.: The EXACT description of biomedical protocols. J. Sci. Math. Bioinform. **24**(13), 295–303 (2006). https://www.ncbi.nlm.nih.gov/pmc/articles/PMC2718634/
12. UNAM: Manual. Manual de prácticas de Laboratorio (2019). http://www.acatlan.unam.mx/licenciaturas/30/
13. UNAL: Laboratorios. Laboratorios Universidad Nacional de Colombia, sede Medellín. Facultad de Minas (2013). http://www.minas.medellin.unal.edu.co/dirlab/index.php/laboratorio
14. W3C: MIBBI. HCLSIG/PharmaOntology/Meetings/2010-05-06 Conference Call (2010). https://www.w3.org/wiki/HCLSIG/PharmaOntology/Meetings/2010-05-06_Conference_Call

Ethics in Big Data: Myth or Reality

A. A. Balyakin(✉), M. V. Nurbina, and S. B. Taranenko

NRC Kurchatov Institute, 1, ac. Kurchatov sq., Moscow 123182, Russia
Balyakin_AA@nrcki.ru

Abstract. We discuss the risks and challenges associated with digitalization process. Big data technology was considered as the most characteristic trait of such a process. Legal aspects of big data regulation are considered. The authors focus their attention on the transition to a new, "impersonal" form of regulation, tied to algorithmic approaches. It is shown that the practice of prohibitions and restrictions is not effective. Authorities act in legal sphere introducing algorithms regulations, and in real world they pursue the localization policy. The process of big data turning into smart content is studied. We show that smart content is in some way an intermediary between digital technology and civil society. Thus it gains an ethical dimension, and introduction of an ethical culture of digital infrastructure implementation becomes crucial for sustainable development. Authors conclude that moral and ethical standards in digital technologies are workable, and ethics should be implemented in core of big data technology in order to assess sustainable development of the society.

Keywords: Big data · Smart content · Legal regulation · Risks and challenges · Scientific data · Ethics

"Ignorance is not the greatest evil. The accumultion of poorly mastered knowledge is worse"

Plato

1 Introduction

Digitalization plays an ever-increasing role in the life of society, and many countries view big data, artificial intelligence and related technologies as breakthroughs, capable of ensuring the country's technological superiority in the future 10–15 years. On the commercial side, the introduction of digital technologies (especially ones connected to the processing of personal data) is an indispensable condition for success: for example, the five largest companies in terms of capitalization in 2019 are exclusively digital (Microsoft, Amazon, Apple, Alphabet and Facebook). They can afford to aggregate data on diverse groups of users and, as a result, become new monopolists in regulating access to information, which leads to serious distortions of the structure and functioning of market mechanisms. To a large extent, such distortions are due to the fact that digitalization carried out by large companies either uses specific technologies for purely instrumental (practical) purposes of promoting specific companies and/or their products, or uses the results of such digitalization for utilitarian marketing and/or political purposes [1, 2].

Á. Rocha et al. (Eds.): ICITS 2021, AISC 1330, pp. 14–22, 2021.
https://doi.org/10.1007/978-3-030-68285-9_2

In general, progress in the field of digital technologies manifests itself extremely heterogeneously in everyday life, and the dynamics of the computer technologies spread in different world regions does not correspond to optimistic expectations [2, 3]. Thus, new socio-economic and political challenges arise, including the emergence of information (digital) spatial inequality, unavoidable in the short term [4].

Despite the fact that now the term "big data" is more often associated with social networks, the financial industry and retail, initially big data was generated in the framework of large-scale scientific projects (more precisely, high energy physics - Large Hadron Collider, or astrophysics research), and rules and approaches of dealing with big data were being formed within the scientific field. This process has not yet been completed, and at present, for example, the European Union is introducing the FAIR principles (Findability, Accessibility, Interoperability, and Reusability). They are designed to create a big data management service through the emergence of many compatible services for the data search, storage and processing within a single ecosystem of scientific data of various thematic origin [5, 6]. In Russia, there are no similar developments, and at present it is only about adapting foreign (primarily European) experience to Russian realities (for example, to the CREMLIN + project, the scientific network of the Union State or the scientific grid system of the BRICS countries). In the European Union, the scientific data management system is incorporated with the e-Infrastructure developed as an integral part of digital science within the ESFRI project [7].

It is important to note that by now, big data technologies have already gone away from a tool for solving a narrow range of applied problems (initially - scientific ones) to an almost independent direction, which for most people is associated exclusively with decision-making systems or with artificial intelligence. This approach seems excessive: rather, big data technology should be understood as a process that includes a large number of technologies that do not have signs of innovation in themselves [8].

In the process of introducing digital technologies, a number of problems arise that require careful assessment and balanced decisions (see, for example, [2, 9]). In addition to purely technical challenges (data centers maintenance, algorithms and methods for working with data, organization of data handling), a number of related difficulties arise that lie in other fields. With this regard we argue that experience gained at scientific facilities and within research teams can be adopted in other fields: scientists were the first to deal with big data and related problems, and have already come through a number of problems the society is now facing.

Therefore in presented paper (supported by the RFBR grant No. 18-29-16130) we investigate the big data technology handling in scientific area, and discuss its probable adoption in other areas. First, we consider the legal regulation of big data, proving it should be based on some ethical principles. Second, we report on current stage of big data development, discussing the phenomenon of smart content. The last mentioned is shown to be also ethically-biased. Thus we conclude that ethics as a set of rules is crucial for big data technology progress and effective implementation.

2 Legal Regulation of Big Data: The Need for Moral and Ethical Basis

In particular, the question of regulating the turnover of big data arises. Nowadays in all countries a reactive approach dominates, focused on solving urgent immediate problems: restrictions and prohibitions are created, which in fact are very easy to get around since there is a lot of gaps in the legal field. As a rule, such actions are justified in the interests of protecting personal data (for example, the European GDPR [10]) or are built in the logic of protecting the interests of national security [11]. It should be noted that this approach, in the form of present or planned legislative restrictions, is designed to maximally protect the human right to privacy. At the same time, it comes into conflict with existing technology (aimed at the full use of all available information) and business practices, since it negates the advantages that Big Data technologies provide [1, 12]. Most likely, in the near future we will observe either the process of overflowing big data and metadata to places with the most friendly control protocols, or an avalanche-like growth of the capabilities of reverse algorithms (for example, deanonymization).

At the same time, the growth of computing power, progress in analytical tools leads to the vanishing (or neglecting) of such a concept as "data protection". The way out of this situation is seen, on the one hand, to classify "data" in general as "natural" (natural data that does not belong to anyone), as suggested in the McKinsey report [1]. On the other hand, management functions, especially on the part of states, are transferred to physical entities: due to the impossibility of controlling algorithms and/or the results obtained, the authorities of all countries are trying to take control of data centers by placing them on their territory [9, 11]. In opposition to these restrictions, a supranational environment for the circulation of big data technologies is emerging, one example of which is the AIPO (OECD AI Policy Observatory), a comprehensive analytical platform for reviewing policies and various national initiatives in the field of artificial intelligence.

In scientific environment another important factor is disinterest of the parties in breaching their obligations, since this may hinder achieving the result for which the research collaboration was created. We should also note that disinterest in breaching the obligations does not completely exclude a possible dispute. Currently, there is already practice of satisfying the claim of violation of moral obligation of a party [13], which gives reason to assert in this case that the soft law becomes binding, and turns into the so-called - hoftlaw (hard + soft law).

Thus, in the scientific field, using the soft law mechanisms, it is possible to settle most of the legal issues that are significant for corresponding projects. It is also important to take into account that implementation of scientific projects based on unique infrastructure affects various areas, and the prospects of such projects can have a wide scope. In light of this, regulating the implementation of such projects by law of obligation seems unlikely to succeed. We can agree with the authors who point out that the use of soft law and hoftlaw, whose main strength lies in the high degree of credibility and effectiveness of inherent scientific logic objectivity [14], seems more relevant, since international cooperation, joint financing, and the need to distribute risks and rights to the work results make it is much easier for participants to agree on the conditions of participation and work at the beginning of the project implementation.

It appears that megascience facilities have played a significant role in obtaining big data, and have logically led the legal community to the idea of establishing the foundations of a new branch of law - research law.

One of the possible consequences of digitalization is the growth of alienation and information inflation. The effects of digital alienation, already clearly manifested in the United States, have been described and analyzed by many authors (see in particular [2]). There is both a dehumanization of society and a disproportionate development of low-tech industries tied to a number of new innovative technologies [15].

There is a so-called digital displacement effect, and we can also see new entrants in the market trying to create their own venues - blue oceans in Rome Club terminology - using digital technology to circumvent the current legislation, working conditions and fiscal systems. Companies like Uber or food delivery services may pose a real danger of uncontrolled development and unethical use of technology, and it is not yet clear how to avoid this [16]. These firms being completely digital do not share common costs (e.g., for the city infrastructure used by drivers, etc.), thereby they distort the existing structure of business entities as a whole and, in their current form, do not meet sustainability criteria.

The data produced by such companies have the initial aim and corresponding algorithms. Thus, the data becomes value-oriented (inclined, biased), gaining the feature not immanent to the information on its origin. An attempt to avoid this contradiction (to judge sine ira et studio) leads to new "impersonal" regulation practice, that deals mostly with algorithmic approaches. A new, "digital" identity emerges, which may differ from the real one [17]. There is an acceleration of the process of shifting the law from the model of "rules" to the model of "operational management" (that is especially characterisitic for non-scientific areas), or, to put it emphasized, to seigniorat. In general, the formation of law in the digital age is characterized by the following features [18, 19]:

- shifting the source of law in favor of the non-public sector;
- withdrawal of this source from the control of the authorities;
- sectoral degradation of law due to the interests of the development of low-tech digitalized industries;
- sectoral and regional defragmentation of legal institutions, dictated by the market power of interested parties;
- forced guild regulation instead of state one as an attempt to maintain state control.

In the applied aspect, the practice of prohibitions and restrictions is not effective. For this reason, as mentioned above, states are moving to regulating algorithms (principles, methods, approaches) in the legal sphere, and to a localization policy in the infrastructure area. The latter means a request to provide control over information and transfer data to the physical jurisdiction of the state [9, 11].

So, the legal regulation of evolving digital economy (process not terminated yet) is the key issue of modern institutional development in the context of a change to new, highly technological world. Current approach and legal traditions are proved to be not effective enough to overcome the emerging challenges. It is necessary to look for other methods and approaches to legal regulation, including informal ones. The latter means relying on practices, customs and traditions that grow out of the basic moral and ethical principles of a particular people and/or country. Thus, there should be the culture of data handling [20].

Otherwise, according to experts (see, for example, Hariri's speech [21]), the philosophical foundations of existence in the world would disappear, decision-making would be left to implicit algorithms. In the long term, this would lead to the emergence of a digital dictatorship and a system of new globalism (obedience to some global rules - algorithms embedded in data processing systems, basics of AI or decision-making systems [2]).

Accordingly, the development of a new law, taking into account the achievements of digitalization, should take place on the basis of some deep principles that are biased (morally and ethically). Competition between these principles has already started, so has between algorithms and approaches. In fact, the commencing race in the field of digitalization leads to a rivalry between both technical characteristics (volume of data storage, processing speed, etc.) and moral and ethical standards that serve as the basis for the development of algorithms and methods (including data access policy). The largest players in this area are the EU (anthropocentric approach), the USA (emphasis on commercialization), and the PRC (values of Confucianism). All other participants (including large corporations that consistently lose their sovereignty in favor of national states or merge/replace with them) will not be able to remain neutral for a number of reasons (see above) and, obviously, would accept the position of some major player.

The digital technology management systems that are being built uphold different values, and, in fact, create different digital worlds with different accompanying technologies, including artificial intelligence technologies (the most promising developments are in the United States and China [22]). In the near future, we will have a competition not just of different methods, but of different minds (AIs) that do not understand each other. They would exist in their own non-intersecting worlds, that in social sphere is represented by so-called filter bubbles [23]. Their presence has a contradictory effect: on the one hand, they facilitate decision-making, standardize procedures, and on the other, they remove all that seem "unnecessary", "doubtful" (i.e. critical to the mainstream). This ultimately leads to the defragmentation of initial "data lakes". In practice, the algorithm forms a system for selecting data from data lakes, which means ignoring and/or neglecting some of the information (believed to be not important). As a result, the filter bubble anchors the walls between communities.

This fact represents a serious threat and challenge for the entire human society, but at the same time it presents a chance for development for third countries, giving them the opportunity to occupy some niche zones.

Here we face the intriguing question whether universal AI is possible? Whether it could be equally strong in all areas? And the answer to this question (in the current reality) is negative.

We note that the automation revolution is not a one-off event. There is a permanent change instead of a one-time transition. A "process" is taking place. According to the authors, the ongoing processes do not mean the creation of new knowledge, therefore, it is proposed to interpret these risks and challenges as not revolutionary, but as another stage in the development of human society. At the same time, the experience of the functioning of international scientific teams, which were the founders of big data per se, can be in demand in society. Methods and approaches for data storing and processing do matter, as well as solving value-related issues.

In practical terms, the following steps are necessary:

First, the harmonization of a single glossary in the field of "big data", taking into account the experience from different fields of activity (from science to business and social media).

Secondly, the rejection of attempts to control "big data", with a shift in emphasis on the result of their use and agents/actors of big data technologies (and not focusing on algorithms and/or methods: they are neither good nor evil).

Third, bridging the gap between the technologies themselves and the consequences of their application, which requires taking into account the socio-economic consequences of high technologies.

Fourth, awareness of the barriers separating different players. And the search for ways of their interaction (perhaps, taking into account the experience of multinational research teams, e.g., CERN, ITER or similar entities). We note that the presence of barriers is not a purely negative point, since it prevents the establishment of a global digital dictatorship (in Hariri's terminology [21]). The existence of barriers contributes to the formation of a unique (one's own) ecosystem. An example is legislative restrictions such as GDPR (it withdraws the EU from the US/RF/PRC space) or requirements for the localization of data centers (Russia and China).

Thus, legal regulation of big data requires solid ethical basis for further development of digital technologies, and experience from scientific field can be in great demand for establishing new legal regulations regarding big data turnover.

3 Big Data Transforming to Smart Content

The above process is superimposed on the evolution of big data itself: it is gradually becoming "smart". One mentions that digital technologies acquire subjectivity. For a long time, big data and AI were perceived as a mechanism for avoiding responsibility, which was shifted to the "mechanical mind" (e.g., an inquire, who is responsible in case of a delict conducted by AI's decision? Owner? Programmer? User?).

The solution for this problems is likely to be the phenomenon of smart content [24]. This approach was first formulated in the business environment as a response to scientific big data, that is characterized by the property of "utility": the formation of scientific data occurs according to given algorithms, supported by models, while big data "in nature" is not structured and is extremely heterogeneous. This phenomenon is often compared with smart grid operation, when by combining heterogeneous data, a qualitative transition from "data" to "knowledge" is carried out.

At the moment, there are no clear rules for the transition from big data to smart, and this process varies greatly depending on the actors involved. It is possible, however, to single out a number of characteristic features of this phenomenon and indicate the existing trends in relation to the processing of large, heterogeneous amounts of information.

First, the task is to get away from a simple search for correlations, which is typical for most big data methods. The results obtained should be categorized as "knowledge" (as a result of data interpretation) and not "new data".

Second, data is thought of as an "active participant" in the processing and decision-making process. The idea was expressed to combine "data" and "code". In particular, in accordance with the FAIR principle [5, 6], algorithms are included in the "circle" of stakeholders and participate in decision-making (as an independent entity, whose "vote" - the decision obtained - is counted with a certain weighting factor).

Thirdly, the social component of big data technologies is taken into account: solutions based on digital technologies are not socially neutral, and must be "adjusted" (i.e. biased) in advance in favor of society preferences. Accordingly, the contradiction between personal data and open data would be removed.

Note that in many countries the problem lies in the fact that for the purposes of "efficiency" and "competitiveness", the issues of digitalization were transferred to the stakeholders of the relevant processes (for example, this happens in the Russian Federation [25]). The interests of the state and society come into conflict with this approach and the question arises - which direction will be given preference: consumerism, anthropocentrism or statism - is still not clear.

Fourth, from the "technical" side, smart content would be characterized by fewer variables than currently big data requires, along with the ability to fold and recover data: the amount of data would decrease without loss of quality.

In our opinion, smart content is the data of the subject area, based on big data, on which, despite their immensity, an adequate interpretation is built due to immanent intelligent procedures.

From the user's point of view, smart content is in some way an intermediary between digital technology and civil society. The current task of the expert community is to make this mediator friendly to society. The solution to this problem is impossible without the formation of an ethical culture of using digital infrastructure (including algorithms applied).

Thus, the logics of technology development in the field of data handling requires ethical approach to big data, and its descendant smart content.

4 Conclusion

We are assured that the main goal of current institutional action is to impose legal regulation of the ongoing process of digitalization, keeping in mind it has not finished yet. The expected rapid development of artificial intelligence (understood today as deep learning networks), big data and related technologies have already been withdrawn from the sphere of science, and belong now to the world of business, society, and governance. Artificial intelligence can help (according to the KTH Royal Institute of Technology, 134 out of 169 UN goals can be achieved by AI) or hinder the achievement of sustainable development goals, depending on how AI technologies are applied in countries with different cultural values and wealth [26].

While evolving, digital technologies would manifest both advantages and risks. This should be taken into account in order to provide sustainable development of the society. Being implemented without ethical control, digital technologies could contribute to political instability, increased inequality, biased election results, and increased nationalism [26]. We conclude that it is necessary to reconcile technically promising approaches with the ethical (not emotionally neutral) development of the digital world.

Understanding of these processes is insufficient to overcome the challenges posed by digitalization. There should be the conception that society (represented by the state) sets norms and rules for business, and not vice versa. That is why it becomes socially necessary to form a well-thought-out and effective system of regulation through the law of relationships between man, science, state, society and nature. Perhaps a part of such a regulatory system belongs to a newly formed complex branch of law - scientific research law (the law of science) [14].

The next step in the study of digitalization processes should be the study of possible socio-economic changes in society caused by the reformatting of ethical norms, and the formulation of proposals to contain negative trends in line with the primacy of social and humanitarian challenges [27]. In particular, the concept of maximum openness of society and the economy (open data policy) is currently being seriously discussed as an alternative to rigidly structured legislation that upholds property rights and the right to protect personal information. The newly proposed form of data (smart content) should rely on some ethical principles and rules in order to produce clear understandable outcomes, conceived by a man, and corrected if necessary.

In general, the authors are confident that moral and ethical standards in digital technologies are real and necessary if humanity aims at sustainable development and prosperity.

Acknowledgments. This work was supported by the RFBR project No. 18-29-16130.

References

1. Big data: The next frontier for innovation, competition, and productivity. Report McKinsey Global Institute (2011). http://www.mckinsey.com/business-functions/digital-mckinsey/our-insights/big-data-the-next-frontier-for-innovation. Accessed 30 Jan 2019
2. O'Neil, C.: Weapons of Math Destruction: How Big Data Increases Inequality and Threatens Democracy. Crown Books, New York (2016)
3. The Global Information Technology Report. The World Economic Forum (2016). http://www3.weforum.org/docs/GITR2016/WEF_GITR_Full_Report.pdf. Accessed 31 Jan 2019
4. Zobova, L., Shcherbakova, L., Evdokimova, E.: Digital spatial competition in the global information space. Fundam. Res. **5**, 64–68 (2018)
5. The FAIR data principles. https://www.force11.org/group/fairgroup/fairprinciples. Accessed 03 Aug 2020
6. FAIR Principles. https://www.go-fair.org/fair-principles/. Accessed 10 Sept 2020
7. ESFRI: The European Strategy Forum on Research Infrastructures. https://www.esfri.eu/. Accessed 10 Sept 2020
8. Hype Cycle for Emerging Technologies 2015. https://www.gartner.com/doc/3100227/. Accessed 15 June 2020

9. Mayer-Schönberger, V., Cukier, K.: Big Data: A Revolution That Will Transform How We Live, Work, and Think. Houghton Mifflin Harcourt (2013)
10. The EU General Data Protection Regulation (GDPR). https://gdpr-info.eu. Accessed 18 Nov 2019, 17 Mar 2020
11. Ivanov, K.V., Balyakin, A.A., Malyshev, A.S.: Big data technologies as a tool for ensuring national security. St. Petersburg State Polytechnical University Journal. Scientific and Technical Sheets. Econ. Sci. **13**(1), 7–19 (2020)
12. Savelyev, A.: The issues of implementing legislation on personal data in the era of big data. Law. J. Higher School Econ. (1), 43–66 (2015)
13. Chetverikov, A.O.: Organizational and legal forms of big science (megascience) in the context of international integration: a comparative study. Part I. Magaziens as a scientific and legal phenomenon. Legal aspects of megascience functioning in the form of international intergovernmental organizations and national legal entities. Legal Science; no. 1 (2018)
14. Kashkin, S.Yu.: The formation of the law of science as a new integrated branch of law. Bull. Univ. named after O.E. Kutafina Moscow (5), 16–27 (2018)
15. Zhulego, V.G., Balyakin, A.A., Nurbina, M.V., Taranenko, S.B.: Digitalization of society: new challenges in the social sphere. Bull. Altai Acad. Econ. Law **9–2**, 36–43 (2019)
16. Von Weizsaecker, E., Wijkman, A.: Come On! Capitalism, Short-termism, Population and the Destruction of the Planet, p. 46. Springer (2018)
17. Kornev, A.V.: Digital technologies, social processes, modernization of law and the possible future of Russian legal education. Lex Russica **4**(149), 23–30 (2019)
18. Andreasyan, A., Balyakin, A., Nurbina, M., Mukhamedzhanova A.: Security issues of scientific based big data circulation analysis. In: Proceedings of the 8th International Conference on Data Science, Technology and Applications, Prague, Czech Republic, vol. 1, pp. 168–173 (2019)
19. Balyakin, A.A., Nurbina, M.V., Taranenko, S.B.: Institutions of law and the digital economy: emerging problems of sustainable development. In: VII International Scientific and Practical Conference “Sustainable Development: Society and Economy”, pp. 71–79. Publishing house of St. Petersburg State University (2020)
20. Balyakin, A.A., Malyshev, A.S., Nurbina, M.V., Titov, M.A.: Big data: Nil Novo Sub Luna. In: Antipova, T. (eds.) Integrated Science in Digital Age, ICIS 2019. Lecture Notes in Networks and Systems, vol. 78, pp. 364–373. Springer, Cham (2020)
21. Yuval Noah Harari: How to Survive the 21st Century- Davos 2020. Yuval Noah Harari Channel. https://www.youtube.com/watch?v=gG6WnMb9Fho. Accessed 10 Sept 2020
22. AI Business. https://aibusiness.com/china-and-the-us-have-the-worlds-most-successful-ai-hubs/. Accessed 03 Aug 2020
23. Pariser, E.: The Filter Bubble: What the Internet Is Hiding from You. Penguin Press, New York (2011)
24. Big data and smart digital environment. In: Kacprzyk, J., Farhaoui, Yo. (eds.) ICBDSDE 2018, Casablanca, Morocco (2019)
25. Order of the Government of the Russian Federation of July 28, 2017 No. 1632-r Program “Digital Economy of the Russian Federation”. http://static.government.ru/media/files/9gFM4FHj4PsB79I5v7yLVuPgu4bvR7M0.pdf. Accessed 02 Feb 2020
26. Vinuesa, R., Azizpour, H., Leite, I., et al.: The role of artificial intelligence in achieving the Sustainable Development Goals. In: Nature Communications, vol. 11, no. 233 (2020)
27. Florio, M., Sirtori, E.: Social benefits and costs of large scale research infrastructures. Technol. Forecast. Soc. Chang. **112**, 65–78 (2016)

Design and Evaluation of a Data Collector and Analyzer to Monitor the COVID-19 and Other Epidemic Outbreaks

Lucas C. de Almeida(✉), Francisco L. de Caldas Filho, Natália A. Marques, Daniel S. do Prado, Fábio L. L. de Mendonça, and Rafael T. de Sousa Jr.

Faculty of Technology, Electrical Engineering Department, University of Brasília, Brasília, DF 70910-900, Brazil
lucas.almeida@redes.unb.br

Abstract. Pandemic situations require analysis, rapid decision making by managers and constant monitoring of the effectiveness of collective health-related approaches. These works can be more efficient with the help of clearer and more representative views of the data, as well as with the application of other measures and projections of epidemiological nature to the information. However, performing such aggregations of data can become a major challenge in contexts with little or no integration between databases, or even when there is no technological core mature enough to feed and integrate technological advances in the workflow of health professionals. This paper aims to present the results of the meeting of project approaches such as the OSEMN framework, a software architecture based on Microservices and Data Science technologies, all tools aligned to make the environment of descriptive and predictive analysis of epidemic data (still dominated by manual processes) evolve towards a context of automation, reliability and application of machine learning, aiming at the organization and addition of value to the results of the data structuring. The project's validation objects were the documents of the situation of the Covid-19 disease pandemic in the region of the city of Brasília, Federal District, Capital of Brazil.

Keywords: Data scraping · Data visualization · Epidemic data analysis · Data automation

1 Introduction

Several areas have gone through revolutions with the application of Data Science methodologies. Among these areas, one has gained space and accelerated the concern with the availability and visualization of data: public health. Recently, a disease has gained the status of a pandemic and, in a short period, has reached all continents and countries around the world, including Brazil, which will be

Á. Rocha et al. (Eds.): ICITS 2021, AISC 1330, pp. 23–35, 2021.
https://doi.org/10.1007/978-3-030-68285-9_3

the test and implementation environment for the system of this work. The disease called Covid-19 [16], because of its infectiousness, requires greater care and short response time in decision making, which can only be made possible with the availability of aggregated data, historical series, projections and other epidemiological measures [18].

Opportunely, Data Science has very widespread and well-known work processes, and one of them, considered almost universal, is the OSEMN framework. good example of implementing this methodology can be seen in [14]). A work in this area, according to the aforementioned framework, must follow five steps, in order, to achieve the kind of results desired by a data scientist: obtain, scrub, explore, model and interpret. In other words and briefly: acquire the data of interest for the study; clean and format them so that machines can understand them; if necessary, find patterns and other measures and build models that exploit those patterns, such as projections; and organize visualizations that allow a complete understanding of the results. Therefore, it is clear that, for health professionals to enjoy the benefits of Data Science applied to collective health, it would be extremely important that they have ways to comply with each of the steps mentioned, which can be challenging in various contexts.

In the context of Emerging Countries, such as Brazil, it is common to verify regions without access to health service management technologies, or, even when access exists, to have jobs marked by the low use and integration of these solutions (as a software to manage patients data that is not fully integrated with central systems and that the professionals don't know how to use properly). In these places, several activities are still done manually, such as control of medical records, recipes and, in the case of epidemics, notifications of active cases and monitoring of the general situation [19]. Therefore, the conclusion is direct that works that demand greater availability and sophisticated data analysis become quite complicated and, in some cases, even unfeasible in these scenarios.

To solve the problem, first of all, it is necessary to understand that even if the data is spread over the Internet, or in documents, it is possible to accumulate this information in a structured way and, later, process it and add value to the analysis. The first step in the case of this work, is to understand the concept of web scraping (a good study on the topic can be seen in [15]): technique in which a program interacts with web pages and seeks information similarly to the way a human user would do. In a complementary way, the so-called "regular expressions" provide research methodologies and filtering capabilities that are extremely powerful, representing almost a standard tool for data scraping over web pages. This is due their flexibility and ability to search for string patterns with high precision and speed. Although there is no standard for the operation of the mentioned "regular expressions", there are efforts to centralize references, and one of them can be seen in [9]. Combining the two concepts, the creation of programs that interact with resources available online and find scattered and unformatted data becomes fast and uncomplicated. This approach is especially interesting when one considers the fact that most information on diseases and epidemics is available in the form of links for downloading files from websites of government agencies and departments, as in the case of Brazil.

This work describes the design and implementation of a system that aims to automate the various phases of a Data Science work in order to aggregate, structure, make available and model public data on the epidemic of Covid-19 disease provided in PDF file format. The application context is related to public health data from the Federal District, capital of Brazil. As a result, it was demonstrated that with a simple project and a lean team, it was possible to convert a scenario of manual readings and analyzes to high availability access to historical series and epidemiological analyzes, in addition to speed in the implementation of tests of new calculations and studies.

It is important to highlight that publishing data in PDF files is as a big challenge to make it accessible by machines. The PDF extension stands for Portable Document Format (PDF) and it defines an ISO standardized document that can contain text, fonts, vector graphics, and figures.

This paper consists of six sections, including this introduction. Section 2 deals with related work, especially on the necessary technical concepts. Section 3 specifies the proposed solution, uniting methodologies widely used in the industry and that integrate seamlessly into the context of users. Section 4 serves to discuss and demonstrate the results obtained in implementing the project with real data. Finally, findings and future work can be found in Sect. 5 and, in the last section, the acknowledgments of the authors.

2 Related Works

The recent arrival of the Covid-19 disease motivated the development of several software in order to inform about the impact of the disease. This work aims to disseminate contagion and death data in a structured and aggregated way, allowing for in-depth analysis, and for that, Big Data, data analysis and Web scraping concepts were used. However, the use of these technologies and methods, together, to integrate and facilitate the work of health professionals is still new, with few proposals with the same purpose. Some works could be related, as described as it follows.

[20] shows a definition of Web scraping, which consists of a set of techniques for obtaining data from an Internet source just like a person would personally do. In the article cited, this tool was implemented with the purpose of collecting images of the company or service to be advertised in order to do a collaborative filtering that aims to suggest suitable ads for a generic web page. The present work uses similar techniques to obtain and filter reports in PDF files from the website of the Secretary of Public Health of the Federal District. The collection of this data is important for the realization of multidimensional modeling of the data extracted about Covid-19 disease.

In the context of Big Data, the article [13] defines several concepts about open data. In addition, their proposed work supports the analysis of a large volume of open government data with the need for rapid processing. However, the proposed solution is barely feasible for technologically immature environments, and its application for cases with different layers of data without proper structuring, for example, documents in PDF format listed as download links in websites, is complex even for advanced users. In contrast, in our work the results and the workflow show that the objective is to integrate completely into the users' routine without requiring new knowledge and concepts.

Due to the large number of cases of malaria contagion in the mountains of East Africa, the work [22] developed an application that processes satellite data for modeling and forecasting epidemics of this disease, which is currently a major public health problem in the region. The cited article uses advanced data science concepts to add value to the predictive analysis of the disease. However, by using past data that is not accessible in real time, it is limited to seeking historical patterns. Similarly, our proposal also deals with an application in which one of the objectives is to make predictions and modeling of the data collected about a disease (Covid-19). However, Data Science is used in the present work to constantly and quickly bring aggregability and integration to public data, even allowing one to trace the start of possible patterns, going beyond the need of a complete organized historic (and often outdated) set of data.

In terms of data analysis, work in [17] makes a significant contribution by addressing relevant methods in the process of linear and non-linear analysis of data by applying efficient techniques of analysis and process diagnosis. These methods can be applied in the present work to perform the test of the data collected from the website of the Public Health Department of the Federal District. However, in a different manner, the modelling is automatically made, gaining time and reducing complexity.

When it comes to the use of Big Data applied to public health, the work [21] brings an explanatory analysis of how Taiwan used a national health insurance database and integrated it with immigration and customs information to analyze possible patients with probability of contagion by Covid-19. It also used new technologies, such as reading online travel history reports and health symptoms to classify travelers' infection risks based on flight origin and travel history in the last 14 days. The use and implementation of these technologies facilitated the country to map the cases of Covid-19. Based on this, it is observed that the work to be proposed in this article represents a major contribution in the area of public health in the sense of mapping the cases of Covid-19 in each region. The major difference is that the present work was implemented aiming to be applied in smaller contexts, generally trying to bring sophistication to less empowered regions.

3 Design of an Epidemic Data Science Project

This work is included in the field of Data Science, which follows, in a generalized way, a sequence of steps very well defined and based on the framework OSEMN, as stated in Sect. 1. Directing the application of this framework to the context of information about diseases implies the choice of certain technologies and preferable procedures. This is due to the fact that the source of the information (health professionals) does not always have knowledge in relation to technological procedures for collecting, aggregating and structuring data storage, therefore, the operation of the project should favor methods that are easy to adhere for these professionals and for existing workflows. In addition, from a technical point of view, the system should use a separate microservices approach, aiming at the low coupling between the different processing steps and the possibility of creating new modules that perform the same steps, but with a different purpose, as it will be addressed.

After describing the application context, it becomes possible to go through the specification of the implementation of the proposed project. Therefore, based on the general work sequence of the Data Science field, it was possible to derive a basic flowchart expected for the project, as shown in Fig. 1. This flowchart describes the basic steps needed for any Data Science project applied to a context of public health. They can be used as a guide to designing other systems and softwares with similar purposes.

Next, the choices for the implementation to be demonstrated will be described for each stage of the proposed flow. Also, from Fig. 1, a diagram of the modules and the general functioning of the project as a validation result of the application of the framework was derived according to Fig. 2.

Obtain: A Python script [8] is executed periodically using the cron scheduler service native to the Linux operating system distributions [5]. In each iteration, the program:

1. Accesses the online repository of the Federal District Health Department;
2. Obtains, by processing the code of the web page, the list of links to the reports in PDF format;
3. Searches and separates, in an ordered list format, the numbers of the available documents using regular expressions to filter variations in the code of the page layout and buttons, as explained in Sect. 1;
4. Checks in a local folder which files (present in the generated list) have not yet been saved and downloads the reports without correspondence;
5. Asks the Operating System to execute the next module.

Scrub: It is also a script developed in Python language. In each iteration, the program:

1. Compares the local folder with files in PDF format with another folder containing files in comma-separated values (CSV) format that is data arranged as tables. The goal is to verify that the two folders are synchronized, so that for each PDF file there is a corresponding CSV one. When reports are found without corresponding table files, it means that the report has not yet been processed, in which case the program proceeds with data extraction;
2. For extraction, using the Python language library Tika-Python [12] (which communicates with the public API Apache Tika [1]), it is sought, using regular expressions that mark the beginning and end of the table of interest, the values separated by administrative regions[1]. Again, with the result of this search,

[1] Administrative regions are government divisions of Brazil's Federal District. For the sake of simplicity, an administrative region can be defined as a district.

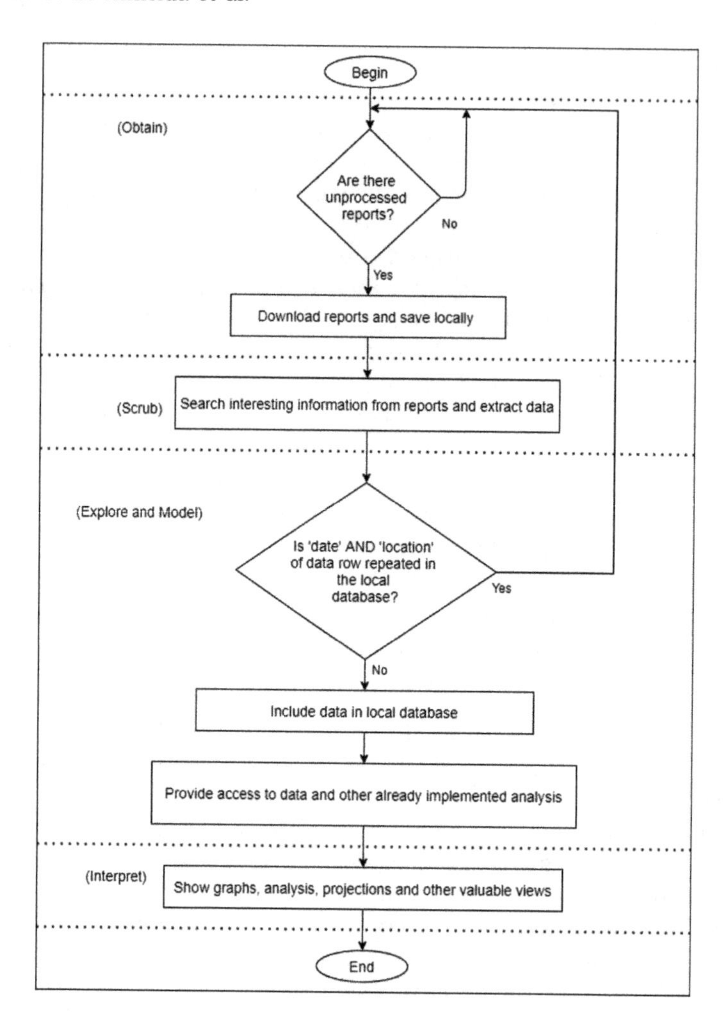

Fig. 1. Flowchart expected from the proposed system.

using regular expressions, each value of each row and column expected for the table structure is inserted into the new CSV file to be created. It is important to note that, as [2], extracting tables from files with the extension PDF can be quite computationally expensive. For this reason, the method used in the aforementioned library does not return tables, only the text available in the document, but without great formatting guarantees. Therefore, it is quite common to check for misalignments, multiple spaces, characters that have been added or incorrectly mapped, among others. However, with the use of regular expressions, it is possible to search the table of interest using the beginning and end markers. It is also possible, by modeling the return with regular expressions for each line, to search the data and to filter the errors of the extraction process quite efficiently;

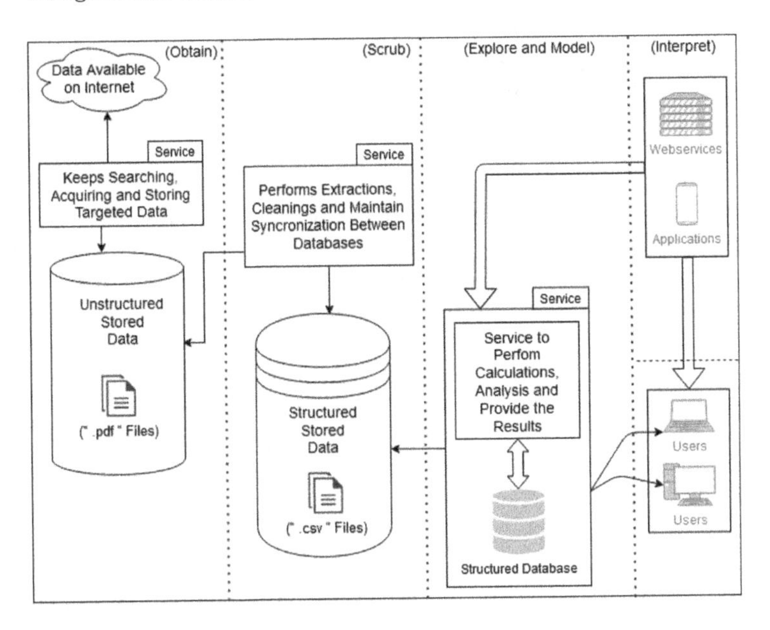

Fig. 2. Diagram of the proposed software system. The arrows point to the targets of actions of persisting or retrieving data. It is worth observing that the users can also be a target of such tasks, for example a researcher that is persisting local data to perform calculations.

3. At the end of the extraction, the program saves the new CSV files in a backup folder (for future references and/or uses). It also deletes all files available in another workbook, used by the next module, and adds the new files extracted there as an output;
4. At the end, it requests the execution of the next module to the operating system.

Explore and Model: For the case of the project, which required agility in the development and approval of features, two versions of this module were made. As explained earlier, the division of tasks was done aiming at the maximum reduction of the coupling, so that there is also no problem in having other versions of the same module creating new branches in the flow of processing of the collected data.

The difference between the versions is related to functionality: one is used in production, for large volumes of requisitions, while the other serves the purpose of accelerating tests and developments. The robust version is based on a Bash language script [3] that automates the inclusion of data in an instance of the MongoDB database [7]. The other version is based on a Python script that includes data in a SQLite database [11] (simple relational database in file format). In this version, data is served using the Flask framework [6], which operates based on Python and has a very low implementation time curve.

A valuable aspect demonstrated at this stage is the flexibility of work once the information is properly organized and categorized after extraction and aggregation. The same workflow can benefit from different types of data storage and access methods (table files, relational databases in file format and high performance non-relational databases served with high availability).

Referring to the term “model”, for further study of the data and tests, a method was implemented in the API supported by the Flask framework which receives an administrative region name and a number of days, and based on the projection model described in [4] and using the library sklearn [10], returns the simple projection of the number of cases from the last available days in the database to the future days, according to the amount passed. This projection is based on the prediction of natural logarithms of the number of cases accumulated per day with an unsupervised linear regression algorithm.

Other models could also be added to the test API to, for example, study the most adherent for each region and, subsequently, implement such calculation in the high performance and availability application interface.

Still on the topic of data modeling, it is important to note that the value of application interfaces is also related to the easy interoperability between different systems, the possible sophistication in data filtering and the speed. It is possible, for example, to request very quickly average values, the time series of a single region, of a set of them, the data of all regions on a given date, the maximum value already collected, among others.

Interpret: This module, as it is only a consumer of processed and structured data, does not work in tandem with the others. In fact, it doesn't even have to be a single program with a single output. Thus, for the commented scenario, the following structured data consuming services were created, which add value to the analysis:

1. A web page that renders time series graphs for the number of reported cases and deaths by region and by region aggregate (for example, all north regions), as well as georeferenced data (based on the name of the administrative region) and with color grading according to the number of cases in relation to the greatest value of the research;
2. Also consuming the data, another sophisticated application was created which, using georeferenced formats and the complete compilation and analysis of the last available historical data from the API, creates short animated videos showing the disease progression in each region for each day of the pandemic.

Several other consumer services could have been created for this stage of the process, such as mobile applications, the use of third-party data analysis software and even the creation of customized report generation software. Again, given the independence between each module and the successful cleaning and organization of data, impractical or extremely repetitive and time-consuming activities for human operators can become fast and even automated.

4 Evaluation of Tests and Results of the Proposed System

In this section, the results obtained with the application of the system will be presented in the context of the data from the Health Department of the Federal District of Brazil made available until August 15, 2020, with 165 reports in the repository.

First, on the efficiency of extraction, it was found that 100% of the reports were downloaded and renamed correctly in PDF format. Of 165 reports correctly obtained from March 23 to August 15, 2020, 138 were processed and resulted in successful data extraction. Also, from reports numbered 1 to 23, the table of interest did not exist yet. Therefore, only 142 reports were real objects of the extraction, and of these, 138 resulted in success. This corresponds to an error rate of about 2.81% and a success rate of 97.19%. This margin of error is due to the various formatting and tabulation problems both in the source code of the repository's web page and in the documents themselves. As stated earlier, regular expressions are used to filter out these flaws, however not all cases can be predicted.

With the data already stored in the database, it was possible to make projections, as explained previously. The most common ways of presenting this information are line and bar graphs, using both ways to create the first visualizations, as shown in Fig. 3.

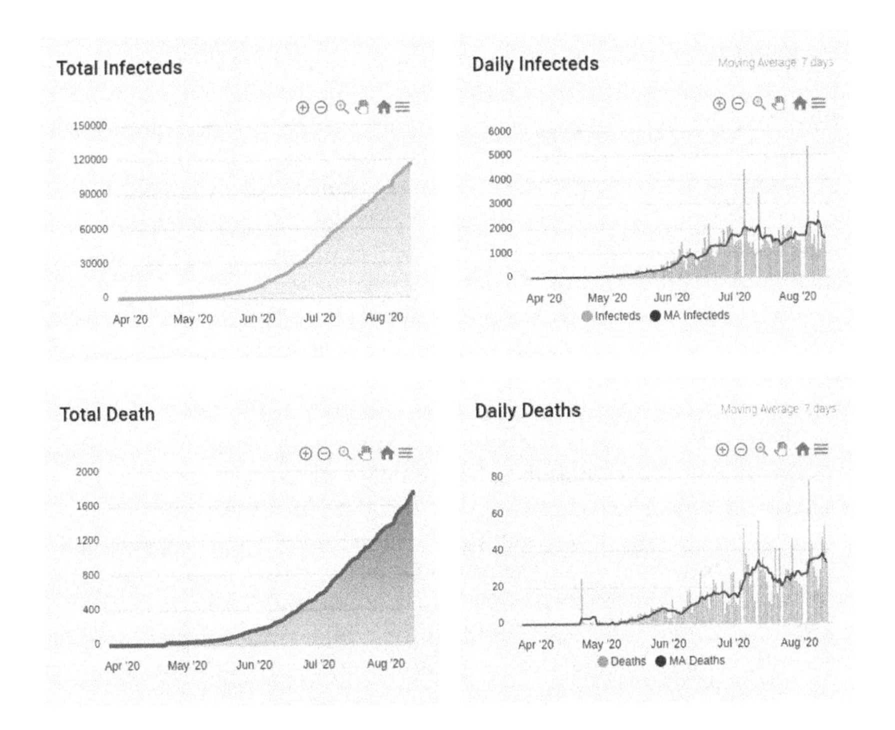

Fig. 3. Line and bar time series graphs with moving averages.

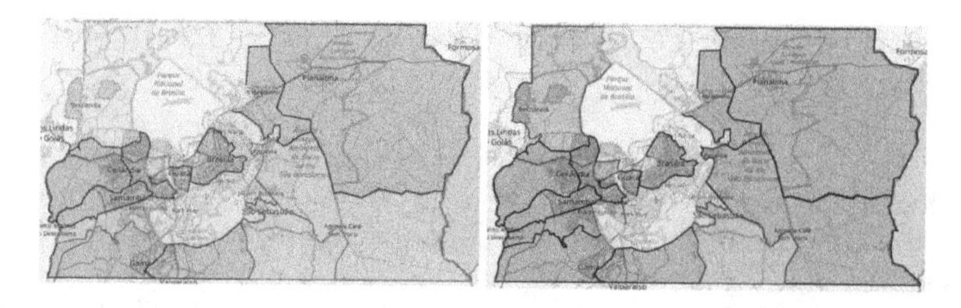

Fig. 4. Deaths and reported cases, respectively. Darker colors represent more affected regions.

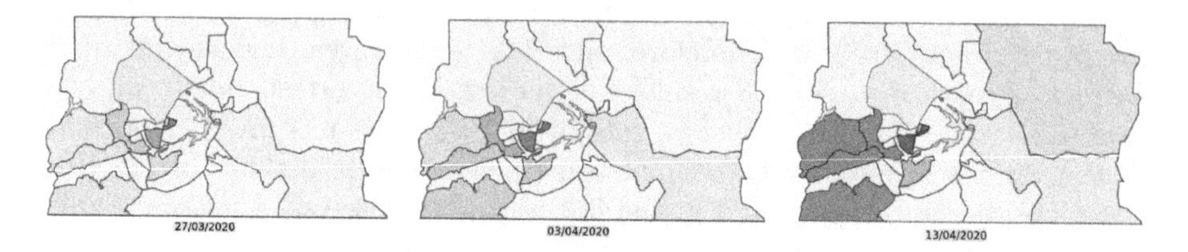

Fig. 5. Frames of a video that shows the spread of the disease over time.

In Fig. 3 it is possible to analyze the evolution of the cases in three different perspectives: accumulated data (figures on the left side); cases per day (bar graphs on the right side); and moving average of the cases (line graphs on the right side). For Covid-19 disease, the period of seven (7) days was used to calculate the averages as shown in the figure. It is also possible to select other periods for the average, such as 15 or 21 days.

Also, in health outbreaks, it is of great importance to geographically identify the spread of the disease. Understanding the evolution in each region, overlapping this data with other factors, such as economic, social, and even population density are of great relevance to understand where possible governmental decisions are most needed. This platform is able to project data by overlaying information on a map based on the names of administrative regions. Figure 4 exemplifies in a georeferenced way the data overlaid in the Federal District region. With this it is possible to analyze, by color, the number of infections and deaths in each region.

Still on the same way of visualization, the platform is also capable of generating short videos showing the spread of the disease over time, generating a historical series of the evolution of the number of georeferenced active cases, according to sample frames from one of the videos in Fig. 5.

Finally, considering the possibility of comparing different regions, it is also possible to create visualization scenarios in which the user chooses the regions of interest. In Fig. 6, three regions are selected: Ceilândia, Taguatinga and Gama. It is possible to observe, for example, that Ceilândia currently holds around 50% in the number of infected and deaths, considering only these 3 options as a total data set.

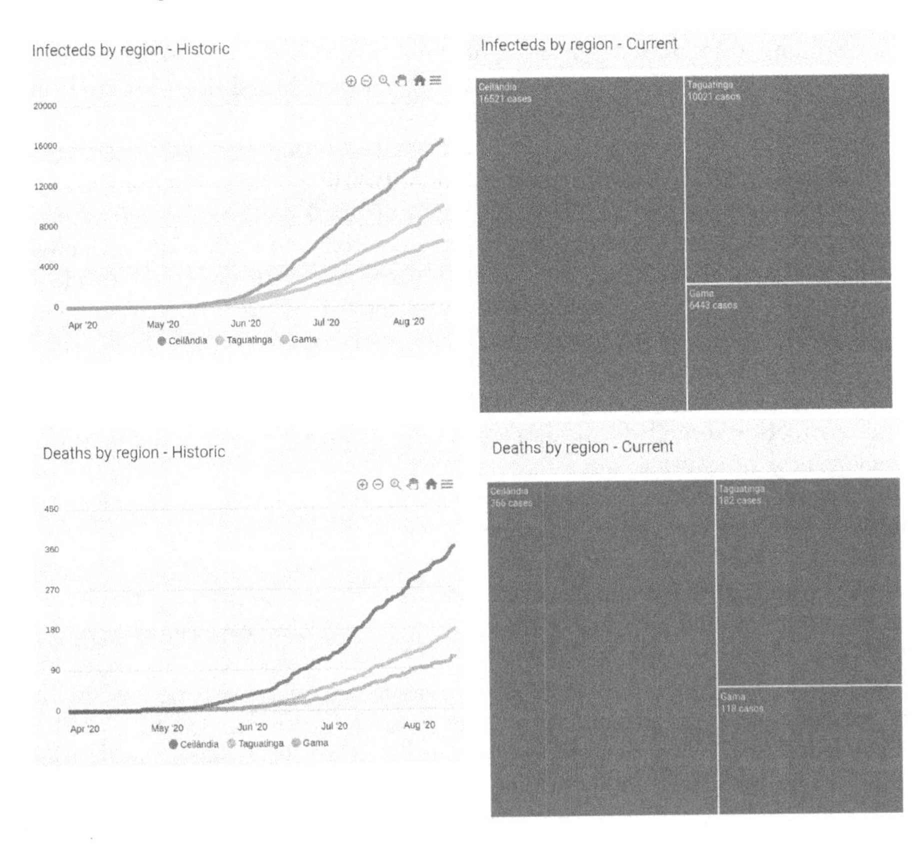

Fig. 6. Multiple analysis are possible with the selection of regions of interest.

There are still several other filters and methods available, especially through APIs, allowing different integrations in third party systems. The malleability of the data resulting from a complete iteration of the system adds new layers of study and decision possible for health professionals without requiring technical knowledge in systems, all because with these approaches, they can continue to use the same human-friendly documents, and still data can be aggregated and modeled with the implementation of different software services in each step of the presented framework.

5 Conclusion and Future Work

In this paper, a system designed to approximate the information and workflows already existing in the field of public health with the most current methodologies of Data Science was presented. It was possible to show that even in teams with precarious conditions for the application and support of data analysis and aggregation technologies, it is possible to produce automated solutions that are integrated into the routine of these teams, being only required a reliable and efficient framework (as the slightly modified OSEMN presented) and associated methodologies (such as microservices architecture principles) for the project design. The system was tested with real data from the government entity responsible for public health in the capital of Brazil and the results of the application reached levels of visualization and analytical modeling as advanced as those performed in technologically mature environments, allowing, ultimately, georeferenced visualization of the spread of the Covid-19 disease epidemic (in the form of a video) from simple, manually edited PDF reports. Still, it should be noted that even in an urban center, situations of lack of integration, problems in the availability and handling of data, among others, are verified, and for this reason, the construction of applications with the objective of automating and abstracting technical processes adds a lot of value to the work scenario and management of health systems.

However, the system, although complex, is still only an inspiration for other initiatives of the same nature. For new advances in the implementation of this proposal, it would be important to create an interface in which the end user would be able, through simple commands that do not require long training, to choose the form and location of the data to be extracted in a model report. In addition, adding the use of OCR (Optical Character Recognition) technology could allow, for example, to register in a structured way past medical records of patients and even medical prescriptions, after the simple scanning of these documents. Still, with the proper training of an artificial intelligence model, it would be possible to add the functionality of checking the level of trust of signatures and stamps, extending the use of the system to pharmacies, audits and transparency portals, for example.

Acknowledgment. The authors would like to thank the support of the Brazilian research, development and innovation agencies CNPq (Projects INCT SegCiber 465741/2014-2, PQ-2 312180/2019-5 and LargEWiN BRICS2017-591), CAPES (Projects FORTE 23038.007604/2014-69 and PROBRAL 88887.144009/2017-00) and FAPDF (UIoT Projects 0193.001366/2016 and SSDDC 0193. 001365/2016), as well as the support of the LATITUDE/UnB Laboratory (SDN Project 23106. 099441/2016-43), cooperation with the Ministry of the Economy (TEDs DIPLA 005/2016 and ENAP 083/2016), the Office of Institutional Security of the Presidency of the Republic (TED 002/2017), the Attorney General's Office (TED 697,935/2019) and the Administrative Council for Economic Defense (TED 08700.000047/2019-14).

References

1. Apache Tika - a content analysis toolkit. https://tika.apache.org/
2. Apache Tika - Class PDFParser. http://tika.apache.org/1.24.1/api/org/apache/tika/parser/pdf/PDFParser
3. Bash Reference Manual. https://www.gnu.org/savannah-checkouts/gnu/bash/manual/bash.html
4. Coronavirus 10-day forecast. https://covid19forecast.science.unimelb.edu.au/

5. Cron(8)—linux manual page. https://man7.org/linux/man-pages/man8/cron.8.html
6. Flask - Web Development. https://flask.palletsprojects.com/en/1.1.x/
7. MongoDB Database. https://www.mongodb.com/
8. Python Programming Language. https://www.python.org/
9. Regular Expressions Reference. https://www.regular-expressions.info/reference.html
10. Scikit-Learn. https://scikit-learn.org/stable/
11. Sqlite database. https://www.sqlite.org/index.html
12. Tika-Python. https://github.com/chrismattmann/tika-python
13. de Carvalho Victorino, M., Shiessl, M., Oliveira, E.C., Ishikawa, E., de Holanda, M.T., de Lima Hokama, M.: Uma proposta de ecossistema de big data para a análise de dados abertos governamentais concetados. Informação & sociedade **27**(1) (2017)
14. Dineva, K., Atanasova, T.: OSEMN process for working over data acquired by iot devices mounted in beehives. Current Trends Nat. Sci. **7**(13), 47–53 (2018)
15. Diouf, R., Sarr, E.N., Sall, O., Birregah, B., Bousso, M., Mbaye, S.N.: Web scraping: state-of-the-art and areas of application. In: 2019 IEEE International Conference on Big Data (Big Data), pp. 6040–6042 (2019)
16. Lima, C.M.A.D.O.: Information about the new coronavirus disease (COVID-19). Radiologia Brasileira **53**, V–VI, April 2020. https://doi.org/10.1590/0100-3984.2020.53.2e1
17. Mogk, M.: Automatic data processing in analyses of epidemics. In: Epidemics of Plant Diseases: Mathematical Analysis and Modeling, pp. 55–77. Springer (1974). https://doi.org/10.1007/978-3-642-96220-2_3
18. Santos, A.D.F.D., Fonseca Sobrinho, D., Araujo, L.L., Procópio, C.d.S.D., Lopes, E.A.S., Lima, A.M.d.L.D.d., Reis, C.M.R.D., Abreu, D.M.X.D., Jorge, A.O., Matta-Machado, A.T.: Incorporação de Tecnologias de Informação e Comunicação e qualidade na atenção básica em saúde no Brasil **33** (2017). https://doi.org/10.1590/0102-311x00172815
19. Santos, S.R., Paula, A.F.A., Lima, J.P.: O enfermeiro e sua percepção sobre o sistema manual de registro no prontuário. Revista Latino-Americana de Enfermagem **11**, 80–87 (2003). https://doi.org/10.1590/S0104-11692003000100012
20. Vargiu, E., Urru, M.: Exploiting web scraping in a collaborative filtering-based approach to web advertising. Artif. Intell. Research **2**(1), 44–54 (2013)
21. Wang, C.J., Ng, C.Y., Brook, R.H.: Response to COVID-19 in Taiwan: big data analytics, new technology, and proactive testing. Jama **323**(14), 1341–1342 (2020)
22. Wimberly, M.C., Chuang, T.W., Henebry, G.M., Liu, Y., Midekisa, A., Semuniguse, P., Senay, G.: A computer system for forecasting malaria epidemic risk using remotely-sensed environmental data (2012)

The #MeToo Movement in Twitter: Fighting Gender-Based Violence

Fátima Martínez[1(✉)], Carolina Pacheco[2(✉)], and Marco Galicia[2(✉)]

[1] Universidad del Rosario, Bogota, Colombia
fatimamargu@gmail.com
[2] Universidad Autonoma de Mexico, Ciudad de Mexico, Mexico
caro.luna26@gmail.com, galprasmarco@gmail.com

Abstract. This paper analyzes the behavior of the Me Too movement in Twitter between December 8, 2018—a little over than one year after actress Alyssa Milano triggered an outcry over sexual harassment, assault and discrimination—and May 8, 2019 through the #MeToo hashtag and numerous other related hashtags. The #MeToo hashtag, representative of a social media trend which seeks to empower female victims of gender-based violence has had a significant international impact. In fact, it viralized beyond the United States in countries where women have turned to social media to raise their voices for women's rights. More than three million tweets were collected during the five-month period previously mentioned using a digital tool called DMI-TCAT to observe the evolution of the #MeToo movement and its consequences inside and outside of Twitter.

Keywords: MeToo · Women · Digital feminism · Twitter · Social media · Digital activism · International movement

1 Introduction

On October 16, 2017, actress Alyssa Milano made a powerful invitation to women who have been sexually harassed or assaulted: replying "Me too" to her tweet [1]. This inspired several women to share their personal stories of sexual harassment and abuse, which led to the revival of a movement. The #MeToo hashtag became then a Twitter trend, which has spread out on other social networks, making Me Too a prominent force on the Internet.

Tarana Burke, an African-American feminist and activist who helped young black girls from disadvantaged neighborhoods who had been sexually abused, started the movement in 2006. The #MeToo movement has grown to include all sorts of victims by offering women an outlet to speak out their truths while promoting necessary conversations on how to stop this kind of violence [2].

Since the end of 2017, the #MeToo movement gained popularity on Twitter. Through multiple hashtags, including #MeToo, people do online activism to support women's rights. This feminist and digital movement has gone viral in numerous countries from the United States to Mexico, Argentina, India, and South Korea. Twitter's hashtags allow for the systematization and categorization of the social

Á. Rocha et al. (Eds.): ICITS 2021, AISC 1330, pp. 36–44, 2021.
https://doi.org/10.1007/978-3-030-68285-9_4

network's information without using intermediaries. Hashtags also highlight the leading topics on the platform through the "trending topics" feature [3].

From a critical discourse analysis [4, 5] along with a feminist perspective [6] we consider the #MeToo movement as a supportive strategy to motivate women—not only public figures—to publicly speak out against gender violence or cyberbullying on social media as a way to strengthen their collective empowerment processes. Follow-ups from tweets with the #MeToo hashtag on Twitter between December 8, 2018 and May 8, 2019 were empirically verified using quantitative and qualitative analysis.

Alyssa Milano's 2017 tweet—"If you've been sexually harassed or assaulted write 'me too' as a reply to this tweet"—had a photo attached with the following message: "Me Too. Suggested by a friend: "If all the women who have been sexually harassed or assaulted wrote 'Me Too' as a status, we might give people a sense of the magnitude of the problem" [1] (See Fig. 1).

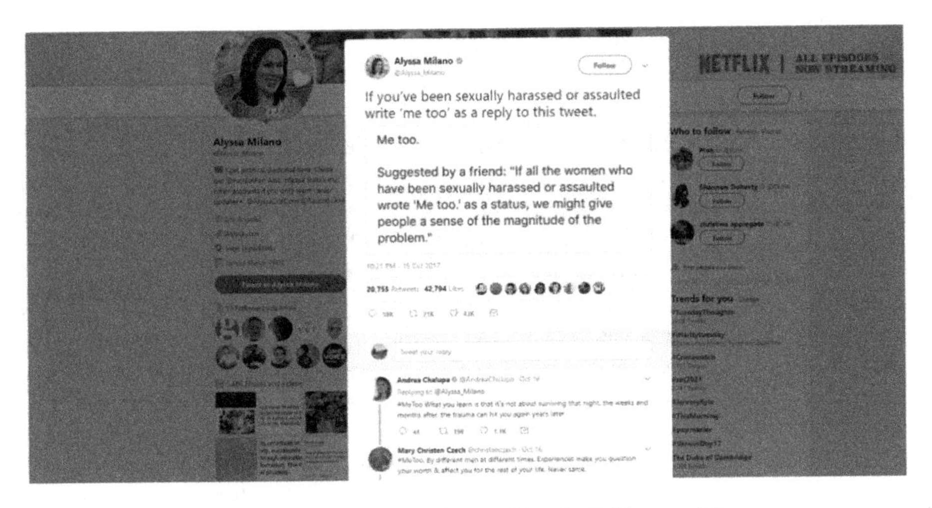

Fig. 1. Screenshot of Alyssa Milano's 2017 tweet [1].

2 Materials and Methods

This research study analyzes the #MeToo movement on Twitter, starting from the year of its existence to the present, through an international viewpoint and the lens of feminist critical theory. The four hypotheses posed below seek to better understand the dynamics of #MeToo:

1. The #MeToo hashtag emerges from a social movement that encourages women to publicly report their sexual harassment stories in social media. Primarily, these are sexual violence survivors.
2. This movement, which gained steam in Twitter with the #MeToo hashtag, sheds light on traumatic experiences women have suffered. It empowers them to report situations they were before silent about. Using a hashtag in a personal and collective sense is empowering. This feminist and activist movement's success lies in the idea

of solidarity for sexual assault or harassment survivors, which not only emerges but evolves over time depending on the different realities in different countries with respect to women who have firsthand experience being sexually harassed, marginalized, or attacked.

3. This movement, which takes place in the environment of social media, encourages online activism from women. While it started in the United States, it spread to different countries where the #MeToo hashtag is used widely, both in English and other languages, such as Spanish and Korean. This hashtag has also been adopted in India as a way for Indian women to report sexual abuse.
4. The #MeToo hashtag does not only remains a trending topic in Twitter. Its effects go beyond social media. It reaches: 1) media (newspapers, radio, television, and online media), where news and editorials make reference to the #MeToo movement, and 2) non-digital or media scenarios where it has important consequences in real life as many of the accusations are brought to trial, resulting in real world changes in support of women.

2.1 Methodology

Both quantitative and qualitative methodologies were used. The qualitative technique involved finding documentation on feminism and online activism related to the Me Too movement using the #MeToo hashtag. The quantitative technique involved gathering more than three million tweets that used different hashtags related to #MeToo between December 8, 2018, and May 8, 2019. All tweets were gathered using the DMI-TCAT tool, which categorizes tweets containing hashtags with the words assigned on Twitter, and it provides statistics and graphs on each of the hashtags gathered [7]. Additionally, the daily news was monitored during the five-month period of the study to monitor the traditional media's response to #MeToo, especially in online newspapers.

Three key #MeToo related-events were identified during the period covered, which we refer to as "milestones" (See Table 1). These events are the #MeToo milestones in the Americas over the five-month collection period, which involved several peaks that made the hashtag a trending topic on Twitter. In general, we considered moments that involved more than 30.000 tweets with the hashtags that were being monitored.

Table 1. Milestones of #MeToo.

Milestone	Date	Country	Topic identified to #MeToo
1	12/8/2018–12/16/2018	Argentina	Harassment allegations from Argentinian actresses on Argentinian social media
2	01/13/2019–01/20/2019	United States	A Gillette ad on social media criticizing the patriarchy's model of hegemonic masculinity
3	03/22/2019–04/08/2019	Mexico	#MeToo goes viral in Mexico among Mexican writers

3 Results

After the five-month period, we nearly observed three million tweets related to the #MeToo hashtag (See Table 2).

Table 2. Data analyzed from Twitter.

Category	Min	Max	Mean	1st Quartile	Median	3rd Quartile
Tweets per account	1	62,058	2.37	1	1	2
Number of account followers	0	54,571,715	3,835.34	75	275	915
Number of friends	0	1,684,339	1,149.3	157	393	988
Number of tweets per account	1	9,434,893	25,373.13	1705	7,525	25,587

From the data presented in Table 2, we conclude that most of the Twitter accounts using the #MeToo hashtag do not come from TweetStars (accounts that belong to stars or public figures having a large number of followers) or people affiliated with media, but rather from personal accounts. This is primarily exemplified by the fact that the average number of tweets per account is 2.37 in addition to the fact that, in general, the average number of followers of Twitter accounts is relatively low at 3,835. Also, when qualitatively analyzing the 200 main accounts, we found that some outlier accounts had skewed the sample's maximums.

In general, however, the #MeToo tweets came from personal Twitter accounts that were not dedicated to tweeting solely with that hashtag. This can be observed from the number of followers in the sample's first and second quartile and seeing that the accounts with more followers are concentrated in the third quartile. Thus, we can conclude that few accounts had many followers in the sample, and most of the accounts tended to be near the median of 275 followers. This corresponds to the first hypothesis posed—that women are the ones who speak out against violence on social media. This leads us to believe that on Twitter, they are primarily posting from personal profiles and not from Twitter profiles affiliated with newspapers, TV stations, broadcasters, or TweetStars.

The three milestones observed during the same time period were the following:

3.1 First Milestone: Harassment Allegations from Argentinian Actresses on Argentinian Social Media

The first milestone occurred between December 8, 2018, and December 16, 2018. During the qualitative analysis of the tweets that were monitored during the study's period, we found that this peak stemmed from harassment allegations by actresses in Argentinian media. This trend went viral in the United States media after being shared

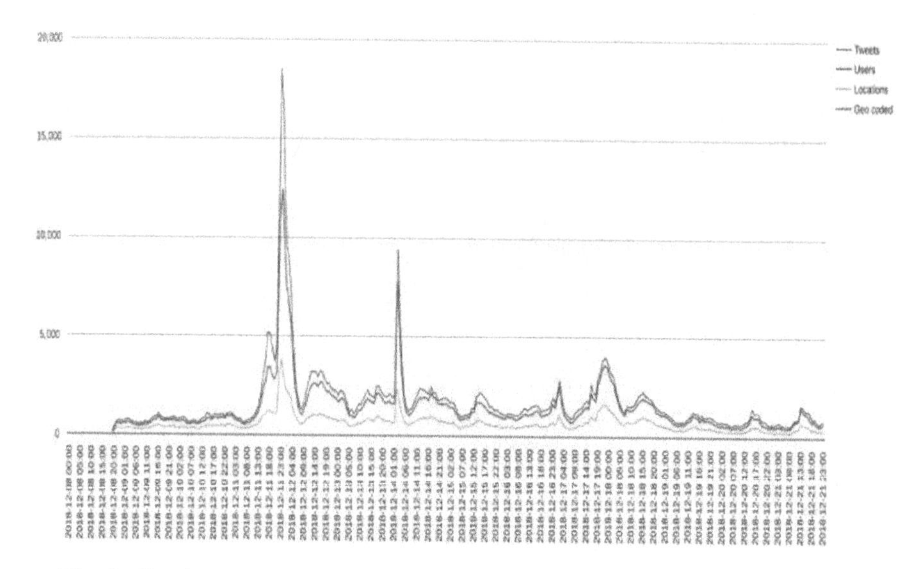

Fig. 2. Graph obtained from the online tool related to the first milestone on Twitter.

by American actresses. While there were accusations made earlier within a time continuum since October of 2017, this milestone's viralization started on December 8.[1]

Beginning on December 9, the first indications of the trend began to come up in the media. This supports the hypothesis that accusations did not come from the media or large news agencies. Nonetheless, acting as analysts, the media disseminated messages posted on Twitter after the accusations of violence became a social media trend. Figure 2 shows this first milestone, where the moments of virality of the accusations can be observed.

3.2 Second Milestone: A Gillette Ad on Social Media Criticizing the Patriarchy's Model of Hegemonic Masculinity

The second milestone involves a Gillette ad that criticizes the model of hegemonic masculinity, which is used as a strategy in United States' social media. The company pushed the #MeToo hashtag to go viral once again, with an online growth from January 13 to January 20, 2019. Specifically, this became a trending topic as part of a commercial media strategy to increase followership from a new market of American women or women living in the United States by aligning the brand to the #MeToo movement and conversation.

Even though the accusations went viral because of the Gillette campaign, they did not belong to an isolated event; they were attached to the list of accusations that had already been made, especially those of the first milestone. Additionally, while this campaign was launched by a company, accusations still occurred in personal Twitter

[1] These conclusions are also outlined in the #MeToo news coverage from several Argentinian newspapers.

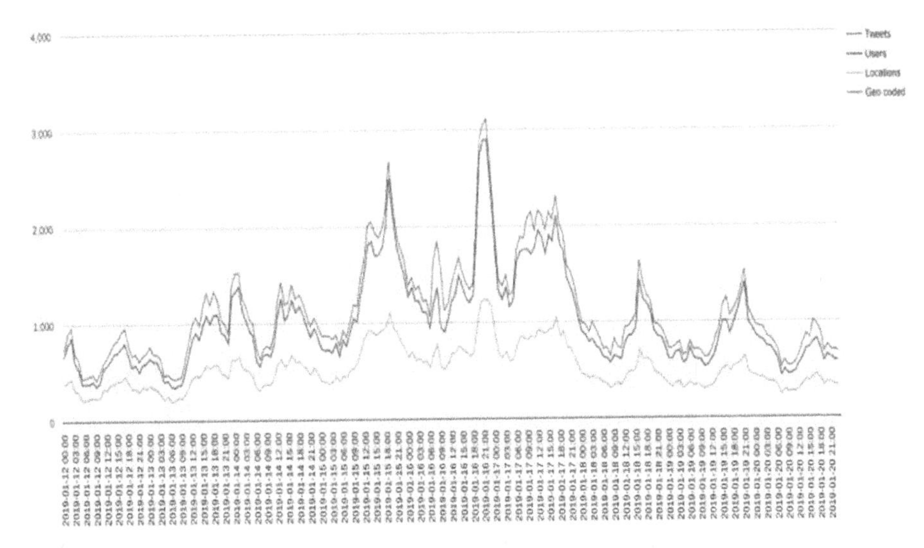

Fig. 3. Online graph of #MeToo tweets during the second milestone.

accounts, from where women spoke out against violence; in this case, in relation to a discussion about the ad and masculinity. Figure 3 shows the growth trend in the use of #MeToo, starting with Gillette launching its ad on January 13, reaching its peak on January 16 and returning to the initial trend on January 20.

3.3 Third Milestone: #MeToo Goes Viral in Mexico Among Mexican Writers

The third milestone pertains to the #MeToo hashtag going viral in Mexico. The hashtag emerged from a personal Twitter account called @MeTooEscritores (In English: @MeTooWriters), which anonymously exposed sexual harassment cases involving Mexican writers. Later, a series of similar accounts related to the same issue emerged (Fig. 4):

- @MeTooMusicaMx (In English: @MeTooMusicMx) promoting #MeTooEscritoresMexicanos (In English: #MeTooMexicanWriters)
- @MeTooActivista1 (In Spanish: @MeTooActivist1) promoting #MeTooActivistasMexicanos (In English: #MeTooMexicanActivists)
- @MeTooPeriodista (In English: @MeTooJournalist) promoting #MeTooPeriodistasMexicanos (In English: #MeTooMexicanJournalists)
- @MeTooTeatroMx (In English: @MeTooTheaterMx) promoting #MeTooTeatroMexicano (In English: #MeTooMexicanTheater
- @MeTooCreativos (In English: @MeTooCreatives) promoting #MeTooCreativosMexicanos (In Spanish: #MeTooMexicanCreatives).

These accounts emerged from March 22 onward. One week later, they went viral and received a great deal of attention, when information about the hashtag was spread by Mexican newspapers and television networks, making it a significant issue in the

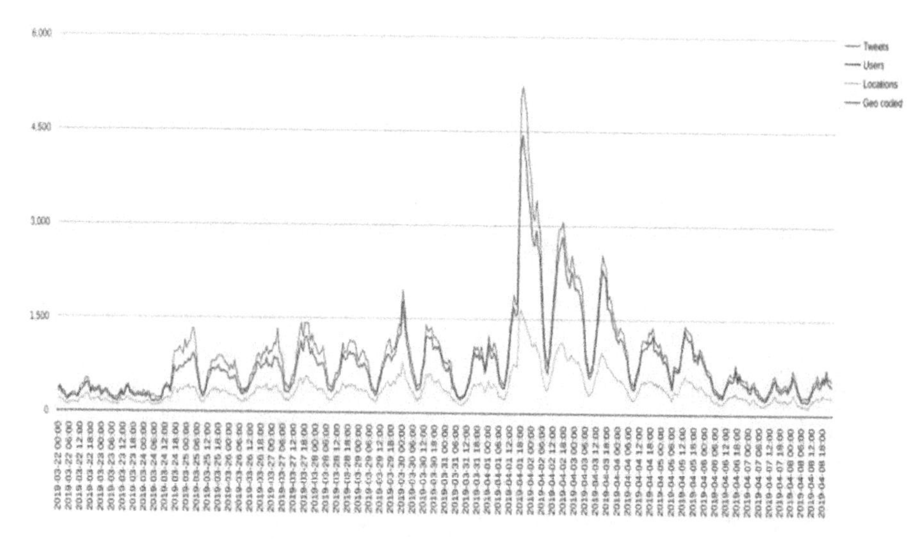

Fig. 4. Graph of tweets with #MeToo during the third milestone.

media. Along with such Twitter accounts, many accusations were spread by Mexican women using personal Twitter accounts to speak out against sexual harassment and violence. They also posted messages of support for the women who made accusations.

The peak occurred in the afternoon of April 1. On that day, the suicide of Mexican musician and band member of Botellita de Jerez Armando Vega Gil was reported. His suicide note mentioned the #MeToo hashtag. This event triggered a discussion on the causes of his suicide, leading to murder charges using the #MeTooMusicosMexicanos hashtag [8]. Then, a new hashtag called #MeTooHombres (In English: #MeTooMen) appeared; it used hateful and reactive language, accusing the #MeToo movement of promoting a breakdown of society. This hashtag was reported and blocked in Twitter. After the musician's suicide and the discussion it generated, the #MeToo movement in Mexico slowed down. On April 2 and 3, the hashtag lost strength again being reduced to previous levels of accusations and it was no longer considered a priority issue in the media agenda.

This milestone concluded on April 8, when a #MeToo forum was held at Mexico City's Commission on Human Rights, which was organized by academics, lawyers, activists, human rights advocates, students, artists, and civil society organizations to support those who started the accounts related to #MeToo to speak out against violence at universities and cultural industries, to affirm that Mexico has a robust regulatory framework to combat cyberviolence, and to encourage the Mexican government to confront violence against women [9]. Even though the #MeToo hashtag was no longer part of the media agenda, it continued to be used to report cases of sexual harassment such as the one from actress Analu Salazar, who reported abuse by a clergyman on May 2 [10]. This news was published by Mexican media outlets.

4 Discussions and Conclusions

The #MeToo movement originated in 2006 in the United States from the efforts of activist Tarana Burke, an African-American woman who fights for women's rights and publicly speaks out against abuse, marginalization, and sexual harassment. This movement was reignited and popularized at the end of 2017 by the American actress Alyssa Milano after she posted a powerful tweet inviting women to reply #MeToo if they were sexual harassment survivors. We have reached the following conclusions from analyzing the evolution of the movement in Twitter during a five-month period (December 8, 2018–May 8, 2019), which starts roughly one year after Milano's invitation:

1. The #MeToo movement remains relevant internationally. In 2019, it was present both in social media and mass media (newspapers, radio, and television), although the former served as a feed for the latter thanks to the usage of the #MeToo hashtag. The three milestones that occurred in three different countries in the Americas (Argentina, the United States, and Mexico) exemplify the peaks and valleys of the movement, which corresponds to social media trending topics. Related hashtags such as #MeTooEscritoresMexicanos (In English: #MeTooMexicanWriters), which was prominent from the end of March through the beginning of April in 2019, gained traction in Twitter and later became part of the Mexican media agenda. Due to the scope of this study, we did not analyze milestones outside of the Americas. However, we recognize #MeToo as a global movement. The usage of #MeToo in South Korea and India, for example, reveals its international impact [11].
2. The #MeToo hashtag is used as a tool for online activism; the more it viralizes on social media, the stronger the movement becomes. This hashtag was not only used by itself; it was also accompanied by other hashtags related to freedom, justice, feminism and democracy, following the course of activist trends on online social media as a mechanism to defend specific rights—in this case, women's rights of women who have been victims of gender-based violence.[2]
3. The #MeToo hashtag encourages women to publicly speak out against abuse, marginalization, and sexual harassment on Twitter (and other online social media networks), and seeks to expand a culture of solidarity, support and listening specially among women who protest against the injustices they have experienced firsthand. Some women may choose too keep their stories privates, yet the brave act of tweeting or commenting on a Tweet with the #MeToo hashtag opens up a door for healing and justice [12].
4. The viralization of the #MeToo hashtag comes primarily from personal accounts—women who are survivors from sexual abuse or have experienced sexual assault, marginalization or harassment, as well as people who stand in solidarity with these female survivors.

[2] This conclusion was reached after a personal interview in Mexico with Alcalá, E. about female empowerment and social causes.

5. The #MeToo hashtag belongs to the collective imaginary, as it is widely used internationally, both in social media and traditional media. This hashtag has been adopted by individuals in different professional areas such as journalism and the arts, creating real-life impact.

References

1. Huffpost. https://www.huffingtonpost.ca/2017/10/16/me-too-rape-culture_a_23245109/. Accessed 10 Dec 2019
2. Nuraddin, N.: The Representation of the #Metoo Movement in Mainstream International Media [Master Thesis]. Jonkoping University, Sweden (2018)
3. Salgado, E.: Twitter se despolitiza? Una exploración a los trending topics en Mexico. Discursos Sociedad **9**(4), 469–491 (2015)
4. Van Dijk, T.: El análisis crítico del discurso. Anthropos **186**, 23–36 (1999)
5. Hodge, R., Kress, G.: Social Semiotics. The United Kingdom, Cambridge (1988)
6. Azpiazu, E.: Conciliación entre trabajo y responsabilidades familiares: una revisión teórica con enfoque de género. Investigium IRE: Ciencias Sociales y Humanas, V **1**, 177–194 (2014)
7. Borra, E., Rieder, B.: Programmed method: developing a toolset for capturing and analyzing tweets. Aslib J. Inf. Manage. **66**(3), 262–278 (2014)
8. El Sol de México. https://bit.ly/2KZ9zOr. Accessed 10 Dec 2019
9. The Broadband Commission for Digital Development: Cyber Violence Against Women and Girls: A Word-wide Wake Up Call. Broadband, ITU, UNESCO, United Nations Development Programme, UN Women, London (2015)
10. Milenio. https://www.milenio.com/policia/analu-salazar-actriz-acusa-abuso-sexual-padre-legionarios-cristo. Accesed 15 Oct 2019
11. Barrera, L., Rodríguez, C.: La violencia en línea contra las mujeres en México: Informe para la relatora sobre las violencias contra las mujeres Ms. Drubavka Simonovic. Luchadoras MX, Fundación Heinrich Böll Stiftung México y El Caribe, Asociación para el Progreso de las Comunicationes (2017)
12. El Universal. https://www.eluniversal.com.mx/espectaculos/farandula/inunda-redes-campana-de-alyssa-milano-contra-abuso-sexual. Accessed 28 Sept 2019

Intellectual Capital: Brief *State-of-the-Art*

Óscar Teixeira Ramada(✉)

Faculty of Science and Technology, Fernando Pessoa University, Porto, Portugal
oscarramada@gmail.com

Abstract. The aim of this *paper* is present a *State-of-the-Art* about the *intellectual capital*. This State is divergent in terms of the definition, measurement and value of *intellectual capital*. A qualitative approach was used specially explaining and interpreting the contributions of 21 authors. Instead of contributing to the progress and expansion of knowledge of the topic, it contributes to the retrogression and retraction of their knowledge. It delays important progress on the knowledge of what its value is and how to obtain it.

In the year 2020, the improvement in his knowledge is very restricted mainly because there is no consensus focused in 3 features: what is, how can be measured and how to know his value. Some approaches can be valuable but, they are complex and difficult to apply in companies which prevents scientific progress regarding these 3 features. The results suggest that there is a lot of work to do, in order to achieve consensus over a definition of *intellectual capital*, ways to measure that can be applied to in all industries satisfactorily scientifically and also methods of knowing his value, in the space and over the time with specific requirements. The *paper* suggests avenues for future research: what features should be considered to measure *intellectual capital* and if it will be possible to build a universal method of measuring it and know his value and comparing results among companies.

Keywords: Intellectual capital · Measurement methods · Value (added) creation · Competitive advantages · Business performance

1 Introduction

In the context of the intangible assets, the literature relevant displays some important topics, namely: information systems, copyright and *intellectual capital*. The *intellectual capital*, perhaps, is the most important because managers can know what is his value and, therefore, what is their expression in the production of goods and services. It is a topic related with the themes proposed for the Conference like *A) Information and Knowledge Management (IKM)*.

Some authors have been emphasized, the *intellectual capital* the *per se*, like [1] [Latvia], [2] [Indonesia] and [3] [Romania]. While others such as [4] [Malaysia], [5] [Romania] and [6] [India], give more attention to the *intellectual capital* as a factor in interaction with knowledge, experience and innovation. It follows effects on business value of the companies.

Á. Rocha et al. (Eds.): ICITS 2021, AISC 1330, pp. 45–52, 2021.
https://doi.org/10.1007/978-3-030-68285-9_5

The same about *how to know* the value of *intellectual capital*. The contribution of [24] [USA] and [22] [Netherlands and Belgium] even if they are valuable, at the same time, is complex and difficult to apply in companies.

Therefore, the absence of a consensus definition about the *intellectual capital*, prevents a way of measure it and know his value, in and about companies. The same about business performance, competitive advantages of the companies and their markets. This is the rationale for doing the *paper*.

The main problem (*research question*) that results from such is to know ***what does it say the relevant literature about the intellectual capital?***

This question or problem arises from make an idea about the State-of-the-Art in the *intellectual capital* (in year 2020) and is also the purpose of this *paper*.

In terms of the contribution for increasing scientific knowledge, is display in an orderly manner, 5 important topics about the *intellectual capital*: **Definitions, Measurement Methods and Empirical Evidence, Literature Review and Insights.** The authors selected followed criteria of relevance, objectivity and clarity of exposure.

The *paper* is organized in 4 Sections: Sect. 1, *Introduction*, Sect. 2, *Intellectual Capital: Definitions, Measurement Methods and Empirical Evidence, Literature Review and Insights*, Sect. 3, *Discussion* and Sect. 4, *Conclusions*.

2 Intellectual Capital: Definitions, Measurement Methods and Empirical Evidence, Literature Review and Insights

[14] [Turkey] say that it was with John Kenneth Galbraith, in 1969, and Tom Stewart in 1991, in an article published by this in *Fortune* magazine, that the expression *intellectual capital* was introduced for the first time.

Although several authors present definitions of *intellectual capital*, [1] [Latvia], [3] [Romania] and [11] [Malaysia], in special, are those who present a view supported by empirical evidence.

2.1 Definitions

[1] [Latvia] define *intellectual capital* as the sum of *Human Capital* (HC) (labor costs), *Structural Capital* (SC) (added value[2] minus HC) and *Capital Employed* (CE) (the money). Such definition is associated with efficiency measures in each component: *Human Capital Efficiency* (HCE) (added value divided by HC), *Structural Capital Efficiency* (SCE) (SC divided by the added value) and *Capital Employed Efficiency* (CEE) (added value divided by CE), respectively. HCE is the added value for each € invested in the company in HC, SCE in SC and CEE in CE.

In turn, [3] [Romania], adopt the same expressions for the first two components but with different meanings. In fact, *Human Capital* consists, according these authors, in

[2] Sum of wages (W), interests (I), dividends (D), taxes (T) and net income (R), being the added value (AV) equal to: $AV = W + I + D + T + R$, according to Riahi-Belkaoui (2003) *in* [1, p. 891] [Latvia].

the stock of knowledge of the workers and *Structural Capital* arises associated with the productivity and organizational culture of the companies. The third component, *Customer's Capital*, different from [1] [Latvia], is defined as the relationships that the companies establish with customers and suppliers, among others. Thus, the same expression, corresponds to different semantic meanings, which gives a different concept of *intellectual capital.*

Finally, [11] [Malaysia], suggest another definition of *intellectual capital* based on six components: *Human Capital* (education and experience of human resources), *Structural Capital* (relationships between customers and suppliers), *Customer's Capital* (patents, computers, …), *Social Capital* (institutions, relationships and norms), *Technological Capital* (information technology, research and development, innovation) and *Spiritual Capital* (effects of spiritual and religious practices). It is observed that, [1]'s [Latvia] meaning of *Human Capital* and *Structural Capital*, does not match with [3] [Romania] and also with [11] [Malaysia]. On the other hand, *Customer's Capital*, is absent in [1] [Latvia] and also have different meanings from the other two authors. As a result of such differences, the underlying idea of *intellectual capital* will be different too, which will give rise to values, performances, value created, different competitive advantages in companies, in the industry and even in the same company!

2.2 Measurement Methods and Empirical Evidence

[1] [Latvia], sum of HCE, SCE and EEC are used to construct the method of measuring *intellectual capital*: *value added intellectual coefficient* (VAIC). This method is no more than the sum of HCE, SCE and EEC. VAIC, originally conceived by [23] *in* [1] [Latvia], measures the efficiency of the value that was created within a company and based on the value added (from accounting), produced by *intellectual capital*, for [1] [Latvia]. The value obtained by this method is interpreted as being the added value for each € invested in the company. The higher (low), the higher (lower) the value of the *intellectual capital* created in a company in a given period of time, per € invested. In a sample of 64 companies in the Baltic Listed Companies (Latvia, Estonia and Lithuania) from 2005 to 2011, covering the sectors of industry, namely, consumer goods (24), industrial (17), consumer services (7), basic materials (5), health care (5) and utilities (4), technology (1) and telecommunications (1), on average, in sample period, for each € invested, companies created a new value in excess of €, the lowest being € 1.67 in 2007, in Lithuania, and the highest, in € 2006, in Estonia, in 2006. The absolute values assumed by VAIC in these three countries follow this decreasing order: Estonia, Latvia and Lithuania.

Regarding [3] [Romania], the base sample is a Romanian company, in the year 2011, of the information technology sector. For each of the three components of *intellectual capital*, defined by them, the authors exhibit indicators, consisting of the measurement method of *intellectual capital*, which leads to the conclusion that, instead of [1] [Latvia], there is no final value of the *intellectual capital*. In *Human Capital*, they showed a close association with labor productivity, level of remuneration, qualifications and use of technologies. In *Customer's Capital*, the focus is on quality management, particularly, in customer relations. These authors concluded by a high level of implementation of *intellectual capital*, according to their definition, based on the satisfaction of the workers, in strategies that motivate them, in the management of knowledge, which increase business *performance*.

For its part, [11] [Malaysia], when studying the relationship between business *performance* and each of the six components of their definition of *intellectual capital* in the context of Malaysian companies, concluded that the components follow the order of importance: *Customer's Capital*, *Technological Capital*, *Spiritual Capital*, *Social Capital*, *Human Capital* and *Structural Capital*. These authors follow a method of measuring *intellectual capital*, by its components, not reaching an overall value of it. Strong endowments of the six components together enable more competitive advantages and better business *performance*.

2.3 Literature Review

[4] [Malaysia] studied the relationship between business *performance* and the components of its definition of *intellectual capital* in companies of the Malaysian Stock Exchange: *Human Capital*, *Structural Capital*, *Relational Capital* and *Spiritual Capital*. They concluded that the strongest linear correlation with business *performance* occurs, first with *Relational Capital* (+0,727), followed by *Spiritual Capital* (+0,62), *Structural Capital* (+0,562) and *Human Capital* (+0,533). In general terms, the authors concluded, that *intellectual capital* is a *sine qua non* factor for business success in companies.

In turn, [13] [Turkey], studied the effects of *intellectual capital*, innovation and organizational strategy on business *performance*. The study focused on 186 insurance companies in Antalya[3] in the year 2013. As main conclusions, they found evidence that the effects are positive, being +0,222, +0,221 and +0,101, respectively, as measured by the linear correlation coefficient.

[5] [Romania] elaborates another analysis in everything similar to that of [13] [Turkey] but, considering contexts of crisis of the economies. The study is limited to the Romanian economy, from 2008 to 2011. It concludes that factors such as knowledge, qualifications and experience, are business development factors in times of crisis and that is translated by the stimulus to adoption of new procedures within companies.

[20] [Lithuania] studied the influence of (open) innovation on *intellectual capital*. It is understood as a movement characterized by inflows and outflows of knowledge from a company. It concludes that it is through *Relational Capital* that the influence is most evident and that (open) innovation, while contributing to better *intellectual capital* (creating competitive advantages), simultaneously, creates greater complexity in companies, which instead of enabling improvements, hinders, and the management of knowledge is forced to respond. But in this context, it also contributes to... More (open) innovation!

[21] [Hong Kong] did a contribution related to the knowledge of the impact of *intellectual capital* on business *performance*, in particular, on companies called MAKE (Most Admired Knowledge Company) and non-MAKE, which were and were not awarded, respectively. MAKE companies have demonstrated efficiency in the

[3] City in southern Turkey, located in the Mediterranean region.

management of *intellectual capital*, creating more value. This translates into a more appropriate use of financial assets, which leads to an increase in income and net income and, therefore, greater ROA and ROE.

[12] [United Kingdom], presents a methodology for measuring the value of *intellectual capital* with the purpose of knowing what are the determinants of it. This author says as the levels of *intellectual capital* value increase, the ranking of business *performance* also increases both for *performance measures* (*Return On Equity* (ROE) and *Return On Assets* (ROA).) and for *intellectual capital* value measures (*Market-To-Book*, *Ratio* Q de Tobin and *Long-Run Value-To-Book* (LRVTB)).

2.4 Insights

The topic of the *intellectual capital* raises some insights. First of all, it is impossible to assess quantitatively, until a consensual definition of what it is, especially with the goal of being expressed in the financial statements of companies. On the other hand, methods of measuring it are intimately linked with the definition, which, in order to be quantified, requires clarification in their general and specific features. Only thus, one can make comparisons and know, better business *performance*, value creation and competitive advantages.

3 Discussion

Regarding the Definitions, it is clear that [11] [Malaysia] has the double of the componentes comparing with [1] [Latvia] and [3] [Romania]. Instead of making it clearer it makes it more confusing. Therefore, the final result in practice is closer to the opposite than to coincide with whoever will be in the authors' intention. The final outcome is make the definition of *intellectual capital* more unknown than known. The problem is even more serious when the content of each component is still more different from author to author.

With regard the Measurement Methods, [1] [Latvia], [3] [Romania] and [11] [Malaysia] we can notice that, the lack of clarity is even more reinforced as well as the methodological contributions to know the value of *intellectual capital* are even more empty of application in daily practice. Specially, [11] [Malaysia], it does not exhibit a guiding thread that can be applied, even more, from one or more companies, based on the notion of *intellectual capital* with six components!

In the Literature Review, [4] [Malaysia], the relationship between *intellectual capital* and business *performance* is made by correlating the components of the first with the second. It turns out that the idea of associating the two factors is reductive as the second does not come exclusively from the first. [13] [Turkey], do the same like [4] [Malaysia] but, in insurance sector. There is evidence that the indicator linear correlation coefficient, it is too generic and based on historical information that adds little or nothing as a method of measuring *intellectual capital*. Regarding [5] [Romania], there is a clear contradiction. Economic crises are unwanted, but in the conclusions of this author it should be desired because they have virtuous effects over the economies. Is it preferable, then, to live permanently in crisis? Of course not. In [20] [Lithuania], through component *Relational Capital* that innovation most influences *Intellectual Capital*. But such, it has the assumption that innovation and *Intellectual Capital* are different from each other. These authors do not prove this, so their conclusions are undermined. [21] [Hong Kong], the authors found that MAKE companies have demonstrated a better use of financial assets in the management of *intellectual capital*, which explains the creation of more value by them. Finally, [12] [United Kingdom], presents a methodology for measuring the value of *intellectual capital* with the purpose of knowing what are the determinants of it. He concluded that the levels of *intellectual capital* value increase, *performance* and *intellectual capital* value measures.

Finally, in the Insights, it can be said that the absence of, first, a definition, second, a quantitative, objective method with universal application and, third, a projective way (not historical), the possibility of knowing the value of *Intellectual Capital* is very tenuous. Diffuse contributions regarding the *Intellectual Capital*, make this objective even more difficult to achieve.

4 Conclusions

This *paper* aims to make a brief *State-of-the-Art*, about *intellectual capital*. In general, it can be said that it is disjointed, incoherent, presents definitions, methods and value for *intellectual capital* that do not contribute to the progress of scientific knowledge but, rather, to an amalgamation of diffuse ideas that little or nothing advance the progress of the topic. Regarding its definition, even though one possesses the idea associated with the mind (not wrong) it is still far from presenting a consensual definition among several authors. On the other hand, several of these, such as [1] [Latvia], [3] [Romania] and [11] [Malaysia] present definitions of *intellectual capital* based on a different number of components and, above all, underlying meanings, that makes it difficult such consensus, leading to nonconsensual measurement methods and non-generalized applications. Avenues for future research are, defining what elements should be included in *intellectual capital*, define them as well as what methods should be adopted to measure, in order to be applicated in companies, which is a task for regulatory authorities. There is another doubt: *there will be a method of measuring intellectual capital equally applicable to any industry?*

References

1. Berzkalne, I., Zelgalve, E.: Intellectual capital and company value. Procedia Soc. Behav. Sci. **110**, 887–896 (2014)
2. Nuryaman, N.: The influence of capital intellectual on the firm's value with the financial performance as intervening variable. Procedia Soc. Behav. Sci. **211**, 292–298 (2015)
3. Gogan, L., Draghici, A.: A model to evaluate the intellectual capital. Procedia Technol. **9** (2013), 867–875 (2013)
4. Abdullah, D., Sofian, S.: The relationship between intellectual capital and corporate performance. Procedia Soc. Behav. Sci. **40**, 292–298 (2012)
5. Sumedrea, S.: Intellectual capital and firm performance: a dynamic relationship in crisis time. Procedia Econ. Finan. **6**(2013), 137–144 (2013)

6. Sekhar, C., Patwardhan, M., Vyas, V.: A delphi-AHP-TOPSIS based framework for the prioritization of intellectual capital indicators: a SME's perspective. Procedia Soc. Behav. Sci. **189**(2015), 275–284 (2015)
7. Gadau, L.: The intellectual capital – a significant, but insufficiently highlighted source in the financial situations. Procedia Soc. Behav. Sci. **62**, 668–671 (2012). World Conference on Business, Economics and Management, WC-BEM 2012
8. Vosloban, R.: The influence of the employee's performance on the company's growth – a managerial perspective. Procedia Econ. Finan. **3**(2012), 660–665 (2012)
9. Delgado-Verde, M., Martín-de-Castro, G., Amores-Salvado, J.: Intellectual capital and radical innovation: exploring the quadratic effects in technology-based manufacturing firms. Technovation **54**(2016), 34–47 (2016)
10. Palacios, T., Galván, R.: Intangible measurement guidelines: a comparative study in Europe. J. Capital Intellect. **8**(2), 192–204 (2007)
11. Hashim, M., Osman, I., Alhabshi, S.: Effect of intellectual capital on organization performance. Procedia Soc. Behav. Sci. **211**, 207–214 (2015)
12. Goebel, V.: Estimating a measure of intellectual capital value to tests its determinants. J. Intellect. Capital **16**(1), 101–120 (2015)
13. Kalkan, A., Bozkurt, Ö., Arman, M.: The impacts of intellectual capital, innovation and organizational strategy on firm performance. Procedia Soc. Behav. Sci. **150**, 700–707 (2014)
14. Örnek, A., Ayas, S.: The relationship between intellectual capital, innovative work behavior and business performance reflection. Procedia Soc. Behav. Sci. **195**, 1387–1395 (2015)
15. Yildiz, S., Meydan, C., Güner, M.: Measurement of intellectual capital components through activity reports of companies. Procedia Soc. Behav. Sci. **109**, 614–621 (2014)
16. Gouveia, L., Couto, P.: A importância crescente do Capital Humano, Intelectual, Social e Territorial e a sua associação ao conhecimento. Atlântico Bus. J. **1**, 28–34 (2017)
17. Gouveia, L., Pinto, P.: Contributo para a discussão sobre a contabilização do Conhecimento e do Capital Humano nas Organizações. Atlântico Bus. J. **1**, 35–37 (2017)
18. Kaufmann, L., Schneider, Y.: Intangibles – a synthesis of current research. J. Intellect. Capital **5**(3), 366–388 (1994)
19. Arvan, M., Omidvar, A., Ghodsi, R.: Intellectual capital evaluation using fuzzy cognitive maps: a scenario-based development planning. Expert Syst. Appl. **55**, 21–36 (2016)
20. Uziene, L.: Open innovation, knowledge flows and intellectual capital. Procedia Soc. Behav. Sci. **213**(1), 1057–1062 (2015)
21. Zhicheng, L., Zhouer, C., Shing, L., Wah, C.: The impact of intellectual capital on companies' performances: a study based on MAKE award winners and non-MAKE award winner companies. Procedia Comput. Sci. **99**(2016), 181–194 (2016)
22. Rodov, I., Leliaert, P.: FiMIAM: financial method of intangible assets measurement. J. Intellect. Capital **3**(3), 323–336 (2002)
23. Pulic, A.: VAIC – an accounting tool for IC management. Int. J. Technol. Manag. **20**(5–8), 702–714 (2000). https://doi.org/10.1504/IJTM.2000.002891
24. Housel, T., Nelson, S.: Knowledge valuation analysis – applications for organization intellectual capital. J. Intellect. Capital **6**(4), 544–557 (2005)
25. Ramada, O.: A proposal to measure intellectual capital using the Andriessen's (2004) method: application to a case study in a national context (Portugal). Ph.D thesis, Faculty of Science and Technology, Fernando Pessoa University, Porto, Portugal, pp. 1–315 (2019)
26. Andriessen, D.: IC valuation & measurement – why and how? In: PMA IC Research Symposium, 1–3 October 2003, Cranfield School of Management, pp. 1–28 (2003)
27. Leedy, P., Ormod, J.: Practical Research – Planning and Design, 11.ª Edição, Pearson, pp. 1–409 (2016)

Digital Transformation in Higher Education: Maturity and Challenges Post COVID-19

Adam Marks[1(✉)], Maytha AL-Ali[2], Reem Attasi[1], Abdellah Abu Elkishk[1], and Yacine Rezgui[3]

[1] American University of the Emirates, Dubai, UAE
adam.marks@aue.ae
[2] Zayed University, Dubai, UAE
[3] University of Cardiff, Cardiff, UK

Abstract. Digital transformation in higher education, especially after COVID-19 is inevitable. This research explores digital transformation maturity and challenges post COVID-19. The significance of the study does not only stem from the critical role of higher education in building the workforce and knowledge economy. This study triangulates the findings of multiple research instruments, including survey, interviews, case study, and direct observation. The research findings show a significant variance between the respondents' perception of digital transformations maturity levels, and the core requirements of digital transformation maturity. The findings also show the lack of holistic vision, digital transformation competency, and data structure and processing as the leading challenges of digital transformation.

Keywords: Digital transformation · Higher education · UAE · COVID-19

1 Introduction

Organizations today are operating in a massive, digitally connected world, and their stakeholders expect seamless and personalized digital services (Bayler and Oz 2018). The increased use and production of knowledge places organizations into a necessary digital transformation. This digital transformation affects the core components of an organization - from its operating model to its infrastructure. Organizations usually do not transform by choice, more often when they fail to evolve and keep up with market changes and technology disruptions (Thompson 2013).

Digital transformation refers to an organizational change realized by means of digital technologies and business models with the aim to improve organization's operational performance. It involves much more than implementing a well-chosen technology solution, it is a close alignment between information technology and business processes that will lead to a substantial outcome for the organization, keeping in mind organizational readiness, change management, and managing key stakeholders (Norton and Shroff 2018). Digital transformation is concerned with transforming organizational processes; build new competencies and models through digital technologies in a profound and strategic way (Mahlow and Hediger 2019).

Á. Rocha et al. (Eds.): ICITS 2021, AISC 1330, pp. 53–70, 2021.
https://doi.org/10.1007/978-3-030-68285-9_6

Chaurasive, 2020 states "Digital transformation is not only the transformation of tool, technologies and process but it is transformation of entire business model. It changes the way a business operates and interacts within itself and with external world. Business transformation is a change in mind set that business is evolving faster than we are adapting". Digital transformation should have a strategic objective defined in an operational architecture with business use cases (Chaurasive 2020).

The impact and magnitude of the COVID-19 pandemic forced many sectors to attempt to do business online. Education and higher education institutions across the globe had to make quick and important digital transformation adjustments to sustain operations. Questions about course delivery, virtual classrooms, seats, capacity, conducting exams and assessments, academic integrity, use of web cams, capacity and quality of video conferencing, and many other questions were raised. Many universities signed up with Zoom, MS Teams, Respondus, and other software systems to sustain operations. Higher education institutions today must integrate digital technologies into their business to a much greater extent than before (Marks and Alali 2016) and (Seres et al. 2018).

Literature in digital transformation maturity and challenges, specifically within higher education, and more specifically within developing nations is scarce. This study aims to address those identified gaps. Given the importance of higher education in today's information society and knowledge economy, this study is significant to higher education institutions, as well as to other stakeholders involved in the hierarchy of higher education, including students, educators, researchers, institutions, and government agencies. Digital transformation in higher education, especially after COVID-19 is seen as inevitable not only to compete, but also to survive and sustain key operations. This research explores digital transformation maturity and challenges within United Arab Emirates "UAE", one of the advanced developing nations in terms of IT infrastructure and digital transformation plans. The significance of the study does not only stem from the critical role higher education is responsible in educating and training future leaders, workers, and citizens; but also from the key role digital transformation plays in today's knowledge economy, which became more evident after the COVID-19 pandemic.

The next section of this paper presents some of the literature related to the main topic of this study, followed by the research methodology, the discussion, conclusion, and recommendation.

2 Related Work

Klaus Schwab, the founder of the World Economic Forum (WEF), describes the emergence of the fourth industrial revolution by linking three fundamental factors. These are: "Speed: New technologies that are connected to each other and are very versatile move quickly at an exponential speed, triggering each other. Width and Depth: Digitization speeds up the industry 4.0. However, the increase in technology diversity in the industry has brought about the change. System Impact: Industry 4.0 is expected to undergo a total change as digital industries, companies, and even countries" (Schwab, 2016).

There are several reasons why organizations undergo digital transformation; however, the main reasons are related to the issues of competitive advantage and survival. Digital transformation of an organization represents an objective process capable at responding to disruption in critical functions and changing organizations environments (Schwertner 2017) (Solis and Anon 2017).

As in other industries, four elements are driving digital transformation in education: customer experience, competitiveness, profitability, and agility (Clark 2016). The impact of digital transformation transcends technology. Through the process of digital transformation, organizations use multiple new digital technologies, with the intent to achieve superior performance and sustained competitive advantage. In such way, they transform different dimensions of business, such as the business model, the customer experience and operations, and simultaneously impacting people and networks (Mahlow and Hediger 2019), (Ismail 2017).

Ismail (2017) presents six dimensions to digital transformation, they are: digital strategy, adaptable collaborative processes in modern business models; Complete automation of business processes; Detailed analysis and research of customers' decision making; Information technology supporting all organizational business processes; Usable and relevant data, use of data analytics as a basis for decision making in line with the organization's goals and strategy (Ismail 2017). Bounfour (2016) talks about four dimensions of digital transformation, namely "the purpose", "degree of strategy", "speed of strategy", and "the value source". Table 1 shows those dimensions in more details.

Table 1. Digital transformation: dimensions, issues, and implications

Dimension of digital transformation	Questions for Managers (Strategy, Organization, Business Models)	Main topics
The purpose of the digitalization strategy	Which analytical methods will be selected in the company? What are the spaces for development and value creation	Determining and analysing the value creation spaces
Degree of Digitization Strategy	What is the relative importance of platforms? What kind of typology? Which governance structures promote innovation?	Defining and analysing the idea of creating new platforms
The Speed of Digitization Strategy	How to define and define innovative offers	Fast and systematic analysis of phenomena
Value sources, creation based on digital strategies	What are the sources of value creation in digital space?	Define the proposed values in the digital space

Source: Bounfour, 2016

Figure 1 outlines the main sectors that have been disrupted by digital transformation. Public Sector related entities, including education, are being disrupted now by digital transformation. The opportunity for the Public Sector is to learn from the previous experiences of other sectors.

In the Middle East and North Africa "MENA" region, the trends of information and communications technologies (ICT) are very diverse due to different levels of development both between and within countries. This gap can be attributed to different aspects including infrastructure, economic conditions, job market, and lack of adequate governance. Nevertheless, nearly all countries in the region are pursuing policies supporting digitalization to further development.

Countries – such as the United Arab Emirates (UAE), and Saudi Arabia, are well equipped for further technological development (Göll and Zwiers 2018). However, the National ICT Index still shows that those countries still lag behind other developed economies in terms of Digital Government capabilities (Deloitte 2019) and (Parlak 2017). To address their own economic, social and environmental challenges, several governments in the MENA region have launched national transformation plans with a focus on enabling ICT and Digital Transformation technologies. Integrating digital technologies to develop smarter cities and become smarter nations is a key outcome of their national visions (Deloitte 2019) and (Limani et al. 2019).

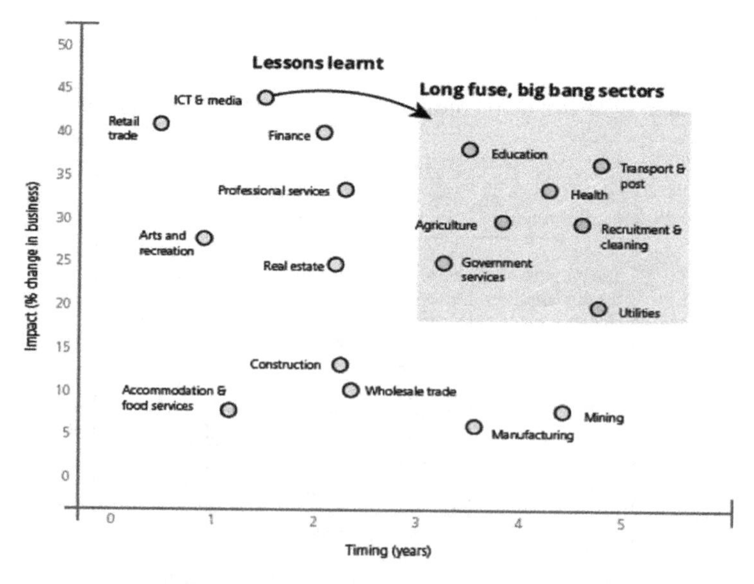

Fig. 1. Sectors affected by digital transformation. Source: (Deloitee 2019)

Globally, higher education digital transformation is highly influenced by government policies and institutional development strategies (Walker, et al. 2016). The aim of the digital transformation process in higher education is to redefine educational services and redevelop operational processes. This can be achieved using three possible approaches. The first approach involves service-first transformation. It focuses on changing and redefining services before making key improvements and changes to operations. The second approach is the operation-first transformation. In this approach, the higher education institution identifies new and amends current digital processes, activities and operations. The third approach is service-operation combination, involves integrated transformation through systematic interrelation of both previous approaches (Sandkuhl and Lehmann 2017). As shown in Table 2, Petkovics et al. 2014 maps higher education business processes into four hierarchical levels: mega processes, major processes, sub-processes, activities and tasks. The mega processes include the learning and teaching process, the research process, the enabling process, and the planning and governance process (Petkovics 2014).

Table 2. Overview of mega and major processes in higher education institutions

Learning and teaching process	Research process
• study programme accreditation	• research planning
• teaching process preparation and realisation	• research preparation
• teaching process outcomes monitoring	• research conduct
• teaching process assessment	• research outcomes monitoring
• student and teacher mobility realisation	• research evaluation
Enabling processes	**Planning and governance processes**
• student administration services	• organization management services
• library services	• change and business process management
• staff provision and development services	• plan development
• finance and accounting services	• budget and funds planning
• marketing, sale and distribution services	• performance assessment
• procurement services	

Source: [Petkovics, 2014]

The Organization for Economic Co-operation and Development "OECD", 2016 study provide a similar presentation of higher education business processes and example of new digital trends, as shown in Table 3.

Deloitte presents a Digital Maturity Assessment Framework, using five key critria:

- Does the organization have the right vision and strategy for digital, and the leadership, communications and focus required to support this vision?
- Does the organization have the right talent, skills and knowledge to support its vision, products, and services?
- Does the organization have the right processes, controls and digital technologies to support the operations of the organization?
- Does the organization have the right technologies and infrastructure as well as the ability to develop, manage and deliver?
- Does the organization have the right approach to understanding and communicating with its customers to succeed in a digital environment? (Deloitte 2019).

Figure 2 visualizes results from Gartner's 2017 CIO Survey, shows where higher education intuitions are in terms of digital transformation. Organizational mind-set is what separates the No digital initiative institutions from those with Desire/Ambition. This transformation requires a framework that is effectively communicated to key stakeholders and decision makers. If employed correctly, digital transformation can play a major role in today's higher education including in the areas of admission tracking, enrollment optimization, and academic advising (Gartner 2017).

Table 3. Categories of educational services and new digital trends

Categories of educational services	Examples of new digital trends
Administration	University College administrative processes, for example, application for enrolment, enrolment for exams, generation of grade mirrors, class schedule, literature download - are increasingly being digitized as part of e-government programs and student requirements.
Communication	Communication between professors, students, and others is also a field with a high level of digitization. With collaboratively accessible collaborative platforms, colleges can save time and provide easier and more flexible online communication - provided that students are ready and have sufficient skills and access.
Teaching and preparing the lesson	With the ongoing development and growth of digitalization, the replacement of traditional literature, books, and prints with online learning resources, as well as the organization and preparations for the exchange of knowledge and practices are also being replaced.
Teaching and Learning	The direct traditional interaction between the professor and the student is developing slowly, while the forms of learning and online learning emerge
Reviews and Examinations	Reviewing letters, tests, and exams is also a part of the lesson time. This area is currently being digitized, though the markets seem to be rooted in the digitalization of teaching materials and teaching in general.

Source: OECD, 2016

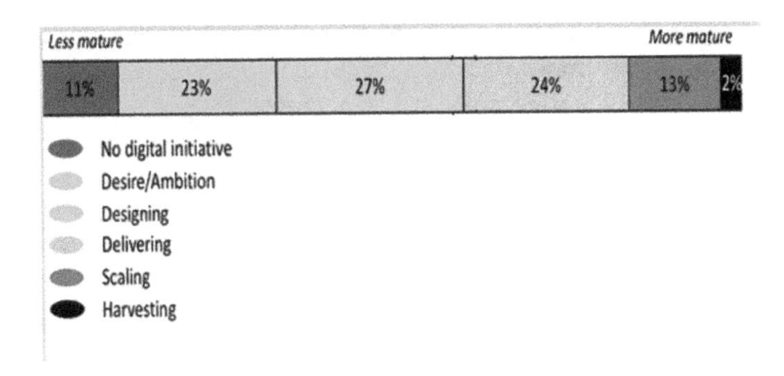

Fig. 2. Digital transformation maturity in higher education

The literature shows that digital transformation is usually faced with a number of challenges. More often than not, those challenges are not listed in any specific order based on criticality, and they are not attached to a specific industry. Challenges reported include: the changing customer experience/expectations, resistance to change, resistance to technology, lack of leadership support, lack of competency and digital transformation skills, failing or poor analytics, lagging and legacy business models and systems, poor planning, misalignment with business strategy, technology and data challenges, lack of clear vision, and digital literacy of stakeholders. (Maltese 2018), (NV 2017), (Chaurasiya 2019).

If implemented correctly, digital transformation tools and technologies such as Artificial Intelligence (AI), Internet of Things (IoT), big data, block chain, social analytics, and cloud services can enhance and change educational system practices, especially in a world where students are constantly interacting with technology in all other aspects of their everyday lives, digital transformation clearly offers opportunities for student engagement that are not always available in the fixed environment of the classroom (Al Tamimi & Company 2019) (SAP 2018). (Xiao 2019).

Seres et al. (2018) discuss four ways where digital transformation may change how education looks in the future, namely smart content, differentiated and personalized learning, global and remote learning, and administrative efficiencies.

In the area of "Smart Content", digital transformation can be attained through e-books, new learning platforms, virtual content such as video lectures and conferences, electronic curricula, distributed educative information across devices.

Similarly, in the area of "Differentiated and Personalized Learning", digital transformation can be attained through personalized electronic tutoring customized to the learning styles and particular needs of the student. The traditional curriculum is designed to suit as many students as possible. For students in the top 10% and the bottom 10%, AI for example can be used to provide testing and feedback to those students to give them challenges they are ready for, identify gaps in knowledge and re-direct them to new topics when appropriate.

In the area of "Global and Remote Learning", digital transformation can facilitate learning from anywhere and at any time. Furthermore, it can be used to support students with homework and exam preparation remotely with advanced tutoring and study programs.

Last, in the area of "Administrative Efficiencies", digital transformation can support higher education to shift from wondering about the future into predicting, forecasting, and shaping the future; making proactive and informed decisions and taking action based on that information. Digital transformation can support universities in using conventional and unconventional (unstructured), internal and external data to discover hidden patterns underlying performance in different areas, track admissions, optimize enrolment, manage grants, enhance academic advising etc. Digital transformation can help higher education institutions to know what is happening (descriptive analytics), what is likely to happen in the future (predictive analytics) and to examine trends, causes and likely forecasts and use that information to make decisions (prescriptive analytics) (Seres, Pavlicevic, Tumbas 2018).

3 Methodology

While the importance of digital transformation is recognized, data about digital transformation maturity and challenges is scarce, especially in developing nations, and specifically within higher education. This study explores digital transformation maturity level in Higher Education Institutions in UAE higher education. The study uses a new framework that is based on the Petkovic's et al. 2014 mega and major processes mapping, and Deloitte's 2019-maturity assessment framework. The researchers believe that Petkovic provides a balanced and comprehensive classification of higher education business processes using four hierarchical levels: mega processes, major processes, sub-processes, activities and tasks. Unlike other classifications, the Petkovic's classification does not contain overlapping, ambiguous, and repetitive processes. The researchers also believe that digital transformation maturity criteria listed in the maturity assessment framework by Deloitte are comprehensive, tangible, easy to understand and reflective. Based on the above, the combination of both models provides a good starting point for higher education institution to assess their digital transformation maturity level, and identify areas that need improvements. The proposed framework in this study is flexible, customizable, and can support further more detailed analysis as required.

The study examines public and private higher education institutions in the UAE. It ranks the criticality of digital transformation challenges using identified pattern codes such as the regulatory and business environment, IT infrastructure, data governance, affordability and budget constraints, personnel competency and IT skills, etc. We consider that the problem identification and related key issues are very important in order to achieve successful implementation of digital transformation.

The goal of this study is achieved throughout several objectives, beginning from the literature study of the state of the art, continuing with the wide-distributed survey, in-depth semi-structured interviews, direct observation of the researchers, and case study. The literature study provided the possibility of identifying and analyzing trends related to the topic, while the survey, interviews, direct observation, and case study provided the possibility of identification and analysis of trends in the field of digital transformation at the national level in public and private higher education institutions. This study is a phenomenological research to determine the views of IT director and

Several authors were identified with *papers* about the *intellectual capital*[1]. By countries, can be seen, in particular, graphically in this Fig. 1.

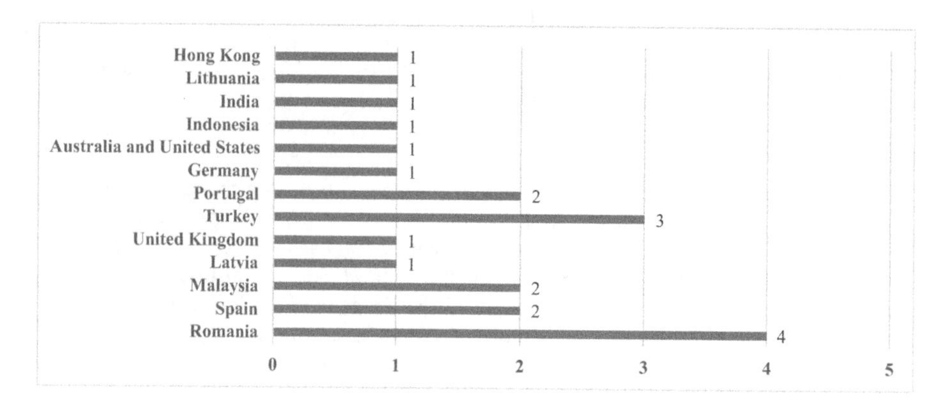

Fig. 1. Number of papers by country

As can be observed the countries are ranked according this order:

- [Romania (4): [5, 7, 8] e [3]],
- [Turkey (3): [13, 14] e [15]];
- [Spain (2): [9] e [10]], [Malaysia (2): [4] e [11]] and [Portugal (2): [16] e [17]]
- [Latvia (1): [1]], [United Kingdom (1): [12]], [Germany (1): [18]], [Australia and United States (1): [19]], [Indonesia (1): [2]], [India (1): [6]], [Lithuania (1): [20]] and [Hong Kong (1): [21]].

About 71% (15/21 of the total *papers*) of the sample are related to the years from 2014 to 2017. The underlying criteria was to know what about *intellectual capital* has been published from more to less recent dates in order to get an idea about the latest *State-of-the-Art* as much as possible. It is also the rationale of doing these research.

According [27, p. 81] [United States], a qualitative approach has the objective to *explore, explain and interpret*, the purpose of the research. So, our methodology is qualitative. The data are textual and were collected in journals regarding the *intellectual capital*, for instance, *Journal of Intellectual Capital*, *Procedia Social and Behavioral Sciences*, *Procedia Computer Science*, and others. In all of these sources, scientific academic information were collected and the *papers* were also selected.

Until 2020, with these authors, there is no consensus, about *what* is *intellectual capital*. For two reasons: there isn't a definition only ([10] [Spain]) and there are differences in the meanings about it – components and number of them ([1] [Latvia], [3] [Romania] and [6] [India]).

Also there is no consensus, about *how can be measured* the *intellectual capital*, as referred by [22] [Netherlands and Belgium], [23] *in* [1] [Latvia] and [24] [USA].

[1] See [25] [Portugal] *in A Proposal to Measure Intellectual Capital Using the Andriessen's (2004) Method: Application to a Case Study in a National Context (Portugal).*

senior academicians on the maturity and challenges of digital transformation. Phenomenological researches may not reveal generalizable situations, but they can provide examples, explanations and experiences that will help to a phenomenon identified and understood better (Yıldırım & Şimşek, 2013), (Limani et al. 2019)

The survey was conducted in both public and private higher education institutions, targeting IT directors, chief information officers, and senior academicians concerned with digital transformation. The survey was sent to 61 individuals. Response was received from 52. The questionnaire design and construction consist of 15 Likert Scale closed-ended multiple-choice-five-pointer questions. Respondents were required to complete the questionnaire indicating the extent to which they agree or disagree with the questions. A room for comments for each question was also available. The survey questions were divided into three sections; the first section measures the respondent's view on the institution's level of digital transformation maturity; the second section verifies the existence/non-existence of key elements of digital transformation maturity; and the third section ask about the respondents rating of the challenges faced

Six in-depth, semi-structured interviews were held with IT directors, and another four interviews were held with senior academic administrators to gain deeper understanding of expected value and the challenges faced during digital transformation; direct observation was used to verify what people do, rather than what people say they do, lastly, a case study was conducted at one of the public universities to validate and triangulate the results of the survey, direct observation, and the interviews.

The importance and the rationale of this research lie in the identification and analyses of the readiness of higher education institutions in the UAE to embrace a meaningful and mature level of digital transformation in higher education academic processes post COVID-19.

The research results can be used as an important input to the design of new academic processes that would be more effective, aligned, efficient, and cost-effective. Moreover, this study points to the key challenges faced by higher education institutions in the UAE in achieving mature digital transformation, turning data into a valuable asset that could be used for prescriptive, corrective and predictive decision-making, using a proposed framework to measure digital transformation maturity, and pinpoint areas of concern and areas of strength. The research also provides the practitioners from the field of digital technologies with the information and knowledge related to their potential market and the related trends.

4 Discussion

In this section, we present the findings of this study, responding directly to the two key research questions

What is the level of digital transformation maturity in the examined UAE higher education institutions?

The data collected in this study shows a significant variance between digital transformations maturity level perception reported by the respondents, and the core requirements of digital transformation maturity. While more than 80% of the examined institutions reported a digital transformation maturity level between "delivering or

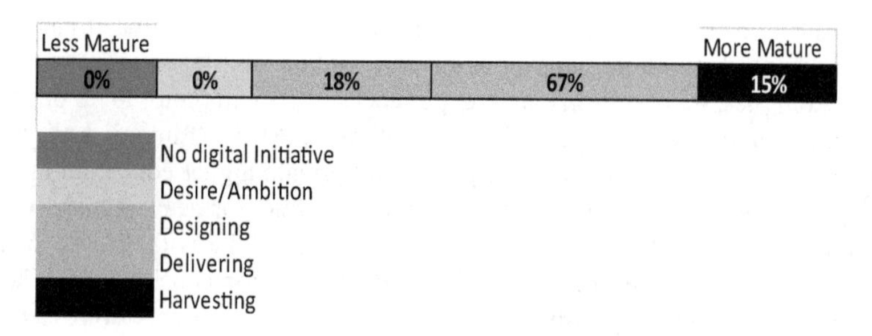

Fig. 3. Digital transformation maturity reported

harvesting" as seen in Fig. 3, none of them had a comprehensive digital transformation plan. A few reported a list of digital transformation initiatives. However, those list of initiatives were more aligned with automation not digital transformation, and they were mostly initiated to comply with external regulatory requirements by the Telecommunication Regulatory Authority TRA. In almost all cases, digital transformation initiatives had no connection to a return on investment, gained value, or a real transformation of a business process. The great majority of cases were concerned with the automation of electronic forms, adding workflow and approvals. None of the initiatives reported was concerned with analytics, machine learning, artificial intelligence, big data and other more-advanced digital transformation techngies.

Based on the proposed higher education digital transformation maturity assessment model, it was noticed that most institution focused their digital transformation effort in the area of enabling processes, while much less in planning and governance, learning and teaching, and research respectively (see Table 2). Respondents attributed this to the availability of third party systems supporting student administration, library services, finance an accounting, etc. Ellucian and Oracle are among the main contenders in this area. Systems supporting accreditation, research, and curriculum management are relatively new in the market, compared to systems supporting enabling processes. Finally, the data findings shows clearly how critical is the role of IT governance in ensuring that all mega and major processes receive the needed support. A segmented structure of data ownership can deliver a segmented vision that can directly affect digital transformation completeness, effectiveness, and alignment.

What are the key digital transformation challenges facing the UAE examined higher education institutions?

The data findings of this study reveals a number of digital transformation challenges. While some of the reported challenges are consistent with some of the challenges reported in literature, some of the challenges listed were specific to the UAE higher education. Table 5 lists digital transformation challenges in UAE higher education, as reported by the respondents.

4.1 Wholistic View

The data supported findings of this study show that the most critical challenge facing digital transformation in UAE higher education is the lack of a wholistic digital transformation vision. The data reveal that none of the examined institutions had a stand-alone digital transformation vision or plan. Two public universities had list of key performance indicators to satisfy government requirements. Most of the other institutions worked on a list of ad-hoc automation initiatives, mainly driven by IT personnel. Some of the respondents reported confusion about digital transformation ownership (Table 4).

Table 4. DT maturity reported by mega and major process

Higher Education Mega and Major Processes	DT Maturity
Learningh and Teaching Processes	
Study Program Accreditation)	32%
Teaching processes preperation and reliazition	74%
Teaching process Assessment	22%
Student and Teacher Mobility Realization	38%
Enabling Processes	
Student Administration Services	92%
Library Services	86%
Staff provosion and development services	65%
Finance and accounting services	83%
Marketing, sales, and distribution services	56%
Procurement services	83%
Research Processes	
Research planning	36%
Research preperation	32%
research conduct	18%
Research outcome monitoring	18%
Research evaluation	24%
Planning and governance processes	
Organziation management services	54%
Change and business process management	42%
Plan development	64%
Budget and fund planning	88%
Performance assets	36%

Table 5. Digital transformation challenges in UAE higher education

Digital Transformation Challenge	
Wholistic Vision	76%
Personnel Competency and IT Skills	54%
Data Structure, Data Processing, and Data Reporting	52%
Redundant Systems	42%
Third Party Reporting Systems	42%
Manual Entries (Middle Man)	38%
Potential Use by Customers	28%
Regulatory and Business Environment	16%
Social and Economic Impact	12%
Privacy and Security Concerns	4%
IT infrastructure	3%
Affordability and Budget Constraints	2%
Other capability constraints	1%

4.2 Personnel Competency and IT Skills

The second critical challenge of digital transformation reported was the lack of critical IT skills. IT Personnel in the UAE are mainly comprised of Asian expatriates. IT personnel lack previous experience in the higher education sector, and support English-based systems, English-based curricula, and operations. Similarly, many IT managers and directors did not have prior higher education or technical experience, which is critical in establishing a wholistic digital transformation vision and plan.

Respondents reported systems that were not used, while annual license is regularly paid; in-house developed systems that were redundant; IT governance, that was not established, requests for new systems driven by individuals based on familiarity; segmentally initiated digital transformation decisions; poor and underdeveloped automation efforts that did not create any real value for the institution or provide services with customer –centric view.

Given the profile of the workforce in the UAE, this digital transformation challenge is more specific to the UAE higher education environment. Experience with systems such as Banner, people soft, campus solution, etc. is more difficult to attract in the MENA region compared to the US and Europe. With few exceptions, many universities are supporting critical operations, including admission, registration, advising, scheduling, and graduation with personnel that have had no prior experience with those systems or higher education.

4.3 Data Structure, Data Processing, and Data Reporting

The Third critical challenge reported was data structure, processing, and reporting (input-process-output). This challenge can be linked to the second reported challenge, as it is also more pertinent to the UAE higher education environment.

Because of the lack of fundamentals such as an IT strategy, IT governance, and/or data governance, data structures and key codes were not setup correctly, and key modules were not utilized. For instance, one institution used Banner Student Information Systems "SIS", but did not utilize Banner workload module. Instead built a stand-alone system to manage faculty workload. No one knew that a small number of forms need to be populated and a fully integrated module will be available in a more efficient way.

Inconsistent college codes, program codes, major codes, etc. used across different creating inconsistent data outputs that is very difficult to verify. Business rule codes were also not well defined, organizations ended up with problems in critical academic, business, and financial areas, including major out of balance issues. Data reporting relied heavily on data extraction and ad-hoc (backend) queries and reporting. System built-in reports were limited or short of customer's specific needs; keeping in mind that most systems are made to align with the US higher education environment including Ellucian, Blackboard, Leepfrog, CurriCUNET, Taskstream, Oracle, etc. This variance in input-processing-output created a major hurdle for integration, consistency, and reporting, resulting in failure to create value and enable an effective digital transformation strategy.

4.4 Redundant Systems

The fourth challenge facing higher education institutions in UAE was the existence of redundant systems. As referred to earlier, due to the lack of proper IT and data governance, several organizations did not have a proper system/software acquisition process in place, subsequently resulting in systems, functions, and data overlap and redundancy, creating major challenges for integration and data consistency, reliability, integrity, availability, timeliness, and confidentiality.

4.5 Third Party Reporting Systems

The fifth challenges cited by higher education institutions attempting to digitally transform their processes was the existence of several external reporting agencies/systems that require different data sets, formats, and requirements, including different accreditation reporting systems, and other compliance reporting systems. Several institutions felt that need to manually extract the required data from different internal systems, then format the data sets as required for reporting purposes. Microsoft Excel formatted files are widely used to support this function.

4.6 Manual Entries

Because institutions were required to report to multiple external entities using third-party reporting systems, a lot of data extraction, data manipulation, data formatting, and data entry was taking place; in several cases the people responsible for data extractions from the organization's different systems, were completely different from those making manual entries into third party systems, potentially leading to system and data and submission errors.

4.7 Customer Adoption

The seventh challenge facing higher education institutions in UAE was the potential use by customers. Several respondents cited resistant to change, resistant to technology, buy-in, awareness, and training as leading causes for adoption challenges. Some of the respondents stated that processes were detached from systems causing loopholes, delays, redundancy, and errors. Other respondents cited off-the shelf systems as generic, while others did not support in-house developed system by IT personnel, and viewed them as temp-cheap solutions, driven by the insecurity of IT personnel to guarantee their jobs.

4.8 Regulatory and Business Environment

The eighth challenge extracted from the data set was the regulatory and the business environment. Private institutions felt more at liberty than government institutions in this category. This is natural, given that government institutions receive full funding from the government, and the government audits their books. In addition, the Telecommunication Regulatory Agency TRA has its own protocols and requirements that must be observed, including what data can or cannot be on the cloud. The purchasing process in government institutions also has its own special requirements that may linger or hinder the process of acquiring certain IT assets that would support digital transformation.

4.9 Social and Economic Impact

Although was only cited by 12% of the respondents, the ninth challenge of digital transformation was concerned with the social and economic impact. Some higher education institutions reported that some of the digital transformation initiatives were not rolled out due to social concerns about how the institution will be viewed, including cases where different genders may have direct communications or use of web cams. Many institutions stressed the importance of showing respect of the UAE culture and norms as one of the main factors for attracting UAE national students.

4.10 Privacy and Security Concerns

Only cited by 4% of the respondents, privacy and security concerns was cited as the tenth challenge to digital transformation. Some universities did not feel that their hardware, security, and network was not ready to manage potential security threats that could come with the expansion of their digital infrastructure.

4.11 Affordability and Budget Constraints

Surprisingly only 2% of the respondents reported affordability and budget constraints as the twelfth challenge facing digital transformation in their organizations. Understandably, those were from small private universities, not government universities.

4.12 Other Capability Constraints

The final and the thirteenth challenge reported was reported by only 1%, and it was concerned with random capability reasons that were not directly identified, but simply reported as capability constraints.

4.13 IT Infrastructure

While the majority of respondents viewed their universities IT infrastructure as mature and ready to support digital transformation, 3% of the respondents expressed concerns about the full readiness of their IT infrastructure in its entirety, and reported it as the eleventh challenge.

5 Conclusion

Digital transformation is one of the biggest catalysts of the business environment today, and higher education is not excluded from this evolution. It is a move that goes beyond the scope of systems and new technologies, while also representing the modernization of organization philosophy, purpose, competition, and patterns that change with emerging audiences. As the business environment, students, and employees change, they do so at an accelerated speed that often exceed an organization's ability to adapt. This disruption causes critical business functions and processes to inevitably be exposed within and outside the organization; subsequently requiring the restoring of new investments in technology, business models and processes to more effectively compete in a continual digital economy shift. Digital transformation is an inevitable choice for higher education institutions everywhere, especially after COVID-19. Digital transformation is a process that can hardly be historically compared to any other process, as it does not exclude the development levels of different countries.

In other words, all countries, regardless of their development level must undergo some level of digital transformation; and while in the developed world, the need for digital transformation has been reinforced and installed, and organizations and governments have developed sophisticated methods for applying digital technology to create products or to deliver certain services, and add value, some developing countries are still attempting to move from desire and ambition to planning, delivering and harvesting.

Despite all the talk about digital transformation in developed and developing countries, and across all industries, the reality is that digital transformation is only as useful as its rate of true implementation and return on investment. Otherwise, organizations will not benefit in terms of efficiency, effectiveness, cost-savings, competitive advantage, and decision-making.

For a number of decades, higher education institutions globally claimed digital transformation maturity, citing students' information systems, learning management systems, etc. The COVID-19 pandemic forced many institutions to use remote teaching, disrupting the regular and normal business environment and operations, subsequently exposing critical functions and their true level of digital transformation maturity and challenges.

The UAE is one of the leading developing nations in terms of IT infrastructure, and the adoption of new technologies. The UAE government has made significant leaps in e-government, e-commerce, e-business, and e-services in general. There are several agencies contributing to this advancement, including the Ministry of Artificial Intelligence, Smart Dubai, and the Telecommunication Regulatory Authority.

Given the importance of digital transformation, higher education, and the role they both play in today's digital/knowledge economy, the aim of this study is to examine the digital transformation maturity level and challenges in UAE higher education institutions in the after math of COVID-19, and the need to provide remote e-service to students, employees, and other customers and stakeholders.

The first research question is concerned with measuring the level of digital transformation maturity in UAE higher education institution using Deloitte's digital maturity assessment framework, and Pekovitcs mega and major processes mapping. The data findings reveal a major variance between the perception and the requirements of digital transformation maturity. The examined institutions did not have a digital transformation vision, leadership, strategy, plan, champions, processes, controls, approach, communication, or proper return on investment. Many of the examined institutions viewed their maturity level at delivering and harvesting, when in fact they were at either designing or ambition.

In addition, digital transformation was more evident in enabling processes such as student administration services, library services, finance and accounting, but not as much in learning and teaching, research processes, and planning and governance processes.

The second research question is concerned with the digital transformation challenges. Leading challenges reported included challenges with wholistic vision; personnel competency and IT skills; data structure, data processing, and data reporting; redundant systems; third party reporting systems; manual entries; and potential use by customers. The challenges cited by the respondents in this study are not mutual exclusive; in fact, they are interrelated in multiple ways. While some of the challenges are more critical than others, the combination of those challenges create an environment that hinders digital transformation and business success by creating dependency, timeliness, integrity, availability, cost, efficiency, effectiveness, and integration issues.

6 Recommendations

Digital transformation in education is inevitable. Higher education institutions should establish a clear vision, policies, strategies, and plans to support mature digital transformation. Institutions should communicate such policies, strategies, and vision, and receive feedback from internal and external customers and stakeholders about business process engineering and return on investment. Such plans should regularly be evaluated. Institutions should hire digital transformation experts in order to align the business strategy with digital transformation. Digital transformation should not be just another task handed down to IT personnel, or segmented among data owners. The difference between automation and digital transformation should be communicated, and training and awareness should be provided. A corner stone to all of this is to show solid management support to combat resistance to change and resistance to technology, and communicate the long term value gained from digital transformation. Digital transformation should be extended beyond the enabling processes to teaching and learning, governance, and research; specifically, the areas of course, program, and student assessment and evaluation.

References

Al Tamimi & Company. Report on Digital Transformation in the Education Space: A Review of the Impact of New Technologies on Middle East Education (2019)

Baydas, O., Küçük, S., Yilmaz, R., Aydemir, M., Göktaş, Y.: Educational technology research trends from 2002 to 2014. Scientometrics **105**(1), 709–725 (2015). https://doi.org/10.1007/s11192-015-1693-4

Bayler, A., Oz, S.: Academicians' views on digital transformation in education. Int. Online J. Educ. Teach. (IOJET) (4), 809–830 (2018)

Bond, M., Zawacki-Richter, O., Nichols, M.: Revisiting five decades of educational technology research: a content and authorship analysis of the British Journal of Educational Technology. British J. Educ. Technol. **50**(1), 12–63 (2019). https://doi.org/10.1111/bjet.12730

Chaurasiya, D.: Digital Transformation: A Case Study (2019). https://www.researchgate.net/publication/340385414_Digital_Transformation?enrichId=rgreq-e65c791fef4fde3c103a4370994ad74e-XXX&enrichSource=Y292ZXJQYWdlOzM0MDM4NTQxNDtBUzo4NzU4ODQyMzE3OTA1OTNAMTU4NTgzODQ4MjI1MA%3D%3D&el=1_x_2&_esc=publicationCoverPdf

Clark, E.: Digital Transformation: What Is It. Educause (2016)

Gartner, Inc.'s free (2017): Creating Digital Value at Scale. Webinar is a special report based on the opening keynote of the 2017 Gartner Symposium (2018)

Göll, E., Zwiers, J.: Technological Trends In The Mena Region: The Cases Of Digitalization And Information and Communications Technology. (Ict), MENARA. (2018)

Ismail, M.H., Khater, M.: Digital Business Transformation and Strategy: What do we know so far? (2017). https://cambridgeservicealliance.eng.cam.ac.uk/resources/Downloads/Monthly%20Papers/2017NovPaper_Mariam.pdf

Limani, Y., Hajrizi, E., Stapleton, L., Retkoceri, M.: Digital Transformation Readiness in Higher Education Institutions (HEI): The Case of Kosovo", Science Direct, IFAC PapersOnLine, pp. 52–25 (2019)

Mahlow, C., Hediger, A.: Digital transformation in higher education—buzzword or opportunity. Special Issue: Paradigm Shifts in Global Higher Education and eLearning, May 2019

Maltese, V.: Digital transformation challenges for universities: ensuring information consistency across digital services. J. Cataloging Classification Quart. **56**(7) (2018)

Marks, A., Alali, M., Reitesema, K.: Learning management systems: a shift toward learning and academic analytics. Int. J. Emerg. Technol. Learn. (2016)

Mehaffy, G.L.: Challenge and change. Educause Rev. **47**(5), 25–42 (2012)

Digital McKinsey. Digital Middle East: Transforming the region into a leading digital economy. Mckinsey (2016)

NV. Digital Transformation in Higher Education. (Online) Navitas Ventures (2017). https://www.navitasventures.com/wp-content/uploads/2017/08/HE-Digital-Transformation-Navitas_Ventures_-EN.pdf

Norton, A., Shroff, S., Edwards, N.: Digital Transformation: An Enterprise Architecture Perspective, Publish Nation Limited, UK (2020)

The Organization for Economic Co-operation and Development OECD. Digital Government Strategies for Transforming Public Services in the Welfare Areas (2016). http://www.oecd.org/gov/digital-government/Digital-Government-Strategies-Welfare-Service.pdf

Parlak, B.: Dijital çağda eğitim: Olanaklar ve uygulamalar üzerine bir analiz [Education in Digital Age: An analysis on opportunities and practices]. Süleyman Demirel University, Journal of Faculty of Economics and Administrative Sciences **22**(15), 1741–1759 (2017)

Petkovics, I., Tumbas, P., Matkovic, P., Baracskai, Z.: Cloud computing support to university business processes in external collaboration. Acta Polytechnica Hungarica **11**(3), 181–200 (2014)

Sandkuhl, K., Lehmann, H.: Digital Transformation in Higher Education – The Role of Enterprise Architectures and Portals, Digital Enterprise Computing (2017)

Schwab, K.: Dördüncü sanayi devrimi [Fourth industrial revolution]. Optimist Publications, İstanbul (2016)

Schwertner, K.: Digital transformation of business. Trakia J. Sci. **15**(Suppl. 1), 388–393 (2017)

Seres, L., Pavlicevic, V., Tumbas, P.: Proceedings of INTED2018 Conference, 5–7 March 2018, Valencia, Spain (2018)

Solis, B.: 8 Success Factors of Digital Transformation. Altimeter. Prophet Thinking (2016). https://www.prophet.com/thinking/2016/02/brief-the-opposite-approach-8-success-factors-of-digital-transformation/

Thompson, J.: Books in the Digital Age. Wiley, New York (2013)

Yıldırım, A., Şimşek, H.: Sosyal bilimlerde nitel araştırma yöntemleri [Qualitative research methods in the social sciences. Seçkin Publishing, Ankara (2013)

Walker, R., Voce, J., Jenkins, M.: Charting the development of technology-enhanced learning developments across the UK higher education sector: a longitudinal perspective" (2001–2012). Interactive Learn. Environ. **24**(3), 438–455 (2016). https://doi.org/10.1080/10494820.2013.867888

Xiao, J.: Digital transformation in higher education: critiquing the five-year development plans (2016–2020) of 75 Chinese universities. Distance Educ. **40**(4), 515–533 (2019). https://doi.org/10.1080/01587919.2019.1680272

Chen, H., Chiang, R., Storey, V.C.: Business intelligence research. MIS Quart. **34**(1), 201–203 (2015)

Daniel, B.: Big data and analytics in higher education: opportunities and challenge. British J. Educ. Technol. **46**(5), 904–920 (2015)

Long, P.D., Siemens, G.: Penetrating the fog: analytics in learning and education. Educause Rev. **46**(5), 31–40 (2011)

Norris, D., Baer, L., Leonard, G., Pugliese, L., Lefrere, P.: Action analytics. Educause Rev. **43**(1), 42–67 (2008)

Sin, K., Muthu, L.: Application of big data in education data mining and learning analytics – a literature review. J. Soft Comput. ICTAC **5**(4), 1035–1049 (2015)

Suhirman, J., Haruna, C., Tutut, H.: Data mining for education decision support: a review. Int. J. Emerg. Technol. Learn. IJET **9**(6), 304 (2014)

Thoronton, G.: The State of Higher Education in 2013. Pressures, Changes, and New Opportunities, 24 March 2014 (2013). http://www.grantthoronton.com/staticfiles/GTCom

Turban, E., Sharda, R., Denlen, D.: Decision Support and Business Intelligence Systems, 9th edn. Pearson Prentice Hall, Upper Saddle River (2011)

Van Barneveld, A., Arnold, K.E., Campbell, J.P.: Analytics in higher education: Establishing a common language. Educause Learning Initiative **1**, 1–11. ELI Paper (2012)

Wixom, B., Ariyachandra, T., Goul, M., Gray, P., Kulkarni, U., Phillips-Wren, G.: The current state of business intelligence in academia. Commun. Assoc. Inf. Syst. **29**(16), 299–312 (2011)

An Information Literacy Framework Through the Conference Paper Format in the Undergraduate Engineering Curriculum

Elizabeth Vidal(✉) and Eveling Castro

Universidad Nacional de San Agustín de Arequipa, Arequipa, Peru
{evidald,ecastro}@unsa.edu.pe

Abstract. This paper shares our experience in developing information literacy skills through a framework based on writing papers in the IEEE format conference. The framework gave students, in an active learning style, a set of activities to identify the need for information, procure the information, evaluate the information and subsequently revise the strategy for obtaining the information, and to use it in an ethical manner to produce a technical paper. We followed the ACRL's Information Literacy Competency Standards for Higher Education in the development of an assessment tool, course content, and exercises. Initial results show that the proposed framework developed information literacy skills in a maturing stage. We believe that this experience and the design of the framework could be replicated or adapted to different Engineering Careers.

Keywords: Information literacy · Technical writing · Searching · Writing in engineering

1 Introduction

Information Literacy (IL) was endorsed by UNESCO's Information for All Programme (IFAP) as a basic human right. IL is described in the Alexandria Proclamation of 2005 as essential for individuals to achieve personal, social, occupational, and educational goals [5]. IL is defined as a set of abilities requiring individuals to "recognize when information is needed and have the ability to find, evaluate, and use effectively the needed information. It is common to all disciplines and it enables learners to master content, extend their investigations, become self-directed, and assume control over their own learning" [8]. The Engineering field in the contemporary environment has recognized the need to develop engineers who have been taught the advances in core knowledge and are capable of defining and solving problems and to be lifelong learners [1, 9, 18]. According to the Association of College (ACRL) [3], IL forms the basis for lifelong learning. Information Literacy has been incorporated in different ways into the curriculum. Most of the literature refers to academic libraries providing first year students with information on how to use their libraries [10, 12–14]. In this paper, we present a framework with five components to develop IL skills. We have considered ACRL's information literacy five standards. The framework has been developed under the course Writing Articles and Research Reports that is part of the new curriculum at

Á. Rocha et al. (Eds.): ICITS 2021, AISC 1330, pp. 71–79, 2021.
https://doi.org/10.1007/978-3-030-68285-9_7

the Escuela Profesional de Ingeniería de Sistemas at Universidad Nacional de San Agustin de Arequipa. The proposal teaches students in an active learning format on how to write conference papers under the IEEE format. While students develop paper-writing skills they develop information literacy skills when refer to recognizing the information they need, locating and evaluating the quality of information, storing and retrieving information, making effective and ethical use of information, and applying the information to write their articles. We have found that the iterative phases of the process of writing an article give a framework to develop information literacy skills. Initial results showed that students became aware of information literacy skills, but they also acknowledged that it is a continuous process. The rest of the paper is organized as follows. Section 2 describes the Information Literacy description, standards, and performance indicators. Also describes the writing requirements under the conference format for the Engineering field. The rest of the paper is organized as follow: Sect. 3 describes the proposed framework. Section 4 describes the initial results and discussion. Finally, we show our conclusions.

2 Background

2.1 Information Literacy

For UNESCO's Information for All Programme (IFAP), Information Literacy is the capacity of people to: recognize their information needs, locate and evaluate the quality of information, store and retrieve information, make effective and ethical use of information, and apply information to create and communicate knowledge [5]. The Association of College, Research Libraries (ACRL) [3] considers that an information literate individual can: determine the extent of information needed, access the needed information effectively and efficiently, evaluate information and its sources critically, incorporate selected information into one's knowledge base, use information effectively to accomplish a specific purpose and understand the economic, legal, and social issues surrounding the use of information, and access and use information ethically and legally. The ACRL determines five standards and twenty two performance indicators that allow faculty to evaluate the assessment of information literacy skills, The description is shown in Table 1.

2.2 Technical and Scientific Writing in Engineering

Considering that scientific research and scientific communication constitute a process indivisible, for a student, publishing scientific articles should be mandatory [16]. Academic writing is a complex task that must be taught. Based on the sociocultural perspective [17] that affirm that writing is social practices that vary depending on the contexts in which it is carried out we have considered regular practices and recommendations of writing in a specific context Engineering context: paper conferences. The Institute of Electrical and Electronics Engineers (IEEE) is a professional association for Electronic Engineering and Electrical Engineering (and associated disciplines); it is the world's largest association of technical professionals that produces over 30% of the world's

Table 1. ACRL Information Literacy Standards and Performance Indicator

STANDARD 1: The information literate student determines the nature and extent of the information needed.
1. Defines and articulates the need for information.
2. Identifies a variety of types and formats of potential sources for information.
3. Considers the costs and benefits of acquiring the needed information.
4. Reevaluates the nature and extent of the information need.
STANDARD 2: The information literate student accesses needed information effectively and efficiently.
5. Selects the most appropriate investigative methods or information retrieval systems for accessing the needed information.
6. Constructs and implements effectively- designed search strategies.
7. Retrieves information online or in person using a variety of methods.
8. Refines the search strategy if necessary.
9. Extracts, records, and manages the information and its sources.
STANDAR 3: The information literate student evaluates information and its sources critically and incorporates selected information into his or her knowledge base and value system.
10. Summarizes the main ideas to be extracted from the information gathered.
11. Articulates and applies initial criteria for evaluating both the information and its sources.
12. Synthesizes main ideas to construct new concepts.
13. Compares new knowledge with prior knowledge to determine the value added, contradictions, or other unique characteristics of the information.
14. Compares new knowledge with prior knowledge to determine the value added, contradictions, or other unique characteristics of the information.
15. Validates understanding and interpretation of the information through discourse with other individuals, subject-area experts, and/or practitioners.
STANDAR 4 The information literate student, individually or as a member of a group, uses information effectively to accomplish a specific purpose.
16. Applies new and prior information to the planning and creation of a particular product or performance.
17. Revises the development process for the product or performance.
18. communicates the product or performance effectively to others.
19. Comunica el producto o el rendimiento de manera efectiva a los demás.
STÁNDAR 5 : The information literate student understands many of the economic, legal, and social issues surrounding the use of information and accesses and uses information ethically and legally.
20. Understands many of the ethical, legal and socio-economic issues surrounding information and information technology.
21. Follows laws, regulations, institutional policies, and etiquette related to the access and use of information resources.
22. Acknowledges the use of information sources in communicating the product or performance.

literature in the electrical and electronics engineering and computer science fields, publishing approximately 200 peer-reviewed journals and magazines.

IEEE publishes more than 1,200 leading-edge conference proceedings every year. IEEE has standards and recommendations for technical and scientific writing [11]. IEEE gives specific recommendations like paper format (two-columns), organization of sections and subsections, relationships among parts of an argument, formatting, and navigational aids. When refers to writing IEEE highlights choosing words and phrases in short sentences, writing in regular sentence patterns, and including linguistic cues to indicate the structure of an argument. But what makes it different from other fields are the recommendations about *the use of* less textually dense alternatives to paragraphs such us: tables, pictures, charts, graphs formulas, and codes. Each of these can be used effectively to make argumentative points, and each requires specialized literacy to access. Alley [2] defines the scientific writing process as art with four states: prepare, write, revise, and finish. Likewise, writing decisions fall into three categories: content, style, and form. The content is what you want to stream. Style is how it converts ideas into sentences and how they connect into paragraphs. The style also refers to how paragraphs are combined with illustrations to create sections and organize them into documents. The form consists of grammatical writing rules, guidelines for inserting references, and citations.

3 The Framework

A framework is a tool for organizing and communicating ideas about program development. It does provide a cohesive structure made up of proven components, but it is adaptable to work with varying teaching styles, content areas, and student needs (while maintaining the core structure of the framework) [4]. We propose a specific instructional framework for Engineering students to develop information literacy skills while they develop writing-paper skills. The Framework proposes four components: (a) course topics, (b) paper assignment, (c) activities, and (d) paper deliverables. The description is shown in Table 2.

Table 2. Framework description

Component	Description
1. Course Topics	Course topics are organized into 3 units: Overview, Planning, and Writing. A detailed description is shown in Table 3
2. Paper assignment	Each student is assigned the title of the paper in the first week. Each paper's title has to been formulated to give students knowledge about their field in a social context and to motivate them. The final paper only will have six pages, with a minimum of 20 references. There will be incremental deliverables of the papers
3. Activities	The activities that lead students to write a paper while they develop information literacy skills are described in Table 4
4. Paper deliverables	There are five reviews throughout the semester. Students deliver incremental drafts of the paper. The description of the deliverables are shown in Table 5

Table 3. Course topics description

Unit	Course topics
1. Overview	Why is Technical writing important/What is a Call for papers/IEEE conference paper format – navigational aids/Parts of the IEEE article
2. Planning	Keywords, synonyms and related terms/Implementing the search strategy and search parameters /Searching information: Scopus, WOS, IEEE Xplore/Credibility and reliability of sources of data/Critical Reading/The writing plan
3. Writing	Writing sentences and paragraphs /Tables, images, diagrams, formulas and codes/References and citation in IEEE format with Mendeley/Developing unity, coherence, emphasis and clarity/Writing Related Works/Writing the Proposal/Writing Abstract, Writing Introduction and Conclusions

Table 4. Activities

Activity/ Standard	Objective/Description
Activity 1 (Standard 1)	*Objective: To determine the nature and scope of information needed* Given the assigned topic, students should identify from which sources, and with which tools they will obtain the information and the relevance of this information
Activitity 2 (Standard 2)	*Objective: access the necessary information effectively and efficiently* Students search for 5 articles on the assigned topic. Use of the selected indexed databases is requested. You must document the keyworkd, search chain and the subsequent refinements made to the chain justifying the reasons are synthesized, as well as stating how this contributes to the topic of your interest
Activity 3 (Standard 1 and Standard 2)	*Objective: Evaluate information and its sources critically* Students must identify if the papers come from a conference or from a journal. If it is a journal they should identify the quartile (Q1,Q2,Q3,Q4) to which they belong. If it is a conference paper identify if it is indexed to SCOPUS, WOS or other
Activity 4 (Standard 3)	*Objective: Incorporate the selected information into your knowledge base* Students summarize the main ideas of five articles and prepare a short text where the similarities and differences are expressed. Indicate how these articles are similar/diference whith your article (Writing related works)
(Activity 5) Standard 4	*Objective: Applies new and prior information to the planning creation of a particular product* Propose a wrinting plan (structure) considerint sections, subsection and sub subsections
Activity 6 (Standard 4 and Standard 5)	*Objective 1: Student applies new and prior information to creation of a particular product* *Objective 2: Student understands the ethical issues surrounding information* *Objective 3: Student acknowledges the use of information sources in communicating the product* *Objective 4: Student revises the development process for the product* The students will write a 6 page paper in a IEEE-Format. The paper will have five incremental deliverables (drafts) (Fig. 1). Students will have permanent feedback

As we have set in the background section, writing academic texts is a process that requires intense cognitive activity. The process of writing requires students to search for information, select the information they consider most relevant after a critical reading, organize this information to propose a writing plan, to write according to the writing plan, and rehearse various formulations on the paper. The process that we have just described involves the stages described by Alley [2]: prepare, write and revise; and, they do not occur in a linear way (Fig. 1).

Table 5. Paper's deliverables

Paper Deliverable description	Week
Basic Information a. Present the keywords and synonymous related to the topic assigned b. Present the Search Strategy used and the refined search strategy c. Share in the course drive at least five related articles of the topics (justify why they were selected) d. Complete the Paul-Elder template for critical reading for the five articles you selected	2
Writing Plan a. Based on the previous activity develop your writing plan; consider IEEE section format b. Use Title, subtitles and sub-subtitles. c. Describe from general concepts to detailed concepts d. Consider at least 5 references from IEEEXplore, ACM Digital Library, SCOPUS, WOS, SJR.	3
Draft 1 a) IEEE paper format b) 1 page and 10 references. c) No abstract, no introduction and no related works	5
Draft 2 a) Make up observations in Draft 1 b) Write 2 pages and 10 references c) No abstract , introduction and related work	7
Draft 3 a. Make up observations in Draft 2 b. Write 4 pages including Related Works and your Technical Proposal. c. No abstract and introduction d. At least 15 references	10
Draft 4 a. Make up observations in Draft 3 b. Write 6 pages with Abstract, Introduction and Conclusions c. At least 20 references	15
Final paper Make up observations in Draft 4	17

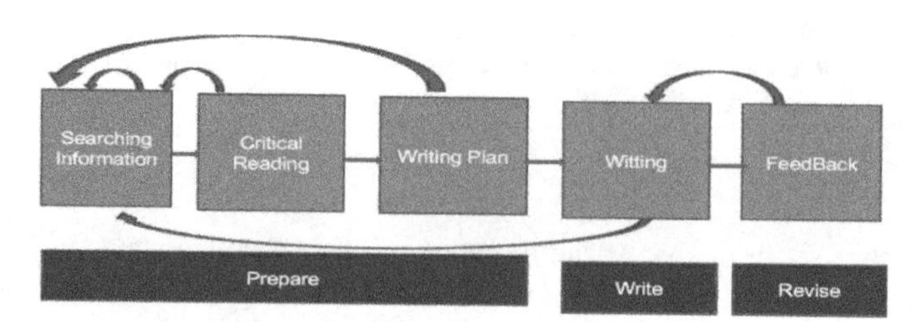

Fig. 1. Phases of the writing process proposed by Alley are shown in purple. The main activities are shown in grey. As we can observe it is a dynamic process. Most of the process is driven by the quality of related articles that students read

4 Methodology

In order to validate if the proposed framework helps in the development of the information literacy skills we carried out a qualitative study. The study population is established in the set of students enrolled in the course Writing Articles and Research Reports. The sample was through a non-probabilistic convenience sampling with 50 students from semester 2020 A. Students were not exposed to the experience of technical writing before the course. The range of ages was between 18 and 20 years old. Students' learning information literacy outcomes were measured using the twenty-two-indicator proposed by ACRL described in Table 1. Each indicator was ranged from 1 to 5: (1) Emerging, (2) Developing, (3) Practicing, (4) Maturing and (5) Mastering. We hypothesized that: "There is an impact in the development of information literacy skills when they are exposed to the writing paper experience".

5 Results and Discussions

The results of the survey are summarized in Table 2. We can see that out of the 22 indicators, the highest values are at level 4: Maturing. Fewer students acknowledge being in the Mastering status. An analysis was made for each standard. Given that each standard has a different number of indicators, a simple average was made in order to make a comparison. From the results obtained, considering only the value 5: Mastering; standard 2 had a score of **18.2**. This standard is referred to as the fact that a literacy information student can access the information he needs more effectively and efficiently. The lowest value referred to as the highest score is given for the standard 4: **14.25**. This standard refers to the use of information for a specific purpose (the paper). Due to the experience in teaching the course, the activities related to the search and validation of information are shorter, precise, and punctual, while the writing process is usually much more challenging for the students. Pimmel [14] expresses that the best way to acquire new skills is by providing students opportunities to practice the skill, giving frequent feedback, and having structured discussion activities. The conference paper format sets the conditions described by Pimmel.

There are other experiences related to informational teaching, most of them oriented to postgraduate [15]. Authors describe experiences like the incorporation of modules from doctoral programs, postgraduate courses, open workshops from doctoral schools, research projects workshops, workshops-seminars in research events. There are other methods for teaching information literacy, some librarians have argued for a "teach the teachers" strategy [7]. The author affirms that librarians concentrate instructional strategies on teaching disciplinary faculty to teach information literacy. The closed work to ours is proposed by Granruth and Pashkova-Balkenhol [6] who also suggest writing assignments that lead to a paper but for Social Work. The main difference with our proposal is that it is oriented to Engineering and makes use of IEEE recommendations. The framework shown is detailed enough to be replicated and general enough to be adapted.

Table 6. Performance indicator ACRL results

	Performance Indicator	Level				
		(1) Emerging	(2) Developing	(3) Practicing	(4) Maturing	(5) Mastering
Standard 1	1		1	5	30	14
	2			13	22	15
	3			7	26	17
	4			9	26	15
	Average	0	0.25	8.5	26	15.25
Standard 2	5		1	10	29	10
	6		1	11	18	20
	7		1	9	19	21
	8			5	25	20
	9		1	8	21	20
	Average	0	4	8.6	22.4	18.2
Standard 3	10	1		7	23	19
	11		1	8	27	14
	12	1		11	23	15
	13		2	9	18	21
	14			8	28	14
	15			11	25	14
	Average	0.3	0.5	9.0	24.0	16.2
Standard 4	16			10	26	14
	17			8	28	14
	18			7	21	22
	19	1	3	10	29	7
	Average	0.25	0.75	8.75	26	14.25
Standard 5	20		2	11	24	13
	21			7	25	18
	22		1	7	24	18
	Average	0	1	8.3	24.3	16.3

6 Conclusions

The process of writing requires students to search for information, select the one that they consider most relevant, perform a critical reading, organize information for proposing a writing plan, to write according to the writing plan, and rehearse various formulations on the paper. The process that we have just described is dynamic and allows students to put into practice information literacy skills. While students develop the writing process, they recognize their information needs by identifying appropriate keywords. They have to use boolean terms 'and' 'or', 'not' to construct a simple search strategy. They identify appropriate resources such as the Digital Libraries. They can use the advanced search facility when it is required. Results show that students have a perception of Maturity Level in the five ACRL standards. From the results showed in Table 6 we can validate our hypothesis: there is an impact on the development of information literacy skills when students are exposed to the writing paper experience. Application of the ACRL's Information Literacy Competency Standards for Higher Education in the process of writing a paper has been proven to be an invaluable mechanism.

As we have set in the background section writing academic texts is a complex cognitive process that requires intense cognitive activity. In a specific field like Engineering, writing a conference paper to show a new design, a prototype, or new software is an important professional skill.

References

1. ABET: Criteria for Accrediting Engineering Programs, 2015 – 2016. http://www.abet.org/accreditation/accreditation-criteria/criteria-for-accrediting-engineering-programs-2015-2016/#outcomes. Accessed 05 Nov 2020
2. Alley, M.: The craft of Scientific Writing (No. 808.0666/A435). Springer, New York (1996)
3. Association for College and Research Libraries: "Information Literacy Competency Standards for Higher Education". http://www.ala.org/acrl/standards. Accessed 05 Nov 2020
4. Carver, R.: Theory for practice: a framework for thinking about experiential education. J. Experiential Educ. **19**(1), 8–13 (1996)
5. Catts, R., Lau, J.: Towards information literacy indicators. (2008). https://dspace.stir.ac.uk/bitstream/1893/2119/1/cattsandlau.pdf. Accessed 05 Nov 2020
6. Granruth, L.B., Pashkova-Balkenhol, T.: The benefits of improved information literacy skills on student writing skills: developing a collaborative teaching model with research librarians in undergraduate social work education. J. Teach. Soc. Work **38**(5), 453–469 (2018)
7. Hammons, J.: Teaching the teachers to teach information literacy: a literature review. J. Acad. Librarianship **46**(5), 102–196 (2020)
8. Horton, F. W.: Understanding information literacy a primer, an easy-to-read, non-technical overview explaining what information literacy means, designed for busy public policy-makes, business executives, civil society administrators and practicing professionals (2008). https://eduq.info/xmlui/bitstream/handle/11515/17980/157020e.pdf?sequence=1
9. Iturbe, C.B., Ochoa, L.L., Castello, M.J., Pelayo, J.C.: Educating the Engineer of 2020: Adapting Engineering Education to the New Century, pp. 1110–1121. IATED, Valenica (2009
10. Julien, H.: Information literacy instruction in Canadian academic libraries: longitudinal trends and international comparisons. Coll. Res. Libr. **61**(6), 510–523 (2000)
11. Kmiec, D., Longo, B.: The IEEE Guide to Writing in the Engineering and Technical Fields. John Wiley & Sons, Hoboken (2017)
12. Lau, J.: Guidelines on Information Literacy for Lifelong Learning. IFLA, Veracruz (2006)
13. Ondrusek, A., Dent, V. F., Bonadie-Joseph, I., Williams, C.: A longitudinal study of the development and evaluation of an information literacy test. Ref. Serv. Rev. **33**(4), 388–417 (2005)
14. Pimmel, R.L.: Student learning of criterion 3 (a)-(k) outcomes with short instructional modules and the relationship to bloom's taxonomy. J. Eng. Educ. **92**(4), 351–359 (2003)
15. Rodríguez, L., Serra T, R., K.: Experiencias internacionales en el desarrollo de habilidades informacionales en la formación doctoral. E-Ciencias de la Información, **8**(2), 159–180, (2018)
16. Unesco-Programa General de Información y Unisist: Guía para la redacción de artículos científicos destinados a la publicación. (2 ed. rev. y act.) por A, pp. 1–13. Martinsson. París (1983)
17. Valery, O.: Reflexiones sobre la escritura a partir de Vygotsky. Educere **3**(9), 38–43 (2000)
18. Ward, D.: Revisioning information literacy for lifelong meaning. J. Acad. Librarianship **32**(4), 396–402 (2006)

Automatic Classification of Research Papers Using Machine Learning Approaches and Natural Language Processing

Ortiz Yesenia and Segarra-Faggioni Veronica(✉)

Universidad Técnica Particular de Loja, Loja, Ecuador
{yaortiz, vasegarra}@utpl.edu.ec

Abstract. This paper shows the automatic classification of research papers published in Scopus. Our classification is based on the research lines of the university. We apply the K-nearest neighbor classifier and linear discriminant analysis (LDA). Various stages were used from information gathering, creating the vocabulary, pre-processing and data training, and supervised classification in this work. The experiment involved 596 research articles published in SCOPUS from 2003–2017. The results show an overall accuracy of 88.44%.

Keywords: K-nearest neighbor classifier · LDA · NLP · Research papers

1 Introduction

The research works carried out by researchers of the university are published as articles in conferences or journals that index in different scientific databases such as Scopus, ISI, DBPL, and others. For this reason, the growth in scientific literature allows exploring through a quantitative analysis that indicates the scientific impact. Scientific research activity is an indicator of universities.

Intending to help the university identify trends in publications, we propose a model to identify research papers according to each research line. This paper presents an automatic learning model based on natural language processing techniques (NLP) and machine learning techniques to retrieve and classify in-formation from scientific publications from Scopus from 2003 to 2017 carried out by university researchers.

In this work, we are interested in evaluating the relationships between research lines of the university and common terms of scientific articles. That is, relevant information will be extracted from words in the abstracts, identifying the research lines related to scientific publications, which allow the search of key resources quickly and efficiently in research. In this work, two supervised learning algorithms were applied: K-Nearest Neighbors (k-NN) and Linear discriminant analysis (LDA). KNN algorithm is instance-based learning, that is, the prediction of a class label to a new datum takes into account the classes associated with the instance closest to it in the space of entry [1]. On the other hand, LDA algorithm solves the problem of semantic similarity measurement in traditional text categorization [2].

Á. Rocha et al. (Eds.): ICITS 2021, AISC 1330, pp. 80–87, 2021.
https://doi.org/10.1007/978-3-030-68285-9_8

This study proposes the implementation of the k-NN algorithm that classifies each new datum in the corresponding group according to the frequency of terms of each document considering the relationship between them.

This paper is organized as follows: Sect. 2 describes related work about k-NN classifier and LDA. Section 3 presents the materials and method applied to this study. Section 4 describes the experimental results and discussion.

2 Related Work

Natural language processing and machine learning techniques are applied to discover knowledge patterns in a group of documents. According to [3], text mining discoveries focus on representation, text level, and world level. Some authors have proposed working methods for document classification. In [4], they applied LSI (Latent Semantic Indexing) algorithm to group the syllabus. Matrix dimensions to determine in vectors the similarity of documents with similar contents. Also, applying NLP techniques to choose the data set is grouped with the previously selected algorithms.

De la Calle Velasco (2004) states that applying NLP techniques to extract and annotate information in scientific publications based on bioinformatics and medical informatics domain [5]. The objective of the study [5] was to design a general model for any scientific discipline. In [6], automatic categorization techniques in biomedical text mining consist of classifying biomedical texts associating entities' biologicals with selected ontology terms. They use text mining techniques to generate knowledge by searching and browsing in various databases where these contents that integrate both topics are stored.

In this work, the algorithms selected are the Knn and LDA. K-nearest Neighbor algorithm is an effective method of machine learning. The k-NN algorithm is a supervised learning algorithm based on the nearest neighbor; it classifies values looking for the "more similar" (by closeness) learned data points in the training set. It is also a method that searches the closest observations trying to predict and classify the point of interest based on most data that surround it. The best value of k depends on the data. Therefore, k-NN calculates a similarity between the document to be classified and the labeled documents in the training set.

The k-NN classification technique is used for supporting keyword search [7, 8]. Some researchers state that the k-NN algorithm's accuracy is high but not competitive compared to better-supervised learning models. Also, K-Nearest Neighbor algorithm is used to guess a real value for an unknown sample [9]. In the study of [10], k-NN is used to Chinese text categorization, and it should be applicable to classification problems for data in other languages. According to [11], LDA is an excellent technique to reduce dimensionality. Thus, LDA is used to reduce the size within a data set for pattern recognition.

Table 1 shows a summary of the related works on automatic document classification:

Table 1. Summary of document classification related work

Author	Study case	Algorithm
Cavillo et al. (2016)	To classify and locate research papers Optimization and precision of information	Bayesian
Caraguay, R. (2016)	Text mining techniques to cluster contents of the syllabus	K – means
Gálvez, C. (2008)	Analysis of the scientific literature in molecular biology and genomics	Text matching
Baoli et al. (2003)	Chinese text categorization	kNN

3 Proposed Work

3.1 Dataset

Scientific research results of professors from Universidad Tecnica Particular de Loja published in journals or conference proceedings are in scientific databases (Scopus, ISI, DBPL). In this study, the dataset was taken from the Scopus database, comprising 596 scientific documents carried out by researchers from the UTPL from 2003 to 2017. Each document article has the author, title, abstract, keywords, and indexed keywords.

3.2 Methods

In this study, we follow the following stages: data collection, pre-processing, information extraction, and classification. Figure 1 shows the stages that were applied in this study to carry out the classification process.

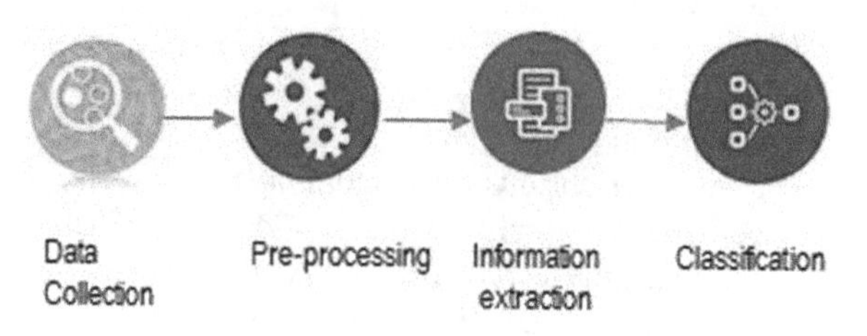

Fig. 1. Methodology overview

We begin applying natural language processing techniques to prepare and clean data, remove common words called stop-words 1; and, lemmatize the dataset to remove inflectional endings and return the base form of a word [12]. In this step, we select the frequent terms by bigrams. To establish a vocabulary, a vector of the matrix of corpus terms is created and then the function is applied to show the frequency of terms in the matrix.

In this study, 15 categories were established based on the research lines described in Table 2. The next step is to classify a text document into 15 categories. These categories were associated with lines of research.

Table 2. Categories based on the research lines

Category	Lines of research	No. Documents
Audiovisual	Communication and Technologies Audiovisual Narratives Organizational Communication Socio-humanistic Literature	77
Biology	Genetic Biology Biochemical Microbiology Biotechnology and production	131
Construction	Structures, Transportation and Construction	10
Economy	Economic theory Economic development	
Education	Basic Education Sciences Pedagogy of Experimental Sciences Basic and Social Educational Psychology	6
Energy	Electronic and energy	14
Geodynamics	Geodynamics, mining, and metallurgy	15
Environmental engineering	Environmental engineering	73
Artificial intelligence	Artificial intelligence	27
Math	Physiochemistry and Mathematics	27
Chemistry	Basic and Applied Chemistry	118
Water resources	Water Resources	7
Health	Public Health and Health Management	14
Software	Software Engineering Information Technology Management	24
Web technologies	Advanced Web Technologies	75

4 Experimental Results and Discussion

In this section, we will detail the construction and validation of the classification model.

4.1 kNN Algorithm Application

k-Nearest Neighbor (k-NN) is among popular classification technique. k-NN is considered to be simple, easy to implement, and the ***k*** value is completely up to user [9, 13]. To carry out the classification model in this study, **75%** of the 596 considered records are used for training, the rest (**25%**) are utilized for test.

k-NN is applied to make the prediction in the confusion matrix. The accuracy of a classification problem result is shown by a confusion matrix. Considering that a matrix of prediction will be compared with the original class of inputs or in other words contains information of actual and predicted value on classification [14].

It is important to perform several tests using the k-NN algorithm, because after some trials a ***k*** value that gives the best result is chosen [9]. The value of ***k*** has a great influence on the classification effect. If the value of ***k*** is too small, it will cause over fitting, which will lead to the classification error of the classifier. Table 3 shows the best accuracy according to the k value. In our study, ***k*** values was ***3, 5,*** and ***20***, in order to get the best accuracy of data classification (See Table 3). In the test ***k*** = 5 by 63.22% of the topics have better performance, while the topics of smaller denomination stay in the margin of error. In the test ***k*** = 20 by 55.48% reflects better precision in large groups of classification. In our study, we chose three (3) as a value of ***k*** parameter with the goal that the closest point to any training data point is in similar classes. Therefore, it allowed to identify the correct classification of articles through the confusion matrix, the test resulted in the highest accuracy of k-NN by 66.89%.

Table 3. k–NN test

k	Accuracy
3	68.89%
5	63.22%
20	55.48%

4.2 Confusion Matrix of the k – NN Classification Algorithm

Once the training percentage is defined, the k-NN model is applied to predict the confusion matrix. The best ***k*** parameter was ***3*** with the average accuracy of **66.89\%.** The prediction does not have a higher percentage because a large number of research articles are related to other research lines, which cause imprecision in the classification carried out. The value of ***k*** is important because the accuracy of the classification is dependent on it [15].

The confusion matrix gives a lot of information. Figure 2 shows the confusion matrix using k-NN as a classifier.

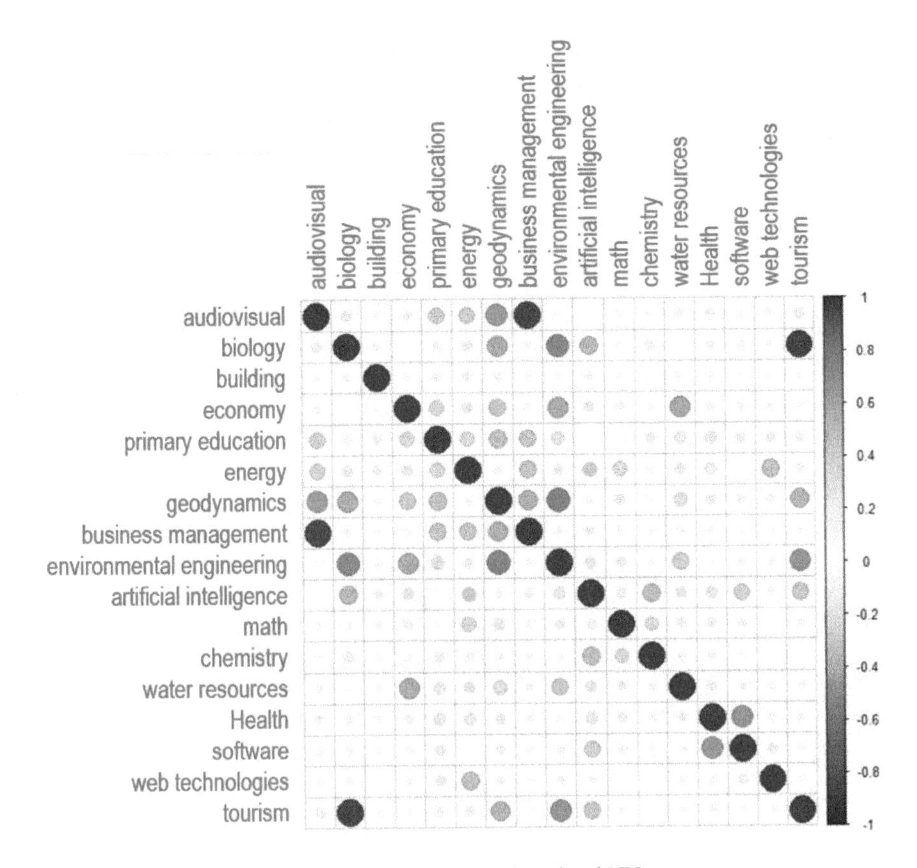

Fig. 2. Confusion matrix using kNN

4.3 Linear Discriminant Analysis (LDA)

After applying the k-NN algorithm, the Linear Discriminant Analysis (LDA) model was used, this linear method performs predictions of qualitative variables by achieving maximum discrimination [16]. The probabilities for each category show a sparse distance. However, there is a close similarity between some lines of research. The comparison result of the training set and test set was **88.44%** of accuracy.

4.4 Discussion

Scientific production reflects the advances of an academic or research activity. Therefore, this work focuses on determining the relations of the research lines with the common terms of research articles to know university research's tendency. Based on the experimental results, the corpus' relevant terms were identified before the classification of scientific documents. Using the algorithm k-Nearest Neighbors (k-NN), the classification articles' accuracy achieves **66.89%** accuracy.

Figure 3 shows the relation between the research lines. We can obtain this graph from the confusion matrix of each line.

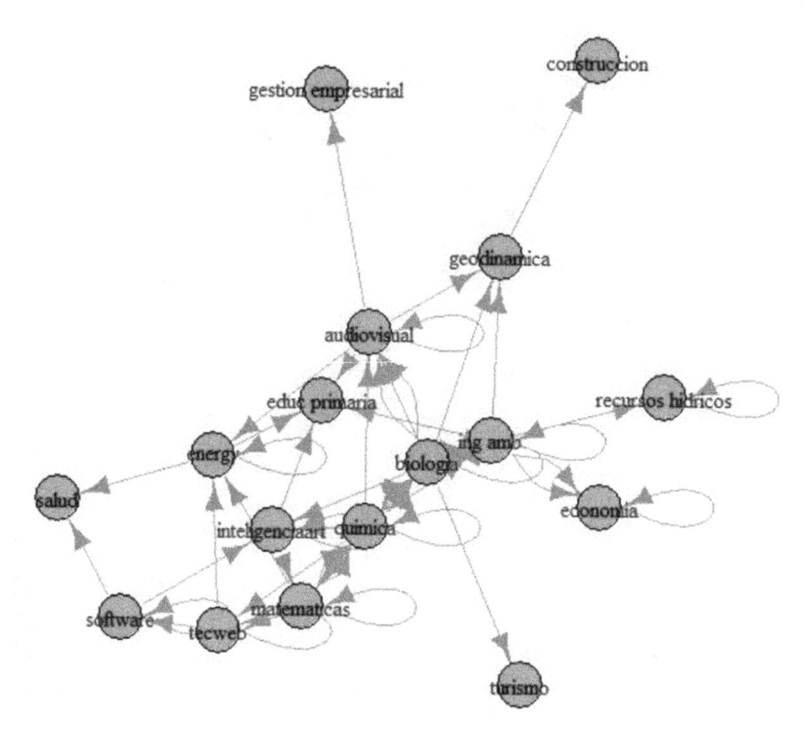

Fig. 3. Relationship between research lines of university

5 Conclusion

In summary, the application of text mining and natural language processing was elementary in the process of this study because it allowed to identify relevant terms of the corpus and eliminate terms and characters that did not provide meaning, these processes are important and should always be carried out prior to the use of an algorithm. The main contribution of our study is to create a validated classifier based on the research lines of the University that improve and speed up the organization of information. In addition, obtaining a glossary of more frequent terms in the research lines that facilitate to search for information.

Finally, this study can be applied in the institutional database system called Scientific and Academic Information System (in Spanish: Sistema de Información Académica Científica - SIAC[1]), to get reliable data for decision-making. Because, in this system is stored the academic–scientific information of the UTPL teaching staff.

[1] sica.utpl.edu.ec.

Acknowledgements. The research team would like to thank Universidad Técnica Particular de Loja, especially to Tecnologías Avanzadas de la Web y SBC Group.

References

1. Nunez, H., Ramos, E.: Automatic classification of academic documents using text mining techniques. In: 2012 XXXVIII Conferencia Latinoamericana En Informatica (CLEI), pp. 1–7 (2012)
2. Chen, W., Zhang, X.: Research on text categorization model based on LDA — KNN. In: 2017 IEEE 2nd Advanced Information Technology, Electronic and Automation Control Conference (IAEAC), pp. 2719–2726 (2017)
3. Montes Gómez, M., Juan de Dios Bátiz, A., Miguel Othón de Mendizábal P, esq C., Gelbukh, A., Aurelio López López, C.-A.: Text mining using comparison of semantic structures. In: CIC-IPN (2005)
4. Caraguay, G.R.: Aplicacion de tecnicas de mineria de texto para el agrupamiento de componentes academicos en base a los contenidos de planes docentes. In: UTPL (2016)
5. De la Calle Velasco, G.: Modelo basado en técnicas de procesamiento de lenguaje natural para extraer y anotar información de publicaciones científicas. In: UPM (2004)
6. Gálvez, C.: Text-mining: the new generation of scientific literature analysis. In: Molecular Biology And Genomics, Encontros Bibli Rev. eletrônica Bibliotecon. e ciência da informaçāo, vol. 13, no. 25, May 2008
7. Cambronero, C.G., Moreno, I.G.: KNN & KMEANS Algorithms (Algoritmos de aprendizaje: KNN & KMEANS) (2016)
8. Draszawka, K., Szymański, J., Guerra, F.: Improving css-KNN classification performance by shifts in training data. In: International KEYSTONE Conference on Semantic Keyword-Based Search on Structured Data Sources, pp. 51–63 (2015)
9. Ananthi, S., Sathyabama, S.: Spam filtering using K-NN. J. Comput. Appl. **2**(3), p. 20 (2009)
10. Baoli, L., Shiwen, Y., Qin, L.: An improved k-nearest neighbor algorithm for text categorization 1, Shenyang, China (2003)
11. Geron, A.: Hands-On Machine Learning with Scikit-Learn & TensorFlow, First. O'Reilly Media Inc, United States of America (2017)
12. Bird, S., Loper, E.: NLTK: the natural language toolkit (2006)
13. Song, Y., Huang, J., Zhou, D., Zha, H., Giles, C.L.: IKNN: informative k-nearest neighbor pattern classification. In: Knowledge Discovery in Databases: PKDD 2007, Berlin, Heidelberg, pp. 248–264. Springer, Berlin Heidelberg (2007)
14. Patil, T.R., Sherekar, S.: Performance analysis of naive bayes and j48 classification algorithm for data classification. Int. J. Comput. Sci. Appl. **6**(2), 256–261 (2013)
15. Ganesh Jivani, A.: The novel k nearest neighbor algorithm. In: International Conference on Computer Communication and Informatics (2013)
16. Wang, Z., Qian, X.: Text categorization based on LDA and SVM. In: 2008 International Conference on Computer Science and Software Engineering, pp. 674–677 (2008)

Blockchain and Government Transformation

Teresa Guarda[1,2,3,4,6(✉)], Maria Fernanda Augusto[1,2,3,6], Lidice Haz[1,3], and José María Díaz-Nafría[1,5,6]

[1] Universidad Estatal Península de Santa Elena, La Libertad, Ecuador
tguarda@gmail.com, mfg.augusto@gmail.com, victoria.haz@hotmail.com, jdian@unileon.es
[2] Universidad de las Fuerzas Armadas, Salinas, Ecuador
[3] CIST – Centro de Investigación en Sistemas y Telecomunicaciones, Universidad Estatal Península de Santa Elena, La Libertad, Ecuador
[4] Algoritmi Centre, Minho University, Guimarães, Portugal
[5] Madrid Open University, Madrid, Spain
[6] BiTrum Research Group, Madrid, Spain

Abstract. In today's digital age, transactions are made increasingly in the virtual universe. Some do all financial transactions online, but some do not even want to think about this possibility since the consideration is unreliable. To make online financial transactions more reliable, many technologies have already been tested. Since passwords, tokens, physical code generation accessories, there have been many attempts, being the most elaborate, the blockchain. Governments, especially in developed countries, often find it difficult to gain the trust of their citizens, especially when it comes to evidence of service provision and improvement of existing services. In developing countries, blockchain requests from governments would be useful in eliminating some important problems, such as corruption, while ensuring more effective deployment and distribution of resources. The adoption of such technologies can also help to facilitate better use of resources. In this paper, we will analyze the impact of the application of blockchain technology on e-government.

Keywords: Blockchain · e-Government · Security · Digital transformation · Competitiveness

1 Introduction

Information and communication technologies provide advances to society, and the role of the State tends to favor the population by offering better conditions of access to information or its services, through more interactive and participatory electronic means with the citizen, from e-Government perspective.

e-Government tends to create a new type of public service, based on an increasingly integrated and effective relationship with society, whose public institutions start to articulate public actions and policies, based on a more modernized service provision, and a an increasingly integrated database and information, in order to offer a better service to the population.

Á. Rocha et al. (Eds.): ICITS 2021, AISC 1330, pp. 88–95, 2021.
https://doi.org/10.1007/978-3-030-68285-9_9

The transformation of government services and governance can be seen as a citizen's right to request, through electronic e-government, at any time, the institution of the State and its public services, in a logic of functioning 24 h a day, in which the request for services and information would be made at any time or instant, via Internet, by the citizen. In this perspective, it is imperative to guarantee the security and privacy of the network, being able to be operationalized through blockchain technologies.

Blockchain is seen as a trust network, which will be executed on the internet data transfer protocols, allowing business to be carried out automatically and at a reduced cost between people, companies and countries. Every record of information or execution results must pass through the network sieve to be validated. This is what makes this technology so precious, as it ensures that all nodes in the network agree and have exactly the same information, avoiding common fraud situations in systems that depend on the validation of a third party.

It is very common to have some confusion with the terms Blockchain and Bitcoin, since Blockchain is the technological platform used for the operation of the Bitcoin network and several other cryptocurrencies; and Bitcoin is the first and most well-known application of Blockchain technology. Blockchain technology emerged in 2009 as bitcoin infrastructure, an indisputable phenomenon, blockchain technology was soon perceived as disruptive and with much greater potential than cryptocurrencies. And it has come to be seen in the last ten years as a high-potential technology [1]. The reality, however, has shown that its maturation process is still ongoing.

With. The P2P network has users who share tasks, work or files without blockchain technology, it is possible to have an anonymous, decentralized internet with guaranteed privacy protection. The system consists of two parts: a peer-to-peer (P2P) network and a decentralized database the need for a central server. All participants have equal privileges and influence on the environment [2]. Each computer in the network is a node and, whenever new data enters the system, all nodes receive it. This information is encrypted and there is no way to track who added it, it is only possible to verify its validity. As a security measure, the method makes the distributed registration of information to decentralize the process [3]. Thus, when one node leaves the network, the others already have a copy of all the information shared. Likewise, if new nodes enter it, the rest create copies of your information for them.

According to the Gartner Group's 2019 CIO Agenda survey, 60% of CIOs said they expected some level of adoption of blockchain technologies in the subsequent three years [2]. This contrasts with the scenario of the same survey conducted in 2018, the result of which indicated that only 22% of CIOs shared this expectation at the time. After a growing interest in 2018 [3], and the appearance of more applications in different sectors in 2019, the forecast was that 2020 would be the year of the blockchain to take off.

2 E-Government as a New Paradigm of Public Administration

In the 1990s, there was great enthusiasm for the development of electronic government or e-Government [4], which was driven by the globalization of the economy, the proliferation of the internet, but also due to a new attitude of the political class towards the use of Information Technologies (IT) in Public Administration (PA).

Over the past few decades, the attention and interest of political power in e-government has evolved, with a widespread recognition today of the importance of IT as an instrument for the transformation and modernization of PA.

In addition to efficiency gains in terms of the internal functioning of the administrative machine, the use of IT has been seen as an opportunity to change the paradigm of a PA centered on the agency, to one more centered on the citizen and companies, capable of providing quality services and information, accessible at any time, and according to the needs of those looking for them.

E-Government consists of the use of information technologies to deliver state products and services to all citizens, and to industry. This type of tool brings the Government closer to all citizens of a given country [5].

The development strategies for e-government are based on three essential aspects (Fig. 1): user-centricity (the focus on the citizen with the intention of facilitating and globalizing citizens' access to public services); transparency and accountability (in order to promote effective public management, the results of which are measurable and more easily scrutinized); and e-participation (consultation or participation of citizens, as a process of co-creating new policies, services or projects).

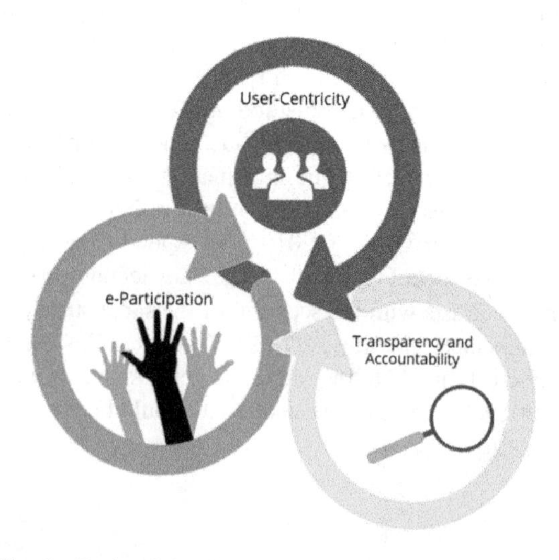

Fig. 1. Basis of development strategies for e-government.

Civil society, increasingly demanding in terms of quality, accessibility, efficiency and transparency in the services provided by the State. The strategic challenge for PA is, therefore, to provide electronic services of high quality and usability, in order to match the level offered by some services in the private sector.

However, improving the services provided to citizens and businesses depends, first of all, on the optimization of internal processes, the reduction of time in the execution of tasks, and greater cooperation between public bodies, which requires the ability to interoperate both between information systems, either between sectors and government organizations.

The realization of e-government and a transparent, effective and rigorous PA is only possible with the use of solid tools and technologies, capable not only of ensuring the security of communications and data privacy, but also of responding to the specificities and requirements of operation various public bodies.

3 Blockchain Technologies

Blockchain emerged as a Bitcoin system, mainly after the publication of the Bitcoin article "A Peer-to-Peer Electronic Cash System". The author was famous for his pseudonym Satoshi Nakamoto [1, 6].

Blockchain terminology refers to a chain of blocks, a series of blocks of data that are chained together cryptographically.

Blockchain came to be known as the system that would conduct the transactions of all Bitcoins and later other cryptocurrencies [7]. The system has come to be widely used by security for those who choose this type of currency in the digital environment [8].

The blockchain is a network that works with very secure chained blocks that always carry content along with a fingerprint. In the case of bitcoin, this content is a financial transaction. The catch here is that the back block will contain the fingerprint of the previous one plus its own content and, with these two pieces of information, generate your own fingerprint [9]. And so on.

Blockchain is not a new technology. It is a combination of proven technologies and applied in a new way (internet, private key encryption and protocol). The result is a system for digital transactions that does not need a third party to intermediate them.

This technology has the potential to change the way we buy and sell, interact with the Government and verify the authenticity of everything. Blockchain technology combines the openness of the Internet with the security of cryptography to offer everyone a faster and more secure way to verify important information and establish trust [10].

Blockchain networks can differentiate into public or private networks (Fig. 2). The public ones have their own rules, functioning independently of legal or regulatory aspects (the case of Bitcoin), the validators of the transactions are anonymous and the entrance to participate in the network of miners is of free access. Private bockchain networks follow regulations, and participants are pre-selected, applications are restricted to closed corporations [11]. That is, the encrypted access keys for carrying out operations on an open network are widely accessible and anonymous, while on a private network, access keys are controlled and there is a need to request permission for transactions [12].

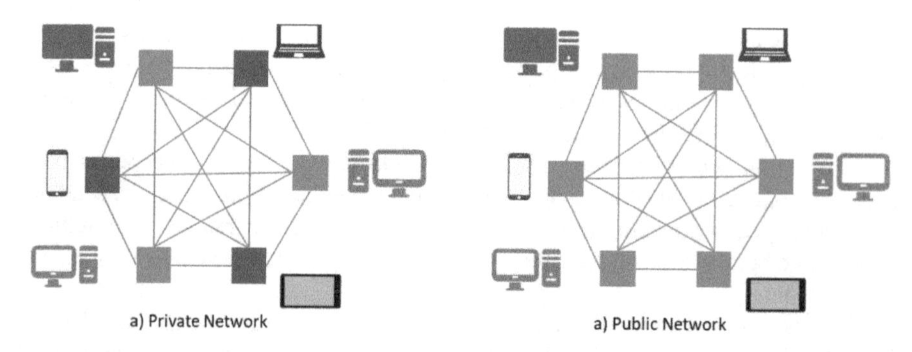

Fig. 2. Blockchain networks: ■Validator node: validates, initiates or receives a transaction; ■ Member node: only initiates or receives a transaction.

The main innovative feature of blockchain technology, is the ability to track transactions in decentralized and public databases, reducing the possibility of counterfeiting and fraud [13]. The distributed ledger provides an almost immutable record and guarantees the traceability of transactions since it will be very difficult to manipulate the data on the blockchain, since the changes are immediately reflected in all copies of the reason by the network and they are linked to the previous transaction [10].

An increasing number of countries are investing in applications of blockchain technology, aiming to improve the provision of public services and improve the governance of their countries [14].

Blockchain can add value to public administration due to its properties of immutability, transparency, traceability, reliability and operational resilience.

4 The Benefits and Applications of Blockchain for Governments

Governments have the task of facilitating transparency, especially in governance, in the distribution of resources and in achieving greater efficiency, among other things. For these reasons, the governments of many countries have expressed an interest in technology.

Blockchain orders for governments can be the missing link in helping governments become fully digitized. The world has been on the path of digitization, as has been seen in many sectors, such as retail and entertainment, among others. Governments have also felt pressure to follow suit, but it is easier said than done. One of the biggest obstacles that have stood in the way of digitization for many governments is the issue of security.

Bringing personal data of millions of people to digital platforms represents a great risk in case the system is hacked. However, blockchain has been touted as safe due to its structure and that means it could offer a viable solution that can help governments finally go digital. The fact that blockchain is virtually non-hackable makes it more attractive for the development of digital systems for governments. However, this is just the tip of the proverbial iceberg compared to what blockchain-based government services can achieve [15].

Blockchain technology offers new opportunities for Governments, different levels: transparency and access to information; control against fraud; the highest quality of public data, control and information security; greater efficiency; and greater confidence in public administration [16].

Blockchain technology can be used successfully by governments in several areas, such as taxes, payments, citizens' digital IDs, legal applications, security, online protection, health services, among many others (Fig. 3).

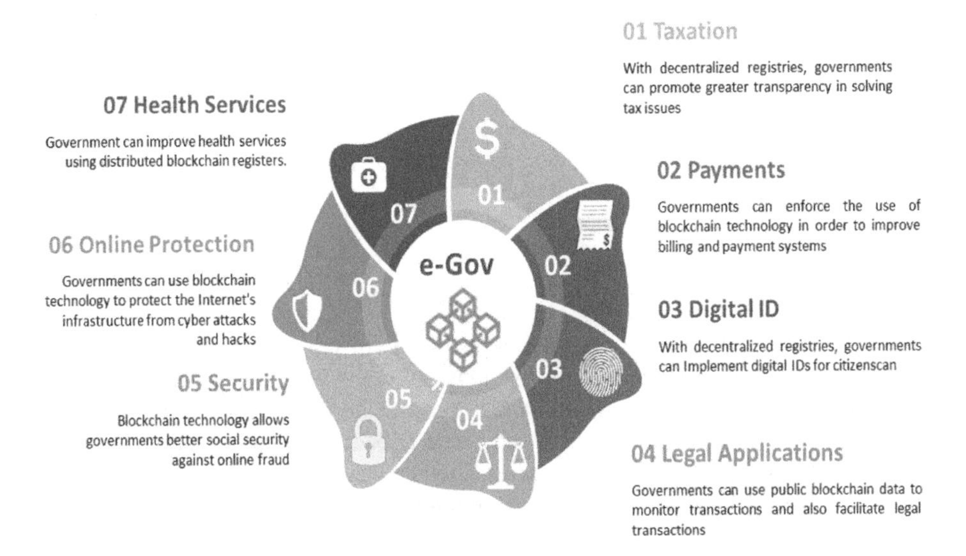

Fig. 3. Government blockchain and its applications.

Taxation, especially regarding the use of digital currencies to avoid taxation, can be controlled through the use of public blockchains and non-private currencies, and can also be the solution for fixing double taxation. In this sense, governments can use blockchain technology to implement protocols that can be used to reduce VAT deficits and reduce the tax burden, eliminating double taxation [17, 18].

The creation of online digital IDs to carry out public services in the self-service modality, will allow a greater number of citizens to have access to services that require the use of digital certificates [19].

Governments can use data from public blockchains to track financial transactions in a similar way to how fiat currency transactions are monitored to ensure that the system does not facilitate illegal transactions. The implementation of this technology for a government can be a fundamental tool to ensure that financial transactions in the digital domain remain legal.

The blockchain decentralizes data and stores it more securely. Thus, it makes it possible to track information and cybersecurity is one of the main results [20]. Blockchain distributed registration technology, due to its secure and immutable character, is ideal to meet the new requirements of governments and function as a reliable repository for identification purposes, and could be the answer to the fight against money laundering and enterprise fraud.

Blockchain applications in the public sector can help governments ensure better protection over their critical infrastructure, thus keeping cyberattacks at a distance. Most of the critical systems used by governments around the world to facilitate service delivery are connected to the Internet. This emphasizes the importance of security for critical systems and like those that cannot be hacked, then it is potentially the best solution. A decentralized registry can also be developed in such a way that it is able to track the integrity of government systems. This would significantly reduce the chances of attacks and data tampering.

Governments can leverage blockchain technology to bring significant improvements to healthcare sectors by developing healthcare systems that offer more efficiency when handling and storing medical data.

5 Conclusions

Blockchain technology has been referred to as one of the most today transformative technologies, being one of the most obvious use cases the application of cryptocurrencies. It is the same technology on which cryptocurrencies are based, but they have many other applications. Blockchain can be described as a highly secure and decentralized accounting system in which information can be stored, but cannot be changed. Instead of storing data on servers as used to through cloud storage, blockchain focuses on using a network of computers that store and verify data. Computers on a given network can be distributed worldwide and the network is not centrally controlled.

Governments are generally criticized for not knowing how to handle most of their processes efficiently, but new technologies, such as blockchain, have the potential to bring about drastic changes in the situation through blockchain solutions for the entire government. Governmental Blockchain may be something that will become a reality in the coming years, as governments begin to adopt this technology.

Blockchain technology can increase access and transparency of information, the predictive capacity of data, control of public data, control against corruption and fraud, and information security [21], enabling the provision of more efficient public services by improving business processes for government actors at any level of government; and enable the creation of fast, cheap and especially safe public records.

Blockchain can be very beneficial for public institutions, making them more transparent, close and secure. However, administrations also have to reposition themselves in this new scenario, since it is not only about its benefit, but also the commitment to promote this technology that will have an impact on the entire economy. These benefits can contribute to the observance of governance principles and, consequently, contribute to good governance in the public sector.

References

1. Nakamoto, S.: Bitcoin: a peer-to-peer electronic cash system, Manubot (2019)
2. Zhang, C., Wu, J., Long, C., Cheng, M.: Review of existing peer-to-peer energy trading projects. Energy Procedia **105**, 2563–2568 (2017)
3. Puthal, D., Malik, N., Mohanty, S.P., Kougianos, E., Das, G.: Everything you wanted to know about the blockchain: its promise, components, processes, and problems. IEEE Consum. Electron. Mag. **7**(4), 6–14 (2018)
4. Gartner: Gartner Group's 2019 CIO Agenda survey, Gartner (2019)
5. Gartner: Gartner Group's 2018 CIO Agenda survey, Gartner (2018)
6. Guijarro, L.: Analysis of the interoperability frameworks in e-government initiatives. In: International Conference on Electronic Government, Berlin (2004)
7. Twizeyimana, J.D., Andersson, A.: The public value of e-government–a literature review. Gov. Inf. Q. **36**(2), 167–178 (2019)
8. Barber, S., Boyen, X., Shi, E., Uzun, E.: Bitter to better—how to make bitcoin a better currency. In: International Conference on Financial Cryptography and Data Security, Berlin (2012)
9. Filippi, P., Loveluck, B.: The invisible politics of bitcoin: governance crisis of a decentralized infrastructure. Internet Policy Rev. **5**(4), 1–32 (2016)
10. Abeyratne, S.A., Monfared, R.P.: Blockchain ready manufacturing supply chain using distributed ledger. Int. J. Res. Eng. Technol. **5**(9), 1–10 (2016)
11. Swan, M.: Blockchain: Blueprint for a New Economy. O'Reilly Media, Inc., Sebastopol (2015)
12. Natarajan, H., Krause, S., Gradstein, H.: Distributed Ledger Technology and Blockchain. World Bank, New York (2017)
13. Li, X., Jiang, P., Chen, T., Luo, X., Wen, Q.: A survey on the security of blockchain systems. Future Gene. Comput. Syst. **107**, 841–853 (2020)
14. Reijers, W., Coeckelbergh, M.: The blockchain as a narrative technology: Investigating the social ontology and normative configurations of cryptocurrencies. Philos. Technol. **31**(1), 103–130 (2018)
15. Jun, M.: Blockchain government-a next form of infrastructure for the twenty-first century. J. Open Innov. Technol. Market. Complex. **4**(7) 1–12 (2018)
16. Boireau, O.: Securing the blockchain against hackers. Network Secur. **2018**(1), 8–11 (2018)
17. Saberi, S., Kouhizadeh, M., Sarkis, J., Shen, L.: Blockchain technology and its relationships to sustainable supply chain management. Int. J. Prod. Res. **57**(7), 2117–2135 (2019)
18. Tasatanattakool, P., Techapanupreeda, C.: Blockchain: challenges and applications. In: International Conference on Information Networking (ICOIN) (2018)
19. Fulmer, N.: Exploring the legal issues of blockchain applications. Akron L. Rev. **52**(1), 162–187 (2019)
20. Alexopoulos, C., Charalabidis, Y., Androutsopoulou, A., Loutsaris, M.A., Lachana, Z.: Benefits and obstacles of blockchain applications in E-government. In: 52nd Hawaii International Conference on System Sciences (2019)
21. Zyskind, G., Nathan, O.: Decentralizing privacy: using blockchain to protect personal data. In: IEEE Security and Privacy Workshops (2015)
22. Per, A.: Blockchain as an anti-corruption tool: case examples and introduction to the technology. U4 Anticorruption Tool **2020**(7), 1–33 (2020)

Impact of GDPR on Access Profile Management in an HR Information System

Pedro Henriques[1(✉)], Paulo Simões[2], and Nuno Santos Loureiro[2]

[1] Military University Institute, Lisbon, Portugal
pjhenriques@emfa.pt
[2] Portuguese Air Force, Lisbon, Portugal
pgsimoes@emfa.pt, nuno.a.loureiro@gmail.com

Abstract. New technologies allow private companies and public entities to use personal data on an unprecedented scale. This study proposes risk mitigation measures for the rights of data owners, associated with the process of managing profiles of access to the personal data of the Human Resources (HR) of the Portuguese Air Force (PoAF), present in an information system, within the scope of the implementation of the General Data Protection Regulation (GDPR). Through deductive reasoning, answers are obtained for a recent and undeveloped area of scientific knowledge, based on a qualitative research strategy. The process of managing the profiles of access to the personal data of HR of the Portuguese Air Force, is characterized and verified as to its compliance with the fundamental principles: the need and proportionality of processing and the protection of the rights of the holders of personal data. The risks associated with the process are assessed for their probability and severity, and consequently, a set of mitigating measures for them is identified.

Keywords: GDPR · Risk · Information system · Human Resources · DPIA

1 Introduction

New technologies allow private companies and public entities to use personal data on an unprecedented scale in the performance of their activities. The Human Resources (HR) of a military organization are an essential pillar in the mission entrusted to the Armed Forces (FFAA) and the military defense of the Portuguese Republic. The Data Protection Act [3], came regulating the General Data Protection Regulation (GDPR) [1], in the national legal order, which, reflects a conceptual change associated with issues related to the security of personal data.

The present investigation aims to manage profiles of access to the personal data of Portuguese Air Force (PoAF) HR, present in the Integrated Management System of National Defense - Human Resources (SIGDN-RH), in the context of the impact evaluation of the GDPR implementation. Santos [4] defines an Information System (IS) as a set of interrelated components that collects or collects, processes, stores and distributes information to support decision-making, coordinate and control work processes. SIGDN-RH is a technological tool and an integrated defense management tool that enhances the adoption of standard procedures within the Ministry of National

Á. Rocha et al. (Eds.): ICITS 2021, AISC 1330, pp. 96–107, 2021.
https://doi.org/10.1007/978-3-030-68285-9_10

Defense (MDN) and Armed Forces (FFAA), in the areas of planning, administrative management and salaries, career management, obtaining and recruiting personnel, training, justice and discipline [5]. In this context, this study aims to propose measures to mitigate the risks to the rights of data owners, associated with the process of managing profiles of access to the personal data of PoAF HR, present in SIGDN-RH, in the context of the implementation of GDPR, through the (1) Characterization of the process of managing access profiles to PoAF HR personal data, present in SIGDN-RH; (2) the analysis of compliance with the fundamental principles, such as the need and proportionality of processing and the protection of the rights of PoAF HR data owners present in SIGDN-RH; and (3) the risk assessment associated with the process of managing profiles of access to the personal data of POAF HR, present in SIGDN-RH.

These objectives are reflected in the following research question: **How can the risks associated with the process of managing profiles of access to PoAF HR data, present in SIGDN-RH, be mitigated, respecting the rights of data owners defined by the GDPR?**

2 Theoretical and Conceptual Framework

The GDPR has defined a new legal framework for the protection of personal data, with direct applicability in all Member States of the European Union from 25 May 2018 [1] The Data Protection Act [3] ensured the implementation of the GDPR, in the national legal order, on the protection of single persons with regard to the processing of personal data and the free movement of such data. The Chief of Staff of the Air Force (CEMFA) defined the responsibilities and tasks to be carried out by PoAF, regarding the processing of personal data, in particular with regard to the approval of internal regulations, changes to computerized processes or manuals for the processing of personal data, the recording of the operations carried out and the information to be provided to the data subject. It also appointed the Data Protection Officer (EPD), responsible to monitor the compliance of PoAF's data protection policies with the GDPR, as well as the realization of Data Protection Impact Assessment (DPIA) [6].

Data Protection Impact Assessment (DPIA). Article 35 of GDPR defines the concept of DPIA, as a process designed to describe the processing, assess the necessity and proportionality of such processing, and help manage the risks to the rights and freedoms of single persons arising from the processing of personal data, assessing them and determining the measures necessary to address those risks. A DPIA is a process aimed at establishing and demonstrating conformity [7]. A DPIA can also be useful for assessing the impact on data protection of a technological product, such as a software, where it is likely to be used by several data processing officers to carry out different processing operations [7].

Access Profile Management Process. Authorization profiles allow control access to all features available in an IS for each of the users, as well as access to the data, according to the functions performed and the respective organizational structure in which it is located [5]. Being a process the way an entity organizes itself to achieve a certain objective [9], it can be considered that, in the context of the protection of data present in an IS, a process of management of access profiles constitutes a set of rules and procedures, defined by the organization, for the management of authorization and profiles of access to information available in an IS, intending to protect the rights of data owners.

Necessity and Proportionality According to GDPR, the need to assess a particular data processing stems from its susceptibility to entail high risks to the rights and freedoms of single persons [7], and is therefore an essential requirement of a DPIA. The proportionality of a process of management of profiles of access to data, aims to assess the adequacy in relation to the objectives defined by the various regulations of information management and protection of personal data. According to GDPR [1], the "processing of personal data should be designed to serve people.

Protection of The Rights of Data Owners. The data owner is a "single person who can be identified, directly or indirectly, in particular by reference to an identifier, such as a name, identification number, location data, electronic identifiers, or to one or more specific elements of the physical, physiological, genetic, mental, economic, cultural or social identity of that natural person" [1]. Also, Chapter III of the GDPR defines several rights of the data subject [1]. Thus, the protection of the rights of data owners should ensure the security of any information relating to an identified or identifiable single person, in particular in the defense of his legally established rights.

Process Risks. The concept of risk is associated with the effect of uncertainty in achieving the objectives [8], as well as a scenario describing an event and its consequences, estimated in terms of severity and probability [7]. As a process, the way an entity organizes itself to achieve a certain objective [9], in this context, the risks of the process encompass the effects, estimated in terms of severity and probability, resulting from the application of the set of rules and procedures defined by the organization with a view to the management of authorization and access profiles to the data available in an IS.

3 Method

The present case study follows a deductive reasoning, based on a qualitative research strategy, supported on documental analysis and elite interviews, carried out between October 2019 and February 2020, to: (E1) Head of Communications and Information Systems Division of the PoAF Headquarters; (E2) Head of the Personnel Information Administration Office of the PoAF Personnel Command; (E3) Head of the Data and Social Protection Office of the Personnel Directorate; (E4) Head of Information Systems Administration Section of the Communications and Information Systems Directorate; (E5) Coordinator of the Technical Human Resources Information Area of the Information Systems Services Directorate of SGMDN; (E6) Head of The Administration Area of Application Systems of the Services Directorate of the Defense Data Center of SGMDN.

The semi-structured interview guide was adapted to the different interviewees, based on an extensive set of controls and questions submitted by Saldanha [10],

ISO/IEC 27001 [11] and the Commission Nationale de l'Informatique et des Libertés (CNIL) [12]. In the data analysis, we used a DPIA, based on the guidelines of the CNIL [12] and GT Art.º 29 [7], also using the Privacy Impact Assessment (PIA) software, developed by CNIL [13, 14].

4 Data Presentation and Results Discussion

4.1 Process of Management of Access Profiles to Personal Data in SIGDN-RH

According to the GDPR, Personal Data consists of any information relating to an identified or identifiable single person (data owner) [1]. The Personal Data of all PoAF HR (or data owners) available in SIGDN-RH "are organized into infotypes and sets of infotypes, with personal and organizational information" (E3). Access to them through an access profile allows the processing of personal data, that is, one or more transactions carried out by automated or non-automated means [6]. Access profiles "are organized by an array that crosses the structure profiles with the authorization profiles" (E2), that is, by "macro-processes, processes, and organizational structure, which then derive to the infotypes". There is a profile structure previously defined by functional area and function type" (E3). On January 16, 2020, SIGDN-RH had 273 PoAF users. "The management of access to SIGDN-RH is based on the occupation of the function/position, in line with the established in Directive 06/2017 of the GEN CEMFA – Organic Management of THE SIGDN in the PoAF, which advocates the management/governance model of that system in PoAF, namely the management of users (…)" (E1), however, "there is no predefined list of roles/positions that can have access to SIGDN-HR. The granting of access is assessed on a case-by-case basis, depending on the information contained in the application" (E3). "PoAF Personnel Data Administrator (ADIAP) is responsible for managing users of applications/modules in the Personnel area (…) in close coordination with members of the Personnel area with responsibilities assigned in this area" (E1). The SIGDN-RH user profiling process is explained in flowchart in Appendix A, which is the result of the various interviews conducted and CEMFA's Directive No 06/2017, however, "there is no policy on the creation of new users" in SIGDN-RH (E3), as this Directive has not yet been updated after the start-up of the HR module in 2018. The request process for changing the profile is the same, however, the cancellations of access authorization are the result of computer procedures controlled by the Administration of Application Systems (E6), as well as periodic audits of users, who have ceased their duties (E3), although it is "determined that whenever the user ceases to perform the functions that gave rise to the assigned profile, the managers must communicate" (E2). Therefore, the process of managing access profiles to the personal data of PoAF HR, present in SIGDN-RH, consists of a mechanism between the user and the data structure of SIGDN-RH, which ensures that the Personal data of PoAF data owners, are only subject to processing by users, according to their functions and position in the organizational structure of PoAF. Through a set of procedures reflected in the flowchart in Appendix A, it is intended to control access to all information and functionalities available in SIGDN-RH, with a view to protecting the rights of data subjects.

4.2 Compliance with the Core Principles of the GDPR

Necessity and Proportionality. Facing the purpose of the processing, the need for data security, as well as the protection against its unauthorized or unlawful treatment and against its accidental loss, destruction or damage, present in the Principle of Integrity and Confidentiality, are the values to retain. Thus, it can be concluded that there is a total need for the process of managing access profiles to PoAF HR personal data, present in SIGDN-RH to exist, in order to protect access, and consequently, the rights of single persons, comply with several fundamental principles and reduce the risk of personal data breach. According to the various interlocutors interviewed, the personal data requested in the authorization request are adequate and proportional to the intended purpose, framing the response time in the requirements. Referring to the degree of proportionality of the process of managing access profiles to the personal data of PoAF HR, present in SIGDN-HR, it is stated that, in view of the data collected, the process is appropriate and proportional to the objective defined for it.

Protection of the Rights of Data Owners. Chapter III of GDPR identifies the rights of data subjects. It is not because someone delivers personal data to an organization that it is no longer the owner or responsible of it; and those who receive them have a duty to respect that ownership. However, rights are not absolute and may be limited by security, defense, justice, social and economic well-being, which constitute the "backbone" of democratic societies [9]. The PoAF is developing several actions leading to compliance with GDPR (E1). It is the opinion of several interviewees, that although GDPR and SIGDN-RH are recent, the users of the system are the same as the previous one, and the rigor of organizational policies and the level of awareness of the sensitivity and nature of personal data were maintained (E1, E2, E3, E5). By the analysis carried out, the rights of data subjects remain protected, given the general level of awareness of the sensitivity and nature of personal data, although there is no document defining the procedures for this process and the PoAF is in a phase of adaptation to the GDPR and SIGDN-RH.

4.3 Risks Associated with the Access Profile Management Process

A risk is a scenario that describes an event and its consequences, estimated in terms of probability and severity. Probability expresses the possibility of a risk, estimated in terms of the level of vulnerabilities of the supporting assets and the level of resources of the sources of risk to exploit them. Severity represents the magnitude of a risk and depends mainly on the harmful nature of possible impacts on data subjects [13].

A DPIA under the GDPR is an instrument of the organization that aims to manage risks to the rights of data subjects and, as such, assesses them from the perspective of the latter [7]. It is important to determine the origin, nature and particularity of these risks, and their probability and severity may vary between Negligible, Limited, Significant or Maximum, in order to define a strategy for prioritizing and defining the implementation of mitigating controls [1]. To assess risks in a dynamic way, PIA software and its Knowledge Base, developed by CNIL [12, 13], divides risks into three categories: illegitimate access (C1), unwanted modification (C2) and disappearance of data (C3), thus enabling the introduction of threats, sources of risks, impacts and mitigation controls, and also to estimate the probability and impact if risks occur. Based on this methodology, as well as the documental analysis and content of the interviews, Tables 2, 3 and 4 present the threats, sources of risk of the process and possible impacts. A threat is a procedure that comprises one or more individual actions on data that supports assets, used intentionally or not, by risk sources and can cause an unwanted event. Table 1 presents the various threats identified during the content analysis of the interviews (E1, E2, E3, E4, E6), linking them with the various risks, taking into account the different consequences in the event of a risk scenario presented by the CNIL [13].

Table 1. Threat by risk

Threat	C1	C2	C3
Outdated access management policy for SIGDN-HR	**X**	**X**	**X**
User training/awareness	**X**	**X**	**X**
Absence of identity management	**X**		
Profile unsuitable for user functions	**X**	**X**	**X**
Lack of information on refused applications	**X**		
Form of communication	**X**	**X**	**X**

CEMFA's Directive 06/2017 has not yet been updated (E1). Although there is a centralized authentication and access management system (Micro Focus iManager), it does not integrate SIGDN-RH (E4). User training and awareness is an identified need (E1, E6) in PoAF and an important factor in any process involving human intervention. The attribution of an inappropriate profile to the user's functions can lead to access or alteration of data (E6). The absence of a file of access requests, with a record of the reason for rejection, may lead to an incorrect analysis of subsequent requests. The form of communication is a threat always present in any situation involving data transmission between different actors or IS, as is the case during the procedures of the process under analysis, identified in Appendix A. Sources of risk may cause a risk deliberately or accidentally, and be of a human nature, internal or external to the organization, or even of a non-human nature [13]. Table 2 presents the various sources of risk identified (E2, E3, E5, E6) in the process under analysis.

The Concurrent Employment feature allows certain HR information to be available for consultation and alteration by other MDN entities, where the military work temporarily, however, "enables unwanted data change, not through the access profile, but by the way functionality is being applied, at the level of procedures" (E3). Access to the information necessary for the correct execution of the user's tasks is guaranteed through an access profile matrix "where all possible profiles, activities and infotypes that each user needs for the performance of their daily tasks are identified" (E2), so,

Table 2. Sources of risk

Source of risk	C1	C2	C3
SIGDN-RH user	X	X	X
Concurrent employement	X	X	X
Entity/Person external to PoAF	X	X	X
Access profile matrix	X	X	X

consequently, that matrix can be considered a source of risk, in case of any incurred. Internal persons (SIGDN-RH users) and external to PoAF can be considered as a source of risk, since, due to insufficient personal training, people facilitate the requirement to perform their duties (E6). The consequences presented in Table 3 in the event of a risk scenario arise from the threats and sources of risk identified above, also based on the various examples presented by the CNIL [13].

Table 3. Risk impacts

Impact	C1	C2	C3
Illegitimate use of personal data	X		
Disclosure of personal data	X		
Violation of personal data	X	X	X
Incorrect information		X	X
Information on SIGDN-HR-dependent systems		X	X
Lack of relevant information			X
Impediment of access to SIGDN-HR			X
Illegitimate use of personal data	X		
Disclosure of personal data	X		

Based on the risk classification, the main stakeholders of the process under analysis were questioned about their estimate of the probability and severity of the risks presented, taking into account the threats and sources of risk identified, and the severity of their impacts, and the answers contained in Table 4 were obtained.

4.4 Risk Mitigation Measures Associated with the Access Profile Management Process

Almeida [2] identified that PoAF intends to adopt, whenever possible, the measures for the protection and security of personal data suggested by the GDPR, namely pseudonymization, minimization and also data encryption. In addition to these legally foreseen, and based on the knowledge developed by CNIL [13] and ISO [8, 11, 15], as well as the analysis of the interviews conducted, a set of measures is presented, some already implemented (not underlined), appropriate to the processing of data in question (Table 5), as well as a description of them.

Table 4. Risk estimation

	Probability		Severity	
	E2	E3	E2	E3
C1	3-Significant	2-Limited	2-Limited	1-Insignificant
C2	3-Significant	3-Significant	3-Significant	2-Limited
C3	4-Maximum	3-Significant	3-Significant	2-Limited

The definition of an EPD, the establishment of a monitoring group in its dependence and the definition of roles, responsibilities and interactions between the main actors in this area (M10, M11), allow the organization to obtain the ability to manage and control the protection of personal data in its possession (M9) [13]. Users should follow the organization's best practices in the use of restricted information for authentication [8]. Also password management systems should be interactive and ensure the use of quality passwords (M12). Information and training sessions are necessary to implement a multi-level structure for risk management (M1, M12) [8]. The procedures for exercising the rights of data subjects must be established and disclosed, and the process should not be discouraging or cost-to-people (M1) [13]. The person responsible for the management of the system should regularly review the access rights of users in order to limit access to information or processing resources (M2) [8]. The authentication method must be appropriate to the expected context, level of risk, and robustness. A strong authentication mechanism requires a minimum of two separate authentication factors, between something tangible and an individual-specific feature (M6) [13]. The management of software, hardware and communication networks should reduce the possibility of adversely affecting personal data stored in a computer system (M5) [13]. An access control policy should be established, documented and revised, based on the organization's business and information security requirements (M10) [11]. The management of privileges and tasks by profile should be appropriate to the required and regularly reviewed (M2 and M13) [13]. A Provisioning system is an automatism that improves "the management of people versus users of the system, that is, while a person is associated with a position, has the inherent accesses" (M15) (E6). To ensure the availability and/or integrity of personal data, while maintaining its confidentiality, backups should be made regularly (M4) [13]. Traceability (log management) ensures that queries and actions performed by data processing users are recorded and assigned so that it is possible to provide evidence during periodic analyses (M3) [13]. A storage period must be defined for each data type and justified by legal requirements and/or processing needs (M14) [13]. Risk management is relevant and transversal to the organization because it allows to control the risks that all personal data processing operations pose to the rights and freedoms of data subjects [13]. Table 6 associates the measures identified by risk.

Table 5. Mitigating measures

	Measure	Description
M 1	Responsibilities and Procedures	Development of actions leading to GDPR compliance; Definition of procedures and responsibilities for the exercise of the rights of the holders
M 2	Control of active users	Implementation of Identity Management system; Periodic analysis of active users in the system; Access lock, after three failed login attempts
M 3	Traceability (Log management)	Control of users' actions in real and deferred time; Log of failed login attempts; Login expiration for no action
M 4	Information backups	Total and partial, daily and weekly backups
M 5	System maintenance	Responsibility for maintaining the system shared between SGMDN and PoAF; Greater autonomy of the PoAF for the management and creation of users
M 6	Internal network security	Rules defined by MDN and PoAF for internal network use; Inclusion of additional authentication factors
M 7	Physical access control	Physical space inserted in military infrastructures, with specific access rules
M 8	Protection against non-human sources of risk	Periodic safety inspection and accident prevention in military areas
M 9	Supervision	Appointment of the EPD; Periodic inspections (GDPR compliance); Constitution of the Coordinating Group of Information Management (GCGI); Annual report of user activity
M 10	Policy Review and Update	Revision and update of Directive 06/2017, CEMFA (Organic Management of SIGDN at PoAF); Definition of the information flow of the management process of access profiles to SIGDN-HR
M 11	Local Information Delegate	Identification of Local Information Delegates for SIGDN-RH
M 12	Training/Awareness raising	Training/Awareness-raising of users (initial and ad hoc); Good practices in the definition of Password
M 13	Access profile matrix	Review of the profile structure by functional area and type of function; Restructuring of the profile matrix
M 14	Order archive	Definition of procedures for the archive of access requests, for the period necessary
M 15	*Provisioning system*	- Integration of HR data with user data (Access inherent to the position)

Table 6. Risk mitigation measures

	Measure	C1	C2	C3
M1	Responsibilities and procedures	X		
M2	Control of active users	X	X	X
M3	Traceability (Log Management)	X	X	X
M4	Information backups		X	X
M5	System maintenance		X	X
M6	Internal network security	X		X
M7	Physical access control	X		X
M8	Protection against non-human sources of risk		X	X
M9	Supervision	X		
M10	Policy Review and Update	X	X	
M11	Local Information Delegate	X		
M12	Training/Awareness raising	X	X	X
M13	Access profile matrix	X	X	X
M14	Order archive	X		
M15	Provisioning system	X		

5 Conclusions

The purpose of this investigation was to manage the process of managing profiles of access to the personal data of PoAF HR, present in SIGDN-HR, in the context of the GDPR implementation impact assessment. Using a DPIA, a set of procedures and responsibilities is identified. It is verified that, the personal data requested in the request for authorization is adequate and proportional to the intended purpose, framing the response time in the requirements. Although there is no document defining the procedures for this process and the PoAF is in a phase of adaptation to the GDPR and SIGDN-RH, the rights of data owners are protected, given the fact that users of this IS are the same as before, and the rigor of organizational policies and the general level of awareness of the sensitivity and nature of personal data have been maintained. Threats, sources of risk of the process and possible impacts are identified, relating them in three categories of risk: illegitimate access, unwanted modification and disappearance of data. Subsequently, based on the threats, sources of risk and impacts identified, the main actors of the process under analysis were questioned about their estimate of the probability and severity of the risks presented, highlighting a maximum level of probability of data disappearance and a negligible impact in case of illegitimate access. In order to mitigate the risks identified, in addition to the measures of protection and security of personal data suggested by the GDPR, based on the knowledge developed by CNIL [13] and ISO [8], as well as the analysis of the interviews conducted, a set of measures, some already implemented, appropriate to the processing of data in question and related by risk typology is presented. These measures include safeguards, security measures and procedures to ensure the protection of personal data, demonstrating compliance with the GDPR, taking into account the rights and legitimate interests of data subjects. In this follow-up, the main contributions to the knowledge arising from this research are the characteristics of the process of managing profiles of access to the personal data of PoAF HR, present in SIGDN-HR; the verification of its compliance with the fundamental principles: the need and proportionality of processing and the protection of the rights of the holders of personal data; and the assessment of the risks associated with the process, and consequent identification of mitigating measures thereof. As regards future studies, and in view of the fact that such conformity assessments and risk analysis are continuous in nature, it is considered appropriate to validate the measures submitted under this DPIA in relation to the financial feasibility of implementing them, and where necessary, their reformulation and adequacy, in the light of new facts or constraints.

Appendix A - SIGDN-RH User Profile Process

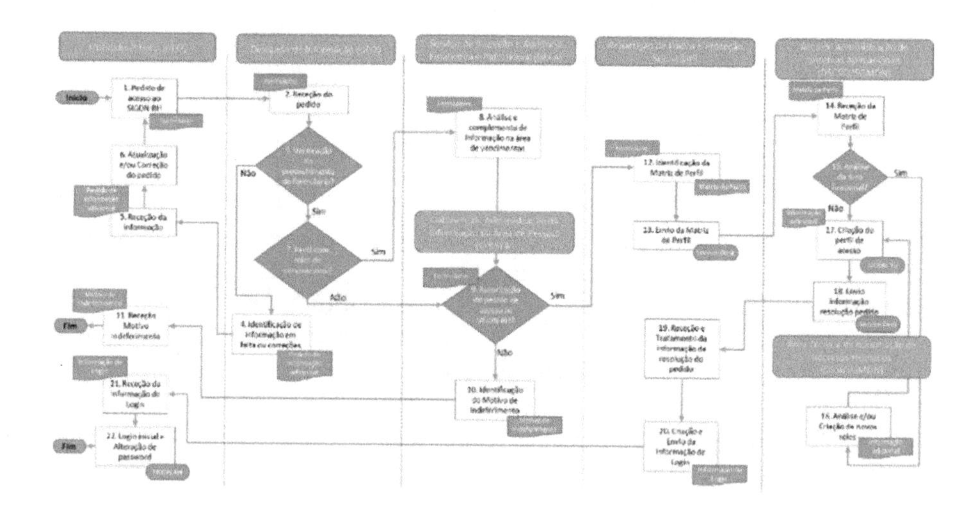

References

1. European Parliament. Regulation 2016/679 of 27 April 2016 on the protection of natural persons with regard to the processing of personal data and the free movement of such data and repealing Directive 95/46/EC (General Data Protection Regulation). European Union: Official Journal of the European Communities (2016)
2. Almeida, I.F.: The General Data Protection Regulation: Impact on the Air Force's Information Management Policy. Air Force Academy, Sintra (2018)
3. Law 58/2019 of August 8th. Data Protection Act. Diário da República, 1st Grade, 151. Three to forty. Assembly of the Republic, Lisbon (2019)
4. Santos, V.: Creativity in Information Systems. FCA, Lisbon (2018)

5. General Secretariat of the Ministry of National Defense: SIGDN Human Resources and Salaries. Business Blueprint, Lisbon (2011)
6. Directive 12/CEMFA/2018 of 14 June. Protection of Personal Data in the Air Force. Air Force, Lisbon (2018)
7. Article 29 Working Group on Data Protection. Guidelines on The Impact Assessment on Data Protection(IPCC) and determining whether the processing is 'likely to result in a high risk' for the purposes of Regulation (EU) 2016/679. European Union (2017)
8. ISO/IEC 27000: Information technology - Security techniques - Information security management systems - Overview and vocabulary. International Organization for Standardization, Switzerland (2018)
9. Antunes, L.: Putting the GDPR into Practice - What Changes for Us? What about Organizations?, p. 98. FCA, Lisbon (2018)
10. Saldanha, N.: GDPR - Guide to a Compliance Audit - Data, Privacy, Implementation, Control, Compliance. FCA, Lisbon (2019)
11. ISO/IEC 27001: Information technology - Security techniques - Information security management systems – Requirements. Switzerland: International Organization for Standardization (2013)
12. Commission Nationale de l'Informatique et des Libertés: Privacy Impact Assessment (PIA) - Methodology. Autor, França (2018)
13. Commission Nationale de l'Informatique et des Libertés Privacy Impact Assessment (PIA) - Knowledge Bases. Author, France (2018)
14. Commission Nationale de l'Informatique et des Libertés: The open source PIA software helps to carry out data protection impact assessment (2019)
15. ISO 31000: Risk Management-Guidelines. Switzerland: International Organization for Standardization (2018)

Agile Application for Innovation Projects in Science Organizations - Knowledge Gap and State of Art

Zornitsa Yordanova(✉)

University of National and World Economy, 8mi dekemvri, Sofia, Bulgaria
zornitsayordanova@unwe.bg

Abstract. The paper examines the process of innovation development in science organizations and its project management. As a methodological approach, the study focuses on Agile method as a technique for improving some old-school practices for innovation, product and project management. The research methodology goes through a systematic review of the literature of the Agile application in project management and innovation in scientific organizations to prove the knowledge gap. The results of the research are of interest to scientists from a wide range of fields and offer knowledge and insight for more effective project management for innovative projects in research organizations. In addition, the methodology applied in this study is based on a specially developed data analytics tool for word and context analysis (based on the principles of literature review) and might be re-used in many other studies which aim at identifying knowledge gap and state of art in interdisciplinary fields of study. The tool is Advanced Analytical Literature Review (AALR) built on Qlik software.

Keywords: Agile · Innovation management · Technology management · Science project management · Process mining · Process breakdown

1 Introduction

Agile as a project, product and innovation management practice and method from one hand and its potential application for innovation projects in science organizations. Applying Agile in different industries, organizations and cases have already existed in the practice and it has described in the literature [1, 2]. This paper presents a systematic literature review of the current Agile application for innovation projects in science organizations if such exists partially, identifies knowledge gap and proposes some recommendations for extending this use since Agile has been already defined as strong tool for managing projects and innovation [3].

The paper addresses the decreasing innovation performance of science organizations and universities [4]. The main reason for this poor performance is not the lack of innovation capacity among scientists, insufficient funding or organizational shortcomings, but the lack of methodology for development and management of innovation projects by science organizations [5]. The theoretical basis of project management and innovation management as part of the management and economics science has not yet found an

Á. Rocha et al. (Eds.): ICITS 2021, AISC 1330, pp. 108–117, 2021.
https://doi.org/10.1007/978-3-030-68285-9_11

integrative approach in between and there is no developed and applied methodology to help and support the management of innovative projects. Agile is a potential tool in this direction so to link project, innovation and product management for strengthening innovation development in science organizations. The paper also discusses some specifics of this type of projects, as well as some features of scientific organizations.

The results of the study are from interest for scientists from wide range of fields and brings knowledge and insight for making project management for innovation projects in science organizations more efficiently. Additionally, the methodology applied in this research is based on a specially developed data analytical tool for word and context analysis (data mining) and might be re-used in many other studies which aim at identifying knowledge gap and state of art of interdisciplinary research areas. The tool is Advanced Analytical Literature Review (AALR) built on Qlik software.

2 Theoretical Background

2.1 Agile

Agile is a project management framework based on the Agile Manifesto values [6]. The Agile manifesto focusses on individuals and interactions over processes and tools, working methods (software in case of IT firms) over comprehensive documentation, customer collaboration over contract negotiation, and responding to change over following a plan [7]. It emphasizes on collaborative work with customer centricity for project success [8].

Agile is best suited for projects that are iterative and gradual [9]. This is a type of process in which demands and decisions are developed through the joint efforts of self-organizing and multifunctional teams and their clients. Originally created for software development, it was created in response to the inconsistencies of the Waterfall method, whose processes do not meet the requirements of the highly competitive and constant movement of today's industry. By having this in mind, the current research is analyzing the potential of applying Agile for innovation management of innovative projects in science organizations. Agile has already been identified as a tool not only for project management but also as an innovation management instrument [10]. Other advantages include easy collection of requirements, retrieval and analysis, effective management [11] which all seem very relevant to innovation projects in science organizations as well. The use of the Agile approach in the management of innovation projects is characterized by greater flexibility, which in practice allows innovators to take one or more steps back from the development and implementation process to eliminate any problems or errors at the moment in their identification [12]. This differs from the traditional methods of innovation development, in which the development and commercialization is planned for the whole project at the beginning and then the plan is followed according to its specifics [13]. Pearson, Costley and Nuttall [14] concluded in their research that despite several decades of dedicated R&D, fusion, a potentially world-changing energy source, remains decades away from commercialization. This research reasoning is based on the same assumption about innovative projects in the scientific organization.

2.2 Project Management in Science Organizations

Project organization and project management handle many activities better than any other organizational structure [15]. In order to meet the needs of modern organization for rapid development of projects and to respond quickly to the globalization of innovation processes, more and more organizations implement Agile as project management approach [16]. The Project Management Institute reported that the Agile approach is widely being used for project management practices, not only in software industry and the result have shown a significant impact on business growth and project performance [17]. However, science organizations and universities in particular do not use Agile for their project management in general although some exceptions exist [18].

Generally, the waterfall model is being used in science organizations and universities for project management activities. There are some research which aimed at developing collaborative university-industry R&D initiatives by presenting a hybrid project management approach, with a set of key distinct project management practices in the context of such collaboration [19].

2.3 Innovation Management and Innovation Projects in Science Organizations and Universities

Innovation management has still not been incorporated efficiently as a major value in universities no matter that universities are with the innovation triangle [20]. University technology transfer is often associated with formal transmission of science-based inventions, for instance through the licensing of patented technology to a firm [21] and it is heavily overloaded process in most countries. No matter that research universities are well-recognized sources of new knowledge, and their contributions to innovation are manifested through the creation and transfer of new technologies originating from academic research, this transfer is still struggling from slow procedures and bureaucracy.

Innovation projects in this research are considered as projects in science organizations with the goal of some kind of commercialization or applied science projects in general. The definition adopted for innovation projects in this research is as follows: 'Innovation projects are systematically managed endeavors that use inputs in order to transform them into outputs with a certain scope and aims at achieving something new, in a new way or at improving something existing' [22]. It originates from both project management and innovation management theory.

3 Research Design

The research design is based on a systematic literature review and the use of a specially developed tool for data analytics. The purpose is the revealing of the currently use of Agile in innovation projects in science organizations.

3.1 Systematic Literature Review

For the purpose of defining comparatively wide and still focused scope of the research, an advanced search in Scopus database has been first undertook. It aims at scoping science papers which are relevant to innovation, product or project management on one hand and on the other, to analyze the use of one of the flexible methodologies, usually used for handling with these three purposes of modern organizations: Lean Startup, Agile, Scrum, Design Thinking, User centricity or User Innovation. This is the formula used in the Scopus advanced search:

TITLE-ABS-KEY ("innovation management" OR "project management" OR "product management") and TITLE-ABS-KEY ("Lean Startup" or "Lean-startup" or "Lean start-up" or "Agile" or "Scrum" or "Design Thinking" or "design-thinking" or "User centricity" or "user-centricity" or "User Innovation" or "user-innovation" or "lead user" or "lead-user" or "led user" or "led-user" or "user-driven" or "user driven"). This data source gives the state of art of the Agile knowledge in the literature. The inclusion criteria with some other flexible methodologies for project and innovation management is because of their frequent research in combination. The dataset of science papers received by these searches has been cleaned (by removing duplicates) and uploaded in the Advanced Analytical Literature Review software as well.

3.2 Tool for Advanced Analytical Literature Review (AALR)

A special tool has been designed and developed to serve the research. The tool is based on Qlik Sense application, which is amongst the best recognized business intelligence tools for data analytics. It is basically used for enterprise data analysis and its application in the current research through AALR is indeed an innovative method for deeper word and contextual analyzing the topic. It aims at revealing hidden connections between science papers with different focus from first impressions and with the traditional approaches. AALR is very useful and extremely appropriate for interdisciplinary research where the analysis includes literature from diverse science fields. The tool AALR integrates all the results from the Scopus research showed in the first step of the research design. AALR is configured with various filters by which the researcher can search and combine different words in order to compare and collate the use of different words within large amount of research papers in their titles, author keywords and abstracts. By doing this, the researcher is able to reveal hidden context, to discuss and analyze the use of words in their context, to make comparison and to identify links between different research. It might be also useful for defining knowledge gaps (Fig. 1).

The tool presents the titles, author keywords and the abstracts and decomposes all the words in these. There are search functionality on each of the sections for titles, author keywords and the abstracts which proposes an advanced search functionality to search by many words of words' fractions.

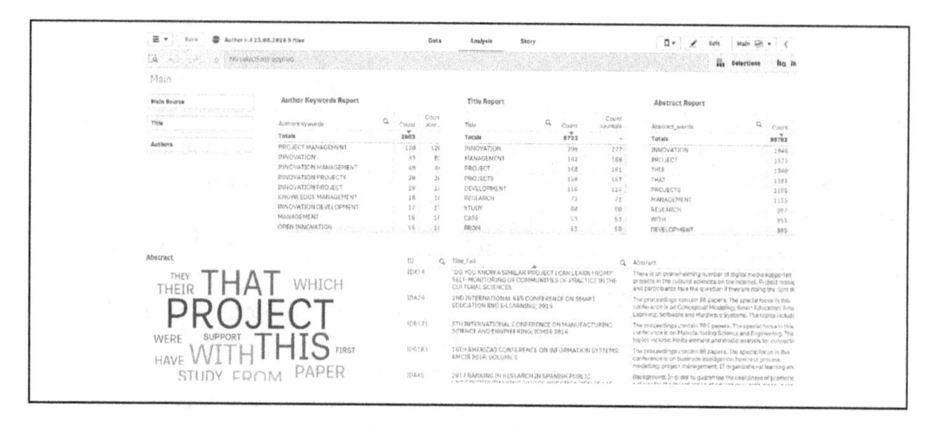

Fig. 1. Data analytics tool AALR for word and context analysis of science papers

4 Results and Discussion

4.1 Systematic Literature Review in Scopus Database

The whole systematic literature review is based on 388 science papers from Scopus database. The first simple filtering search for existence of both ‘Agile’ and ‘Science’ in these 388 papers showed only two papers examining their connection and mentioning both words in the title. They both are from one and the same author - Senabre Hidalgo and were published in 2019 [23, 24]. Both the papers are dealing with adapting the scrum framework for agile project management in science and presented a case study of a distributed research initiative. The papers reasoned the knowledge transfer of Agile practices to science with providing proofs about Agile success in many other types of organizations. The same filter search of both ‘Agile’ and ‘Science’, but in the abstracts of the golden source of 388 science papers, reveal 28 documents, mentioning in their abstracts both the words. By systematically going through these 28 documents, it is turned out that only few of them are actually dealing with Agile for science and university issues. Obviously, Agile in science organizations and universities is s a very unexplored topic, but still has gained the research interest in the recent years.

The secondly done series of searches, identifying the literature achievements about project and innovation management in science organizations and universities are only analyzed with the tool, because of their broad scope and complexity. Only one of the searches showed results that are worth to be discussed separately without the combinatory approach of the tool. They are not Agile related. This is the search about science papers (4 described in the research design section), analyzing project and innovation development on one hand and science organization or university on the other [25, 26].

In the same source, handling with both project and innovation management in science organizations and universities, Agile does not exist at all.

4.2 Results from Word and Context Analysis with the Tool for Advanced Analytics AALR

The first dataset is analyzing Agile for project and innovation management and it reasonably gives a lot of proofs for such a synergy and already existing application. In the search functionality, the tool has options for word search in titles, abstracts and author keywords of the uploaded dataset. All the words with less than four symbols are removed. All words with four and more symbols which are non-relevant for the context analysis are also removed (for example more, than, with, where, etc.). The figure below presents a word analysis of all the words and their frequency used in the abstracts of the analyzed 388 science papers, based on the first Scopus advances search, presented in the research design (Fig. 2).

Fig. 2. Word analysis of science papers, dealing with project and innovation management and flexible methodologies

The results from the word analysis show that when it comes to project and innovation management and flexible methodologies, Agile is the most frequent used word in science papers' abstracts. In combination with it, topics and words such as: 'development', 'design', software', projects', 'user', etc., are also very common. This analysis gives insight about the context of Agile research and the relevant topics usually researched along it.

Analyzing the top 10 words, used in abstracts of these 388 research papers (based on the first shown search in the research design), these are: 'project', 'management', 'agile', 'development',' design', 'software', 'process', 'innovation', 'information', 'scrum'. Looking at titles, the results show the same words, but with additions to 'construction', system', 'engineering' and 'robotics'. The analysis of the author keywords gives clearer analysis with these top 10 used words/expressions: 'project management', 'design thinking', 'agile project management', 'scrum', 'agile', 'innovation management', 'innovation',' robotics', 'lean startup' and 'construction management'.

Exploiting the AALR feature for filtering a particular word and funneling the results for all science papers, the next performed analysis is based on filtering papers which contains Agile in their abstract, title and author keywords. The assumption is that these would be those papers, dealing extremely with the topic of Agile. The results show 70 science papers which are predominantly published in the period between 2016 and 2020. Removing the filter from the author keywords, but keeping it for titles and abstracts, the analysis of the top used words shows what are the topics related to the research of Agile. These are mostly related to project management, Scrum, IT/software development, product development, innovation management and agile methods. A different approach has been performed so as to reveal what are the topics discussed in science papers with Agile in the author keywords and the results are again related to projects, management, Scrum, product development.

Analyzing the second series of advanced searches in Scopus database (9 searches, described in the research design section), the analyses performed aim at revealing how science organizations and universities are dealing with project and innovation management/development at the moment. The tool has automatically cleaned all he ninth datasets from the Scopus searches from duplicates. A fast conclusion might be done about the evident knowledge gap in the literature about studies which research innovation development in science projects and also innovation development in science organizations. The rest of the datasets are uploaded in the AALR tool and the analyses performed and presented below are results from them. The reduced number of the nine searches presents 360 science papers, which answer to the inclusion criteria.

First analysis is based on filtering amongst abstract for the word Agile, which basically means using Agile in science organizations and universities in the context of project and innovation management. The results showed three papers, which titles only are presented below for giving the scope of the literature coverage: The Long Way from Science to Innovation – A Research Approach for Creating an Innovation Project Methodology [27]; Agile Project Management in University-Industry Collaboration Projects [28]; Project Management Practices for Collaborative University-Industry R&D: A Hybrid Approach [19]. All three of the research papers were published in the last two years (2018 and 2019), which again emphasized the relevance and topicality of the researched subject matter. All three of them are discussing interdisciplinary research which indirectly also frames the research question for application of Agile and flexible methodologies in general, in the context of innovation projects in science organizations as a suitable approach, inviting and asking for much further research.

The following figure presents a wave analysis on the papers, dealing with Agile in science organizations and universities in the context of project and innovation management. The wave analysis shows other relevant topics discussed alongside with Agile (Fig. 3).

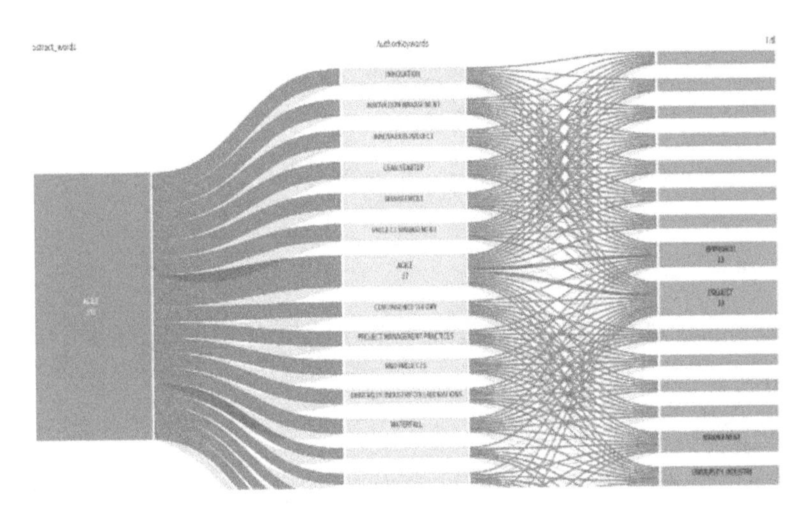

Fig. 3. Wave analysis of science papers, dealing with Agile in the context of project and innovation management in science organizations and universities

Analyzing the results of the wave analysis based on Agile word in the abstract, the most discussed topics in the context and presented in the title are: ‘innovation’, ‘innovation management’, ‘innovation project’, ‘lean startup’, ‘project management’, ‘contingency theory’, ‘R&D projects’, ‘university-industry collaboration’, ‘waterfall’.

The results show that papers, researching Agile are categorized by their authors with the following keywords: ‘agile project management’, ‘ICT’, ‘project-based learning’, ‘R&D learning development’ and ‘Scrum”. These results give the scope of the state of art of Agile in science organizations for the purposes of project and innovation development. Analyzing further the results from the AALR tool, some other topics appeared as well: ‘interdisciplinary’, ‘knowledge’, ‘competence’, ‘collaboration’, ‘hybrid’, ‘creation’, etc.

5 Conclusion

In conclusion, several research results and contributions might be highlighted. The research paper first presents an innovative approach with a specially designed and developed tool for advanced analytics for performing deep systematic literature review of science papers called AALR. Secondly, the paper analyzes a hot topic related to strengthen project management of innovation projects in science organizations and universities which based on the results of a systematic literature analysis on Scopus datasets, is under analyzed and underestimated by researchers. Thirdly, the paper provides some well identified knowledge gaps for further examination and research as: Agile in specific research projects and its impact on innovation performance of a science organization in general. Many co-topics or close topics have also been identified by performing word. Wave and context analyses of the words used in the scoped science papers’ titles, abstracts and author keywords. Further research of the author will be developing a new project management framework for innovation projects in science organizations and universities which can be used easily from all scientists no matter of their management skills and knowledge.

Acknowledgments. The paper is supported by the BG NSF Grant No KP-06 OPR01/3–2018.

References

1. Conforto, E., Salum, F., Amaral, D., Silva, S., Almeida, L.: Can agile project management be adopted by industries other than software development? Proj. Manag. J. **45**(3), 21–34 (2014)
2. Ciric, D., Lalic, B., Gracanin, D., Palcic, I., Zivlak, N.: Agile project management in new product development and innovation processes: challenges and benefits beyond software domain. In: 2018 IEEE International Symposium on Innovation and Entrepreneurship (TEMS-ISIE), Beijing, pp. 1–9 (2018). https://doi.org/10.1109/TEMS-ISIE.2018.8478461
3. Špundak, M.: Mixed agile/traditional project management methodology – reality or illusion? Procedia Soc. Behav. Sci. **119**, 939–948 (2014)
4. Tseng, F., Huang, M., Chen, D.: Factors of university–industry collaboration affecting university innovation performance. J. Technol. Transf. **45**, 560–577 (2020). https://doi.org/10.1007/s10961-018-9656-6
5. Benneworth, P., Pinheiro, R., Karlsen, J.: Strategic agency and institutional change: investigating the role of universities in regional innovation systems (RISs). Reg. Stud. **51**(2), 235–248 (2017). https://doi.org/10.1080/00343404.2016.1215599
6. Beck, K., et al.: Manifesto for agile software development (2001). https://agilemanifesto.org. Accessed 10 Feb 2009. Accessed 01 Jan 2020
7. Hoda, R., Murugesan, L.K.: Multi-level agile project management challenges: a self-organizing team perspective. J. Syst. Softw. **117**, 245–257 (2016)
8. Vanhala, E., Kasurinen, J.: The role of the customer in an agile project: a multi-case study. In: Hyrynsalmi, S., Suoranta, M., Nguyen-Duc, A., Tyrväinen, P., Abrahamsson, P. (eds.) Software Business. ICSOB 2019. Lecture Notes in Business Information Processing, vol. 370. Springer, Cham (2019)
9. Oorschot, K., Sengupta, K., Wassenhove, L.: Under pressure: the effects of iteration lengths on agile software development performance. Proj. Manag. J. **49**, 78–102 (2018). https://doi.org/10.1177/8756972818802714
10. Cooper, R., Sommer, A.: The agile–stage-gate hybrid model: a promising new approach and a new research opportunity. J. Prod. Innov. Manag. **33**(5), 513–526 (2016). Virtual Issue: Collection of Robert G. Cooper's JPIM Articles
11. Lin, Y., Markowsky, G., Liu, S., Markowsky, L.: Adapting American ERP systems for China: cross-cultural issues and a case study. In: 2013 IEEE 7th International Conference on Intelligent Data Acquisition and Advanced Computing Systems, vol. 2, pp. 572–577 (2013)
12. Misra, S., Singh, V., Bisui, S.: Characterization of agile ERP. Softw. Qual. Prof. **18**(3), 39–46 (2016)
13. Bianchi, M., Marzi, G., Guerini, M.: Agile, stage-gate and their combination: exploring how they relate to performance in software development. J. Bus. Res. **110**, 538–553 (2020)
14. Pearson, R.J., Costley, A.E., Nuttall, W.J.: Technology roadmapping for mission-led agile hardware development: a case study of a commercial fusion energy start-up. Technol. Forecast. Soc. Change **158**, 120064 (2020)

15. Hobday, M.: The project-based organisation: an ideal form for managing complex products and systems? Res. Policy **29**(7–8), 871–893 (2000)
16. Tidd, J., Bessant, J.: Managing Innovation: Integrating Technological, Market and Organizational Change. Willey, Hoboken (2018)
17. Raharjo, T., Purwandari, B.: Agile project management challenges and mapping solutions: a systematic literature review. In: Proceedings of the 3rd International Conference on Software Engineering and Information Management (ICSIM 2020), pp. 123–129. Association for Computing Machinery, New York (2020). https://doi.org/10.1145/3378936.3378949
18. Akhmetshin, E.M., Romanov, P.Y., Zakieva, R.R., Zhminko, A.E., Aleshko, R.A., Makarov, A.L.: Modern approaches to innovative project management in entrepreneurship education: a review of methods and applications in education. J. Entrepreneurship Educ. **22**, 1–15 (2019)
19. Fernandes, G., et al.: Project management practices for collaborative university-industry R&D: a hybrid approach. Procedia Comput. Sci. **138**(2018), 805–814 (2018)
20. Etzkowitz, H., Zhou, C.: The Triple Helix: University–Industry–Government Innovation and Entrepreneurship. Routledge, Abingdon (2018)
21. Hayter, C.S., Rasmussen, E., Rooksby, J.H.: Beyond formal university technology transfer: innovative pathways for knowledge exchange. J. Technol. Transf. **45**, 1–8 (2020). https://doi.org/10.1007/s10961-018-9677-1
22. Yordanova, Z.: Innovation project tool for outlining innovation projects. Int. J. Bus. Innov. Res. **16**(1), 63–78 (2018)
23. Hidalgo, E.S.: Adapting the scrum framework for agile project management in science: case study of a distributed research initiative. Heliyon **5**(3), e01447 (2019)
24. Hidalgo, E.S.: Corrigendum to "adapting the scrum framework for agile project management in science: case study of a distributed research initiative." Heliyon **5**(4), e01542 (2019)
25. Mylnikov, L., Fayzrakhmanov, R., Kurushin, D.: Information support of project management processes in the network of research organizations and technological companies. Astra Salvensis **6**, 603–612 (2018)
26. Kossukhina, M.A.: The interaction among universities and industrial enterprises as the main factor of innovative development. In: 2015 4th Forum Strategic Partnership of Universities and Enterprises of Hi-Tech Branches (Science Education Innovation), pp. 75–76 (2016)
27. Yordanova, Z., Stoimenov, N., Boyanova, O., Ivanchev, I.: The long way from science to innovation – a research approach for creating an innovation project methodology. In: Abramowicz, W., Corchuelo, R. (eds.) Business Information Systems. BIS 2019. Lecture Notes in Business Information Processing, vol. 353. Springer, Cham (2019)
28. Säisä, M.E.K, Tiura, K., Matikainen, R.: Agile project management in university-industry collaboration projects. Int. J. Inf. Technol. Proj. Manag. (IJITPM) **10**(2) (2019). https://doi.org/10.4018/IJITPM.2019040102

Managing, Locating and Evaluating Undefined Values in Relational Databases

Michal Kvet[(✉)] and Karol Matiasko

Faculty of Management Science and Informatics, University of Žilina, Univerzitná 8215/1, 010 26 Žilina, Slovak Republic
Michal.Kvet@fri.uniza.sk

Abstract. Database systems provide complex and robust solutions for dealing with data with emphasis on the performance aspect covered by the index access methods. Undefined values can provide significant processing drop, whereas NULLs are not part of the B+tree indexes. In this paper, we discuss the existing techniques based on the data transformations and function-based indexes. The main contribution of the paper is based on our own techniques, mapping, and ensuring partially undefined data states to be present in the index. Thanks to that, data can be easily located inside the index by shifting the necessity of sequential data block scanning into the index range scan.

Keywords: Undefined state · NULL · Relational database · Indexing · Mapping

1 Introduction

Database systems are the inseparable part of the information technology. Information systems need to manage, evaluate, and complexly process the wide amount of the data, which are commonly stored in the databases ensuring robustness, consistency, reliability, security, and data independence [10]. Thus, any change in the physical database storage does not influence the interface and communication principles of the information systems and applications. The relational paradigm is still often used and widespread. The main aspect supplying the strength of the solution is based on the relational algebra describing the access principles and evaluation criteria to compose the complex query.

A relational database system consists of two modules specifying the data structure – entities and relationships between them [2]. The task is to define the data structure suitable for the application system with an emphasis on data normalization. Thanks to that, the data structure is optimized reducing data amount to be stored as well as many problems caused by inconsistency, reliability issues, and improper management [13].

Data to be stored and evaluated can origin from any system or segment. Nowadays, the data amount is significantly rising and the efficiency of the processing is significant [3]. Physically, data are stored in the data blocks located in the tablespaces. Each tablespace consists of at least one physical data file split into individual blocks. Blocks are associated with the data segment – mostly representing the table or index. Blocks

Á. Rocha et al. (Eds.): ICITS 2021, AISC 1330, pp. 118–127, 2021.
https://doi.org/10.1007/978-3-030-68285-9_12

are, however, not allocated separately due to the processing and system demands. Therefore, blocks are grouped into extents and allocated in that manner [11].

Tablespace forms the interlayer between the physical database and memory forming database instance. It is a significant module, by which the physical storage separation can be done. The user specifies just the query to be processed and evaluated, without any information about the process of data obtaining, as well as the data structures, in which the data reside [11]. Therefore, the database system must be autonomous to ensure the correctness of the results with emphasis on the processing time and resource consumption.

This paper deals with the process of data evaluating and accessing. It provides a brief summarization of the index structures and methods based on the index and query suitability. The core part of the paper deals with the undefined states characterized by the undefined values. Data as the input for the database system can be produced by various subsystems, modules, and external systems, which can have various quality, reliability. The sensor data environment does not need to produce the required data in the specified time point, data can be corrupted, inapplicable or just not obtained. All such cases produce undefined values to be handled by the database system. Object state usually consists of various data attributes, not just the one. Thus, the state is only partially defined and attributes, which do not hold the data in the specified state must be marked. With the rise of temporality, in which the system does not manage just current valid data, but the whole spectrum is delimited by the time validity, the problem of undefined value management is even sharper [9]. This paper aims to propose own solution for dealing with undefined values with an emphasis on the NULL definition, which is, however, not part of the index. Thanks to that, if the query references at least one attribute, which can hold undefined value, particular index managing such attribute cannot be used [2, 12], even if there is no physical NULL value. As a consequence, the specification can therefore cause a significant performance drop. Therefore in this paper, we propose and discuss various techniques to ensure indexing and accessing undefined data more properly and effectively.

2 Analysis

An index is an optional structure associated with the data table to propose effective access to the data based on the indexed values. It aims to improve the performance of the *Select* statement, as well as to ensure the consistency of the data. Thus, its significant advantage can be felt in other data manipulation statements (*Insert*, *Update*, and *Delete*) as well, mostly when dealing with the constraints and referential integrity. By using an index, all data blocks do not need to be scanned completely sequentially, but the pointer to the particular block is used. Such block is then loaded into the memory to the consecutive processing and evaluation. If the query condition is unique, the index produces no more than one pointer to the block [2]. Thus, after its loading and evaluation, processing can end. Vice versa, if the condition can produce a various number of data rows, several pointers and produced to be evaluated. However, the amount the be processed is always significantly lower in comparison with the total number of data blocks [8]. In the memory, the block is loaded and located in the *Buffer cache* – matrix structure of the block granularity, to which database block can be loaded. Similarly, any data change is firstly applied to the memory blocks and consecutively loaded into the database on demand. The memory *Buffer cache* block can hold a *clean* or *dirty* (occupied) mark. If the block is *clean*, it can be rewritten at any time, whereas the data are stored in the physical storage, thus there is no possibility for data loss. Vice versa, if the block is *dirty*, before replacing, its image must be copied into the database.

Data block loading is part of the data access and it forms the crucial performance aspect. The granularity of the *Buffer cache* is a block, thus the whole block from the database must be loaded for the evaluation, regardless of the relevant amount inside it to be processed.

As stated, the data index can provide significant performance improvement in terms of data locating. From the perspective of the commands modifying the database, the benefit may no longer be any, or minimal reflecting the constraints [3]. To emphasize the validity of the index, each data change must be applied to each index, therefore the set of the indexes must be strictly limited to provide just a slight impact on the performance.

The index can be defined either explicitly or in an implicit manner by the primary key, respectively any unique constraint.

A – Index Structures

In individual database systems, the architecture of the index can vary depending on the layout and access to the data. In principle, the index can consist of the pointers to the data mostly defined as *ROWID* or it can hold the direct data there – index-organized table structure [10]. *ROWID* is the locator for the data row by specifying its position in the database – data file, data block, and position of the row inside the block. It is unique among the standalone database.

In the database systems, various index structures and techniques can be identified. The common and most often used structure is a *B+tree*. Its main advantage is the robustness and balance. It does not perform degrade with the increase of the data holding inside. The structure consists of the tree in which the path from the root to the leaf layer has the same length [7]. Three node types are present: *root*, *internal node*, and *leaf node*. *The root* and *internal node* contains pointers S_i and values K_i, the pointer S_i refers to nodes with lower values the corresponding value (K_i), pointer S_{i+1} references higher (or equal) values. Leaf nodes are directly connected to the file data (using pointers) [2].

Other indexes used in database systems are *bitmaps* (characterized mostly by the warehouse environment) and *hash* indexes available just in some systems [1, 6].

B – Index Access Methods

The importance of the index definition is underlined by the sequential scanning of all associated data blocks, loading into memory followed by the parsing, if no suitable index is available. In that case, the *Table Access Full* (*TAF*) method is used. On the opposite side of the corridor, index access methods can be identified. In that case, the index is scanned resulting in obtaining all required data or by providing relevant *ROWID* values to locate data rows in the database blocks. If the condition is based on the unique value, mostly delimited by the primary key identifier, an *index unique scan* is used. Interval data conditions are evaluated by the *index range scan*. The specific method category is covered by the full index scans, by which the whole index is scanned fully, whereas its processing is far easier in comparison with sequential data block scanning. Its activity uses the fact, that the index is usually already located in the memory. Moreover, the size of the index is significantly smaller compared to the whole table [8].

DBS Oracle has proposed another access method – *index skip scan*, method, which is based on skipping the leading attribute for the evaluation. Physically, it is implemented by the index highlighting the first attribute. In the layers, there are pointers to the secondary structure covering the rest indexed attributes [1, 5].

3 Undefined Values – Existing Solutions

Undefined values form an important aspect, that must be handled to ensure the performance of the system. First of all, each stored attribute is characterized by the domain, to which it belongs. Column integrity maintains the values, which can be present with emphasis on the uniqueness, duplicities management, and undefined values. Marking attribute specification with the *NULL* flag means, that it can hold undefined value. Although such attributes can be part of the index, many times, a particular index cannot be used, whereas undefined values are not indexed in the default *B+tree* index structure at all. Thus, for using the index, such missing values must be taken away inside the query. If not, the optimizer will choose sequential scanning of all data blocks. It can have huge impacts if the data fragmentation is present [2].

As already mentioned, the default B+*tree* index provides a robust background for the access methods. Its strength is, however, limited by the impossibility to deal with *NULLs*, which are not present there. Although database statistics contain the information about the number of *NULL* values for each table attribute, it just reflects the estimation, whereas statistics are not generated after any data change, but regularly based on the specification. Thus, the database optimizer cannot rely on them to evaluate the position of the *NULL* with regards to the index usage. In principle, nowadays, two principled solutions are present to be served. The first solution (**model 1**) is based on the value transformation, so the undefined value is not physically present in the database storage [10]. Transformation can be done either explicitly, or automated by the trigger:

```
Create or replace trigger trig_null
 Before insert or update on table_name
  For each row
begin
  if :new.atr_val is null  then :new.atr_val:=null_replacement_val; end if;
end;
/
```

Such a trigger is specified for each table separately. In the body, any attribute, which can hold a *NULL* value, can be processed by checking its value and transforming it into *null_replacement_val*, if suitable. Thanks to that, *NULL* values are not stored physically and the index covers all data tuples. On the other hand, this implementation brings additional demands on size and storage, whereas such transformed values must be physically stored in the database. Moreover, a particular user must have the privilege to create a trigger. Finally, adding a new trigger for the evaluation brings new costs dealing with *Insert* and *Update* data operation.

The preceding model removes the *NULL value* impact on the physical layer. The attribute itself cannot hold NULL from the definition, which is ensured by the checking during a data operation. The second existing solution removes the impact of *NULL* on the logical layer (**model 2**) by using a transformation function inside the index forming function-based index [4]. From the performance perspective, it does not load the physical storage by any additional demand. Vice versa, evaluating is enhanced by the function call for each data row.

4 Undefined Values – Own Contribution

Based on the previously defined solutions and performance analysis, to propose a more effective and robust solution, it is necessary to ensure these two requirements:

- do not increase the size demands of the database storage,
- do not increase performance drops caused by the function calls and evaluations.

As evident, these requirements cannot be passed completely. Note, that physical undefined value management can be located either directly in the table data, or the index definition, or both.

In this paper, we propose four own models, which principles are described in the following section. The first solution (**model 3**) is based on the *mapping* executed during the data loading from the database into instance memory. It is relevant just for the table data, whereas in the index, particular references are not present. Thanks to that, the *NULL* value with no storage demand is located in the database, but after the loading into the instance memory, the undefined value is replaced by the defined default value. The principle is shown in Fig. 1.

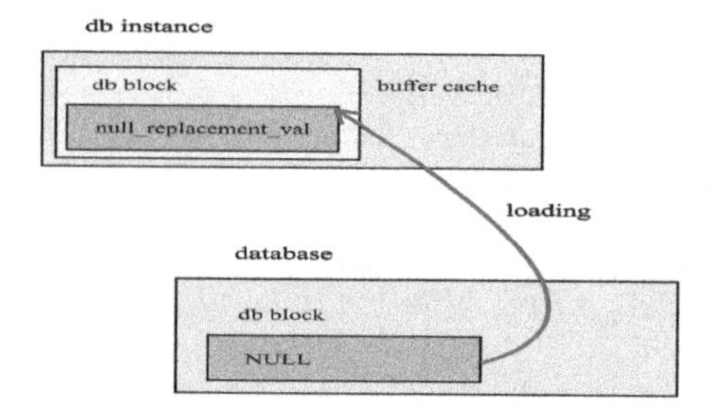

Fig. 1. Model 3 – mapping

The execution is done during the data loading, before the block parse operation. The proposed *model 3* has the following properties and limitations: There are no extra size demands dealing with physical storage. Additional demands are present when the data are to be loaded. Note, that each data block must be parsed and evaluated to identify undefined values. Such activity is necessary to be done, even, if the block is loaded as the result of the *ROWID* value obtaining, by which the direct data row can be located. That is the consequence of ensuring data consistency. There is, moreover, one extra significant problem. The *NULL* value itself does not require storage capacity. However, if the original *NULL* value is to be replaced, a new value needs to be stored in the *Buffer cache*, formed in the block shape, as well. The change during the transformation may no longer fit in the original block if it is full. In that case, the transformation cannot be done. The workaround of the problem is to reflect consistency between the database and instance memory is based on splitting the original data block in the database storage, as well as memory. As a consequence, after the processing, from one block, now, in the database and memory, one additional block needs to be added. The principles are in Fig. 2. Note, that the original block is invalidated in the last phase.

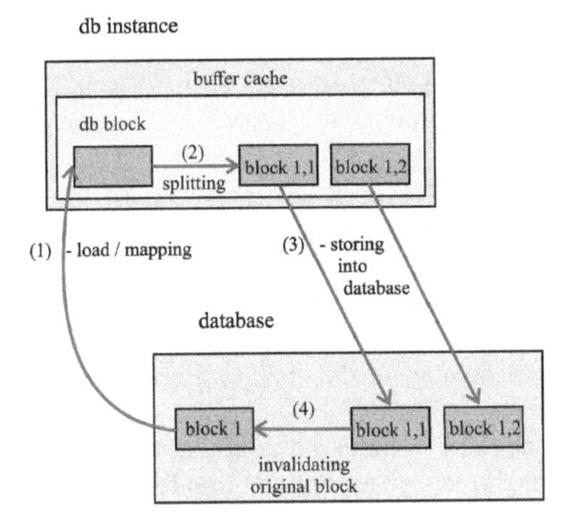

Fig. 2. Model 3 – block splitting

Model 4 improves the principles of parsing. Each block now consists of the list of undefined values with positional data in the header. Thanks to that, identification of the undefined (*NULL*) values during the loading and mapping is straightforward. It does not, however, solve the problem of data block splitting necessity. **Model 4b** fills the block just up to 90% to provide an additional size for memory mapping. Reflecting the performance evaluation section, such a decision automatically increase database storage demands, even if the no-undefined value is present in the table. Thus, a partial model (*4b*) does not bring the relevant solution.

Model 5 does not influence the database storage in terms of the table itself, at all. Instead of that, the *B+tree* indexing strategy is enhanced. In our proposed solution,

NULLs are present in the *B+tree* and evaluated based on the defined criterion. The solution was inspired by the usage of the *Order by* clause of the *Select* statement. In a conventional environment, *NULL* values cannot be mathematically compared, so they are not present in the index. By specifying the principle of the traversing, additional nodes dealing with undefined values can *B+tree* be there. In the physical implementation, two clause options are available for handling and locating undefined values: *{nulls first|nulls last}*.

NULLS FIRST option instructs the system to locate undefined values in the left part of the index. Vice versa, if the *NULLS LAST* option is used, invalid values are in the right part. By default, *NULLS FIRST* definition is used, however, when dealing with the index followed by data sort, the correct locator of the undefined value can bring additional power resulting in lowering time processing consumption, as well as lowering resource consumption.

Physical implementation depends on the index specification. One way or another, the index itself can hold multiple values of the *NULL*, which cannot be sorted – these values are always transcoded into the same value. To solve such limitations, our proposed solution uses *ROWID* value as the extension for dealing with *NULLs*. *ROWID* value is unique and easily sortable.

Model 5 holds the original structure of the index. Undefined values are transformed directly in the index structure, therefore in the physical data or the table, no change in comparison with the ordinary *B+tree* index can be identified. As a consequence, multiple nodes dealing with undefined values can be there, differentiated by the *ROWID* value.

Introduced **model 6** creates a separate segment for the undefined values. It is formed just by one node dealing with undefined values stored in the dynamic table consisting of the pointers (*ROWID*) to the data tuples. The dynamic table holds unsorted data, based on the assumption of the direct impossibility of *NULL* values comparing and sorting. In comparison with *model 5*, the depth of the B+tree index is not influenced, whereas pointers of the undefined values are stored in a separate segment structure.

Extension of the *model 6* is based on replacing the dynamic table by the *B+tree* index, where the sorting key is either *ROWID* (**model 6b**) or primary key (**model 6c**).

5 Performance

Performance characteristics have been obtained by using the *Oracle 19c* database system based on the relational platform. For the evaluation, a table containing *10* attributes originated from the sensors were used, delimited by the composite primary key consisting of two attributes. The table contained *1 million rows*, 10% of them contained undefined values. Two index structures were defined, one implicit covered by the primary key, the second index extends the primary key by covering attribute, which can hold *NULL* value. *The select* statement was evaluated covering all attribute values. The condition limited the amount of data to *10%* of the whole data set.

The reference model for the evaluation and comparison was **model 0**, which two *B+tree* indexes (primary key and user-defined) with no undefined values special support. Thus, in the results, this model reaches 100% of the processing costs and size demands.

The first evaluation deals with the total size of the structures stored in the database – table itself and index. If the value was replaced and stored physically (*model 1*), an additional 6,2% was added, whereas *null_replacement_val* was located directly. 2,7% was added as the result of the index extension, which now covers the undefined values, as well. Logical transformation ensured by the function calls (*model 2*) inside the index does not add a significant extension requirement – 1,9% caused by the index covering the whole data set. *Model 3* and *model 4b* have risen demands, however, they are caused by the data block splitting necessity. *Model 4* adds a list of undefined values position for the table, the index itself is not influenced. The results are shown in Fig. 3. Size requirements are in percentage to express the costs on the disc storage. *Model 6* category (*models 6, 6b,* and *6c*) have almost the same demands, whereas the difference is just based on the *NULL* value locating inside the index delimited by the sorting principles.

The second part to be evaluated is reflected by the *Select* statement performance and index usage. Similarly, *model 0* is a reference model providing 100%. In that case, the

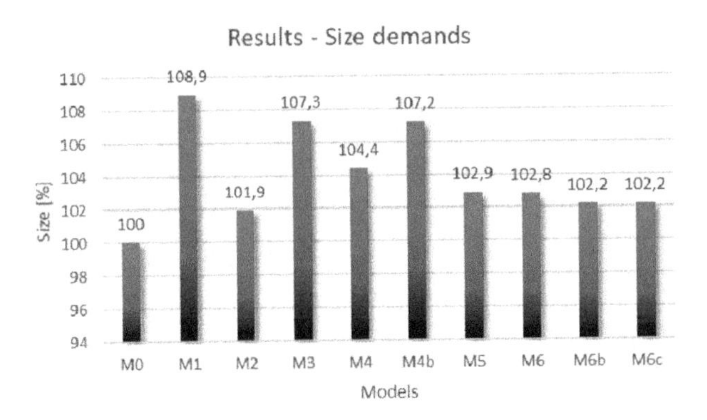

Fig. 3. Results – size demands

whole table must be scanned sequentially (by using the *TAF* method), whereas undefined values cannot be located in the index structure, at all. The highest processing time demands have *model 1*. It is caused by the necessity of the transformed value evaluation. Whereas the size demands are increased, the loading process requires more blocks to be transferred, as well. *Model 3* is influenced by the mapping and block splitting necessity during the loading process. Thus, the processing lasts 39,1% in comparison with the *TAF method*. In comparison with *model 2*, there are additional demands, whereas physically, *NULL* values are stored followed by the transformation on demand. Thus, comparing *model 2* and *model 3*, approximately 44% of the processing time is increased. Proposed models based on the direct covering of the undefined values provide the best solution. To be specific, indexing *NULL* values directly in the index requires 20% of the processing time. By extracting such undefined values, even better results can be reached. Locating *NULLs* in the unsorted table (*model 6*) offers the 17,7% of the processing time, extraction into *B+tree* based on the *ROWID* (*model 6b*) reflects 16,5% and the sorting option defined by the primary key (*model 6c*) requires only 15,0%. The difference between *model 6b* and *model 6c* is based on the layout. Data to be produced to the result set should be mostly sorted based on the unique identifier of the object (primary key), which can be user decoded. Vice versa, information about the physical position of the row in the database does not provide any additional power to the user, although such information can benefit, whereas it is unique, as well. Thus the processing can be shifted into a *unique scan* instead of a *range scan*. Figure 4 shows the experiment results – processing time expressed in percentage.

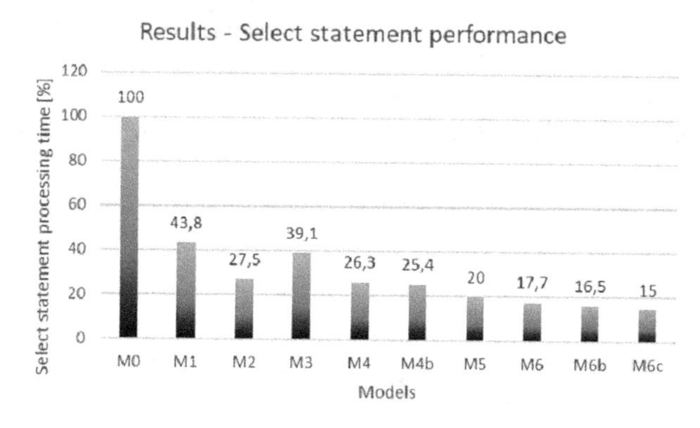

Fig. 4. Results – processing time

6 Conclusions

Database technology is a complex environment proposing a robust solution for data management. The relational paradigm is based on the relational algebra. Many times, the Spatio-temporal model core is covered by relational theory, as well. In that environment, data evolve originating by various modules, sensors, and systems. Each data tuple to be evaluated is delimited by the quality and reliability resulting in storing partially undefined or untrusted values. These values are not part of the index, thus their identification and access are complicated, too time and resource consuming. This paper deals with the existing solutions based on the physical and logical undefined value identification and transformation secured either by the trigger or function-based index. These approaches have additional demands for the Select statement evaluation and size of the storage structure. Therefore, we proposed several layers to cover undefined values in the index.

The core solution is based on the B+tree extension by locating undefined values using the ROWID pointer identifiers. NULL values can be located and sorted either as the first or last value based on the specification (nulls first/nulls last). Reflecting the reached results, the improvement was done by segregating the structure into separate modules covered either by the unsorted table or B+tree with the ROWID, respectively primary key reflection. The best solution is delimited by the unique identifier (primary) as the traverse path.

In the future, we would like to deal with the Spatio-temporal enhancement of the undefined status, to propose complex model monitoring evolution based on the analytical perspective. We would like to shift the solution into the distributed environment, as well. The specification of such an environment is delimited by the local and global indexes in the partitioned structures.

Acknowledgment. The work is also supported by the project VEGA *1/0089/19 Data analysis methods and decisions support tools for service systems supporting electric vehicles* and by the *Grant System of University of Zilina* No. 9/2020.

References

1. Abdalla, H.I.: A synchronized design technique for efficient data distribution. Comput. Hum. Behav. **30**, 427–435 (2014)
2. Bryla, B.: Oracle Database 12c the Complete Reference. Oracle Press (2013). ISBN 9780071801751
3. Burleson, D.K.: Oracle High-Performance SQL Tuning. Oracle Press (2001). ISBN 9780072190588
4. Delplanque, J., Etien, A., Anquetil, N., Auverlot, O.: Relational database schema evolution: an industrial case study. In: IEEE International Conference on Software Maintenance and Evolution, ICSME 2018, Spain, pp. 635–644 (2018)
5. Eisa, I., Salem, R., Abdelkader, H.: A fragmentation algorithm for storage management in cloud database environment. In: Proceedings of ICCES 2017 12th International Conference on Computer Engineering and Systems, Egypt (2018)
6. Ivanova, E., Sokolinsky, L.B.: Join decomposition based on fragmented column indices. Lobachevskii J. Math. **37**(3), 255–260 (2016)
7. Kvet, M., Matiaško, K.: Concept of dynamic index management in temporal approach using intelligent transport systems. In: Recent Advances in Information Systems and Technologies: Volume I. Advances in Intelligent Systems and Computing, vol. 569, pp. 549–560. Springer, Cham (2017). ISBN 978-3-319-56534-7. ISSN 2194-5357
8. Kvet, M., Matiaško, K.: Temporal flower index eliminating impact of high water mark. In: Innovations for Community Services: Proceedings, pp. 85–98. Springer, Cham (2018). ISBN 978-3-319-93407-5
9. Moreira, J., Duarte, J., Dias, P.: Modeling and representing real-world spatio-temporal data in databases. In: Leibniz International Proceedings in Informatics, LIPIcs, vol. 142 (2019)
10. Smolinski, M.: Impact of storage space configuration on transaction processing performance for relational database in PostgreSQL. In: 14th International Conference on Beyond Databases, Architectures and Structures, BDAS (2018)
11. Ochs, A.R., et al.: Databases to efficiently manage medium sized, low velocity, multidimensional data in tissue engineering. J. Vis. Exp. (JoVE) (153) (2019)
12. Vinayakumar, R., Soman, K., Menon, P.: DB-learn: studying relational algebra concepts by snapping blocks. In: International Conference on Computing, Communication and Networking Technologies, ICCCNT 2018, India (2018)
13. Zaitseva, E., Levashenko, V., Rabcan, J., Kvassay, M., Rusnak, P.: Reliability evaluation of multi-state system based on incompletely specified data and structure function. In: ISAACS 2019, pp. 741–746 (2019)

Usability Analysis of the Concordia Tool Applying Novel Concordance Searching

Rafał Jaworski[1(✉)], Ivan Dunđer[2], and Sanja Seljan[2]

[1] Faculty of Mathematics and Computer Science, Adam Mickiewicz University in Poznań, Poznań, Poland
rjawor@amu.edu.pl

[2] Faculty of Humanities and Social Sciences, University of Zagreb, Zagreb, Croatia

Abstract. This paper describes a novel tool for concordance searching, named Concordia. It combines the capabilities of standard concordance searchers with the usability of a translation memory. The tool is described in detail with regard to main applied methods and differences when compared to already existing CAT tools. Concordia uses three data structures, i.e. hashed index, markers array and suffix array, which are loaded into memory to enable fast lookups according to fragments that cover a search pattern. In this new concordancing system, sentences are stored in the index and marked with additional information, such as unique ids, which are then retrieved by the Concordia search algorithm. The usability of the new tool is analysed in an experiment involving two English-Croatian human translation tasks. The paper presents a detailed scheme and methodology of the conducted experiment. Furthermore, an analysis of the experiment results is presented, with special emphasis on the users' attitudes towards the usefulness and functionalities of Concordia.

Keywords: Concordance searching · Computer-assisted translation · Approximate searching · Suffix array · Human evaluation

1 Introduction

In order to bridge the gap between the industry and the research, various studies have been conducted regarding the usability of computer-assisted translation (CAT) tools in the translation process. CAT tools can be considered as stand-alone systems, or tools that are integrated with electronic dictionaries, machine translation (MT) engines, concordancers, terminology managers, full-text search tools etc. CAT tools and MT systems, along with integrated plug-ins and resources can offer quick gisting translation, but still lack the quality. Numerous experiments have been conducted in order to assess the usability of CAT and/or MT systems. While CAT technology is used to find matching sentences from sentence-aligned translation memories (TM), translators often need translations of sub-sentence units, e.g. phrases, expressions etc.

Á. Rocha et al. (Eds.): ICITS 2021, AISC 1330, pp. 128–138, 2021.
https://doi.org/10.1007/978-3-030-68285-9_14

One of the key requirements is to have high-quality aligned parallel corpora, see [7,14]. Bilingual concordancers are still often used by translators. For a given query, the system retrieves a source-target translation unit pair containing the queried sequences of characters. Bilingual concordancers represent an extension of dictionaries, allowing for searching of multi-word units, collocations or idiomatic expressions (e.g. "look forward to"), phrases or even entire sentences.

CAT systems typically use sets of previously translated sentences, called translation memories. For a given sentence, a CAT system searches for a similar sentence in the translation memory. If such a sentence is found, its translation is used to produce the output sentence. This output sentence is then used as a suggestion for translation, while a human translator carries out the post-editing.

This technique is applied in many leading CAT platforms, such as SDL Trados [12] or Kilgray memoQ [11]. Its main advantage is the fast detection of situations, in which a translator is presented an identical or almost identical sentence to the one previously translated. In this case the old translation can be reused with a minimal need for post-editing. However, the main drawback of translation memory searching lies in the fact that the situations described above happen relatively rarely.

Another mentioned technique used in the mentioned CAT platforms is concordance searching – a technique of looking up single words or multi-word units from the translated sentence in a translation memory. Occurrences of these words are then presented to the translator with the appropriate contexts.

It is crucial to know which of these techniques can prove valuable in the translation process. Therefore, evaluation of the translation productivity is conducted in order to obtain or maintain a suitable translation quality and/or reduce work time and costs. Translation productivity, analysed through post-editing of CAT/MT-translated text, is often performed in combination with a survey of the users' skills, cognitive efforts and the quality of the translated text.

Human evaluation of usefulness of a CAT tool mostly takes into account the impact on post-editing speed and effort, usability of the interface, ease of translation spotting, autocompleting of translations etc. On the other hand, automatic evaluation is commonly analysed with the help of human-targeted translation edit rate (HTER) as shown in [16], the BLEU metric and, in more recent works, fuzzy matching measures: [2,3].

This paper presents a new CAT tool, i.e. a novel concordance searcher named Concordia, and evaluation results regarding its usability. The term usability in this paper reflects to what extent the CAT tool is helpful during the complex translation workflow. It refers to the user perception of the tool's usefulness in terms of functionalities and the possibilities of integrating it into the translation process, with various aspects being measured, such as time needed to translate segments and the number of lookups in the CAT tool. Concordia uses a combination of well established algorithms and data structures to facilitate fast queries and combines the advantages of standard concordancers with the capabilities of a translation memory. Subsequent sections describe related work in the field, the details about the Concordia search algorithm, followed by the description of the experimental evaluation with the corresponding results, whereas conclusions are given in the final section. The authors intend to implement refinements in the CAT tool, according to the comments and findings of human evaluators in this experiment.

2 Related Work

Usability and productivity studies of various CAT tools have recently emerged due to the interest of industry leaders, software engineers, computer and information scientists, translators, localisers and data scientists. Numerous assessments taking into consideration different aspects, ranging from human evaluations up to automatic metrics, have been conducted.

The paper [3] assessed the user productivity of a commercial CAT tool with the publicly available MyMemory plug-in and an integrated commercial machine translation engine. Twelve translators participated in a real translation project. The productivity was measured by human and automatic evaluations. The machine translation engine was analysed through the rate of words per hour, fuzzy match, productivity gain, BLEU and TER score. The results showed that post-editing efforts significantly decreased when using a combination of translation memories and machine translation. The post-editing speed implied significant differences across translators, languages, and domains. In another research (see [13]) the machine translation post-editing productivity with regard to speed and needed effort was measured. The results were obtained with the help of an eye-tracking system.

An interesting experiment is described in [4]. It involved eight professional translators who were given a task to translate ca 800 source words from scratch, using a glossary, a translation memory with mainly 80–90% fuzzy matches and a commercial statistical machine translation (SMT) engine trained on the translation memory content. The productivity was measured in terms of speed and quality of translated texts. Relative time improved 4% to 52%, with an average of 27%. Short (1–10 words), medium (11–20 words) and long (>20 words) segments were used, whereas the highest quality increase of machine translations was observed on medium to long segments.

The paper [1] described a searchable translation memory relying on statistical machine translation, using word-alignment and phrase-based SMT with the possibility to search for all possible substrings, i.e. unseen phrases. The authors recommended the Linear B system available for Arabic, Chinese and seven European languages. The evaluation was done with regard to precision and recall.

A bilingual concordancing system which displayed occurrences of a specific word or an expression is presented in [15]. The tool can also be accessed over the internet and performs thousands of user queries per day. It searches through a large database of bitexts (sentence-aligned text) – Hansard and Court Decisions, which mostly contains bigram expressions, followed by 3-g, 4-g and unigrams. The authors proposed a word-processor add-on functionality which would allow to submit queries to the TransSearch system directly from a word-processor.

Also, the paper [8] presented a system for term extraction by extracting the contexts and combining word alignment and concordancing. The aim was to develop the Terminology Management System (TMS) of legal phraseology and terminology for French, Dutch and German. In an experimental tool called FragmALex, links between the source and target text were created using lexical resources (lemmas and their translations) borrowed from dictionaries, terminology bases, documents and cognates, i.e. words with common etymological origin which are similar in the source and target language.

3 Concordia: A Concordance Search Algorithm

This section presents a novel solution for concordance searching. It differs significantly from the standard concordance searchers. The most important difference is that Concordia tries to search sequences of words in the translation memory, instead of single word occurrences. In order to carry out the search procedures efficiently, an offline index, based on the suffix array [9,10] and other auxiliary data structures are used.

3.1 Operations on Index

Main operations performed on the index are stated subsequently.

- *void addToIndex(string sentence, int id)* – this method is used to add a sentence to the index along with its unique id. The id is treated as additional information about the sentence and it is then retrieved from the index by the search algorithm. This is useful in a standard scenario, when sentences are stored in a database or a text file, where the id is the line number. Within the *addToIndex* method the sentence is tokenised and from this point forward treated as a word sequence.
- *void generateIndex()* – after adding all the sentences to the index the generateIndex method should be called in order to compute the suffix array for the needs of the fast lookup index. This operation may take some time depending on the number of sentences in the index. Nevertheless, its length rarely exceeds one minute (during experiments with 2 million sentences the index generation took 6–7 s).
- *concordiaSearch (string pattern)* – the main concordance search method returns the longest fragments from the index that cover the search pattern.

3.2 Index Construction

The index incorporates the idea of a suffix array and is aided by two auxiliary data structures – the hashed index and markers array. The first serves as the "text" (in terms of approximate string search algorithms) and the second facilitates the process of retrieving matches from the memory.

During the operation of the system, i.e. when the searches are performed, all three structures (hashed index, markers array and suffix array) are loaded into

RAM. For performance reasons, hashed index and markers array are backed up on the hard disk. When a new sentence is added to the index via the aforementioned addToIndex method, the following operations are performed:

1. tokenizing of the sentence
2. stemming of each token
3. converting each token to a numeric value according to a dynamically created map (called dictionary)

The coded stems are stored in the index. Stemming each word and replacing it with a code results in a situation, where even large text corpora require relatively few codes. For example, a research of this phenomenon presented that a corpus of 3593227 tokens coming from a narrow domain (the JRC-acquis corpus) required only 17001 codes (see [6]). In this situation each word could be stored in just 2 bytes, which significantly reduces space complexity.

3.3 Concordia Searching

The Concordia search is aimed at finding the longest matches from the index that cover the search pattern. Such a match is called "matched pattern fragment". Then, out of all matched pattern fragments, the best pattern overlay is computed.

The pattern overlay is a set of matched pattern fragments which do not intersect with each other. Best pattern overlay is an overlay that matches the most of the pattern with the fewest number of fragments.

Additionally, the score for this best overlay is computed. The score is a real number between 0 and 1, where 0 indicates, that the pattern is not covered at all (i.e. not a single word from this pattern is found in the index). The score 1 represents a perfect match – pattern is covered completely by just one fragment, which means that the pattern is found in the index as one of the examples. The formula (1) is used to compute the best overlay score:

$$score = \sum_{fragment \in overlay} \frac{len(fragment)}{len(pattern)} \cdot \frac{log(len(fragment)+1)}{log(len(pattern)+1)} \tag{1}$$

According to this formula, each fragment covering the pattern is assigned a base score equalling the relation of its length to the length of the whole pattern. This concept is taken from the standard Jaccard index [5]. However, this base score is modified by the second factor, which assumes the value 1 when the fragment covers the pattern completely, but decreases significantly when the fragment is shorter. For that reason, if one considers a situation where the whole pattern is covered with two continuous fragments, such overlay is not given the score 1.

An example illustrating the Concordia search procedure is given hereafter. Let the index contain the sentences from Table 1.

Table 2 presents the results of searching for the pattern: "Our new test product has nothing to do with computers".

Table 1. Example sentences for Concordia searching.

Sentence	Id
Alice has a cat	56
Alice has a dog	23
New test product has a mistake	321
This is just testing and it has nothing to do with the above	14

Table 2. Concordia search results.

Pattern interval	Example id	Example offset
$[4, 9]$	14	6
$[1, 5]$	321	0
$[5, 9]$	14	7
$[2, 5]$	321	1
$[6, 9]$	14	8
$[3, 5]$	321	2
$[7, 9]$	14	9
$[8, 9]$	14	10
Best overlay: $[1, 5][5, 9]$, Score = 0.53695		

These results list all the longest matched pattern fragments. The longest is $[4, 9]$ (length 5, as the end index is exclusive) which corresponds to the pattern fragment "has nothing to do with", found in the sentence 14 at offset 7. However, this longest fragment was not chosen to the best overlay. The best overlay are two fragments of length 4: $[1, 5]$ "new test product has" and $[5, 9]$ "nothing to do with". It should also be noted, that if the fragment $[4, 9]$was chosen to the overlay, it would eliminate the $[1, 5]$ fragment. The score of such an overlay is 0.53695, which can be considered as quite satisfactory to serve as an aid for a translator.

4 Experiment and Evaluation

The aim of the proposed experiment involving a human translation task is to get an insight into the users' perspective on the usefulness, functionalities of the system Concordia and the possibilities of integrating it into the translation workflow.

4.1 Methodology

At first, the Concordia tool was fed with the SETimes2 corpus consisting of approximately 200k sentences from the news domain (for corpus description see [17]). Then, the evaluation of Concordia is performed for the English-Croatian language pair. Each of the 14 evaluators, which were separated into two groups (group A and B), was given 20 sentences, also from the news domain (but not present in the SETimes2 corpus).

Those 20 sentences were divided into two test sets: while evaluators from group A translated the sentences without Concordia, but could use other internet resources, group B translated the same test set with Concordia and with the possibility to use other internet resources. Then, in the second task, group A translated with Concordia (possibility to use other internet resources) and group B without Concordia (but with the possibility to arbitrarily use resources on the internet).

Evaluators were graduate students at the Faculty of Humanities and Social Sciences, University of Zagreb, fluent in English. Before starting with the translation tasks, a pre-translation survey was carried out in order to acquire more information on the evaluators' background (study group and familiarity with translation tools).

Prior to starting with the translation tasks the evaluators were shown three examples on how to use Concordia. Each evaluator was then asked to record the total time needed for translating a sentence.

The translators, i.e. evaluators, were not interrupted during the work in any way and could take any time they needed. Apart from the total translation time, the best Concordia, i.e. overlay, score for each of the sentences from the test sets was recorded.

After translating the given sentences with and without the help of Concordia, a post-task questionnaire was given to the evaluators. The questionnaire contained various questions regarding the usefulness of the tool, existence of necessary functionalities for effective translation, purpose of using Concordia during translation (single words in a dictionary-like style, multi-word units or entire phrases), intuitiveness of design etc. The list of questions of the questionnaire is presented in Table 3.

The translation times, automatic best overlay scores and the survey results were then analysed, as valuable user feedback with regard to usability and user-friendliness can be utilised to upgrade the new CAT tool.

4.2 Results and Discussion

In total, 14 evaluators (9 were male, 5 female) participated in this experimental study. They were randomly selected and split into two groups (7 per group). 10 evaluators were studying informatics and 4 were students of translation study groups of various languages. Figure 1 shows the distribution of digital language resources according to the students' familiarity.

Both groups were allowed to use any other preferred language resource, when translating with or without Concordia. Among 14 evaluators, 2 used only Google Translate, whereas 2 evaluators used both Google Translate and Bing Translator. Each test set consisted of 10 sentences from the same news domain regarding

Table 3. Post-translation questionnaire.

Question	Type
Did the system help you in the translation task?	Yes/no
Please rate the intuitiveness of the system	Score 1–5
How many times did you look up hits suggested by the system?	Number
After looking up a hit suggested by the system, did you find its translation easily?	Always/sometimes/never easy
What did you look up most often in the system?	Single words/multi-word units/entire phrases
Please list suggestions for improvement	Short comment

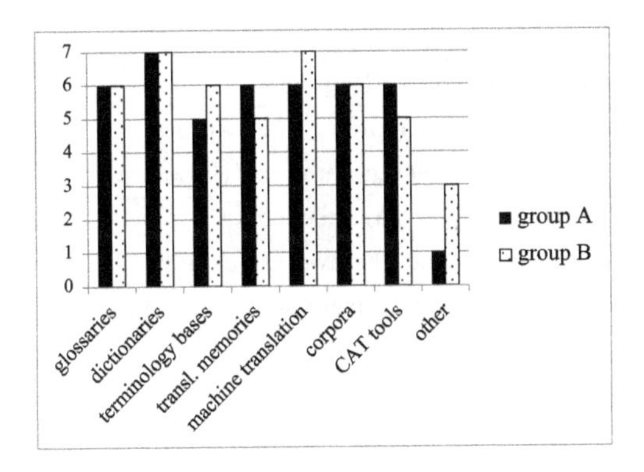

Fig. 1. Familiarity with digital language resources.

traffic accidents in the region. Minimum length was 5 words, whereas the maximum sentence length was 38 words. Average sentence length for the first test set was 21.8 and for the second 18.2 words, an in total 218 and 182, respectively.

Table 4 shows the total time needed for the translation tasks. Time needed for translation was very similar, but the main advantage would be consistent translation of the specific abbreviation when using Concordia. However, due to relatively small test sets, differences in time and quality are not clear. The reason for a little longer time with Concordia was that students were not said to be relatively quick with translation task, but on the contrary, they took more time to investigate all possible doubts. The test set 1 contained the specific abbreviation and a few more specific terms, which caused longer time for the translation task. The total number of lookups for the test set 1 was 69, and for the test set 2

of the students (57%) answered that after looking up a hit suggested by the system, it was "sometimes easy" to find its translation, whereas 29% of the students stated that it was "always easy" to find the corresponding translation.

When answering the question what did they look up most often in the system, 7 students answered "multi-word units" and 4 students stated "single words".

When asked to list suggestions for improvements or remarks, evaluators suggested that the system should be fed with much larger corpora. Also more interactivity with the end-users would be desirable. Furthermore, design improvements with regard to displaying long sentences that stretch across the screen were proposed, as displaying of longer sentences is not user-friendly.

5 Conclusions

Concordia combines capabilities of standard concordance searchers with the usability of a translation memory. In the pre-translation task the familiarity with language resources was evaluated, indicating that on average 85% of students of mainly non-translation study group were acquainted with various language resources.

In the translation task, two experiments were conducted without and with Concordia tool, using both tests sets from the news domain. 71% of users declared the Concordia system useful in the translation task, with the average score for the system intuitiveness 3.5. The system was mostly used to find multi-word expressions, followed by single words. 57% of users indicated that it was "sometimes easy" to find its translation, whereas 29% of the students stated that it was "always easy". However, there are no firm conclusions regarding the speed of translation, due to relatively small test sets and specific terminology and abbreviation which has caused longer time for the translation task. However, the Concordia system was more useful with longer and more complex sentences where translation time decreased for 17–25%.

The future research would focus on improvements of interface design, better interactivity, corpus enlargement and more extensive experiments.

References

1. Callison-Burch, C., Bannard, C., Schroeder, J.: Searchable translation memories. In: Proceedings of ASLIB Translation and the Computer, vol. 26 (2004)
2. Escartín, C.P., Arcedillo, M.: A fuzzier approach to machine translation evaluation: a pilot study on post-editing productivity and automated metrics in commercial settings. In: Proceedings of the ACL 2015 Fourth Workshop on Hybrid Approaches to Translation (HyTra), pp. 40–45 (2015)
3. Federico, M., Cattelan, A., Trombetti, M.: Measuring user productivity in machine translation enhanced computer assisted translation. In: Proceedings of the Tenth Conference of the Association for Machine Translation in the Americas (AMTA) (2012). http://www.mt-archive.info/AMTA-2012-Federico.pdf

4. Guerberof, A.: Productivity and quality in MT post-editing. In: Proceedings of the 12th Machine Translation Summit (MT Summit XII) Workshop: Beyond Translation Memories - New Tools for Translators, p. 9 (2009)
5. Jaccard, P.: Étude comparative de la distribution florale dans une portion des Alpes et des Jura. Bull. de la Société Vaudoise des Sci. Nat. **37**, 547–579 (1901)
6. Jaworski, R.: Anubis – speeding up computer-aided translation. In: Computational Linguistics – Applications. Studies in Computational Intelligence, vol. 458, Springer (2013)
7. Jaworski, R., Jassem, K.: Building high quality translation memories acquired from monolingual corpora. In: Proceedings of the Intelligent Information Systems Conference, pp. 157–168 (2010)
8. Kockaert, H.J., Vanallemeersch, T., Steurs, F.: Term-based context extraction in legal terminology : a case study in Belgium. In: Fóris, A., Pusztay, J. (eds.) Current Trends in Terminology: Proceedings of the International Conference on Terminology (Terminologia et Corpora Supplementum 4), pp. 153–162 (2008)
9. Makinen, V., Navarro, G.: Compressed compact suffix arrays. In: Proceedings of the 15th Annual Symposium on Combinatorial Pattern Matching (CPM). LNCS, vol. 3109, pp. 420–433 (2004)
10. Manber, U., Myers, G.: Suffix arrays: a new method for on-line string searches. In: First Annual ACM-SIAM Symposium on Discrete Algorithms, pp. 319–327 (1990)
11. Multiple: Kilgray Translation Technologies: memoQ Translator Pro. http://kilgray.com/products/memoq/
12. Multiple: SDL Trados translation solution. http://www.sdl.com/en/sites/sdl-trados-solutions
13. O'Brien, S.: Towards predicting post-editing productivity. Mach. Transl. **25**(3), 197–215 (2011). https://doi.org/10.1007/s10590-011-9096-7
14. Seljan, S., Gašpar, A., Pavuna, D.: Sentence alignment as the basis for translation memory database. In: INFuture 2007–The Future of Information Sciences: Digital Information and Heritage. Odsjek za informacijske znanosti, Filozofski fakultet, Zagreb (2007)
15. Simard, M., Macklovitch, E.: Studying the human translation process through the TransSearch log-files. In: Knowledge Collection from Volunteer Contributors, Papers from the 2005 AAAI Spring Symposium, Technical Report SS-05-03, Stanford, California, USA, 21–23 March 2005, pp. 70–77. AAAI (2005). http://www.aaai.org/Library/Symposia/Spring/2005/ss05-03-011.php
16. Specia, L., Farzindar, A.: Estimating machine translation post-editing effort with HTER. In: AMTA Workshop Bringing MT to the User: MT Research and the Translation Industry, Denver, Colorado (2010). http://www.mt-archive.info/JEC-2010-Specia.pdf
17. Tiedemann, J.: Parallel data, tools and interfaces in OPUS. In: Calzolari, N., Choukri, K., Declerck, T., Doğan, M.U., Maegaard, B., Mariani, J., Moreno, A., Odijk, J., Piperidis, S. (eds.) Proceedings of the Eight International Conference on Language Resources and Evaluation (LREC 2012). European Language Resources Association (ELRA), Istanbul (2012)

Intelligent and Decision Support Systems

Computer System Based on Robotic Process Automation for Detecting Low Student Performance

María Guacales-Gualavisi[1], Fausto Salazar-Fierro[1], Janneth García-Santillán[2], Silvia Arciniega-Hidrobo[1], and Iván García-Santillán[1](✉)

[1] Faculty of Engineering in Applied Sciences, Universidad Técnica del Norte, Ibarra, Ecuador
{mmguacalesg, fasalazar, srarciniega, idgarcia}@utn.edu.ec
[2] Unidad Educativa Juan Pablo II, Ibarra, Ecuador
janneth.garcia@educacion.gob.ec

Abstract. Robotic Process Automation (RPA) is the technology-oriented to the use of robots based on software robots to automatically execute repetitive processes performed by humans but without their limitations, such as exhaustion, stress, time limits, etc. For this reason, RPA has been applied in different fields as in the educational context, improving the academic work of teachers by engaging to other more relevant academic activities. Thus, the objective of this study was to develop and evaluate an RPA system for an Educational Unit using the SCRUM agile software methodology and the UiPath Studio tool, adapting to the academic system used in the institution. The quality evaluation performed using the GQM (Goal Question Metric) methodology, showing adequate levels of effectiveness (system functionality), efficiency (task execution times), and satisfaction (ease of use).

Keywords: RPA · UiPath Studio · SCRUM · GQM · Academic performance

1 Introduction

Robotic Process Automation (R.P.A) is technology-oriented to the use of software-based robots that perform manual and repetitive tasks in existing digital environments and works 24/7 without rest. The bot follows a set of rules programmed by a person, simulating the human execution of a repetitive task. In this way, it reduces human error and allows staff to generate value for the institution. RPA carries out the replication of easy actions that performed on a computer-based on simple rules and business logic. Some tasks conducted by RPA are moves, records, and copies of digital information; periodically extract, synthesize, processes and reports data; enters data into multiple non-integrated systems; monitors detect, and reports operational performance, etc. RPA applies in several areas of the business and some benefits are: quality, time, continuous improvement, costs, customer and employee satisfaction, productivity, and competitiveness [1, 2].

Á. Rocha et al. (Eds.): ICITS 2021, AISC 1330, pp. 141–150, 2021.
https://doi.org/10.1007/978-3-030-68285-9_15

RPA applies in several domains, such as telecommunications [3], robotic process mining [4], and university contexts [5, 6]. However, during the research process, we have not found RPA implementations applied in secondary education, automating repetitive tasks, relieving the teacher of administrative work, and giving the teacher valuable time to focus on other more relevant academic activities, such as: implement new pedagogical strategies in their remedial classes, improve the knowledge imparted, grades and contribute to reducing year losses and dropouts [7].

In this context, an academic RPA is important for the Educational Unit "San Juan de Ilumán" (Otavalo-Ecuador) which has a remarkable rate of growth of year losses and desertions in each school cycle [8]. Therefore, the objective of this study is to develop and evaluate an RPA system for the automatic detection of low-performance high school students using the SCRUM agile software methodology [9, 10] and the UiPath Studio development tool [11]. The system is coupled to the existing academic system which works with a SQL Server 2008 manager [12]. The quality assessment of the RPA system (efficacy, satisfaction, and usability) carries out with the G.Q.M. methodology (Goal Question Metric) [13]. This research allows the teachers to send the grades by email and obtain timely reports of poor student performance. This facilitates to the academic staff to make appropriate decisions and actions.

Some RPA-related works in several applications domains that served as the basis for this study include the following:

In [14] analyzed the application of RPA carried out by a service outsourcing company in which they issue invoices and send their customers in PDF via email was analyzed, increasing their ability to serve users by 20% and at run times by 2%. The company Opus Capita [15] analyzed that companies dedicated to the development of RPA, for each robot created charge between € 3000 a € 5000, showing that it is a profitable service for the developer company and its traders. In [16] showed that RPA processes do not need to start from scratch, since RPA action flow can be developed by visually recognizing existing system interfaces, called agile RPA development. The analysis of tools for RPA development [5], highlighted UiPath Studio as the leading tool for bot development for the versatility of working with other programs, unattended robots, visual processes development, the development community, and free information. In the University education system [6], it was pointed out that with the use of the Automation Anywhere tool, for the development of a bot that extracts information from PDF files it gets a time reduction of 94.44% compared to the work done manually. In [17], they proposed a method of how and when adopting relevant RPA processes that should be proposed in a company's business logic to increase its profitability, productivity, data security, reduction in times and decrease of routine work in employees, resulting in 83% of RPA strategies are viable. In [3] a study was analyzed in a telecommunications company to obtain zero complexities in the execution of routine tasks through RPA processes. The results obtained with RPA were fast and efficient, zero complexity and effort, as well as agility in service. In robotics process mining [4], the technique to identify processes to automate called Pipeline RPM was proposed, concluding that the task of detecting processes to be developed as RPA cannot be automatically detected with any technique, but rather have to be interpreted by the developer with customer experience support.

What follows from the manuscript is divided as follows: Sect. 2 indicates the modular structure of the proposed RPA System; in Sect. 3 the results obtained from the system and the evaluation based on GQM; and in Sect. 4 the main conclusions of the work and future work.

2 Materials and Methods

The general architecture of the proposed RPA System is shown in Fig. 1. The System was developed, administered, and published with technology shared by the RPA UiPath service company [11], in its free version Community. The UiPath Studio tool was chosen because it turned out to be a good choice for developing RPA processes [5]. UiPath Studio is a software-based robot development environment using activity flows designed by the developer. Once the development of each RPA robot is completed, these are published, using the credentials of a developer account, on the RPA bot management server named UiPath Orchestrator, which allows accessing and executing RPA robots from a client machine, through an application called UiPath Robo. This application shows the user the bots uploaded in the UiPath orchestrator and allows them to run on a local computer. The proposed RPA System interacts with (i) the SQL Server 2008 database manager of the existing academic system in the institution, and (ii) a prerequisite JSF system that allows the entry of information to the database.

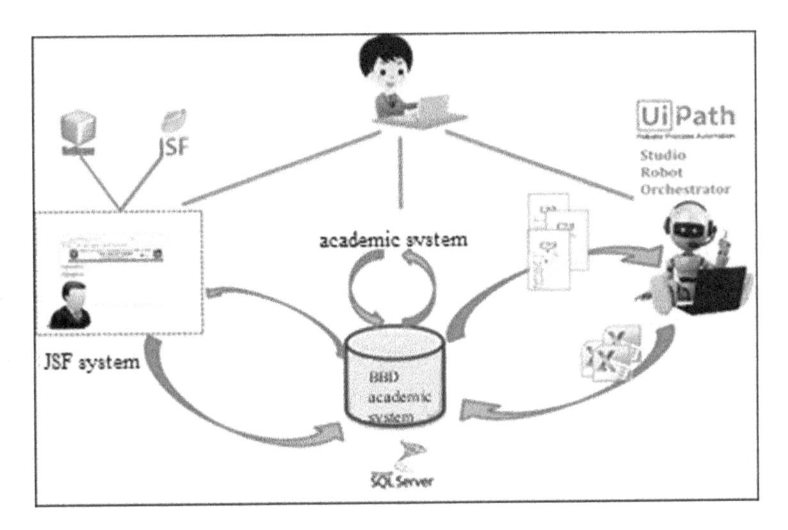

Fig. 1. General architecture of the RPA system

2.1 RPA System Modules

The JSF prerequisite module was developed in Apache NetBeans IDE [18] with the JSF Framework [19], which allows the entry of information to the SQL Server 2008

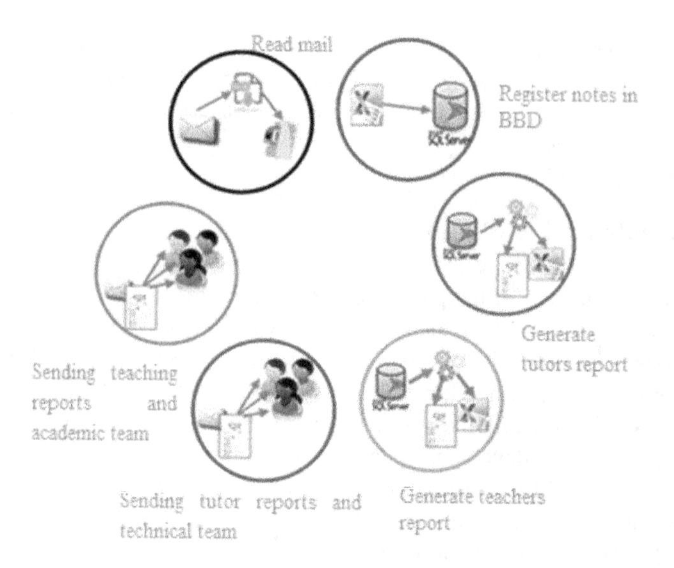

Fig. 2. RPA system modules

database [12] of the existing academic system, and later it allows to work the 6 RPA modules (Fig. 2) that were developed with UiPath Studio technology.

The RPA system modules correspond to the 6 RPA robots that have a development and commissioning cycle, starting from (i) developing RPA processes and publishing to UiPath Studio, (ii) Managing RPA in UiPath Orchestrator y (iii) the visualization of the RPA services on the client computer using UiPath Robot. The components involved in the RPA system architecture are the following (Figs. 1 and 2):

- **JSF prerequisite system:** It allows the entry of information, in the database of the existing academic system, regarding storage paths in the client computer of notes downloaded from emails and reports generated by the RPA system.
- **UiPath Orchestrator:** Internet server containing developed RPA.
- **UiPath Robot:** UiPath client, installed on the client computer, which enables you to activate existing RPA processes in UiPath Orchestrator.
- **RPA 1.** Download Email Account Notes: the robot that downloads Excel files from the institutional email account containing student notes.
- **RPA 2.** Register Notes from Excel files to the BDD of the Academic System: robot that extracts the notes from the Excel file previously downloaded and records in the database of the academic system of the institution.
- **RPA 3.** Generate reports for tutors: the robot that generates PDF reports of students with grades less than 7 (low academic performance), filtered by course, and saved on the client computer.
- **RPA 4.** Generate reports for teachers: the robot that generates PDF reports of students with grades less than 7, filtered by the teacher and their subject taught and then stored on the client computer.

- **RPA 5.** send by email reports for tutors and academic team: the robot that sends the reports generated for tutors and work team to the personal email of each one.
- **RPA 6.** send by email reports for teachers and academic teams: the robot that sends the reports generated for teachers and the work team to the personal email of each one.

2.2 Development of the RPA System Using the SCRUM Methodology

The RPA system was developed using the SCRUM methodology [10] in which the equipment is organized as follows: Scrum Master (Development manager system), Product Owner (The representative of the Educational Unit), Development Team (System developers), and Users (24 Teachers, ICT manager, and Academic Team). The Scrum Master, Product Owner, and Development Team organized the requirements requested by the educational unit, shown in Table 1. Afterward, user history was established for each functional requirement.

Table 1. Functional requirements of the RPA system

Order no.	Functional requirements
1	Teachers note entry template in MS. Excel
2	RPA 1 Download notes from the institutional email account
3	RPA 2 Register notes from Excel files to the BDD of the academic system
4	RPA 3 Generate reports for tutors
5	RPA 4 Generate reports for teachers
6	RPA 5 send reports by email to tutors and academic team
7	RPA 6 send reports by email to teachers and academic team
8	Integration and management of RPA processes in UiPath orchestrator
9	Documentation of the developed project
10	Socialization of the project to technical staff and end-users

A daily meeting was held to verify compliance with each story user to verify the meeting progress at the end all user stories called *Sprint Review* to make sure they know the user's needs, and a final meeting of the process development called *Sprint Retrospective* in case you need to make any improvements.

2.3 Software Quality Evaluation Using GQM (Goal Question Metric)

Because the metrics proposed by the ISO/IEC 25000 [20], standards can be complex to use efficiently by people with little experience in the quality area. [21], it was decided to perform the software quality analysis with the GQM (Goal Question Metric) methodology [13], through a set of simple questions, for the feasibility of working with measurable parameters to measure the effectiveness, efficiency, and user satisfaction. The Quality Assessment was developed as follows:

- **GQM planning**
  Phase 1: Creation of the evaluation team consisting of the Scrum Master, Scrum Owner, and Head of ICT.
  Phase 2: Selection of areas to evaluate: effectiveness, efficiency, and satisfaction of use.
  Phase 3: Assigning responsibilities to GQM team members.
  Phase 4: Evaluation planning (objectives, hypotheses, and metrics to evaluate).
- **GQM metrics measurement plan**
  Effectiveness: It will be measured by the functionality, in percentage, of the fulfilment of the objectives proposed in Phase 4.
  Efficiency: The direct rule of three is applied as a metric to measure the percentage of time improvement presented by the system in each RPA module.
  Use satisfaction: the methodology of the SUS scale [22] is applied to measure the usability of the system and the Cronbach Alpha coefficient [23, 24] for the reliability of the results. Although there are other questionnaires to measure system satisfaction [25], the SUS questionnaire was used due to its ease of application and interpretation of results.
- **Development of the GQM plan**
  Objectives, hypotheses, questions, and metrics were raised to evaluate the efficiency, effectiveness, and usability of the RPA system.
- **GQM analysis plan**
  It determines the acceptability that each criterion must have assessed if the institution requests the following: exceed 90% functionality, times must improve by a minimum of 100%, and in usability, the system must have intuitive functionality and an easy-to-use interface.
- **Data collection** through an evaluation carried out when running the system and interpretation of the results.

3 Results and Discussion

3.1 Efficacy Results

Efficiency is measured by the operation of the UiPath Orchestrator server with the RPA processes developed, uploaded, and running. It involves the client with the processes active in UiPath Robot, and the Excel files downloaded from the institutional email account containing the teachers' grades, and reports generated (*.pdf) with low student performance and stored in the client computer for tutors, teachers, and academic team.

The test execution of the RPA system, carried out by the SCRUM team, analyzes the fulfilment of the objectives as indicated in Table 2, having a result of 100% system efficiency in the development of the 6 RPA processes required by the institution. To obtain the percentages of the achieved objective, the server functionality, client, and previously described results were verified.

Table 2. Percentage of project effectiveness using GQM

Study objectives focused on the functioning of	% Goal fulfilled	% Project effectiveness
Servers	100%	14.29
Clients	100%	28.57
Module: "Entering information with prerequisites"	100%	42.86
RPA 1 "Read Email"	100%	57.14
RPA 2 "Notes entry"	100%	71.43
RPAs 3, 4 "Generating grade reports for academic staff"	100%	85.71
RPAs 5, 6 "Sending report cards for academic staff"	100%	**100.00**

3.2 Efficiency Results

System efficiency results performed by comparing manually recorded time against automatically using the RPA system, run on a personal computer with modest features (Intel Core i7 processor, 4 GB RAM, Windows 64-bit). After the test, the results of Table 3 obtained, showing (column 4) that the time improved by 995.43% (for the 6 RPA processes), since, the time occupied for tasks manually is approximately 708 min, compared to the time used by the R.P.A. system of 71 min. The improvement in times of automatic execution of email reading activities, entry of notes to the system, generation, and timely distribution of reports with low academic performance to tutors, teachers, and academic authorities is evident.

Table 3. Percentage of increased efficiency at run times using RPA processes.

Activity	Time (s) used manually (A)	Time (s) used by the RPA system (B)	Improving efficiency by % = (A*100)/B
RPA 1 Read email	2500	500	500.00
RPA 2 BDD notes entry	8089	1300	622.23
RPA 3 Report course notes	15000	850	1764.71
RPA 4 Report teacher notes	15000	1000	1500.00
RPA 5 Send email tutors	958	270	354.81
RPA 6 Send emails teachers	958	350	273.71
Total in seconds	42505	4270	995.43
Total in minutes	**708.41**	**71.17**	**995.43**

3.3 Satisfaction and Usability Results of the RPA System

After testing the RPA system (Tables 4 and 5) the predefined S.U.S. survey [22] was applied to 24 secondary teachers in the educational unit to measure system satisfaction. The survey consisted of 10 well-known questions and was evaluated using the 5-point Likert scale. Then, the Cronbach Alpha coefficient [24] was calculated to determine the level of internal reliability of the SUS scale applied. After the resulting calculations with the support of the MS program. Excel, we have that Cronbach's Alpha was 0.999 reliability, that is, 99.9%. This demonstrates the extent to which the instrument produces consistent and consistent results.

Using the SUS methodology [22], the analysis and interpretation of the results obtained carried out, indicating that the system has a 92.175% acceptability from users, which allows affirming that the satisfaction and usability of the RPA system are at an Excellent level with grade A, according to the interpretation of the SUS Scale, Fig. 3.

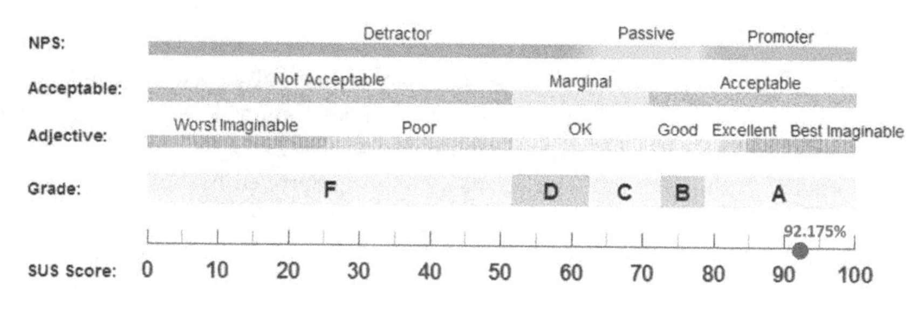

Fig. 3. Interpretation of the S.U.S scale [22]

3.4 Discussion

This research addresses the future work suggested in [5] that considers important to implement an RPA-based education system, taking into account its ability to interact with other programs, unattended robots, and free development tools. The costs of developing and implementing RPA processes, using UiPath Community technology for free, allows you to have a high-value system, without having to make an economic investment for licensing, that in conjunction with other items, these software-based robots can be onerous [15]. In this way, like [3] the goal of zero complexity and effort in the use of the RPA system is achieved, resulting in less user intervention in the processes to be automated and agility in the services to enter and report student notes.

Regarding the performance of the proposal, in the study [14] where RPA processes apply for PDF generated invoice resulted in a 2% improvement in time performance (single process) that compared to this study has an improvement, in the 6 RPA processes, of 995.43% (Table 3). RPAs 3 and 4 for reporting low-performance students represent the main efficiency improvement. This turns out to be important, as well as the case study developed for reading PDF files. [6] where a time reduction of 94.44% is determined compared to the same work performed manually. On the other hand, in the study carried out on the agile development of RPA [16] It is determined that the construction of automatic RPA by interface recognition is faster but subject to errors if such an interface changes. Thus, in this study, activity flows for automation are performed manually to avoid errors if interfaces such as SQL Server, Excel, or Windows Operating System change.

Finally, unlike [17] where they calculate metrics to identify the processes to automate, this study uses the SCRUM methodology that involves the user as an active member of the team and helps to quickly identify the required and feasible automation processes. This is corroborated with [4] which concludes that there is no specific automatic technique for identifying processes to be automated and the best option is the use of experience provided by the same user.

4 Conclusions

This work develops and evaluates an RPA system for low academic performance detection at an educational institution using the SCRUM agile development methodology and the UiPath Studio tool, adapting to the institution's existing academic system. The system quality assessment (efficiency, efficiency, and satisfaction) carries out with the G.Q.M. methodology. The RPA system achieves a 100% efficiency level in meeting the goals set (Table 2), an efficiency level regarding execution times of 995.43% for the 6 RPA processes (Table 3), and an excellent level of satisfaction (Fig. 3) using the SUS questionnaire. Thus, the RPA system enables automatic detection of low-performance high school students, obtaining timely reports which facilitate the teachers, course tutors and authorities involved in making appropriate decisions and actions. As future work, the scalability of the project is proposed for the primary section of the same educational institution.

References

1. Innovasys: automatización robótica de procesos – RPA. https://www.innovasyscorp.com/automatizacion-robotica-de-procesos. Accessed 2020
2. Valois, J.: Qué es RPA. Conoce la solución para automatizar procesos. https://www.pragma.com.co/blog/que-es-rpa-conoce-la-solucion-para-automatizar-procesos. Accessed 11 July 2019
3. Schmitz, M., Dietze, C., Czarnecki, C.: Enabling digital transformation through robotic process automation at deutsche telekom. In: Urbach, N., Röglinger, M. (eds.) Digitalization Cases. Management for Professionals (2019)
4. Leno, V., Polyvyanyy, A., Dumas, M.: Robotic process mining: vision and challenges. Bus. Inf. Syst. Eng. (2020)
5. Issac, R., Muni, R., Desai, K.: Delineated analysis of robotic process automation tools. In: Second International Conference on Advances in Electronics, Computers and Communications (ICAECC), pp. 1–5 (2018)
6. Suryakan, P., Vinod, M., Preeti, P.: Social innovation in education system by using robotic process automation (RPA). Int. J. Innov. Technol. Explor. Eng. (IJITEE) **8**(11), 3757–3760 (2019)

7. Vila, D., Cisneros, S., Granda, P., Ortega, C., Posso, M., García-Santillán, I.: Detection of desertion patterns in university students using data mining techniques: a case study. In: Communications in Computer and Information Science, vol. 895, pp. 420–429 (2019)
8. Unidad Educativa San Juan de Ilumán: Unidad Educativa Intercultural Bilingüe San Juan de Ilumán. https://www.infoescuelas.com/ecuador/imbabura/unidad-educativa-san-juan-de-iluman-en-otavalo/. Accessed 2020
9. Schwaber, K., Sutherland, J.: La Guía Definitiva de Scrum: Las Reglas del Juego. https://www.scrumguides.org/docs/scrumguide/v2017/2017-Scrum-Guide-Spanish-SouthAmerican.pdf. Accessed November 2017
10. Scrum: The home of Scrum. https://www.scrum.org/. Accessed 2020
11. UIPath: Plataforma de RPA Empresarial de UiPath. https://www.uipath.com/es/producto/plataforma. Accessed 2019
12. Microsoft: SQL Server 2008. https://www.microsoft.com/es-es/download/details.aspx?id=1695. Accessed 2008
13. Basili, V.: Software modeling and measurement: the goal/question/metric paradigm. University of Maryland, CS-TR-2956, UMIACS-TR-92-96 (1992)
14. Aguirre, S., Rodríguez, A.: Automation of a business process using robotic process automation (RPA): a case study. In: Communications in Computer and Information Science, vol. 742 (2017)
15. Asatiani, A., Penttinen, E.: Turning robotic process automation into commercial success – case OpusCapita. J. Inf. Technol. Teach. Cases **6**(2), 67–74 (2016)
16. Cewe, C., Koch, D., Mertens, R.: Minimal effort requirements engineering for robotic process automation with test driven development and screen recording. In: 15th International Conference on Business Process Management, vol. 308, pp. 642–648 (2018)
17. Leshob, A., Bourgouin, A., Renard, L.: Towards a process analysis approach robotic process automation. In: IEEE 15th International Conference on e-Business Engineering (ICEBE), pp. 46–53 (2018)
18. Apache: Apache NetBeans IDE. https://netbeans.org/. Accessed 2020
19. JSF: JavaServer Faces. https://www.javaserverfaces.org/. Accessed 2020
20. ISO: ISO 25000: calidad de software y datos. https://iso25000.com/. Accessed 2020
21. Calabrese, J., Muñoz, R.: Asistente para la evaluación de productos de software segun la familia de normas ISO/IEC 25000 utilizando el enfoque GQM. Tesis de grado. UNLP, Facultad de Informática. https://sedici.unlp.edu.ar/handle/10915/67212. Accessed 2018
22. Brooke, J.: SUS: a quick and dirty usability scale. In: Usability Evaluation in Industry. CRC Press (1996)
23. González, J., Pazmiño, M.: Cálculo e interpretación del Alfa de Cronbach para el caso de validación de la consistencia interna de un cuestionario, con dos posibles escalas tipo Likert. Publicando **2**(2), 62–77 (2015)
24. Cronbach, L.: Coefficient alpha and the internal structure of tests. Psychometrika **16**(3), 297–334 (1951)
25. Juma, A., Rodríguez, J., Naranjo, M., Caraguay, J., Quiña, A., García-Santillán, I.: Integration and evaluation of social networks in virtual learning environments: a case study. In: Communications in Computer and Information Science, vol. 895, pp. 245–258 (2019)

Services Based on the Enriched Profile of a Person Within a Smart University

Viky Julieta Arias Delgado(✉) and Enrique González(✉)

Pontificia Universidad Javeriana, Cra. 7 #40-62, Bogotá, Colombia
{viky.arias,egonzal}@javeriana.edu.co

Abstract. This paper aims to review the literature on context-aware services inside a smart university by using a person's profile enrichment. An exploratory investigation was conducted through a systematic state of the art review, to find relationships among a selection of context-aware services with profile-based services such as location, likes, preferences, behavior, and applicability. Also, conceptual pillars of smart university, such as wellbeing, knowledge, environment, governance, and society, are used to find relationships with context-aware services. Subsequently, different features of a profile were identified in order to enrich the design of the services through the non-intrusive acquisition of data from emerging technologies. After analysis, research opportunities in context-aware services through profile enrichment are presented to guide further work as well as experimental validation protocols.

Keywords: Smart university · Smart campus · Context-aware services · Enriched profile · Personalization

1 Introduction

Currently, there is a trend for higher education institutions to become a smart campus. This transformation includes innovative campus strategies towards the pursuit of sustainability. Some universities carry out projects associated with the vision of a smart campus. For instance, the University of Covenant, Nigeria, implemented a strategy for daily network traffic scanning to monitor resource consumption [1]. At the University of Aveiro, Portugal, a measurement of the level of intelligence centered on people and oriented to the learning ecosystem was implemented to find, through the use of data analytics, features of people's satisfaction related to service administration, infrastructure, and changes in the environment [2]. At the University of Brescia, energy quality improvement was achieved in a classroom building, performing real-time monitoring to increase confidence and awareness about energy consumption, as well as monitoring the behavior of people aligned to technical and economic criteria [3]. Other studies carried out at the Wuhan University of Technology indicate that by embracing concepts such as the internet of things (IoT), cloud computing, and virtualization technology based on the existing network, a smart campus can be achieved [4].

The services designed within the context of an intelligent campus are supported by emerging technologies to obtain data in real-time. This data acquisition is used to control, optimize, and to predict different actions in different interconnected

Á. Rocha et al. (Eds.): ICITS 2021, AISC 1330, pp. 151–163, 2021.
https://doi.org/10.1007/978-3-030-68285-9_16

it was 35. However, the time needed to translate longer sentences having from 23–38 words decreased by 17–25%.

Table 4 presents the total times needed for performing translation tasks.

Table 4. Total time needed for the translation tasks.

Test set	Without Concordia + add. resources	With Concordia + add. resources	No. of words	No. of look-ups
Test set 1	Group A: 67 m 10 s	Group B: 66 m 42 s	218	69
Test set 2	Group B: 49 m 31 s	Group A: 50 m 38 s	182	35

Table 5 shows various Pearson correlation results.

Table 5. Pearson correlation results.

Correlation	Test set 1	Test set 2
Concordia score and the number of fragments	**−0.78**	**−0.69**
Concordia score and the average number of lookups	−0.17	−0.49
Number of fragments and the average number of lookups	0.59	0.72
Concordia score and the average time needed	−0.73	−0.50
Average number of fragments and the average time needed	**0.86**	**0.84**
Average number of lookups and the average time needed	0.52	**0.96**

The most indicative are the following correlations:

- correlation between the average number of fragments and the average time for completing the translation tasks (0.86 and 0.84) indicating that more fragments ask for more verification and longer time.
- correlation between the Concordia score and the number of fragments (−0.78 and −0.69), where higher values of Concordia scores imply lower numbers of fragments. Higher Concordia scores are obtained with smaller number of fragments.
- in the second translation task there was a high correlation (0.96) between the average number of lookups and average time that was needed

Other correlations, such as "Concordia score and average number of lookups", "number of fragments and average number of lookups", "Concordia score and average time needed" were relatively low. Still, more extensive human evaluation on a larger test set is planned. Evaluators were asked whether the system helped them in their assigned translation tasks. 10 students confirmed that the system was helpful during translation, whereas 4 students did not find the system useful. Average score of the system intuitiveness given by the evaluators was 3.5. Most applications. Technologies such as the internet of things, wireless sensor networks, computing paradigms such as cloud, edge or fog, and big data solutions allow the connection of objects and people for future changes and services, raised on the vision of a smart campus [5]. The context-aware services of a smart-campus are focused on improving people's experience by extracting features of the behavior in a non-intrusive way, treating large amounts of data, and contributing to personalized intelligent services. When conducting a review of latent context-aware services, we found that authors refer to different groups of services.

This paper aims to answer the research question associated with context-aware services (CAS) that can potentially be supported by enriching the person's profile by capturing information through emerging technologies in the context of an intelligent university. In this article, services based on location, likes, and preferences, people's behavior are considered within the framework of a smart campus. When presenting the systematic review of the literature, there are CAS that contribute to the enrichment of a person's profile using information collected by emerging technologies, which in turn contributes to the development of the Wellbeing, Knowledge, Environment, Governance, and Society pillars of a smart university.

2 Materials and Methods

The question defined for the purpose and scope of the present study is: ¿What are the context-aware services that can potentially be supported by the person profile enrichment, by capturing information through emerging technologies in the context of a smart university? According to the question posted previously, the state of the art review was carried out through exploratory research as well as a systematic review of the literature (SRL) [6]. The SLR process is composed of three stages. In the first stage, database sources are selected to create information search equations. Then, in the second stage, selection criteria are defined to filter out non-related literature. Finally, in the third stage, data extraction and synthesis are performed to build final references and literature sets. During these stages, the strategy used in the search for information was associated with primary sources, using papers and conferences from bibliographic databases (ScienceDirect, Scopus, and Web of Science) and secondary sources (catalogs).

2.1 First Stage: Paper Search

In the first stage, the search was carried out in the bibliographic databases, using the keywords listed in Table 1. The logical operators act as functions that represent conjunction or disjunction among different keywords related to a specific topic. The logical operator used to set topics relationship is the "and" operator, which forces the result of the search to comply with all keywords. Reserved identifiers such as "title" (to set a search inside document's title) or "body" (to set a search inside documents' content) will not be listed, as they differ to each bibliographic search engine. The final search equation for the SRL is a conjunction of T1 (for Smart university), T2 (for Context-Adapted Services), T3 (for Emerging Technologies), and T4 (for Person Behavior).

Table 1. Keywords and operators used for bibliographic search

Id	Topic	Keywords	Logical operator
1	T1	Smart University, Smart University, Smart City, Smart Cities	OR
2	T2	Context-Aware, Adaptative Services, Services, Personalization, Recommender System, Adaptative System	AND
3	T3	Emerging Technologies, Ambient Intelligence, IoT, Big Data, Sensor, RFID, Wi-Fi	OR
4	T4	Person Behavior, Person Profile, Preferences, Enriched Profile	OR

During the searching process, multiple examples were found related to different topic keywords, having a strong relationship with the standard search term. For instance, "smart university" and "smart campus" are closely related, but the smart university keyword is the standard search term. Additionally, search equations were adjusted to comply with scenarios that favor smart-campus concept pillars such as wellbeing, knowledge, environment, governance, and society. These pillar concepts were identified from different works such as Pagliaro et al. [7] and Hirsch & Ng [8].

2.2 Second Stage: Paper Selection and Assessment

In the second stage, the search equation was used and complemented with a document's publication date filter. The selected period was from 2001 to 2019, and after the process, the search yielded 973 articles. Subsequently, a selection criteria such as relevance in the search results, number of reported citations provided by the bibliographic databases, h-index indicator, English and Spanish language, scientific articles in national and international journals and average citations per item per year, were applied to obtain a final subset of forty (40) papers. With the reduced subset of papers, a quality assessment was carried out to extract a final subset of documents. Table 2 describes the quality assessment criteria (QAC) used in this process. Each of the QAC was codified with one (1) for compliance and zero (0) for non-compliance. Subsequently, criteria 1–4 is to be met by any paper to accept it for further analysis. When the quality assessment criteria were carried out over the selected articles, all the subset had full compliance with the requirements.

Table 2. Quality assessment criteria description

Id	Keywords
1	The article is diverse and significant in the context of the SRL
2	The article is related to the keywords of the context of the present investigation
3	The article clearly defines the methods, technologies and aspects that should be considered in solving the question asked for the SRL
4	The article presents conclusions and future work clearly concerning the question asked for the SRL

2.3 Third Stage: Information Extraction and Synthesis

During this stage, the analysis of the information was conducted to identify content. Synthesis result in a set of subtopics that will be mentioned in Sect. 3 and is listed as follows:

- Use of the person's profile: The usage of a regular or enriched person profile for adaptative services design.
- Emerging technology that refers to tools, approaches in the usage of emerging service technology such as IoT, ambient intelligence elements and analytics.
- Smart-campus vision: concepts related to smart-campus such as wellbeing, knowledge, environment, governance, and society.
- Service group: classification of the service adapted to the location, likes, preferences, behavior, and applicability.
- Challenges: limitations and ethical considerations of the adapted service

2.4 Limitations of the Study

The study presents some limitations in terms of the ability to select state-of-the-art research and information selection criteria. Regarding the ability to choose the information; free primary sources available and raised by the Pontificia Universidad Javeriana were used. Regarding the information selection criteria, the relevance, number of citations, and diversity were used, concerning the research question of this study.

3 Results

This section is intended to present results after the SRL process. Section 3.1 will discuss the main aspects of a person's profile. Section 3.2 will mention CAS groups applications, and Sect. 3.3 will enunciate potential CAS, which could be supported by an enriched profile.

3.1 Person Profile

According to the literature review, different authors have focused their research on aspects that are associated with a person's profile (PPA). One of these aspects is demographic features, which refer to variables that do not change frequently [9] like the name, address, or marital status. An example of this data usage can be evidenced in systems like MAIPU [10]. This system suits shopping website features to users' preferences by recognizing them after login. Another related feature is the socio-cultural features that enunciate the beliefs, values, and meanings of the symbols that a person has, as part of a group or society. As is mentioned in works as Aballay et al. [11], culture is considered as a frame of reference that influences the way of how people think, perceive, or interprets things in a social environment. Based on this concept, a use case could be an adaptive service for students which interface would be adjusted according to a place of origin, language, terminology, idioms, among others.

The psychological aspect of a person's profile defines the emotional dimension by assessing mood or interpersonal behavior. Previous studies analyzed the physical comfort perceived by the subject in the workplace as a significant predictor of stress syndrome and low personal performance. As Goldberg et al. [12] shows that personality defines aspects of the profile in which a person has the aptitude for new experiences, performs activities under his responsibility, develops actions with other people and maintains emotional stability in daily tasks. An example of personality features usage could be people who participate on a virtual academic collaboration platform, achieving tasks of and objectives and earning skills, supported by the goals and discipline of the person.

Interest and preferences are highly related, but their definitions are different. The preferences of a person correspond to the activities that are decided to perform at a given time, facilitating service customization. Sutterer et al. [13] showed a notification system in an office according to the high-level interpretation of the user's current situation. On the other hand, interests refer to features that vary according to surrounding behavior or environmental conditions. Both aspects could closely related to likes which refers to senses that allow persons to perceive the world. An example of the like usage is the smart-emotional bracelet iGenda [14] that collects likes information allowing discreet service to users. Also, it facilitates the management of daily activities through alerts and notifications.

Learning styles refer to the characteristics of a person to acquire new knowledge in an active, reflexive, theoretical or pragmatic way. For example, a person with a pragmatical learning style could earn skills easier through experimentation and practice. Multiple intelligences refer to how a person can express their abilities in verbal, logical, mathematical, spatial, kinesthetic, musical, interpersonal, intrapersonal or linguistic forms [15]. For example, the skills of a plastic artist are assessed by the way the artist makes graphics or drawings in a defined spatial environment.

Business context refers to the representation of the physical and virtual environment where the environment behavior dictates the models' adjustment. Previous work has shown adaptation strategies, considering components such as architectural models, which are in turn of reliability, performance, and safety [16].

3.2 CAS Context-Aware Services

To design systems that include scenario features, the concept of context must also be analyzed. Context is any information that can be used to characterize the situation of an entity. An entity is a person, place, software service or object that is considered relevant for the interaction between a person and an application [17]. Context-awareness was defined by Dey as the use of context to provide information and/or relevant services to the person, where relevance depends on the person's task [18]. The services adapted to the context of a smart university, are aimed to improving the experience of people, by knowing the behavior of people, treating large amounts of data and contributing to decision-making by campus directives [19]. When conducting a review of potential CAS that suits the context, it was found that different authors refer to four CAS groups. In this work, this group classification of CAS is used, according to previous research on location, likes and preferences, behavior, and applicability.

CAS focused on the location: these are computer applications that provide information about the location of the device and the person, mainly through mobile devices and mobile networks. Weng et al. [20] proposed a method to quickly extract urban landmarks from spatial databases on large scales, based on the dual aspects of spatial knowledge acquisition and public spatial cognition rules. Qiu and Wang [21] focused on location-based tracking data, proposing a framework for segmentation and grouping, for inference of road maps from GPS traces.

CAS focused on likes and preferences: these rely on the person and their environment in each situation. The enrichment of the context allows personalizing various products or services to individual persons, based on the knowledge of the custodians, the characteristics of the persons, and the past activities of the persons, such as previous purchases [22]. For example, a system can customize a list of nearby museums that are currently open, according to the location of the person and the list of selections previously stored on the smartphone.

CAS focused on the behavior of a person: a sequence of poses defines an action, and a sequence of acts define an activity [23]. For example, the CAVIAR system [24] focuses on city monitoring and market surveillance. The available video clips included an actor walking, a meeting of actors with other actors, shop windows, actors entering and leaving the store, and finally, leaving an object in a public place.

The fourth CAS group on a smart university focuses on applicability, which includes characteristics of the person's perception of the service, what is analyzed by emerging technologies and considers concerns about ethical and legal issues of smart home technologies regarding privacy, security and confidentiality, due to the highly identifiable nature of the data [25]. For example, challenges such as smart homes for health care services for the elderly include both technical and experimental points of view, integrated with ubiquitous computing and IoT technology, home automation, communication and security network, home intelligence and remote control, real-time and long-term health monitoring, disease prevention by detecting anomalies and discreet support for activity that does not interfere with people's daily activities, reducing care costs and improving the experience of the elderly.

An additional analysis allowed the association with sixteen (16) context-aware services tagged as potential for a smart university: virtualization of services (S1) [26], content recommendations (S2) [27], detection and treatment of anomalous events (S3) [28], Agent-based services and IoT (S4) [29], services that contribute to decision-making (S5) [30], based on Intelligent Tutoring Systems (ITS) and Artificial Intelligence in Education (AIED) (S6) [31], based on information security (S7) [32], based on interaction with the campus (S8) [33], based on the use of space (S9) [20], based on quality and satisfaction (S10) [34], based on research (S11) [3], based on socio-cultural aspects (S12) [35], oriented towards ergonomics (S13) [36], time-based (S14) [37], based on the state of the person (S15) and based on the ethical perception of the service (S16) [2]. Table 3 shows the relation between the mentioned CAS potential services (CASPS) and CAS groups (CASG). Each cell is qualified with the correspondence and relationship between the service and the group of services at the following scales: strong (+++), medium (++), low (+) and not applicable (-).

Table 3. CAS potential services for each CAS service group

CASPS	CASG			
	Location	Taste and preferences	Behavior	Applicability
S1	+	-	++	-
S2	+++	+	-	-
S3	+	-	++	-
S4	++	+++	+	-
S5	-	-	++	+
S6	+	+	+++	-
S7	-	-	-	+++
S8	++	+++	++	-
S9	++	+	-	-
S10	+	-	+	+++
S11	-	-	+	++
S12	-	+	++	-
S13	-	-	+++	+
S14	-	-	+	+++
S15	-	-	+++	+
S16	-	-	-	+++

When analyzing the potential CAS concerning the service groups in a smart university, the research opportunity is emphasized on the behavior of a person given the strong relationship between the features evaluated. Nevertheless, the behavior observation should not be conducted in a controlled environment, but in a natural one, as the behavior of people can drastically change if they feel monitored or even controlled. For this reason, a non-intrusive way to perform observations should be considered to avoid any fail or bias in the data collection and analysis. For this analysis, the term non-intrusive is considered as the ability to observe and capture the action of a person in a transparent way, using emerging technologies. Also, it contributes to ambience intelligence, through the integration and the context awareness according to likes and preferences [27]. For example, from the observation of mobile device connections to Wi-fi networks, it is possible to design personalized services, considering features of the context such as time and space.

3.3 Context-Aware Services from the Enriched Profile

The design of CAS should consider the pillars of a smart university, as well as the features of the person profile. Table 4 shows the relationship between the CAS and the specific pillars in the vision of a smart university (SCP). Each cell is qualified with the correspondence and relationship between the service and the group of services at the following scales: strong (+++), medium (++), low (+) and not applicable (-).

Table 4. CAS groups vs smart university pillars relationship

CASG	SCP				
	Wellbeing	Knowledge	Environment	Governance	Society
Location	+++	++	+	++	-
Preferences	++	+	-	++	-
Behavior	-	+++	-	++	+
Applicability	-	-	++	+	+++

The potential CAS are supported by the PPA considered in Sect. 3.1: demographic features (PPA1), likes (PPA2), preferences (PPA3), interests (PPA4), business context (PPA5), socio-cultural (PPA6), personality (PPA7), learning style (PPA8), multiple intelligences (PPA9) and psychological aspects (PPA10). Table 5 shows the relationship between these PPA, which can be used to contribute to the design of the CAS of a smart university. Each cell is qualified with the correspondence and relationship between the service and the profile feature with the following symbols: (Y) It is used, (N) It is not used, (-) if it does not apply.

4 Discussion

Different author's proposals were explored related to the tendency of higher education facilities transformation into smart universities. This process includes innovative campus strategies towards the search for sustainability. A smart university is supported then by emerging technologies for the design of adapted services, with stable features that can be accessed from any moment from any device. However, few authors have proposed to address a person's features as a fundamental axis, in the design of the adapted service. In this scenario, a huge amount of information of a person should be captured in a non-intrusive way, facilitating the enrichment of the profile and the provision of potential people-centered services.

Table 5. Relationship among potential services and person profile aspects

CASPS	PPA									
	PPA1	PPA2	PPA3	PPA4	PPA5	PPA6	PPA7	PPA8	PPA9	PPA10
S1	Y	Y	Y	Y	Y	Y	-	Y	Y	-
S2	Y	Y	Y	N	Y	N	-	Y	Y	-
S3	Y	Y	Y	Y	Y	Y	Y	Y	Y	Y
S4	Y	Y	Y	Y	Y	Y	-	Y	Y	-
S5	Y	Y	Y	Y	Y	Y	Y	Y	Y	Y
S6	Y	N	Y	Y	Y	Y	-	Y	Y	-
S7	Y	N	N	N	Y	N	-	N	N	-
S8	Y	Y	Y	Y	Y	Y	-	N	N	-

(continued)

Table 5. (*continued*)

CASPS	PPA									
	PPA1	PPA2	PPA3	PPA4	PPA5	PPA6	PPA7	PPA8	PPA9	PPA10
S9	Y	Y	Y	Y	Y	N	-	N	N	-
S10	Y	N	Y	Y	Y	N	-	N	N	-
S11	Y	Y	Y	Y	Y	Y	-	Y	Y	-
S12	Y	Y	Y	Y	Y	Y	-	N	N	-
S13	Y	N	N	N	Y	N	-	N	N	-
S14	Y	Y	Y	Y	Y	N	-	N	N	-
S15	Y	N	N	N	Y	Y	Y	N	N	Y
S16	Y	Y	Y	Y	Y	N	-	N	N	-

The SRL allowed to find sixteen different types of potential CAS for a smart university. Some of these services could be accessed or activated by involuntary (gestures or emotions) or intentional human expressions (voice commands). For example, by observing Wi-Fi connectivity, it is possible to design services that consider features of the context such as time and space. A recommendation for a new book can be sent to a user located in a bookstore, based on the history of activities and themes previously consulted. Also, the SRL allowed the identification of potential services classified in one of the four groups. The applicability group can be extended to other study fields such as security, ethics, quality of service and perception by the people who are using CAS.

Different research opportunities are presented, to enrich the profile of a student and offer a base for services adapted to the context, which contribute to the solution of different problems of a smart university such as security, efficient management of environmental resources, detection and the treatment of anomalous events, support for academic processes in real-time and the intelligent analysis of large volumes of information that support decision-making. Some examples of feasible CAS are listed as follows:

- CAS that contributes to the wellbeing pillar could be materialized to show advertising or recommendations in a cafeteria.
- CAS that contributes to the knowledge pillar could be used to prevent academic dropout.
- CAS that contribute to the environment pillar could be used for monitoring the use and saving of natural resources on campus.
- CAS that contributes to the governance pillar could be used to measure the frequent use of the different spaces.
- CAS that contributes to the pillar society could support training strategies in sustainable human development.

5 Conclusions

Through the systematic review of the literature, we found different approaches related to trends of higher education institutions transformed into smart universities. This process includes innovative campus strategies for the pursuit of sustainability. The analysis performed in Table 4 describes the relationship between potential CAS and the features of a person's profile, indicating that the greater the number of features included in the CAS, the greater the profile's enrichment. A wide range of services could be offered, as they get close to the user's behavior and needs. Examples of these are the recommendation of location-based services or preferences-based services.

Some limitations should be considered during the design and implementation of CAS. Among the most significant, we found the number of resources of the IoT devices (sensors, actuators, controllers, and embedded systems), intermittent network connectivity due to external factors, and the accuracy of the sensor which records a person's location on campus and synchronizes the information in real-time. Reviewing other research approaches, we confirm the diversity and importance of CAS in sixteen (16) types of services. Also, we found a strong relationship with four (4) groups of services, such as location, preferences, behavior, and applicability. In turn, the correspondence of the types of services is aligned with the pillars of a smart-campus, well-being, knowledge, environment, governance, and society.

After selecting, evaluating, and synthesizing the evidence found in the state-of-the-art review, important aspects were found for the selection of a CAS supported by the enrichment of a person's profile. These aspects are the virtualization of services that allows the monitoring of the campus infrastructure; the recommendation of services adapted to the user and to the context that contribute to the experience; the detection of anomalies based on statistical analysis, behavior pattern recognition, coverage criteria and simulation that contribute to environmental intelligence; the use of agent approaches that include interaction, planning, organization and regulations of some services; the use of machine learning techniques to support the improvement of educational systems; the strengthening of smart campus security, by monitoring other access security platforms; the orientation in the interaction with the campus that help and facilitate the daily tasks and the communication between the university community; the features of the product or service that are key drivers of community satisfaction; support for research and intellectual production processes that are important for strengthening research networks and associations of people with similar interests, for managing campaigns, events in favor of social and experiential values of the city; the aspects that seek the well-being of people, considering characteristics of ergonomics in face-to-face or virtual modalities of campus services; the analysis and appropriation of the temporal approach, as one of the most valuable contextual factors in many domains of recommender systems and the feedback of the intelligent service by the user, that is in fact significant for the application and appropriation of the pillars of the smart-campus.

The selection of a person's profile features through emerging technologies in a non-intrusive way allows real-time data management. Likewise, the before mentioned capabilities, added to the transactional information systems, contribute to the design of

CAS using context to provide information or services relevant to the person. As future work, a research opportunity is proposed to select a CAS based on some of the relevant aspects found in the systematic review of the literature. Our approach is based on the enrichment of the profile of the person considering likes, preferences, and observation of behavior through sensors. As validation is conducted through an experimental protocol and using emerging technologies such as IoT or edge compu-ting, we should integrate security components, privacy, and ethics aligned with the country's law and restrictions.

References

1. Adeyemi, O.J., Popoola, S., Atayero, A., Afolayan, D., Ariyo, M.: Exploration of daily Internet data traffic generated in a smart university campus. Data Brief **20**, 30–52 (2018)
2. Galego, D., Giovannella, M., Mealha, O.: Determination of the smartness of a university campus: the case study of aveiro. Procedia Soc. Behav. Sci. **223**, 147–152 (2016)
3. De Angelis, E., Ciribini, A., Tagliabue, L., Paneroni, M.: The Brescia Smart Campus Demonstrator. Renovation Toward Zero Energ. Classroom Build. Procedia Eng. **118**, 735–743 (2015)
4. Guo, M., Zhang, Y.: The research of smart campus based on Internet of Things and cloud computing. In: 11th International Conference on Wireless Communications, Networking and Mobile Computing (WiCOM 2015), Shanghai, China (2015)
5. Uskov, V., Bakken, J., Howlett, R., Jain, L.: Smart Universities: Conceptsm Systems, and Technologies, vol. 70. Springer, Cham (2018)
6. Gough, D., Oliver, S., Thomas, J.: An Introduction to Systematic Reviews, Second Edition ed., SAGE Publications Ltd, London (2017)
7. Pagliaro, F., Mattoni, B., Gugliermenti, F., Bisegna, F., Azzaro, B., Tomei, F., Catucci, F.: A roadmap toward the development of Sapienza smart campus. In: 2016 IEEE 16th International Conference on Environment and Electrical Engineering (EEEIC), Florence, Italy (2016)
8. Hirsch, J., Ng, W.P.J.: Education beyond the cloud: anytime-anywhere learning in a smart campus environment. In: 2011 International Conference for Internet Technology and Secured Transactions, Abu Dhabi, United Arab Emirates (2011)
9. Carrillo, A., Aragon, F., Cardenas, J., Cristancho, J., Higuera, M., Marin, D., Niño, L., Nova, J., Orozco, A., Rico, A., Romero, A.: IAM: integrated adaptation model. Avances En Sistemas E Informática **6**(3), 145–162 (2009)
10. Orozco, A., Cardenas, J., Florez, L., Carrillo, A.: MAIPU: adaptative information model based in user profile data for customizing product selling throughWeb sites. Avances En Sistemas E Informática **5**(3), 93–100 (2008)
11. Aballay, L., Aciar, S., Collazos, C., Gonzalez, C.: Adaptation model content based in cultural profile into learning environment. IEEE Latin Am. Trans. **13**(2), 490–495 (2015)
12. Goldberg, K., Foster, K., Maki, B., Emde, J., O'Kelly, M.: Improving Student Motivation through Cooperative Learning and Other Strategies, Dissertations/Theses (2001)
13. Sutterer, M., Droegehorn, O., David, K.: UPOS: User profile ontology with situation-dependent preferences support, In: First International Conference on Advances in Computer-Human Interaction, Sainte Luce, Martinique (2008)
14. Costa, A., Rincon, J., Carrascosa, C., Julian, V., Novais, P.: Emotions detection on an ambient intelligent system using wearable devices. Future Gener. Comput. Syst. **92**, 479–489 (2019)

15. Kezar, A.: Theory of multiple intelligences: implications for higher education. Innovative High. Edu. **26**, 141–154 (2001)
16. Garlan, D., Cheng, S.-W., Schmerl, B.: Increasing system dependability through architecture-based self-repair. Architecting Dependable Syst. **2677**, 61–89 (2007)
17. Abowd, G., Dey, A., Brown, P., Davies, N., Smith, M., Steggles, P.: Towards a better understanding of context and context-awareness. Handheld Ubiquit. Comput. **1707**, 304–307 (2001)
18. Dey, A.: Understanding and using context. Pers. Ubiquit. Comput. **5**, 4–7 (2001)
19. Rohs, M., Bohn, R.: Entry points into a smart campus environment - overview of the ETHOC system. In: 23rd International Conference on Distributed Computing Systems Workshops, 2003. Proceedings., Providence, Rhode Island, USA (2003)
20. Weng, M., Xiong, Q., Kang, M.: Salience indicators for landmark extraction at large spatial scales based on spatial analysis methods. Int. J. Geo-Inf. **6**(3), 72 (2017)
21. Qiu, J., Wang, R.: Road map inference: a segmentation and grouping framework. Int. J. Geo-Inf. **5**(8), 130 (2016)
22. Mori, M., Li, F., Dorn, C., Inverardi, P., Dustdar, S.: Leveraging state-based user preferences in context-aware reconfigurations for self-adaptive systems. In: Barthe, G., Pardo, A., Schneider, G. (eds.) Software Engineering and Formal Methods, SEFM 2011. Lecture Notes in Computer Science, Springer, Berlin (2011)
23. Cao, Z., Simon, T., Wei, S.-E., Sheikh, Y.: OpenPose: realtime multi-person 2D pose estimation using part affinity fields. In: IEEE Transactions on Pattern Analysis and Machine Intelligence (2019)
24. Saad, M., Hussain, A., Kong, W., Yasmin, F., Tahir, N.: SESRG-InViSS: image and video data set for human pose, action, activity and behaviour detection. In: 2013 IEEE 3rd International Conference on System Engineering and Technology, Shah Alam, Malaysia (2013)
25. Nebeker, C., Lagare, T., Takemoto, M., Lewars, B., Crist, K., Bloss, C., Kerr, K.: Engaging research participants to inform the ethical conduct of mobile imaging, pervasive sensing, and location tracking research. Trans. Behav. Med. **6**(4), 577–586 (2016)
26. Baccarelli, E., Amendola, D., Cordeschi, N.: Minimum-energy bandwidth management for QoS live migration of virtual machines. Comput. Netw. **93**, 1–22 (2015)
27. Moon, A., Kim, H., Lee, S.: Context-aware active services in ubiquitous computing environments. ETRI J. **29**(2), 169–178 (2007)
28. Roba, A., Katina, M., MG, M.: The regulatory considerations and ethical dilemmas of location-based services (LBS) a literature review. Inf. Technol. People **27**(1), 20 (2014)
29. Pico, P., Holgado, J.: Agentification of the internet of things: a systematic literature review. Int. J. Distrib. Sens. Netw. **14**(10), 1550147718805945 (2018)
30. Villegas, W., Palacios, X., Luján, S.: Application of a smart city model to a traditional university campus with a big data architecture: a sustainable smart campus. Sustainability **11** (10), 2857 (2019)
31. Tsai, I., Yeh, C.: Integrating SERVQUAL and importance-performance analysis for assessing smart campus service quality: a case study of an english training programme in Vietnam. In: IEEE 15th International Conference on Advanced Learning Technologies, pp. 436–440 (2015)
32. Kwon, D., Kim, H., An, D., Ju, H.: Container based testbed for gate security using open API mashup. Procedia Comput. Sci. **111**, 260–267 (2017)
33. Villegas, N., Müller, H.: Managing dynamic context to optimize smart interactions and services. In: Chignell, M., Cordy, J., Ng, J., Yesha, Y., (eds.) The Smart Internet. Lecture Notes in Computer Science, vol. 6400, pp. 289-318. Springer, Berlin (2010)

34. Chacon, I., Pinzon, A., Ortegon, L., Rojas, S.: Scope and management of carbon footprint as a driving force of branding for companies implementing these environmental practices in colombia. Estudios Gerenciales **32**(140), 278–289 (2016)
35. Benerecetti, M., Bouquet, P., Bonifacio, M.: Distributed context-aware systems. Hum. Comput. Interact. **16**(2), 213–228 (2001)
36. Zhe, C., Gines, H., Tomas, S., Shih, W., Yaser, S.: OpenPose: real-time multi-person keypoint detection library for body, face, hands, and foot estimation. In: de IEEE transactions on pattern analysis and machine intelligence (2018)
37. Adomavicius, C., Dietmar, J.: Preface to the special issue on context-aware recommender systems. User Modeling User-Adap. Inter. **24**, 1–5 (2014)

Multimedia Systems and Applications

Players Attitudes Towards In-Game Advertising

Luis F. Rios-Pino[1](✉), José E. Mejía-Perea[2], and Eliana E. Gallardo-Echenique[2]

[1] Communication and Advertising Program, Universidad Peruana de Ciencias Aplicadas, Prolongación Primavera 2390, Lima 15023, Peru
u201313256@upc.edu.pe

[2] School of Communications, Universidad Peruana de Ciencias Aplicadas, Prolongación Primavera 2390, Lima 15023, Peru
pcpujmej@upc.edu.pe, eliana.gallardo@upc.pe

Abstract. Users are actively protesting against invasive digital ads by downloading adblocking software. Recent games developers introduce dynamic and interactive advertising into virtual worlds. In this context, video games like NBA2K18 launched different platforms like billboards and other advertising media within the game. This study aims to analyze how the user's experience in NBA2K18 is regarding the participation of brands in it. It is positioned in the interpretive paradigm and semi-structured interviews were conducted. The participants were 15 NBA2K18 players, considered "hardcore players" whose ages ranged from 18 to 25. MyCareer, a game mode in NBA 2K18 will be studied. Findings show all participants agreed that brands are necessary in this game because they help the players fulfilled every fantasy requirement. Young people's perceptions revolved around how the brands help them pursuing their dreams. Brands in NBA2K18, do not undergo the experience of the game; rather, they enhance it, so that players feel it much more like real life.

Keywords: Videogames · Virtual world · Player · In-game advertising

1 Introduction

Users generally follow brands with social media presence; because they want an ad-free experience. Faced with people's rejection of advertising [1, 2], agencies and brands decided to respond with increasingly creative ideas and with more emotional and relevant messages [3]. Brands decide to migrate and expand to other media; for example, digital advertising investment reached 342 million soles in Peru [4]. As result of this ad bombardment, consumers have become increasingly resistant to traditional forms of advertising. This is how the term prosumer born. This term was coined by Toffler (1980), and refers to a combination of the words "producer" and "consumer" [5]. A prosumer tries to stay as far as possible from advertising; he is unwilling to receive messages that interrupts what he is doing and seeks for something that complements his experience. Prosumer has the ability to participate and generate shareable content with his network; especially experiences that connect with him [6, 7]. Today's young prosumers tend to (a) skip or

Á. Rocha et al. (Eds.): ICITS 2021, AISC 1330, pp. 167–176, 2021.
https://doi.org/10.1007/978-3-030-68285-9_17

avoid looking at ads [1]; click outside paid video ads online; and (c) discard e-mail ads without even reading the announced details [2]. They live in an era of immediacy, of "TV on demand", of "Here and Now" [8].

In this context, users are actively protesting against invasive digital ads by downloading adblocking software [9]. Ad blockers are various software tools that monitor browsers' requests for editorial and advertising content, which makes it possible for more internet users to watch the content provided without ads [10]. A good example of an ad blocker is Blockthrough [11], which by early 2017 had 1.3 million active users per month. Another solution was premium streaming services [12]. The first example is Netflix, which does not have any type of advertising and started 2019 with more than 130 million users worldwide [13]; the second one is Spotify® Premium, which is the ad-free version of music streaming and that has more than 124 million users in the premium services [14].

Faced this new ad-free mechanisms, different techniques such as branded content evolved and adapted to digital media platforms. Branded content is the fusion of advertising and entertainment into a product or service that is integrated with the organization's global brand strategy to be distributed as high-quality branded entertainment content for the users [1]. In order to prevent the content to be immediately perceived as advertising in a negative way, brands must: (a) create, distribute and share relevant, compelling and timely content [15]; (b) implement communication strategies capable of emotionally connecting with the daily lives of consumers [16]. Some examples are Johnnie Walker "Dear Brother", Nike "What girls are made of", and Adidas® "Original is never finished".

Several brands explored the world of video game as an alternative that is gradually getting a good share in the market every day: Video game advertising. This mechanic, which started as part of branded content, now has its own terms according to the type of advertising. Therefore, this study analyzed the advertising in NBA2K18, a basketball video game series developed by Sega Dreamcast. The NBA2K franchise has always innovated its format and gameplay experience over the years; so, they released a new space within the virtual world format in 2017 [17]. The question that guide this study is: How is the user's experience in NBA2K18 regarding the participation of brands in it?

1.1 The Advertising Crusade in Videogames

There are two types of advertising in video games: advergaming and in-game advertising [18, 19].

Advergaming, a term coined by Anthony Giallourakis in 2000 [19], is the process by which the advertising message, logo or other brand information is incorporated into online or offline video games [3]. This technique takes advantage of the game's market share by turning players into allies of the brand [18, 20, 21]. The goal of advergames is to immerse the player in the video game by continuing the narrative without leaving the brand aside [21]. Advergames encourage users to do a little more research on the brand and the product [3, 22]. When this technique is properly used, the brand can stick in people's mind [22] and can even go beyond borders to become viral to achieve the state of flow. A mental state in which the player completely omits the environmental information and achieves a high level of concentration in a specific process [20].

For example, "Dumb Ways to Die" [23] is a public service announcement campaign by Metro Trains in Melbourne (Australia) that promotes railway safety. The idea shows many dumb ways to die in a jingle with cartoons.

Nevertheless, a problem with advergames is that users, as soon as they got bored or frustrated, they can uninstall them [18]. Not every advergame will have the same success as Dumb Ways to Die. Therefore, video game developers created another video game advertising technique called in-game advertising. This model consists of paying for advertising spaces in video games that already exist. Advergames are tailor-made for the advertiser, differing from in-game advertising, which follows a more traditional format of placing products within a gaming environment [24, 25].

Recent game developers introduce dynamic and interactive advertising into virtual worlds [19] which are platforms that recreate a past, present, future and fantasy versions of the world [26]. By materializing our bodies in the way that the games reality (space–time) allow us to interact with different objects, environments and people we meet along the way [27]. People need to feel like they are present in this world (sense of presence) [27] and could handle decisions of all kinds. In this way, the virtual world should generate a user experience in which they become content producers since they can shape their fate and keep hard players (so-called hardcore players/gamers) within the game and the franchise [26–28]. A hardcore player is an individual who spend a significant part of his time (approx. 19 h a week) playing or learning about games and immersive games [28, 29].

Fig. 1. Foot locker and gatorade inside the world of NBA2K18

Virtual worlds such as GTA or The Sims, have spaces that are like our physical world with advertisements [19]. As it is in the real world, the spaces created by video game developers could be rented by brands, not following the static advertising model where it was already embedded in the game, but rather with the dynamic advertising model. This model consists in a dynamic banner (as a billboard) that will change depending on which advertiser rents the space, becoming a source of income for developers and giving movement to the space by making the place have a refresh [30]. In this context, video games like NBA2K18 launched different platforms like billboards and other advertising media within the game to achieve better results [21]. For example, Gatorade decided to pay the developers in order to place its own gym within the virtual world; just like Foot Locker did to set up his own store (see Fig. 1).

2 Methodology

This study is positioned in the interpretive paradigm with a qualitative methodology, where the researcher investigates situations, trying to make sense or interpret the phenomena in terms or meanings that the informants give [31, 32]. For data collection, semi-structured interviews were conducted, a technique that, unlike formal interviews, follows a rigid format with a series of established questions; these were adjusted according to the interviewees, allowing to collect richer and more nuanced information than in the structured interview [33]. The semi-structured interview guide was prepared with open questions that, depending on each interviewee, can be expanded or reformulated, and more information can be requested to collect the data in the best possible way.

The selection of participants was through theoretical and convenience sampling where the informants participated voluntarily until reaching the theoretical saturation; this means that no additional information was found in the topics or categories addressed in this study [31, 34]. The participants were 15 NBA2K18 players, considered "hardcore players" whose ages ranged from 18 to 25. At the same time, all of them are at an age capable of deciding which sports brand they prefer and decide to buy, unlike minors who may be subject to what their parents decide. It should be noted that the NBA separates its leagues in NBA for men and WNBA for women, creating both types of games. For the purpose of this research, MyCareer, a game mode in NBA 2K18, will be studied.

The analysis process consisted of encoding data according to different categories derived from the literature, adjusting to the new ones that emerge during the interviews [34]. The basic analytical procedures were asking questions about the data and making comparisons for similarities and differences between codes and concepts that are labeled and grouped to form categories and themes [34]. "Brief quotations represented each of these themes" [35, p. 153]. Multiple participant voices (hard players) add richness and a more effective explanation on the findings [35, 36].

3 Results and Discussion

Due to space constraints, we will focus on presenting the results of the most relevant questions of the interview guide. The identity of the interviewees is presented through alphanumeric coding in order to maintain their confidentiality and anonymity [37]. When asked about which sneaker brands they decided to sign with in the game, all the respondents answered having signed with the one they wore that year.

I signed with Under Armour because I have always used it, that is, it is the brand of shoes that I wear, in a way I had to sign with them, right? (E01, 21)
Why wouldn't I sign with Nike; they sponsor Lebron. (E03, 20)
I signed with Nike, and all my basketball shoes are Nike. Like I said (…) I just trust them. (E05, 19)

The interviewees signed with the brands they already knew, these have been inserted into the game in the perfect space to talk with people without disturbing them [18] and help people become the players they want to be [26]. It should be noted that in the contract, the player can negotiate with them and the payment they will receive is agreed prior negotiation; thus, giving them the power to decide [25].

Yes, it's true that they offered me a little more money than the others. But I still would have signed with them, not that I don't use it, but it made me make an easier decision. (E05, 19)
Obviously I considered the money, I signed with Under Armour, but not without asking for a little more… I wanted them, but the VC will be useful to me. (E08, 22)

According to the interviewees, brand interaction with people is the main reason why the three sneaker brands (Under Armour, Nike and Jordan) are the most remembered among all the interviewees. In contrast, Evans and Bang [38] mentioned that this kind of interaction would increase the level of brand recall. These three sponsoring brands are not only in participants' minds, there are also Gatorade and Foot Locker, brands that were placed in the virtual world. Being physically placed, these two brands managed to locate themselves in the minds of the interviewees, in addition to the fact that they are completely relevant brands for the context of the game [18] because they are involved in basketball in real physical life. To Herrewijn and Poels [30], virtual games offer brands "the opportunity to become an integral part of the game experience, reaching out to players in a vivid, interactive and immersive environment" (p. 87).

Obviously I have to enter the Gatorade's gym, otherwise my character had less energy. (E07, 22)
I entered Foot Locker to buy shoes, well… I have to, it's the only option and if I don't wear the shoes of the brand that sponsors me to play online, it complains. (E02, 20)

With regard to the question if they would prefer all game brands disappear and returning everything brand-free and generic, all the interviewees shouted "No". Brands should not only choose a good location for the advertising message, the ad should be visually consistent and resemble a real online display ad, so the player could be more receptive to advertising messages within the game [39]. The ads must be well integrated into the game experience, as well as be conceptually correct with the story of the game [38].

"It is impossible to imagine the game without brands… especially the sponsors, it gives it more realism, which is… the billboards do make it feel more real, but the main thing is the brand that sponsors you. Those are the ones we want." (E09, 19)
"The brands that sponsor us are the ones that help us achieve what we want, that is… I want my own sneaker and my TV commercial." (E02, 20)
"To remove the brands of the game would be like going back to 2K11 where you just played. There was nothing else." (E12, 21)

Players are looking to get sponsored by the same brand that in real life sponsored NBA superstars. Many of the brands help the player to accomplish a goal in the game

[25, 38], and to obtain points that can then be exchanged for prizes (special discounts and offers for real-life products) [19].

In addition, the authors were interested to find out if players saw any specific campaigns that use billboard ads within their neighbourhoods. At this point, they do remember several brands that were present in these dynamic advertising formats [40], but they do not remember specific campaigns that exist in the game. For example, none of the respondents saw the Curry 4 campaign; however, some saw the billboards when they belonged to Under Armor (see Fig. 2) as a brand. Due to new media and technology affordances of modern games, advertisers do not utilize static and unchanging images anymore, they update ads in-game based on more accurate settings depending on socio-demographic and gaming characteristics of gamers [30].

Very few interviewees recalled seeing the "Want it all" campaign by Nike (see Fig. 2); however, they were quite mindful of the fact that Nike owned several of those advertising spaces for quite some time. Finally, the most remembered brand was Mountain Dew, because the brand has one of the main billboards in the game.

Fig. 2. Nike and under armour dynamic billboard ads in NBA 2K18

"I did not really notice the Under Armour's campaign about the launch of Curry 4. What I did see were the Mountain Dew posters and a pair of Nikes." (E01, 21)
"I only saw the Mountain Dew one. Oh, I also saw mine from Gatorade." (E08, 19)

In real life, advertising is so annoying, so why interviewees did talk excitedly about what the brand does for them in MyCareer game. Everyone responded that there was no point of comparison. For example, in Facebook, they said that advertising is obstructing them from the goal by not letting them watch the video without finishing the advertisement. In this game, the brand is part of the goal, it is what helps players grow and live the life of a basketball player. Interviewees perceive advertising as something positive; making the game more realistic and improving brand perception [19]. When brands collaborate in users' goals, they will remember the brand over his competitors [19, 38]. Brands are part of the real world, and if they are not in the virtual world, this will not be so real [21]. In line with Morillas and Martín [19], in-game advertising lends a particular realism with real-time interactivity in the game environment.

"What I want is to live up to Lebron and Curry, I want to take all the awards and have people talk about me." (E16, 19).

4 Conclusions

In regard to research question, all participants agreed that brands are necessary in this specific game. In particular, players perceive that brands are helping them not only to go through the game [26], but also fulfill every fantasy requirement (e.g. they are professional basketball players). According to the findings, there is a space in which the brand is well received in gamers' lives, and this occurs in virtual life [19]. All the young people's perceptions revolved around how the brand helps them pursuing their dreams. Brands in NBA2 K18, that only participate in the visual space of the game, do not undergo the experience of the game; rather, they enhance it; players feel it's more realistic, much more like real life [22].

Brands are well received in the MyCareer game by the users, taking into consideration the fact that they must be associated to the game [22]. Also, some brands like Nike, Under Armour and Jordan, which have a higher level of participation in the game's plot and in the individual goals, are the most remembered brands [19, 38]. These brands were engraved in participant's minds followed by Gatorade with the gym and Foot Loocker with the shoe store, which had significant spaces within the virtual world. In both cases, being monopolies of the activity they carried out, and having a certain level of participation in the improvement of the players, these two brands have applied affordances that provide users more freedom of choices and more opportunities to express themselves. According to Xi and Hamari [41], when the player has the freedom to pursue an optimal activity without any external control, the player's sense of autonomy is high and thus increases his motivation. When users interact with immersion-related features, they are more likely to perceive higher feelings of freedom, involvement and engagement in the game [41].

Among the brands present in the game experience, these brands decided to apply the dynamic in-game advertising [42] that could be very important if publishers and developers should be able to offer more effective in-game advertising. In the game experience, brands must help every player to fulfil his true goal: to be a basketball player [25]. Another factor to highlight to ensure and improve the in-game advertising experience, is the ability of the game to give players autonomy, freedom and the ability to decide the development of the story [41]. Therefore, any presence in a game that allows the user to explore these possibilities and encourage the fact that he is free to make decisions will make his experience much more enjoyable [38]. If these games also give people positive feedback, they can increase the users' experience much more [42].

4.1 Limitations and Implications for Future Research

The main limitation of this qualitative research is that the generalizability of the results is limited. The findings may only be applicable to similar players, especially those considered "hardcore players". Much of this study needs to be replicated on various sociodemographic groups (in terms of age, gender, income, race and ethnicity) before more profound conclusions on in-game advertising are drawn. In this regard, the authors encourage researchers to investigate the role of gender since women of nearly all ages are a fast growing segment within the game industry. The future of in-game advertising as an advertising tool looks promising for brands, although it remains a terrain yet to be exploited to create an experience that engages the gamer. There remains much to be done in Peru and other countries in South America.

Acknowledgments. The authors would like to express their gratitude to the professional players who participated in this study. Thanks to the Research Department at Universidad Peruana de Ciencias Aplicadas (UPC) for the support provided in this research study.

References

1. Kim, T.Y., Shin, D.H.: The survival strategy of branded content in the over-the-top (OTT) environment: Eye-tracking and Q-methodology approach in digital product placement. Telemat. Inform. **34**, 1081–1092 (2017). https://doi.org/10.1016/j.tele.2017.04.016
2. Tuchman, A.E., Nair, H.S., Gardete, P.M.: Television ad-skipping, consumption complementarities and the consumer demand for advertising. Quant. Mark. Econ. **16**, 111–174 (2018). https://doi.org/10.1007/s11129-017-9192-y
3. Sharma, M.: Advergaming–the novel instrument in the advertising. Procedia Econ. Financ. **11**, 247–254 (2014). https://doi.org/10.1016/s2212-5671(14)00193-2
4. Perú, I.A.B.: GFK: Informe de Inversión Publicitaria Online del 2017. IAB Perú, Lima, Perú (2018)
5. Ritzer, G.: Automating prosumption: the decline of the prosumer and the rise of the prosuming machines. J. Consum. Cult. **15**, 407–424 (2015). https://doi.org/10.1177/1469540514553717
6. Arbaiza, F., Huertas, S.: Comunicación publicitaria en la industria de la moda: branded content, el caso de los fashion films. Rev. Comun. 17, 9–33 (2018). https://doi.org/10.26441/RC17.1-2018-A1
7. Ritzer, G., Dean, P., Jurgenson, N.: The coming of age of the prosumer. Am. Behav. Sci. **56**, 379–398 (2012). https://doi.org/10.1177/0002764211429368
8. Ron, R., Álvarez, A., Núñez, P.: Bajo la Influencia del Branded Content: Efectos de los Contenidos de Marca en Niños y Jóvenes. ESIC Editorial, Madrid, España (2014)
9. Aguado, J.M.: La publicidad como problema. El impacto de los bloqueadores de anuncios en la industria del conflicto digital. TELOS. Rev. Pensam. sobre Comun. Tecnol. y Soc. pp. 6–9 (2016)
10. Redondo, I., Aznar, G.: To use or not to use ad blockers? The roles of knowledge of ad blockers and attitude toward online advertising. Telemat. Inform. **35**, 1607–1616 (2018). https://doi.org/10.1016/j.tele.2018.04.008
11. Blockthrough: PageFair 2017 Adblock Report, https://blockthrough.com/
12. Oyedele, A., Simpson, P.M.: Streaming apps: what consumers value. J. Retail. Consum. Serv. **41**, 296–304 (2018). https://doi.org/10.1016/j.jretconser.2017.04.006
13. Netflix Inc.: Netflix Releases Fourth-Quarter 2018 Financial Results., California, USA (2019)
14. Spotify: Quick Facts, https://investors.spotify.com/home/default.aspx
15. Müller, J., Christandl, F.: Content is king–But who is the king of kings? The effect of content marketing, sponsored content & user-generated content on brand responses. Comput. Hum. Behav. **96**, 46–55 (2019). https://doi.org/10.1016/j.chb.2019.02.006
16. Roma, P., Aloini, D.: How does brand-related user-generated content differ across social media? Evidence reloaded. J. Bus. Res. **96**, 322–339 (2019). https://doi.org/10.1016/j.jbusres.2018.11.055

17. NBA 2K: Run The Neighborhood, https://youtu.be/3EPHnjHu77g
18. Méndiz Noguero, A.: Advergaming: Concepto, tipología, estrategias e evolución histórica. Rev. científica Comun. y Tecnol. emergentes **15**, 37–58 (2012)
19. Morillas, A.S., Martín, L.R.: Advergaming: an advertising tool with a future. Int. J. Hisp. Media. **9**, 14–31 (2016)
20. Gurău, C.: The influence of advergames on players' behaviour: an experimental study. Electron. Mark. **18**, 106–116 (2008). https://doi.org/10.1080/10196780802044859
21. Moral, M.E., Del, Villalustre, L., Neira, M., del R.: Estrategias publicitarias para jóvenes: advergaming, redes sociales y realidad aumentada. Rev. Mediterránea Comun. **7**, 47–62 (2016). https://doi.org/10.14198/MEDCOM2016.7.1.3
22. Ismail, H., Nasidi, Q.Y.: Adver-games and consumers : measuring the impact of advertising on online games. Int. J. Law, Gov. Commun. **3**, 35–40 (2018)
23. Metro Trains Melbourne, D.W. to D.: Dumb Ways to Die 2 - The games, http://www.dumbwaystodie.com/
24. Ghosh, T.: Winning versus not losing: exploring the effects of in-game advertising outcome on its effectiveness. J. Interact. Mark. **36**, 134–147 (2016). https://doi.org/10.1016/j.intmar.2016.05.003
25. Roettl, J., Terlutter, R.: The same video game in 2D, 3D or virtual reality – How does technology impact game evaluation and brand placements? PLoS ONE **13**, 1–24 (2018). https://doi.org/10.1371/journal.pone.0200724
26. López de Anda, M.M.: Relatos sobre el origen de los mundos virtuales. Virtualis **5**, 92–106 (2014)
27. Girvan, C.: What is a virtual world? Definition and classification. Educ. Technol. Res. Dev. **66**, 1087–1100 (2018). https://doi.org/10.1007/s11423-018-9577-y
28. Ramírez-Correa, P.E., Rondán-Cataluña, F.J., Arenas-Gaitán, J.: A posteriori segmentation of personal profiles of online video games' players. Games Cult. **15**, 227–247 (2018). https://doi.org/10.1177/1555412018766786
29. Poels, Y., Annema, J.H., Verstraete, M., Zaman, B., De Grooff, D.: Are you a gamer? a qualititive study on the parameters for categorizing casual and hardcore gamers (2012). https://lirias.kuleuven.be/retrieve/203496
30. Herrewijn, L., Poels, K.: The effectiveness of in-game advertising: examining the influence of ad format. In: Cauberghe, V., Hudders, L., and Eisend, M. (eds.) Advances in Advertising Research IX Research IX. pp. 87–100. Springer Gabler, Wiesbaden, Wiesbaden, Germany (2018). https://doi.org/10.1007/978-3-658-22681-7_7
31. Creswell, J.W.: Qualitative Inquiry and Research Design: Choosing Among Five Traditions. SAGE Publications Inc, Thousand Oaks, CA (2013)
32. Lincoln, Y.S.: The ethics of teaching in qualitative research. Qual. Inq. **4**, 315–327 (1998). https://doi.org/10.1177/107780049800400301
33. Díaz-Bravo, L., Torruco-García, U., Martínez-Hernández, M., Varela-Ruiz, M.: La entrevista, recurso flexible y dinámico. Investig. en Educ. Médica. **2**, 162–167 (2013)
34. Strauss, A.L., Corbin, J.M.: Bases de la Investigación Cualitativa: Técnicas y Procedimientos para Desarrollar la Teoría Fundamentada. Inc. y Editorial Universidad de Antioquia, Antioquia, Colombia, Sage Publications (2002)
35. Patton, M.Q., Cochran, M.: A Guide to Using Qualitative Research Methodology. Med. San Front. 1–36 (2002). https://doi.org/10.1109/PROC.1978.11033
36. Peterson, J.S.: Presenting a qualitative study: a reviewer's perspective. Gift. Child Q. **63**, 147–158 (2019). https://doi.org/10.1177/0016986219844789
37. Cohen, L., Manion, L., Morrison, K.: Research Methods in Education. Routledge, London; New York (2007)

38. Evans, N.J., Bang, H.: Extending expectancy violations theory to multiplayer online games: the structure and effects of expectations on attitude toward the advertising, attitude toward the brand, and purchase intent. J. Promot. Manag. 1–20 (2018). https://doi.org/10.1080/10496491.2018.1500411
39. Dardis, F., Schmierbach, M., Sherrick, B., Luckman, B., Dardis, F., Sherrick, B.: How game dif fi culty and ad framing in fl uence memory of in-game advertisements. (2018). https://doi.org/10.1108/JCM-07-2016-1878
40. Herrewijn, L., Poels, K.: Exploring player responses toward in-game advertising: the impact of interactivity. In: Rodgers, S., Thorson, E. (eds.) Digital advertising: theory and research, pp. 310–327. Routledge, New York, USA (2017)
41. Xi, N., Hamari, J.: Does gamification satisfy needs? a study on the relationship between gamification features and intrinsic need satisfaction. Int. J. Inf. Manage. **46**, 210–221 (2019). https://doi.org/10.1016/j.ijinfomgt.2018.12.002
42. Hassan, L., Dias, A., Hamari, J.: How motivational feedback increases user's benefits and continued use: A study on gamification, quantified-self and social networking. Int. J. Inf. Manage. **46**, 151–162 (2019). https://doi.org/10.1016/j.ijinfomgt.2018.12.004

Retinal Image Enhancement via a Multiscale Morphological Approach with OCCO Filter

Julio César Mello Román[1,2], José Luis Vázquez Noguera[1(✉)], Miguel García-Torres[1,3], Veronica Elisa Castillo Benítez[4], and Ingrid Castro Matto[4]

[1] Universidad Americana, Asunción, Paraguay
jose.vazquez@ua.edu.py

[2] Facultad de Ciencias Exactas y Tecnológicas, Universidad Nacional de Concepción, Concepción, Paraguay
jcmello@facet-unc.edu.py

[3] Division of Computer Science, Universidad Pablo de Olavide, 41013 Seville, Spain
mgarciat@upo.es

[4] Departamento de Retina, Catedra de Oftalmología, Hospital de Clínicas, Facultad de Ciencias Médicas, Universidad Nacional de Asunción, San Lorenzo, Paraguay
vcastillo@med.una.py, incamatt@hotmail.com

Abstract. Retinal images are widely used for diagnosis and eye disease detection. However, due to the acquisition process, retinal images often have problems such as low contrast, blurry details or artifacts. These problems may severely affect the diagnosis. Therefore, it is very important to enhance the visual quality of such images. Contrast enhancement is a pre-processing applied to images to improve their visual quality. This technique betters the identification of retinal structures in degraded retinal images. In this work, a novel algorithm based on multi-scale mathematical morphology is presented. First, the original image is blurred using the Open-Close Close-Open (OCCO) filter to reduce any artifacts in the image. Next, multiple bright and dark features are extracted from the filtered image by the Top-Hat transform. Finally, the maximum bright values are added to the original image and the maximum dark values are subtracted from the original image, previously adjusted by a weight. The algorithm was tested on 397 retinal images from the public STARE database. The proposed algorithm was compared with state of the art algorithms and results show that the proposal is more efficient in improving contrast, maintaining similarity with the original image and introducing less distortion than the other algorithms. According to ophthalmologists, the algorithm, by improving retinal images, provides greater clarity in the blood vessels of the retina and would facilitate the identification of pathologies.

Keywords: Retinal images · Contrast enhancement · Mathematical morphology · OCCO · Top-Hat transform

Á. Rocha et al. (Eds.): ICITS 2021, AISC 1330, pp. 177–186, 2021.
https://doi.org/10.1007/978-3-030-68285-9_18

1 Introduction

Eye diseases are detected from an eye fundus study. This test is very important to detect pathologies such as choroid melanoma, diabetic retinopathy, glaucoma, age-related macular degeneration, toxoplasmosis or other diseases. An image with a high visual quality is very important for ophthalmologists, because it guarantees a more precise diagnosis. However, image quality can be affected for different reasons at the time of acquisition. Some of the common problems that images often suffer at the time of acquisition are low contrast, poor detail, insufficient lighting or artifact generation. To solve these problems, digital image processing techniques are used to enhance the visual quality of the images [7].

Digital image processing has different techniques that are used to improve the visual quality of the images [1,3–6,8,9,12,14–20,23,24,26]. Among these are algorithms based on histograms and algorithms based on mathematical morphology. Also, some of these techniques have been used to enhance the visual quality of retinal images. For example, in [23] the Contrast Limited Adaptive Histogram Equalization (CLAHE) algorithm was used; in [14] the multiscale top-hat transform with the contrast stretch was used sequentially; and in [17] the multiscale top-hat transform by reconstruction was used. Although these algorithms improve the contrast of retinal images, new alternatives continue to be sought to better highlight the anatomical structures of the retina. This will help ophthalmologists to better diagnosis and follow up the eye disease that the patient.

This paper presents a novel retinal image contrast enhancement algorithm. The proposal is based on mathematical morphology operations. The morphological operation to reduce noise in images is called Open-Close Close-Open (OCCO) [2]. The morphological operation called Top-Hat transform is used in the contrast enhancement technique [22]. This technique improves the contrast of the images, but it also enhances the noises that may be present in the images. First, the OCCO filter is used to remove the noise present in the images. Second, multiple bright and dark features are extracted from the filtered image by the Top-Hat transform. Finally, the original image is enhanced by adding the bright areas and subtracting the dark areas previously adjusted by a weight.

The article is organized in 4 sections. Section 2 presents the morphological operations and the proposed algorithm for retinal image enhancement. Section 3 presents the visual and numerical results and Sect. 4 presents the conclusions of the work.

2 Proposed Algorithm for Retinal Image Enhancement

In this section some concepts of mathematical morphology are presented. The details of the algorithm are given below.

2.1 Basic Concepts

Top-Hat transform is defined from morphological operations of dilation, erosion, opening and closing. Such operations are defined as follows:

i *Dilation and Erosion*: These are the basic operations of mathematical morphology [21]. They are defined as:

$$\delta_B(I)(u,v) = \max_{(x,y)\in B}(I(u-x, v-y)), \tag{1}$$

$$\varepsilon_B(I)(u,v) = \min_{(x,y)\in B}(I(u+x, v+y)), \tag{2}$$

where I is the original image, B is the flat structuring element and (u,v), (x,y) are the spatial coordinates of I and B.

ii *Opening*: To make the opening, first the image is eroded and secondly it is dilated using the same structuring element [22]. Morphological opening is defined as:

$$\gamma_B(I) = \delta_{\check{B}}(\varepsilon_B(I)), \tag{3}$$

where $\check{B}$ is the reflection of B.

iii *Closing*: To make the closing, first the image is dilated and then it is eroded using the same structuring element [22]. Morphological closing is defined as:

$$\phi_B(I) = \varepsilon_{\check{B}}(\delta_B(I)). \tag{4}$$

A structuring element is symmetrical if it is equal to its reflection, i.e. $B = \check{B}$.

Based on previous definition two Top-Hat operations can be defined:

i *White Top-Hat* (WTH): This is used to obtain the bright regions lost in the morphological opening [22]. WTH is defined as:

$$WTH_B(I) = I - \gamma_B(I), \tag{5}$$

where I is the original image, B is the structuring element and $\gamma_B(I)$ is the morphological opening.

ii Black Top-Hat (BTH): This is used to obtain the dark regions lost in the morphological closing [22]. BTH is defined as:

$$BTH_B(I) = \phi_B(I) - I, \tag{6}$$

where I is the original image, B is the structuring element and $\phi_B(I)$ is the morphological closing.

Finally, the Open-Close Close-Open [2] (OCCO) is introduced. It is applied intro an image to reduce noise. It is defined as:

$$OCCO_B(I) = \frac{1}{2}\gamma_B(\phi_B(I)) + \frac{1}{2}\phi_B(\gamma_B(I)). \tag{7}$$

2.2 Proposed Algorithm

The proposal, which is based on mathematical morphology operations, strategically combines OCCO with multiscale Top-Hat. Such combination allows to perform image enhancement without introducing artifacts into the process. The pseudocode is shown in Algorithm 1. As it can be seen, it first applies OCCO filter to the original image. OCCO is used with a small structuring element to reduce noise. Then, features are extracted from the image by applying Top-Hat iteratively. Finally, the weighted maximum values of the bright scales are added to the original image and the weighted maximum values of the dark scales are subtracted from the original image. In this way, image enhancement is performed.

Algorithm 1. Open-Close Close-Open - Multi-scale Top-Hat for retinal image enhancement (OCCO-MTH)

Input: I: Original image, B: Structuring element B, G: Structuring element G, n: Number of iterations, ω: Contrast adjustment weight.
Output: I_E *(Enhanced image)*
Initialisation : B, G, n, ω
1: *Noise removal with OCCO filter* (Equation 7)
$OCCO = OCCO_B(I)$
2: **for** $i = 1$ to n **do**
3: *Multi-scale Top-Hat transform*
$TH_i = WTH_{G_i}(OCCO)$,
$BH_i = BTH_{G_i}(OCCO)$,
4: **end for**
5: *Calculation of the maximum areas of brightness and darkness.*
$MTH = \max_{1 \le i \le n}\{TH_i\}$,
$MBH = \max_{1 \le i \le n}\{BH_i\}$.
6: *Image enhancement calculation.*
$I_E = I + \omega \times MTH - \omega \times MBH$,
7: **return** I_E

---

3 Results

In this work, two different evaluation approaches are considered. The first one is a visual evaluation performed by two ophthalmologists. The other one is based on some performance metrics that allow to assess the quality of the processed images. In order to quantify the performance of the proposed algorithm, 397 color images from the public STructured Analysis of the Retina (STARE) [10] database were used. In the Fig. 1 it can be seen some retinal images from the STARE database.

OCCO-MTH was compared with the Histogram Equalization (HE), Contrast-limited adaptive histogram equalization (CLAHE) [26] and Multi-scale Top-Hat transform (MTH) [5] algorithms.

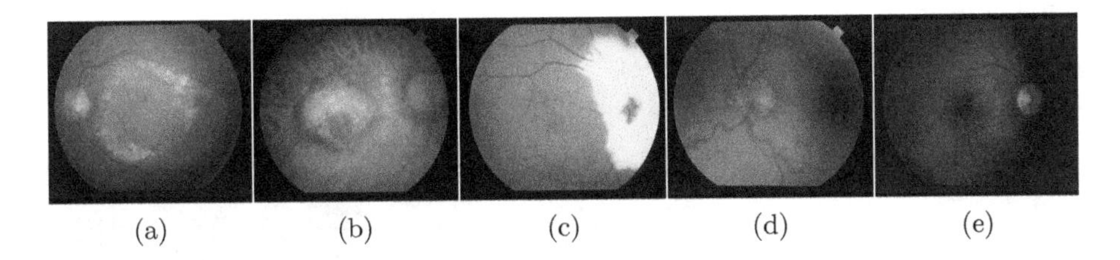

Fig. 1. STARE database retinographies.

The received and configurable parameters of the algorithms are: NumTiles: Number of rectangular contextual regions, clipLimit: Contrast enhancement limit, I: Original image, B: Radius r of the disk structuring element B, G: Radius r of the disk structuring element G, n: Number of iterations, ω: Contrast adjustment weight.

The initial parameters of MTH and OCCO-MTH are: the original image I, the structuring element B disk with radius $r = 1$, the initial structuring element G disk with radius $r = 1$ which increases in a range of $i = \{1, ..., n\}$, the number of iterations $n = 10$ and $\omega = 1.5$ is the contrast adjustment weight. All the algorithms were implemented in MATLAB R2014a. HE does not use parameters and CLAHE uses a default configuration. The RGB images are first converted to the HSV color space, then the algorithms are applied in the V channel, and finally the enhanced image is converted back to RGB.

3.1 Visual Analysis

The visual analysis was performed by two ophthalmologists who evaluated the images processed. Despite the subjectivity of this approach, it is the most common method nowadays.

In the Fig. 2 it can be seen images of diabetic retinopathy improved with the algorithms HE, CLAHE, MTH and OCCO-MTH. When the improvement is made, the choroidal vessels can be seen to be more enhanced. At the level of the retina, intraretinal hemorrhages, and microaneurysms can be seen more clearly, since they become darker, which facilitates the staging of the pathology. Also, the details of the retinal arterial and venous vascular wall can be better observed. The lipid exudates at the macular level are slightly more opaque in relation to the original image. However, in comparison with the other algorithms OCCO-MTH better preserves the original structure of the retinal image and distorts less in the enhancement process.

In the Fig. 3 it can be seen macular scar retinographies improved with the HE, CLAHE, MTH and OCCO-MTH algorithms. By enhancing the image, greater contrast of the choroidal vessels can be observed; the pigment in the macular scar of chorioretinitis due to toxoplasmosis can be seen more defined and with clear limits at the edges of the lesion. Also, the pigment is enhanced at the central level of the scar and at the temporal edges of the optical papilla. The upper and lower temporal vascular arches are seen more clearly than in the original

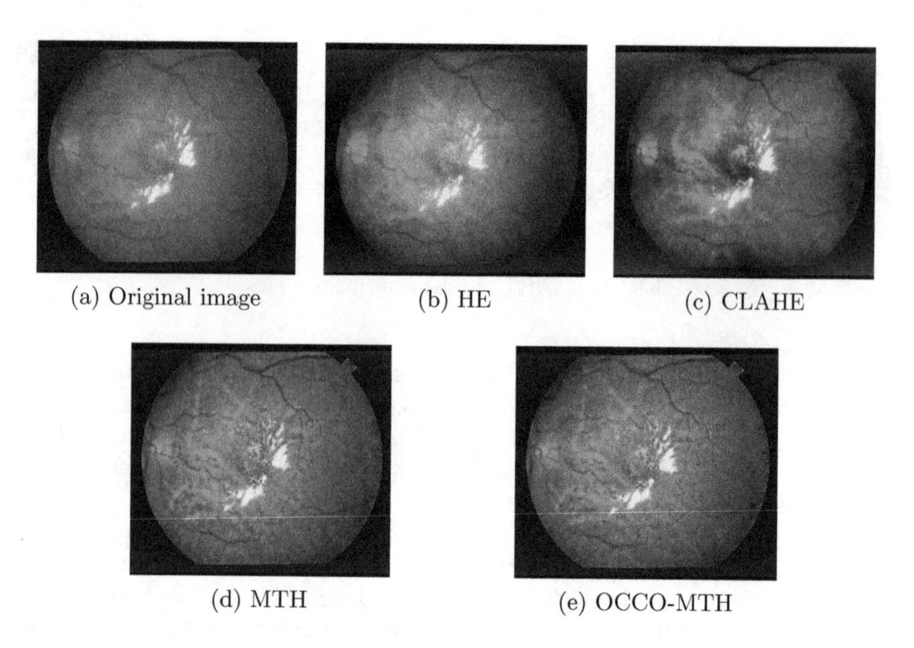

(a) Original image (b) HE (c) CLAHE

(d) MTH (e) OCCO-MTH

Fig. 2. Images with severe diabetic nonproliferative retinopathy.

image. However, HE and CLAHE enhance the brightness in areas not seen in the original retinal image, MTH and OCCO-MTH improve retinal images, although the proposal introduces less distortions.

In the Fig. 4 it can be seen age-related macular degeneration retinographies, improved with the HE, CLAHE, MTH and OCCO-MTH algorithms. By improving the image, the contrast of the choroidal background and the enhancement of the vascular walls can be better appreciated. The macular lesion presents greater definition of the faint pigment at the edges of the lesion, as well as the retinal vessels. However, for this case the CLAHE introduces dark colors around the lesion.

In the Fig. 5 it can be seen age-related macular degeneration images improved with the HE, CLAHE, MTH and OCCO-MTH algorithms. By enhancing the image, the choroidal vessels are more clearly visible. At the level of the retinal vasculature, the vascular walls have greater clarity. The lipid exudates take on a dark color and at the end of the lower temporal arch there is an area of scar atrophy and fibrosis at the central level of the lesion. The target-like intraretinal bleeding located in the lower temporal arch has a darker coloration. However, OCCO-MTH preserves the structures of the image and visually it is appreciated a more smoothed image.

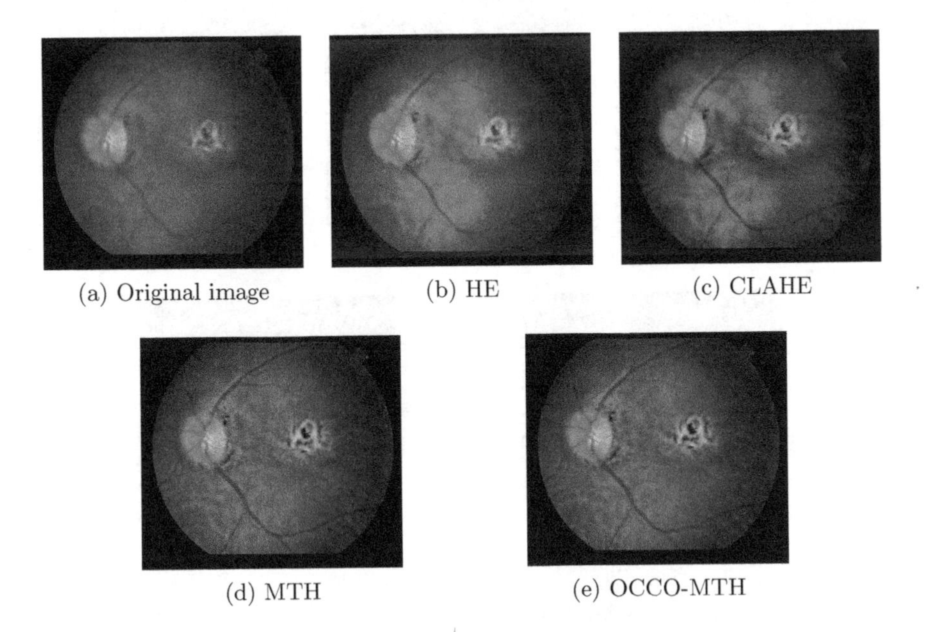

(a) Original image (b) HE (c) CLAHE

(d) MTH (e) OCCO-MTH

Fig. 3. Images of the macular scar of chorioretinitis by toxoplasmosis.

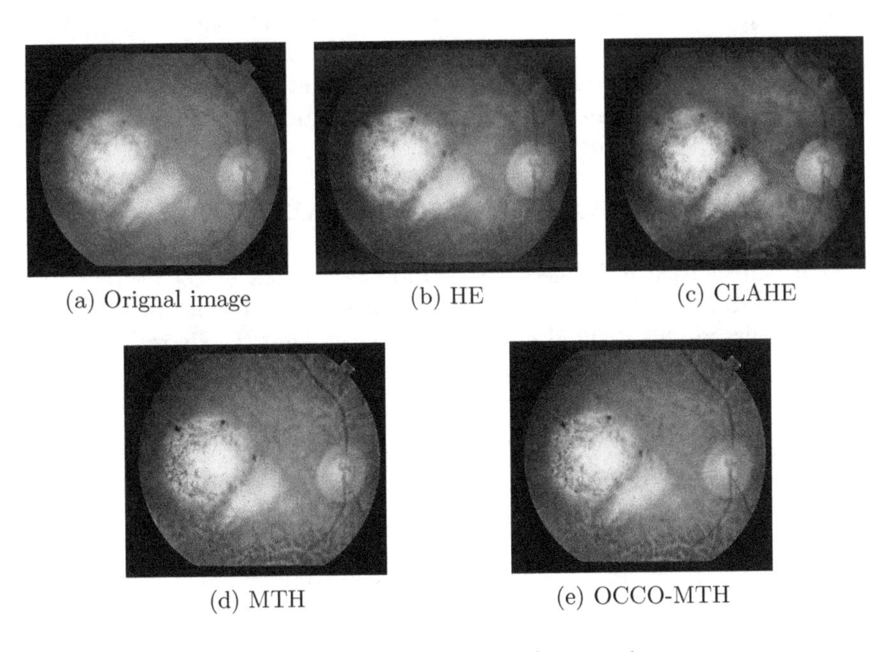

(a) Orignal image (b) HE (c) CLAHE

(d) MTH (e) OCCO-MTH

Fig. 4. Age-related macular degeneration.

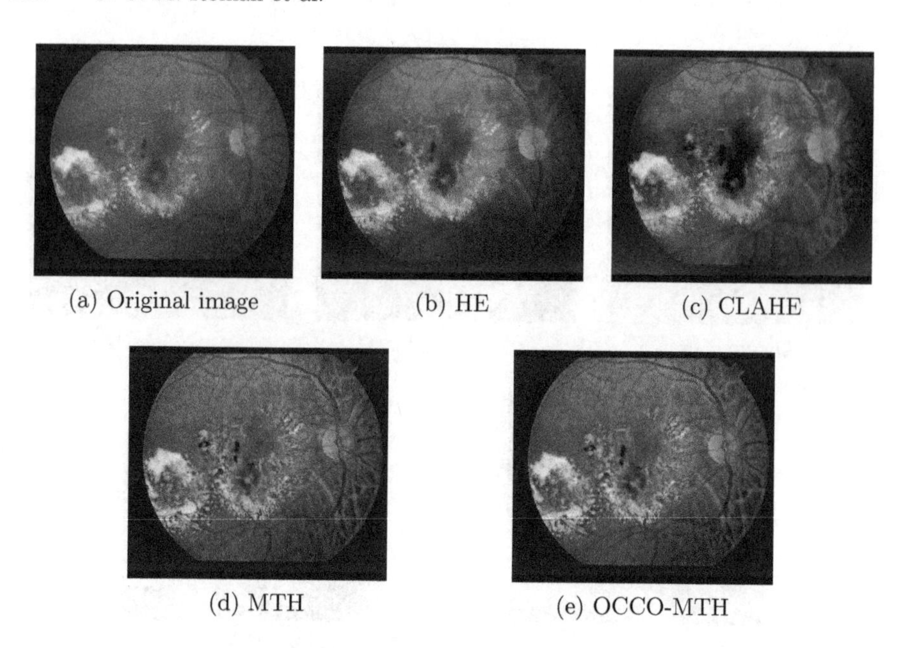

(a) Original image (b) HE (c) CLAHE

(d) MTH (e) OCCO-MTH

Fig. 5. Age-related macular degeneration images.

3.2 Numerical Results

The image quality metrics used to evaluate the results are: Entropy (E) is used to quantify the amount of information contained in the image [13,18]. Peak Signal-to-Noise Ratio (PSNR) is used to quantify the amount of distortion introduced to the image in the enhancement process [4,11]. Structural Similarity Index (SSIM) is used to quantify the similarity between the processed image and the original image [19,25].

The metrics were applied to the V channel of the image obtained by the HE, CLAHE, MTH and OCCO-MTH algorithms.

As it can be seen in Table 1, OCCO-MTH presents better average results in PSNR and SSIM. However, CLAHE obtains better results in E. The best average results are highlighted in bold. This means that the algorithm improves the contrast without distorting the original image too much.

Table 1. Average results of the 397 retinal images processed with the HE, CLAHE, MTH and OCCO-MTH algorithms.

Algorithms	E	PSNR	SSIM
HE	5.821	17.128	0.777
CLAHE	**7.705**	18.175	0.820
MTH	7.259	29.552	0.771
OCCO-MTH	7.175	**32.115**	**0.895**

4 Conclusions

In this work, a novel retinal image enhancement algorithm was proposed. Through this algorithm, the image is improved by increasing the differentiation between the lesions, such as intra- and sub-retinal hemorrhages, and the choroidal background. This is due to the enhancement of the red-yellow spectrum of the image. The proposed algorithm enhances the color of the pigment in the retinographies, which facilitates the staging of lesions such as chorioretinitis scars, subretinal Drusen's accumulations, lipid or cottony intraretinal exudates, and provides greater clarity in the blood vessels of the retina and evaluation of their parietal profile.

OCCO-MTH was compared to algorithms used in improving retinal images. The numerical and visual results show contrast enhanced retinographies, similarity to the original image and low distortion with respect to the compared algorithms.

Acknowledgment. This research was funded by CONACYT, Paraguay, grant number PINV18-846.

References

1. Alharbi, S.S., Sazak, Ç., Nelson, C.J., Alhasson, H.F., Obara, B.: The multiscale top-hat tensor enables specific enhancement of curvilinear structures in 2D and 3D images. Methods **173**, 3–15 (2020)
2. Aptoula, E., Lefèvre, S.: A comparative study on multivariate mathematical morphology. Pattern Recogn. **40**(11), 2914–2929 (2007)
3. Bai, X.: Image enhancement through contrast enlargement using the image regions extracted by multiscale top-hat by reconstruction. Optik **124**(20), 4421–4424 (2013)
4. Bai, X., Zhou, F., Xue, B.: Image enhancement using multi scale image features extracted by top-hat transform. Opt. Laser Technol. **44**(2), 328–336 (2012)
5. Bai, X., Zhou, F., Xue, B.: Noise-suppressed image enhancement using multiscale top-hat selection transform through region extraction. Appl. Opt. **51**(3), 338 (2012)
6. Bai, X., Zhou, F., Xue, B.: Toggle and top-hat based morphological contrast operators. Comput. Electr. Eng. **38**(5), 1196–1204 (2012)
7. Cao, L., Li, H.: Enhancement of blurry retinal image based on non-uniform contrast stretching and intensity transfer. Med. Biol. Eng. Comput. **58**(3), 483–496 (2020)
8. Gayathri, S., Jawhar, S.J.: Enhancement in the vision of branch retinal artery occluded images using boosted anisotropic diffusion filter – an ophthalmic assessment. IETE J. Res., pp. 1–9 (2020)
9. Hassanpour, H., Samadiani, N., Salehi, S.M.: Using morphological transforms to enhance the contrast of medical images. Egypt. J. Radiol. Nucl. Med. **46**(2), 481–489 (2015)
10. Hoover, A., Kouznetsova, V., Goldbaum, M.: Locating blood vessels in retinal images by piecewise threshold probing of a matched filter response. IEEE Trans. Med. Imaging **19**(3), 203–210 (2000)
11. Hore, A., Ziou, D.: Image quality metrics: PSNR vs. SSIM. In: 20th International Conference on Pattern Recognition. IEEE (2010)

12. Li, D., Zhang, L., Sun, C., Yin, T., Liu, C., Yang, J.: Robust retinal image enhancement via dual-tree complex wavelet transform and morphology-based method. IEEE Access **7**, 47303–47316 (2019)
13. Li, P., Yang, X., Yin, G., Guo, J.: Skeletal muscle fatigue state evaluation with ultrasound image entropy. Ultrasonic Imaging p. 016173462095268 (2020)
14. Liao, M., Qian Zhao, Y., Hong Wang, X., Shan Dai, P.: Retinal vessel enhancement based on multi-scale top-hat transformation and histogram fitting stretching. Opt. Laser Technol. **58**, 56–62 (2014)
15. Mukhopadhyay, S., Chanda, B.: A multiscale morphological approach to local contrast enhancement. Signal Process. **80**(4), 685–696 (2000)
16. Pineda, I.A.B., Caballero, R.D.M., Silva, J.J.C., Román, J.C.M., Noguera, J.L.V.: Quadri-histogram equalization using cutoff limits based on the size of each histogram with preservation of average brightness. Signal Image Video Process. **13**(5), 843–851 (2019)
17. Román, J.C.M., Escobar, R., Martínez, F., Noguera, J.L.V., Legal-Ayala, H., Pinto-Roa, D.P.: Medical image enhancement with brightness and detail preserving using multiscale top-hat transform by reconstruction. Electron Notes Theoret. Comput. Sci. **349**, 69–80 (2020)
18. Román, J.C.M., Noguera, J.L.V., Legal-Ayala, H., Pinto-Roa, D., Gomez-Guerrero, S., Torres, M.G.: Entropy and contrast enhancement of infrared thermal images using the multiscale top-hat transform. Entropy **21**(3), 244 (2019)
19. Singh, N., Kaur, L., Singh, K.: Histogram equalization techniques for enhancement of low radiance retinal images for early detection of diabetic retinopathy. Eng. Sci. Technol. Int. J. **22**(3), 736–745 (2019)
20. Singh, N., Bhandari, A.K.: Image contrast enhancement with brightness preservation using an optimal gamma and logarithmic approach. IET Image Process. **14**(4), 794–805 (2020)
21. Soille, P.: Erosion and dilation. In: Morphological Image Analysis, pp. 63–103. Springer Berlin Heidelberg (2004)
22. Soille, P.: Opening and closing. In: Morphological Image Analysis, pp. 105–137. Springer Berlin Heidelberg (2004)
23. Sonali, Sahu, S., Singh, A.K., Ghrera, S., Elhoseny, M.: An approach for de-noising and contrast enhancement of retinal fundus image using CLAHE. Opt. Laser Technol. **110**, 87–98 (2019)
24. Vijayalakshmi, D., Nath, M.K., Acharya, O.P.: A comprehensive survey on image contrast enhancement techniques in spatial domain. Sens. Imaging, **21**(1) (2020)
25. Wang, Z., Bovik, A., Sheikh, H., Simoncelli, E.: Image quality assessment: from error visibility to structural similarity. IEEE Trans. Image Process. **13**(4), 600–612 (2004)
26. Zuiderveld, K.: Contrast limited adaptive histogram equalization. Graphics gems, pp. 474–485 (1994)

Implementing a Web Based Open Source Tool for Digital Storytelling

Juan-Bernardo Tenesaca, Andres Heredia, and Gabriel Barros-Gavilanes(✉)

Universidad del Azuay, LIDI, Cuenca, Ecuador
{juan.tenesaca,andres.heredia,gbarrosg}@uazuay.edu.ec
http://www.uazuay.edu.ec

Abstract. Digital Storytelling is a technique used in education and inclusion. While most systems are not Open-sourced and generate video, our implementation offers a web-based service for generating two types of stories without any additional software than a web browser. This Open source implementation offers scene-based and audio-centered digital storytelling. This system offers scenes consisting of image, text, audio, or video. Additionally, after uploading an audio file, the system allows annotation of text and images. Then, stories are reachable from internet through a URL after agreement on publishing under the Creative Commons license. This module is developed in the context of project Smart Ecosystem for Learning and Inclusion with the main goal of providing support for storytelling workshops for inclusion.

Keywords: Digital storytelling · Web systems · Web-based platform · Open-source software · Multimedia systems

1 Introduction

Digital StoryTelling (DST) is a tool used in fields like education (e.g. folklore [2], English Second Language (ESL) [2]), Inclusion (e.g. feminism [2,6]), and Heritage [5]. From the point of view of teachers or instructors, in some cases DST can be used to assert some kind of competence, or just provide a voice to a specific person or group.

DST could be defined as the integration of audio, image, text, and in some cases video too [2] with the objective of allowing end-users or students to create a story about a topic selected by themselves. Under this description, any video edition or generation software is a viable solution for a DST activity.

However, access and manipulation barriers exist for users suffering from digital divide or digital exclusion. The access barrier comes from the cost of video editing software. Thus, manipulation or handling barrier comes from the lack of

Thanks to funding agency SENESCYT Ecuador, for the financing provided to research, through the ERANet-LAC projects, especially the ERANet17/ICT-0076 SELI project; Smart Ecosystem for Learning and Inclusion. PIC-2019-ERANET-UDA-001.

Á. Rocha et al. (Eds.): ICITS 2021, AISC 1330, pp. 187–196, 2021.
https://doi.org/10.1007/978-3-030-68285-9_19

experience from users never been required to generate a basic video software. For this reason, this work proposes a web-based open source implementation as an alternative to proprietary software and without installing any additional software than a web browser. Additionally, this open source software is meant to be possibly used as a complement tool in the context of DST workshops. Take as example the six steps workshop methodology proposed by Simsek in [6]. In that context, this system support step 5: "Putting the digital story together" and step 6: "Screening" of the story. Additional steps in these workshops are: step 1 forming the story circle, step 2 or scripting, step 3 or sound recording and editing, step 4 or image selection and editing.

This work aims to answer the research question: How feasible is the use of free software for creating digital stories through the use of scenes and timelines? and relates specifically to the software implementation stages of two systems: Scene based DST and Audio based DST. Both modules are integrated in the Learning platform of the Smart Ecosystem for Learning and Inclusion (SELI [7]) project: https://seli.uazuay.edu.ec/.

This work is organized as follows: In Sect. 2 we present a set of methods used in similar applications. Then, Sect. 4 provide information about the modules and sub-modules of both systems. Finally, Sect. 5 presents the conclusions and future works.

2 Related Works

Table 1. Software tools for Digital StoryTelling in literature.

Study	Tools					Topic
	Text	*Image*	*Audio*	*Consolidation*	Presentation	
Chatterjee [2] 2019	MS Word	MS Photo Story, Adobe Premiere Elements, MS PowerPoint	Digital voice recorder	Windows Movie Maker, Apple's iMovie	WeVideo	Folklore in college education
Jumail [3] 2010	Adobe Action Script 2.0				Adobe flash player 6.0	Primary children education
Cao [1] 2010	-	Java upload App	-	Java web application	Preview in browser	Film story creation support
Russo [5] 2013	ASPX web application				In browser	Museum interactive material

There exist many software able to produce high quality audio-visual material. These tools can be used to generate world quality stories in the context of DST. However, as expressed in Sect. 1, our design parameters include: i) web based tool, ii) open source policy, iii) shareable through internet.

The IEEExplore database has been queried using the following keyword to refine the search. First word is "storytelling" with 782 results. Then, word "digital" reduces the number of articles to 242. In this case, only web-based solutions are considered representative; including the refine search term "web" throws 19 articles. Some articles have been removed due to be considered off topic, this is mainly because are related to data storytelling or virtual exploration of heritage sites. Only four articles [1–3,5] from this search are considered relevant for comparison. Works have been classified accordingly to declared tool used for audio, image, and video. Additionally, information about the type of source code license is presented too.

Chatterjee in [2] enumerates existing software for the different task in generating a story: Microsoft Photo Story, Windows Movie Maker, Wevideo, Web 2.0 etc. Additionally, that work reports tools declared in other related studies, for example: Adobe Premiere Elements, MS MovieMaker, Photostory, and WeVideo which are available for free versions. Experiences of application of DST are provided, an interesting case is when DST was used with 23 engineering students for ESL, inclusion and feminism for college students. In this specific case, the following tools were used: Microsoft Word, PowerPoint, Microsoft's Photo Story 3, Apple's iMovie,

Jumail in [3] presents a web based DST system for primary level children through the use of generic flash cards. It is based on proprietary software like Adobe flash player and Action Script. Thus, the number of stories is determined by the number of cards. No further details are provided about concurrency and max number of users.

Authors in [1] present the design and implementation of "YouTell TE" system, intented for multi-user non-linear storytelling. Stories are stored as graphs using GraphML to store part of stories as nodes and edges. Each node registers only images through uploading the image file or using an URL. Then, users can connect nodes to provide desired story flow. Additionally, the system provides templates for fitting stories to specific patterns (e.g. the hero's journey). System uses Java technology in the server and JavaScript with Ajax interactions to enhance user experience in the client. For visualization, a browser-based preview screen presets images and allow navigation if many paths are available.

Homm platform [5] is a network of stories for non-linear and open multimedia narrative. Each node could be: documents, collection of images, clips of video. This implementation has three types of users, namely users, authors, and administrators. Every user can explore and upload multimedia material about specific points in the museum. Authors could create new story networks and organize them in activities. In turn, these activities could also be organized in personal space of users. Administrators deal with user and authors creation and maintenance. From the technical point of view, the web system uses MS ASPX. However, no information is provided about license of code and ways to access it.

Image Selection. Additionally, some works are only dedicated to specific parts of DST systems. Take as example, Liu [4], where a recommendation system for helping users finding images through click or query is presented. The implementation uses semantic technology to help user to find images on the web. Presentation images are labeled accordingly to criteria such as: relevance, semantic and visual diversity, and visual quality. Thus, authors use Conditional Random Field (CRF) based image suggestion. This functionality could be very beneficial for DST systems, however, it is out of the scope of this work.

Non Open Source Software. After the revision of literature, proprietary software like *Adobe spark* www.spark.adobe.com have important functionality by slide, and to handle image, text, audio, and video. Thus, it allows the use of templates and even the use of background music and equalizer.

Another proprietary web software is *weVideo*, because its free version allows only 5 min videos and up to 1 Gb storage with resolution of 480px. However, pay per access is required for accessing copyrighted databases. Users can record their voice or choose from free music clips. Thus, web cam can be used to record video and additional functionalities for effects, scale, and rotation of media are available. It offers a drag and drop interface, however record screen is not available in the free version.

3 Methodology

The methodology applied to the development of the digital stories tool is based on the agile methodology eXtreme Programming [8]. eXtreme Programming is the simplest and easiest software development methodology to implement. It is a methodology based on "continuous feedback between the client and the development team". This methodology was considered due to the fact that the suggestions of different teachers who used and tested the tool during the development process were taken into account in the development process.

Considering the opinion of the teachers, two digital story tools were created, one based on scenes and the other based on a timeline.

The digital history models consider that the user is in charge of uploading the multimedia content to the creation of the story, no tool is contemplated for automatic validation of content or copyright of the material used, which is the responsibility of the teacher.

3.1 Architecture

Story Editor Based on Timeline. The story editor design ensures a proper scalability and visualization of scenes. The editor allows to include stories with tons of scenes without losing the organization of the story. Figure 1a shows the flow of the story editor. This schema describes the process that the storyteller must follow to create a stage of the story. In a first instance, the mandatory audio content must be uploaded, this audio creates the base timeline on which one or more images can be distributed, in the same way several subtitles can be added in several languages. Once the content of a stage has been completed, the editing loop breaks and ends at the "Story materials" phase, and continues with the creation of the story which can contain more than one stage.

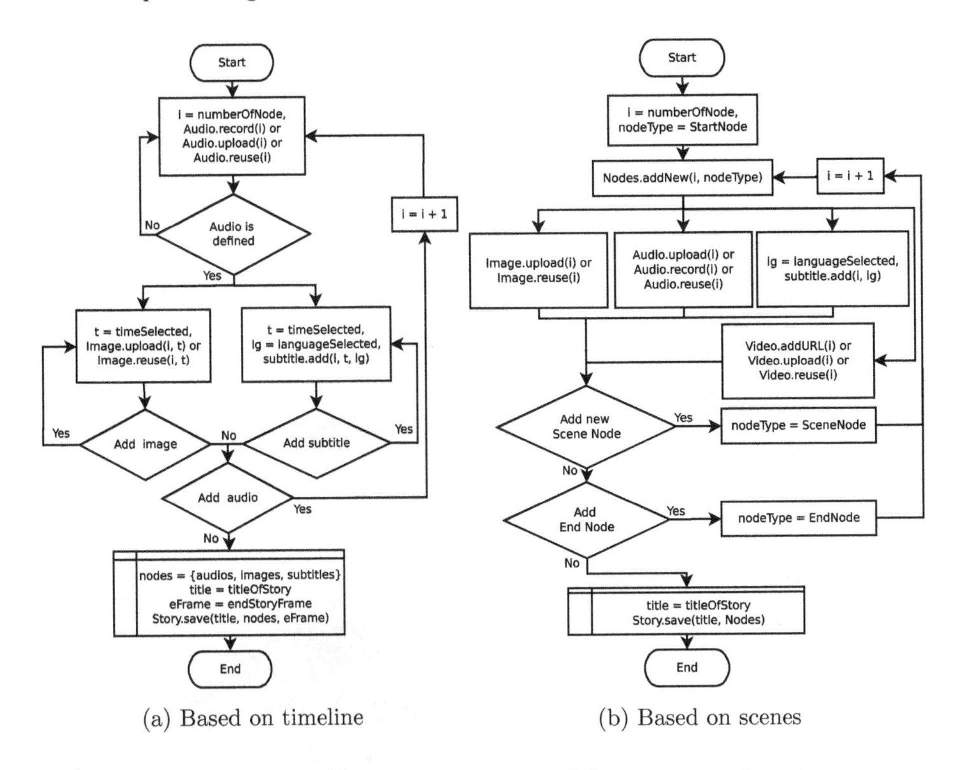

(a) Based on timeline (b) Based on scenes

Fig. 1. Flow diagrams for the creation of the two types of stories.

Story Editor Based on Scenes. The second structure showed in Fig. 1b is based on scenes or nodes, where each of them can contain only one image, audio, video and subtitle. This architecture is less complex than the previous one, creating a fast environment for users that want to create stories in a easier way.

3.2 Multimedia Edition

The story editor is the space to create the story structure. The timeline story is made up of: audio, images and subtitles (different languages) organized in a temporal axis through the audio. The system flow is shown in Fig. 1.

- Audio component: Where the storyteller can record, upload or reuse any audio from the platform. If the storyteller does not select the audio, he cannot build the story. However, if the storyteller selects one or more audio files, he will be able to add images and subtitles through the sound tracks. Figure 2a describes this step on the platform.

- Image step: In this step the application allows upload, change or reuse any image in the platform. Also, the storyteller can rotate the images to build a more fun story. This step is not mandatory and the storyteller decides whether to add or not the image. Figure 2b shows the functionality of this step in the platform.
- Subtitle component: This step allows the storyteller to add a subtitle to the story, it can be used to describe the image at the story stage, or as a descriptive alternative to audio content. This is an optional step in the story. Figure 3 shows this component.

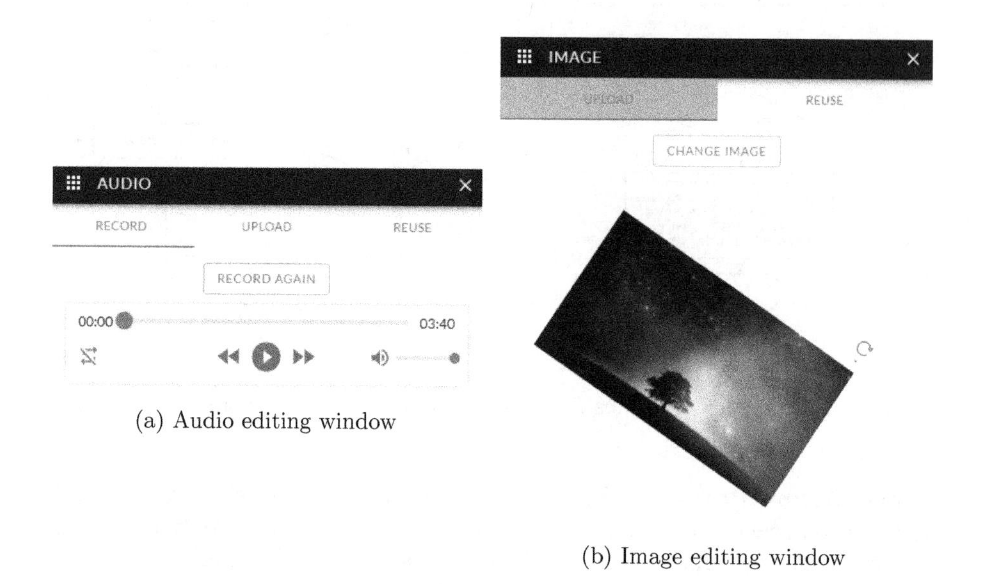

(a) Audio editing window

(b) Image editing window

Fig. 2. Windows for adding multimedia in storytelling.

Fig. 3. Window for adding subtitles in the storytelling tool.

4 Results

The stories should be saved in order to store the content in the database and in consequence would be shared to the public. The story can be saved specifying the end frame (will be showed at the end of the presentation of the story) and the title of the story. This is a mandatory step in the process to create the story. Figure 4 shows the edition window.

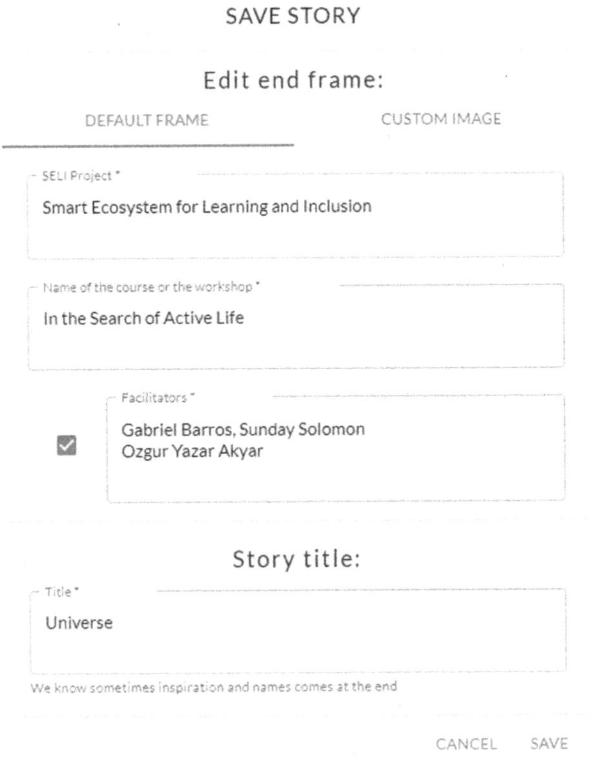

Fig. 4. Step to save the state in the storytelling tool.

Finally, digital story is put together and can be saved or published on a social network. Social behavior is important, because storytellers can share their memories, experiences, and feelings. Stories can be shared on Facebook, Twitter or LinkedIn, see Fig. 5.

Fig. 5. Step to publish the story in social networks.

The use of the story editor is intuitive and allows users represent the information within a timeline and give a narrative twist to the story. The complete dashboards of the two storytelling tools can be appreciated in Fig. 6 and Fig. 7.

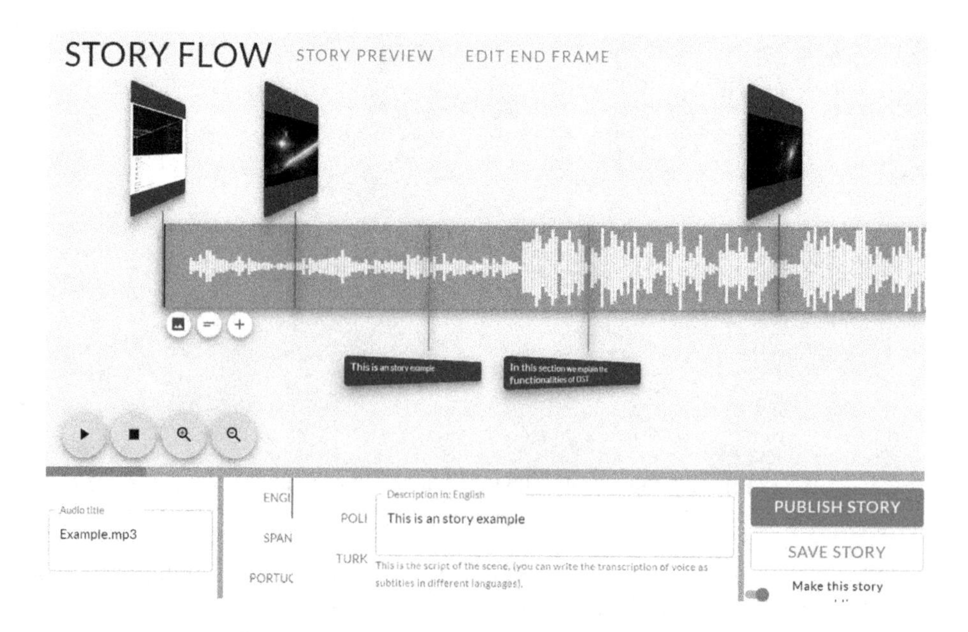

Fig. 6. Dashboard of the storytelling tool based on timeline.

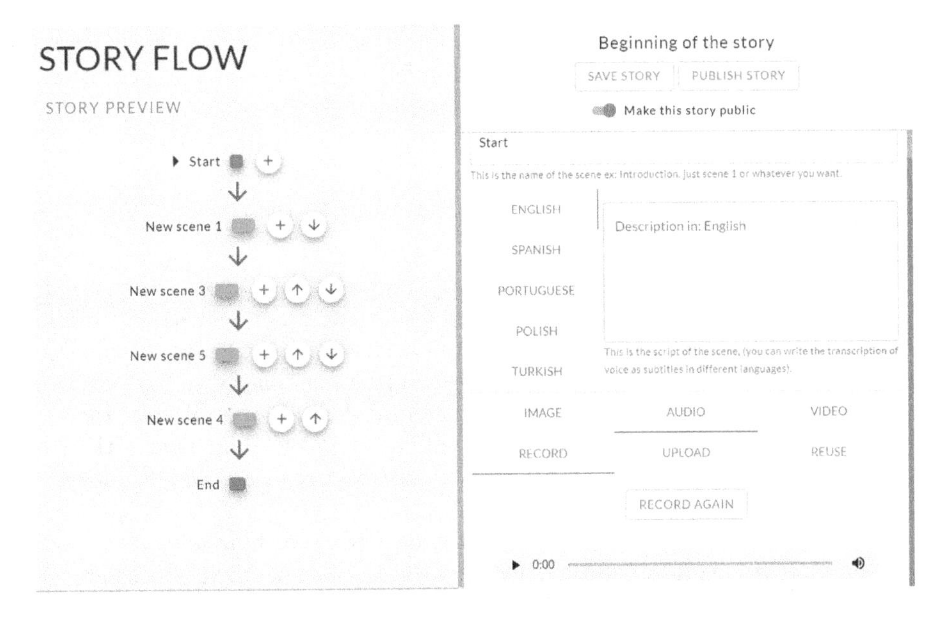

Fig. 7. Dashboard of the storytelling tool based on scenes.

4.1 Manual for SELI Access

To get access to the digital storytelling tools, it is necessary to create a user account in [7] in the SELI platform. If tutor account is selected, it will ask for

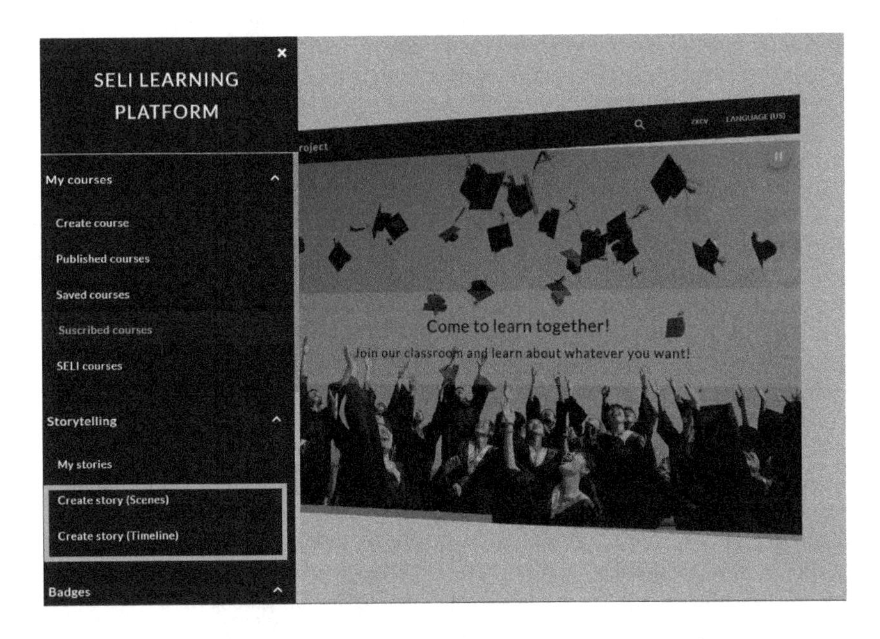

Fig. 8. Access for the storytelling tools in SELI Platform.

more information about the user, but a tutor account has more tools than a student account. Once into SELI, open the left menu in the left up corner and in Fig. 8 the blue box shows the buttons to open the two different storytelling tools.

5 Conclusions

This work presents the implementation of two web based open source system for Digital StoryTelling. To the best of our knowledge, no other open source solution is available for preparing a DST activities on the web. A Scene organization allows a bins-like structure, where each scene contains images, audio, text, and video. Another implementation is based mainly around an audio file, there, it is the user who decides the appearing timing of images and text with more freedom. However, inclusion of video is not possible because of considerations of video synchronization.

Future works include the generation of widget-like screening software (embedded scripltet) to avoid generation of video files, and less bandwidth consumption to visualize the story on the client side.

References

1. Cao, Y., Klamma, R., Hristov, M.: The hero's journey – a stortytelling template engine for ubiquitous multimedia. In: 3rd IEEE International Conference on Ubi-Media Computing, pp. 118–123 (2010). https://doi.org/10.1109/UMEDIA.2010.5543915
2. Chatterjee, P., Mishra, D., Padhi, L.K., Ojha, J., Al-Absi, A.A., Sain, M.: Digital story-telling: a methodology of web based learning of teaching of folklore studies. In: 21st ICACT, pp. 573–578 (2019). https://doi.org/10.23919/ICACT.2019.8702047
3. Wastam, J., Awang Rambli, D.R., Sulaiman, S.: A guided digital storytelling prototype system using illustrated flashcards. In: International Symposium on Information Technology. vol. 1, pp. 1–6 (2010). https://doi.org/10.1109/ITSIM.2010.5561377
4. Liu, Y., Mei, T., Chen, C.W.: Automatic suggestion of presentation image for storytelling. In: IEEE ICME, pp. 1–6 (2016). https://doi.org/10.1109/ICME.2016.7552974
5. Russo, M., Ghose, R., Mattioli, M.: Homm-sw: networks-of-stories to value tangible and intangible heritage in museums. In: Digital Heritage. vol. 2, pp. 463–463 (2013). https://doi.org/10.1109/DigitalHeritage.2013.6744813
6. Simsek, B., Erdener, B.: Digital visual skills education for digital inclusion of elder women in the community. In: Procedia - Social and Behavioral Sciences. 4th WCES, vol. 46, pp. 4107–4113 (2012). https://doi.org/10.1016/j.sbspro.2012.06.208
7. team, S.: Smart Ecosystem for Learning and Inclusion - SELI project. https://seli.uazuay.edu.ec. Accessed 20 May 2020
8. Zhou, Y.: Unix process, merging unified process and extreme programming to benefit software development practice. In: First International Workshop on Education Technology and Computer Science. vol. 3, pp. 699–702 (2009)

Software Systems, Architectures, Applications and Tools

Performance Comparison: Virtual Machines and Containers Running Artificial Intelligence Applications

Jack D. Marquez[1(✉)] and Mario Castillo[2]

[1] Universidad Autonoma de Occidente, Cali, Valle del Cauca 760030, Colombia
jdmarquez@uao.edu.co
[2] Servicio Nacional de Aprendizaje SENA, Cali, Valle del Cauca 760004, Colombia
mgcastillor@sena.edu.co

Abstract. With the continuous growth of data that can be valuable for companies and scientific research, cloud computing has shown itself as one of the emerging technologies that can help solve many of these applications that need the right level of computing and ubiquitous access to them. Cloud Computing has a base technology that is virtualization, which has evolved to provide users with features from which they can benefit. There are different types of virtualization and each of them has its own way of carrying out some processes and of managing computational resources. In this paper, we present the comparison of performance between virtual machines and containers, specifically between an instance of OpenStack and docker and singularity containers. The application used to measure performance is a real application of artificial intelligence. We present the obtained results and discuss them.

Keywords: Cloud computing · Virtualization · Virtual machines · Containers · Machine learning

1 Introduction

With the continuous development of society, technology and the recent advances in different scientific disciplines, enormous volumes of data or called Big Data is being generated. Related to [1] the total global data accumulated in the year 2020 is going to be 44 times more than the value in the year 2009, that is, 40ZB a real big number. They also describe this era like the Data Technology one. As a potential business, big data is forcing the different companies and IT leaders to obtain the value from all kind of data [2].

The use of Cloud Computing (CC) systems is a good choice to deal massive data. Analytic capabilities and efficient processing power are necessary to get value of the massive data. These data must be stored, processed and analyzed efficiently, resulting in a processing problem [3].

The use of CC is getting higher, because it is becoming a new paradigm that can offer flexibility and scalability for every resources requested by the person

Á. Rocha et al. (Eds.): ICITS 2021, AISC 1330, pp. 199–209, 2021.
https://doi.org/10.1007/978-3-030-68285-9_20

who is using the service. Users can ask for a lot of IT resources through CC like storage servers, networks, software, among others. They can connect to those resources from everywhere using internet and only pay what they are using thanks to CC load balancing service and payment model.

Containers are increasingly becoming more popular than VMs even latter have been predominant for managing a lot of applications. Containers like Docker or Singularity are arising as a good option because they theoretically offer low-overhead and better performance than Vms. In this work we present a performance comparison between Virtual Machines using OpenStack, Docker containers and Singularity containers. All of them running an Artificial Intelligence application that through a Convulotional Neural Network is able to identify different pars of the body like hands, face and the body itself.

The rest of the paper is organized as follows. In Sect. 2, we present a related work, including some previous works that have compared these technologies and we show our main difference with them. In Sect. 3, we describe the application we use to test both of technologies and show some background about Convolutional Neural Networks (CNN). In Sect. 4 we describe in detail all the infrastructure used to do this study, and the experiments we did, with their specifications and some other details to guarantee the reproducibility. In Sect. 5, we discuss the experiments performed and show the results obtained by them. Finally, we conclude in Sect. 6.

2 Related Work

Due to the great boom and all the use that has been given by Cloud Computing and all its components, some works have been dedicated to evaluating many of each of them. In this case, as Cloud Computing can be applied in many areas, it can be seen that the components have been evaluated to perform different operations just as in the works presented in [4–18]. It is in our interest to know the works in which previous comparisons have been made regarding virtual machines and containers, so we present the most relevant ones below. In order to provide efficient resource sharing and run concurrently on a virtualized infrastructure, the Big Data Platforms are offering NosQL distributed databases [13]. This is what Shirinbab et al. want to evaluate on the cloud computing systems. In [13], authors want to compare the performance of Apache Cassandra in both virtualization environment, Virtual Machines, and Containers. Besides, they also measured the performance of Cassandra in a non-virtualized environment. They used the Cassandra-stress tool as a benchmark, and they discovered that Docker containers accomplish the virtualization challenges because of its dependencies packing. Docker consumed fewer resources and had less overhead than the virtual machines, but they are less isolated than virtual machines; therefore, a bug in the kernel could affect the entire system.

In [14], authors recognize containers as an essential technology that is increasing its use in cloud systems, due to the light operations, efficiency, dependencies encapsulation, and resource sharing flexibility. They also know that there are some lacks and decide to test some spark applications on virtual machines and containers. Finally, they obtained that Docker containers are more efficient in deployment, boot stages, better scalability, and achieves higher CPU and memory utilization.

One of the most used frameworks to deal with Big Data in Cloud Computing is Hadoop due to its scalability and processing power. In [15], authors want to test this framework running over Virtual Machines and Docker containers. They validate both environments and run some tests with the Teragen benchmark, which put some workloads for the Hadoop Distributed File System (HDFS). After those tests, they discover that the performance of Hadoop on Containers outperforms the performance in Virtual Machines because the cluster gains some speed in containers due to its use of the same kernel.

Containers have shown an excellent solution for most of the previous works. The significant difference with all the mentioned works to our study is that we are going to test a real-word machine learning application using Virtual Machines, Docker containers, and Singularity containers. The application will be explained in the next section.

3 Machine Learning Application

Artificial Intelligence (AI) and its subfield Machine Learning are being widely used, and the technologies to support these applications are improving too. Generally, this kind of applications are data, memory, network, GPU, or CPU-intensive, due to the number of operations they have to do in a short period. Hardware and language programming models are evolving to reduce the execution time of these applications.

To test the performance of these platforms (Docker, singularity, and OpenStack virtual machine instance), we use a Machine Learning application that is Disk and CPU-intensive. We decide to use this application as a benchmark for the platforms because research literature showed that it had not been done before. This test, with this application help us to show one of the best platforms to run further machine learning applications.

The application source code can be found in https://github.com/mariocr73/ArtificialIntelligence_python.git. This application is written in Python and the source file can be run in Jupyter Notebook.

After train the application with more than 1300 images of the body parts we are going to recognize, we provide another image and the application output should be something like this:

face: 0.002709% - Body: 0.0014576% - Hand: 99.995768% - Nothing: 6.204666%

In this case, the identification of one hand by the application was successful.

4 Experiments

Two different architectures are tested in this paper, Virtual Machines and Containers. Each one of them has its advantages and disadvantages because of their configurations and the way to execute every task.

In order to run the previously section, we did some configuration and we use some physical resources to test the three environments. Each environment has the same resources looking for guarantee the homogeneity and do not give any advantage or disadvantage to any of them. For the container tests we use a Dell Workstation with 16 GB of RAM, four Intel Xeon @ 3.4 GHz processors, one 240 GB Solid State Drive (SSD) and a Red Hat 4.8.5-39 linux version with the kernel 3.10.0-1062. For the virtual machine we create an OpenStack instance with the same resources than the machine where we test the containers.

For each one of the platforms we did ten tests, and each test with 100 iterations.

4.1 Virtual Machine - OpenStack Instance

As mentioned by Pepple in [19], OpenStack *"was created with the audacious goal of being the ubiquitous software choice for building public and private cloud infrastructures"*. This platform is one of the most used for open source cloud computing due to its architecture, characteristics, and it supports most virtualization technologies of the market: XEN, KVM, LXC, QEMU [20]. For that reason, we selected it to create a virtual machine instance and test the performance of the machine learning application. Being one of the most used technologies, people can discover how it is going to be the performance compared to containers' performance.

4.2 Containers

Containers are continuously getting more popularity in HPC and Cloud Computing platforms due to their countless benefits, so we decided to test the performance of two of the most used linux containers currently We use two types of containers, Docker [21] and singularity [22]. Each one of them have some main characteristics and that is the reason we also decided to test the performance of both.

The Docker containers used were 1.13.1 version with python 3.6.8, jupyter core 4.5.0 and tensorflow 1.14.0.

The singularity containers used were 3.4.1-1 version with python 3.7.4, jupyter core 4.5.0 and tensorflow 1.14.0.

The Dockerfile and Singularity Receipe for reproducibility of this experiment can be found in https://github.com/mariocr73/ArtificialIntelligence_python.git.

4.3 Extra Configurations

In order to make the experiments of this study, we had to take into account the infrastructure mentioned before. Therefore we configured an entire environment for containers and virtual machines.

During the configuration of the experiments, an important point was to ensure that the machine which was running every experiment would not be executing another task that could affect or generate some noise to the performance of the machine learning application running on it. Also, in this stage we did some preliminary test looking for guarantee all the compatibility of the platforms with the software that supported the application. Throughout this preliminary test we had to downgrade the tensorflow version used in singularity, because it had tensorflow 2.0 and some of the functions were deprecated, consequently we fix this installing the 1.14 version.

5 Results

We tested the machine learning application in each platform (Docker, Singularity, and OpenStack instance), and then we compared the performance of each one of them. For that end, we performed ten tests, as we mentioned before, to detect variations and possible factors that could be affecting this performance. We did each of those tests strictly following the previously mentioned constraints, specifications, and points.

Figure 1 shows the values obtained for the machine learning application in each platform for all the tests done. In general, we observe very similar behavior for each of the variables that occur in each of the technologies tested. In each of the ten tests that were done, all the platforms (Containers and Virtual Machines) maintain equal application values without affecting the outputs of its calculations. However, it can be observed that experiment number 8 of Docker containers presents an atypical behavior to others. We note that for the error rate, it shows the highest peak of the 30 experiments that are offered. Consequently, all the accuracies of that same experiment are affected, presenting, in turn, the lowest values that can be found of all the tests performed. We did not have a record of why it could have generated this, but it may have been due to the type of image provided to the application, perhaps it was not very clear, or the object to be identified was not well located.

For the Error Rate (See Fig. 1a), if experiment number eight of Docker would not have been an atypical value, this would have been the most constant because almost do not have a variation during the ten tests. Singularity had the most lower rate for Error but also had the second and third highest values, and OpenStack had a real variety, getting the lowest rate in the first, third, and ninth experiment.

One of the first steps, when we are using a CNN, is training the network to allow recognizing some patterns than later are going to identify the image we provide to it. In Fig. 1b, we can observe that, again, the unique value which is different than the others is experiment number eight from Docker containers.

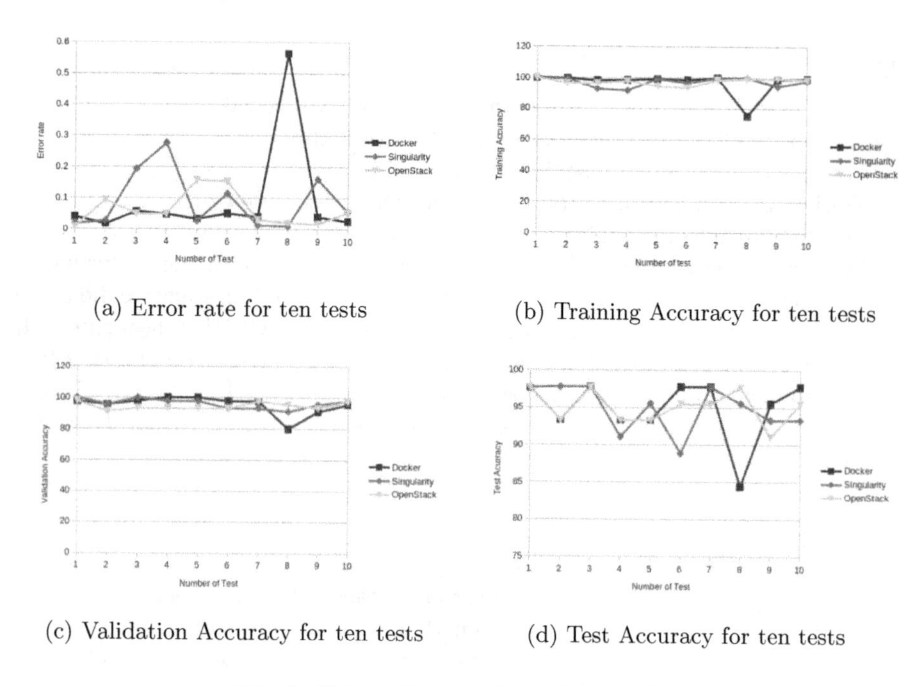

(a) Error rate for ten tests

(b) Training Accuracy for ten tests

(c) Validation Accuracy for ten tests

(d) Test Accuracy for ten tests

Fig. 1. Obtained results for each technology

After training the network, the next step is to validate it. Validation Accuracy values (See Fig. 1c) for Docker, Singularity, and OpenStack were pretty close one each other. Still, at some moments, one technology was better than the other two, and then this was the worst. Docker container almost always had one of the best values, if it was not the best, it was the second-best, except for the experiment number eight, that how we already know is an atypical one. OpenStack, in most of the cases, had the worst validation accuracy value.

When the neural network has been trained and validated, it is the moment to test it with different images and get the final results (See Fig. 1d). For this variable, unlike previous ones, there is variation in values obtained during the execution of ten experiments for each of the technologies tested. Docker's experiment number eight still has the worst value; however, Docker and Singularity had the same accuracy in the first five experiments. OpenStack was not constant in its test accuracy values; it even had the second and third-worst accuracy values in its experiment numbers six and four, respectively.

We could see that, in most cases, the values obtained by the application were not influenced by the technology that was supporting the process. Docker, Singularity, and OpenStack did not affect the outputs that get the application because these are generated by models created by the application itself using the inputs for the training and the learning process it does with it.

Containers and virtual machines do affect the performance of the application or its execution time (see Fig. 2), due to the processes that each one of them has to do to guarantee the execution and the way they provide the computational resources to the applications.

Figure 2, shows that the worst performance was from OpenStack, due to the overhead that creates the hypervisor in the virtual machines. This hypervisor is the one in charge of take the physical resources and offer them virtual resources for every hosted virtual machine. This process takes too long, and it is no very efficient compared to bare metal performance or even with the container's performance [23]. Table 1, 2, and 3, shows the best results for Singularity, Docker and OpenStack instance, respectively. As we mentioned before and as can be seen in Table 3 and 2, OpenStack had the worst performance with an execution time of 32 min and 12 s. Comparing containers, we observe that Docker was better than Singularity, finishing in 16 min and 36 s, while Singularity finished in 16 min and 40 s. This might be a small difference, but if we are running an application that has to work in real-time, those four seconds are going to be a great amount of time. Nevertheless, Singularity in six of the ten experiments was better than Docker. This is possible because Singularity was designed to improve some processes like resource management and the way it uses them, giving a little advantage over Docker containers. The main difference between this two container technologies is that Singularity transfer the memory contents to user namespaces while Docker control its resources using cgroup namespaces, what brings an overhead for this kind of applications.

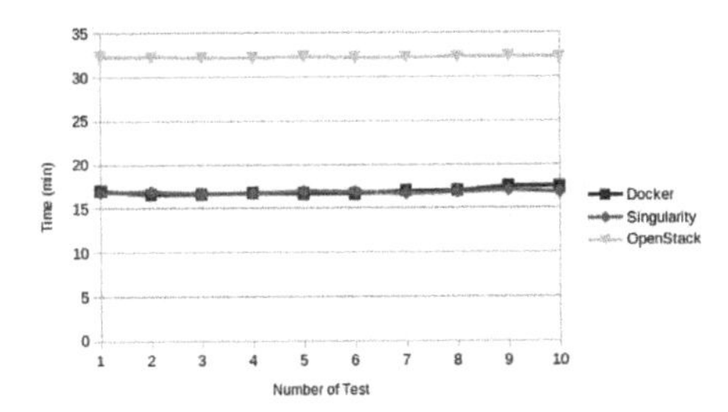

Fig. 2. Training Time

Even Docker had the best performance for one of its experiments, as we mentioned before, Singularity was better in more number of tests. In Table 1, we see that the accuracy values for Singularity outperform the values obtained from OpenStack instance and Docker containers, no matter that had a higher Final Error.

Table 1. Results for Singularity containers

Results	
# Iterations	100
Final rate	0.00066
Final error	0.19381
Train accuracy	92.7%
Validation accuracy	100%
Test accuracy	97.7%
Time	16'40"

Table 2. Results for Docker containers

Results	
# Iterations	100
Final rate	0.00066
Final error	0.01744
Train accuracy	99.6%
Validation accuracy	95.5%
Test accuracy	93.3%
Time	16'36"

Table 3. Results for OpenStack instance

Results	
# Iterations	100
Final rate	0.00066
Final error	0.04975
Train accuracy	97.9%
Validation accuracy	93.3%
Test accuracy	93.3%
Time	32'12"

In Fig. 3, we show for the best Docker result, how CNN was improving in each iteration, starting from a high Mean Square Error (MSQ) and a low percentage of accuracy, to a low level of MSQ and a High level of accuracy. This figure shows

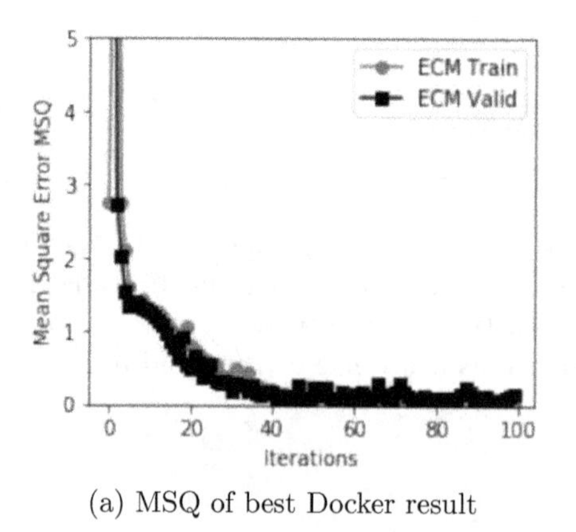

(a) MSQ of best Docker result

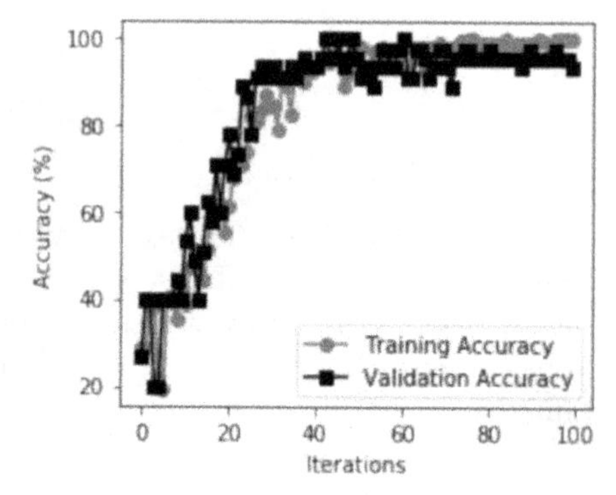

(b) Accuracy of best Docker result

Fig. 3. MSQ and Accuracy for best Docker result

the behavior for the training and validation processes. During iterations, we can see that some values are high, and then there are some low; that is because CNN is always learning and trying to be more precise.

Although OpenStack instance had the worst performance, we would like to test this kind of application to another virtualization types such as Xen or XenServer, to determine if any of them or KVM has more overhead for machine learning applications.

6 Conclusions

In this paper, we have shown a performance comparison of two virtualization types, Virtual Machines, and Containers, precisely an OpenStack instance for Virtual Machine and Docker and Singularity for containers. According to our results, Docker had the best performance with the lowest execution time in all experiments done to all the technologies. Nevertheless, Singularity was better than Docker and OpenStack in six of ten times, due to Docker has an overhead compared to Singularity, because of the use of cgroups namespaces and because Singularity makes more efficient use of the libraries that need the application.

On the other hand, the OpenStack instance had the worst performance of the three technologies in every test. This is because of the overhead caused by the hypervisor, which allows having the virtual resources for each virtual machine. The time used by the hypervisor to put the physical resources of our machines to put them as virtual makes this process less efficient than other virtualization types that do not need this layer in their architecture.

Machine Learning applications find another technology where they can be executed, obtaining good results. Each platform got good results for the outputs of the form without considering the execution time. As future work, we want to try these machine learning applications using GPU and the access that can give any virtualization to this resource.

Acknowledgment. We want to show our gratitude to Oscar Eduardo Castillo for his help during some experiments, and we are also immensely grateful to Armando Uribe Churta, Instructor of SENA, who was the person that provided us the application to make the comparison between these three platforms.

References

1. Zhou, K., Fu, C., Yang, S.: Big data driven smart energy management: from big data to big insights. Renew. Sustain. Energy Rev. **56**, 215–225 (2016)
2. Li, H., Li, H., Wen, Z., Mo, J., Wu, J.: Distributed heterogeneous storage based on data value. In: 2017 IEEE 2nd Information Technology, Networking, Electronic and Automation Control Conference (ITNEC), pp. 264–271 (2017)
3. Bezerra, A., Hernandez, P., Espinosa, A., Moure, J.C.: Job scheduling in hadoop with shared input policy and RAMDISK, pp. 355–363 (2014)

4. Bokhari, M.U., Makki, Q., Tamandani, Y.K.: A survey on cloud computing. In: Big Data Analytics. Springer, pp. 149–164 (2018)
5. Abdelfattah, A.S., Abdelkader, T., EI-Horbaty, E.S.M.: Rsam: an enhanced architecture for achieving web services reliability in mobile cloud computing. J. King Saud Univ. Comput. Inf. Sci. **30**(2), 164–174 (2018)
6. Balmakhtar, M., Persson, C.J., Rajagopal, A.: Secure cloud computing framework. US Patent App. 10/243,959 (2019)
7. Akherfi, K., Gerndt, M., Harroud, H.: Mobile cloud computing for computation offloading: issues and challenges. Appl. Comput. Inform. **14**(1), 1–16 (2018)
8. Lehrig, S., Sanders, R., Brataas, G., Cecowski, M., Ivanšek, S., Polutnik, J.: Cloudstore-towards scalability, elasticity, and efficiency benchmarking and analysis in cloud computing. Fut. Gener. Comput. Syst. **78**, 115–126 (2018)
9. Marvasti, M.A., Harutyunyan, A.N., Grigoryan, N.M., Poghosyan, A.: Methods and systems to manage big data in cloud-computing infrastructures. US Patent 9,948,528 (2018)
10. Arango, C., Dernat, R., Sanabria, J.: Performance evaluation of container-based virtualization for high performance computing environments. arXiv preprint arXiv:1709.10140 (2017)
11. Seo, K.T., Hwang, H.S., Moon, I.Y., Kwon, O.Y., Kim, B.J.: Performance comparison analysis of linux container and virtual machine for building cloud. Adv. Sci.Technol. Lett. **66**(105–111), 2 (2014)
12. Sharma, P., Chaufournier, L., Shenoy, P., Tay, Y.: Containers and virtual machines at scale: a comparative study. In: Proceedings of the 17th International Middleware Conference, p. 1. ACM (2016)
13. Shirinbab, S., Lundberg, L., Casalicchio, E.: Performance evaluation of container and virtual machine running cassandra workload. In: 3rd International Conference of Cloud Computing Technologies and Applications (CloudTech), **2017**, 1–8. IEEE (2017)
14. Zhang, Q., Liu, L., Pu, C., Dou, Q., Wu, L., Zhou, W.: A comparative study of containers and virtual machines in big data environment. In: IEEE 11th International Conference on Cloud Computing (CLOUD), pp. 178–185. IEEE (2018)
15. Singh, A., Gouthaman, P., Bagla, S., Dey, A.: Comparative study of hadoop over containers and hadoop over virutal machine. Interantial J. Appl. Eng. Res. **13**, 4373–4378 (2018)
16. Auliya, Y., Nurdinsyah, Y., Wulandari, D.: Performance comparison of docker and lxd with apachebench. In: Journal of Physics: Conference Series. vol. 1211, p. 012042, IOP Publishing (2019)
17. Gillani, K., Lee, J.H.: Comparison of linux virtual machines and containers for a service migration in 5g multi-access edge computing. ICT Express (2019)
18. Poojara, S.R., Ghule, V.B., Birje, M.N., Dharwadkar, N.V.: Performance analysis of linux container and hypervisor for application deployment on clouds. In: International Conference on Computational Techniques, Electronics and Mechanical Systems (CTEMS), pp. 24–29 (2018)
19. Pepple, K.: Deploying openstack. "O'Reilly Media, Inc." (2011)
20. Endo, P.T., Gonçalves, G.E., Kelner, J., Sadok, D.: A survey on open-source cloud computing solutions. In: Brazilian Symposium on Computer Networks and Distributed Systems. vol. 71 (2010)
21. Merkel, D.: Docker: lightweight linux containers for consistent development and deployment. Linux J. **2014**(239), 2 (2014)

22. Kurtzer, G.M., Sochat, V., Bauer, M.W.: Singularity: scientific containers for mobility of compute. PloS one **12**(5), e0177459 (2017)
23. Li, Z., Kihl, M., Lu, Q., Andersson, J.A.: Performance overhead comparison between hypervisor and container based virtualization. In: IEEE 31st International Conference on Advanced Information Networking and Applications (AINA), pp. 955–962. IEEE (2017)

EasyBio: A Bioinformatics Web Platform to Analyze Families of Genes

Federico Agostini[1(✉)], Pilar Hernandez[2], and Sergio Gálvez[3]

[1] Facultad de Ciencias Exactas y Naturales y Agrimensura, Universidad Nacional del Nordeste, 3400 Corrientes, Argentina
fagostini@conicet.gov.ar

[2] Instituto de Agricultura Sostenible (IAS), Consejo Superior de Investigaciones Científicas (CSIC), 14004 Córdoba, Spain
phernandez@ias.csic.es

[3] Departamento de Lenguajes y Ciencias de la Computación, ETSI Informática, Campus de Teatinos, Universidad de Málaga, 29071 Málaga, Spain
galvez@uma.es

Abstract. Next Generation Sequencing (NGS) techniques have facilitated genome sequencing in a cost-effective way and have impacted the work of geneticists and bioinformaticians. In particular, the study of gene families and transcription factors in plants is among the fields affected by this revolution, and remains a task involving many manual steps and the use of differently located computing resources. From a Bioinformatics point of view, the first steps to take in these studies are well-defined and imply the management of a complex set of computer tools difficult to install and setup. We introduce EasyBio, a web platform whose aim is to strengthen the field of Bioinformatics providing user-friendly tools to biologists and allowing them to focus on the biological meanings of the results instead of applying efforts to the computing technical aspects. EasyBio facilitates the user to interact with one of the main databases of plants, EnsemblPlants, to retrieve gene data and to incorporate it into an easy to use pipeline of tools. At the same time, EasyBio is versatile enough to allow a wide range of customization and output possibilities, displaying results in real time and interactively. The user does not need to learn about complex procedures, purely technical, or to deal with involute command line consoles. A clear web interface, a good user experience (with high quality results) and a good design of the backend are at the heart of this web platform. EasyBio is also free-to-use.

Keywords: User experience · Web services · Tools integration

1 Introduction

Nowadays, the work of a biologist is not only limited to tangible resources, like flasks and chemical products, but also includes the use of a wide range of computer programs that must be correctly managed. In particular, geneticists use many computer tools in tasks ranging from discovering the genes expressed under particular abiotic conditions (drought, salinity levels, etc.) to functionally annotate proteins using large open databases. High-Throughput Next Generation Sequencing methods [1] have drastically

Á. Rocha et al. (Eds.): ICITS 2021, AISC 1330, pp. 210–219, 2021.
https://doi.org/10.1007/978-3-030-68285-9_21

increased the number of bases obtained per sequencing run while, at the same time, decreasing the costs involved. This has led to a flood of data that need to be processed by new computer programs as efficiently as possible.

In this work, we address Gene-Wide analyses in plants, i.e. the study of a particular family of genes or transcription factors. These studies are recently based on pipelines that use computer tools for the analysis, characterization and visualization of genes. This is the case of the analysis of aquaporins in Malus *domestica* [2], the GRAS gene family in barley [3] or the Squamosa Promoter Binding Protein-like (SBP) transcription factors in wheat [4]. The results of the pipeline may include phylogenetic trees, structure of genes, conserved motives, cis-regulatory elements in flanking regions of genes and orthology relations between a species and its ancestors.

Creating a pipeline requires many pieces of different technologies that must be installed and properly configured. This is a time consuming effort that, in many cases, involves a research group to carry only a single analysis. In addition, the parameterization of the programs requires deep knowledge of their behaviors and capabilities. This problem may be addressed by using web platforms that allow biologists executing programs already configured in a server by simply uploading input data. Many web applications have been developed following this approach. In [5] is presented PathwayExplorer, a platform to map the profiles of genes' expressions against the pathways of type regulatory, metabolic and cellular available at KEGG [6], BioCarta [7] and GenMAPP [8]. In [9] is presented a bioinformatic platform that gathers the information currently available on Elaeis *guineensis* (oil palm) and related organisms like those involved in diseases or pathologies. This gives support to researchers interested in the systematic management of huge amounts of data typically used in pathology, physiology, biochemistry, breeding, etc. Other platforms focus on the usage of HPC hardware hard to obtain and makes it available through the web; this is the case of like MC64-Cluster [10], that allows the users executing long pairwise alignments and other time-consuming operations. More recently, the Curio platform (www.curiogenomics.com) includes options to compare genetic data, perform analytic adjustments, apply filters, and so on, obtaining intermediate diagrams in real time that can be fine-tuned until reaching the required result. Finally, Galaxy [11] is an open web platform to design workflows for data intensive bioinformatics analysis by means of a wide range of tools ready to use.

In this work we present EasyBio, a web platform with tools incorporated to manage sets of genes to accomplish the initial analysis of families of genes and transcription factors, displaying results in real time and interactively. Also, it integrates the access to public databases with genetic information of plants so standard data can be managed in a seamless way. The aim is to strengthen the field of Bioinformatics providing user-friendly tools to biologists and allowing them to focus on the biological meanings of the results instead of applying efforts to the computing technical aspects.

1.1 Operations Incorporated in the Platform

The main capabilities of EasyBio can be summarized as follows:

- **Data selection**. A user may upload his/her own data (genes) or import it from a BioMart [12] database (plants.ensembl.org) in a transparent way by means of a genes cart. This approach is similar to that used in other web sites like Phytozome [13] (phy-tozome.jgi.doe.gov).
- **User Experience**. The prime aim of EasyBio is to give to the user an easy way to interact with complex bioinformatics tools. This implies the utilization of clear web pages to parameterize their execution, comparison of genes through different types of diagrams, etc. Not all the capabilities of these tools are available in EasyBio but the most used ones have been included.
- **Open Database Access**. The Ensembl Plants datasets [12, 14] is the main source of genomic information for EasyBio. It contains data from more than 90 species of plants, including important crops for human feeding like wheat, barley, corn and rice. Gene names, DNA sequences, gene structure, functional annotation and protein translation can be retrieved seamlessly by the user. In addition, EasyBio connects two Cis databases, PlantCARE [15] and New PLACE [16] with no effort.
- **Usage of bioinformatic tools**. Data selected by the user can be used as the input for several tools. Selected tools have been chosen to describe the preliminaries of a set of genes functionally related: genes family, transcription factor, etc. The next are included in the current version of EasyBio (they appear in the usual order in which they are used):

a. **Multiple alignment**. ClustalW [17,18] is the tool used to group DNA or proteins sequences calculating the best match among them using the NJ (Neighbor-Joining) or UPGMA methods (Unweighted Pair Group Method with Arithmetic Mean). The result can be drawn as a phylogenetic tree and improved with confidence values through bootstrapping.
b. **Ortholog/paralog diagrams**. Circos [19] is a program written in Perl that allows representing several chromosomes (from the same or different species) in a circular way and to connect specific positions (gene locations) between them. This is useful to draw orthologs (homologous genes found in different species related by linear descent) and paralogs (genes in the same species that derive from the same ancestral gene).
c. **Genes and Cis diagrams**. AMCharts is a JavaScript library to create interactive charts inside a web page. This tools is fed up with data coming from the gene cart and configured through the user interface to show the detailed structure of genes and the cis-regulatory elements in their upstream (the non-coding region of the chromosome just before the gene).

2 Materials and Methods

EasyBio has a robust architecture based on internal web services hidden to the user. Currently, it is executed on an HP Proliant 585 G6 with 24 AMD cores and 256 GiB of RAM located at the Department of Languages and Computer Sciences (LCS) of the

University of Málaga, Spain. This server has installed, among others, the next tools and packages:

- R framework v3.6.0.
- Circos v0.69–9.
- amCharts v3 and v4.
- msa R package v1.18.0 (ClustalW).
- ape R package v5.3 (Phylogenetic tree).
- phangorn R package v2.5.5 (Bootstrapping).

2.1 Programming with Java, PHP, Laravel and Oracle

The data managed by the user in the EasyBio web platform are usually imported from the datasets of Ensembl Plants and it is included in the so called User Cart. In turn, the User Cart is used as input to the tools included into the platform. The results given by these tools, once parametrized and executed, are stored temporarily into session objects private to each user. The languages, frameworks and databases used in the development of EasyBio are mainly:

- **Java**: The internal execution is fully controlled by programs developed in this popular object oriented programming language. SOAP web services have been deployed using NetBeans 11.3.
- **PHP**: This language has been used to create DHTML content. This includes JavaScript routines to improve the user experience.
- **Laravel**: It is a web application framework for PHP. Its expressive and elegant syntax has allowed to increase the productivity in the implementation of EasyBio.
- **Oracle**: EasyBio is supported by an Oracle database instance whose stability and reliability ensures the smooth operation of the platform.

2.2 Front End and Back End

A web application is usually divided into front end and back end. The front end interacts with the user and it is also known as the client side. The back end includes all the technologies executed in the server and, therefore, is known as the server side. In EasyBio, the connection between these two sides is carried out by web services, an easy and standard way of communication, as can be shown in Fig. 1.

Fig. 1. Languages and Tools used in the development of the platform.

At the same time, the use of web services allow a decentralized design, easy extensibility, good maintainability and standardization.

Front End. It covers everything the user sees on the screen, including fonts, colours, responsiveness and visual effects; it also covers what the user can do on the screen: pressing keys, mouse clicks and movements, scrolls, etc. This set of actions and operations is known as the User Experience (UX). EasyBio uses extensively HTML5, CSS3 and JavaScript to provide an easy, engaging and functional user experience.

Back End. The web pages are provided by an Apache server that, in turn, communicates with a Tomcat server through SOAP web services. Due to the restrictions of a Tomcat server to execute commands and applications already installed in the operating system, Tomcat acts as an intermediary against a proprietary server developed in Java that actually communicates with the tools. R applications (multiple alignment, generation of phylogenetic trees, accesses to Ensembl Plants through BioMart, etc.) are executed using REngine (github.com/s-u/REngine), a general Java interface to R, with JRI (Java/R Interface, www.rforge.net/JRI). Perl applications like Circos are executed by launching a custom script. The rest of applications are programmed in Java inside the server; this is the case of Cis search that, in the case of PlantCARE, requires automatic access to the HTML form located at bioinformatics.psb.ugent.be, fulfill of its fields and retrieve the results via e-mail. Figures generated by these programs are returned directly to the front-end that is in charge to show them to the user. In other cases, like the genes figure (including Cis or not), the backend server returns the tabulated data required to build up the graphics through AMCharts, a JavaScript API to create interactive diagrams.

3 Results and Discussion

The main goal of EasyBio is to give support to the biologists by means of a clear and intuitive user interface to create graphics and charts in the analysis of sets of genes. The rationale is to avoid tiresome training or reading vast amounts of documentation. EasyBio provides visual tools that allow the researchers managing data in a smooth and efficient way. As figures are generated in real time, users can afford to do many changes in the parameterization of the tools: comparisons, filtering, adjustments, etc. The main features are summed up as follows:

- **Usability**. Straightforward usable for non technical users, including biologists.
- **Interactive**. DHTML (Dynamic HTML) are shown, and can be manipulated, inside EasyBio.
- **Responsive**. EasyBio adapts itself automatically to any device: laptop, tablet, smart phones, etc.
- **Incremental**. Obtaining intermediary results in real time allows the user to carry out many experiments.
- **Replicable**. Parameterization of tools is recorded in JSON files to be replicated at any time.

In the next section are described the main features of EasyBio, including the integrated tools.

3.1 User Cart

In order to interact with any tool, formerly the user must select the set of genes to work with. To do this, genes can be imported from Ensembl Plants (preferred method) or uploaded in a text file. A query against Ensembl Plants is composed of a taxa (species), a filter and the terms to search for. Once obtained a resulting set of genes, the user may select only those to include in a cart that, in turn, will be used throughout the pipeline of tools; it is the User Cart of genes, as can be shown in Fig. 2.

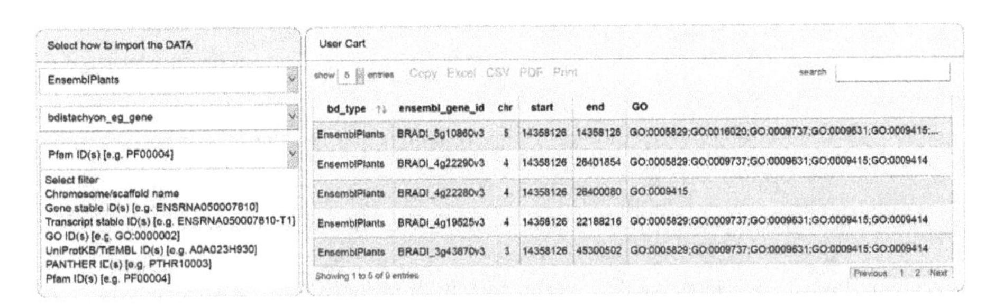

Fig. 2. Main screen to import user data.

3.2 Multiple Alignment: ClustalW

This tool allows generating not only a multiple alignment among the genes in the User Cart, but also obtaining a fully customizable tree: sequence type, gap penal-ties, etc. The result is displayed inside EasyBio and can be downloaded in several formats:.-fasta,.pdf,.tree and, even,.json, a JSON file to replicate the same con-figuration in another experiment. The phylogenetic tree can be aesthetically customized: show or hide labels, colours, sizes, and many other options. In addition, bootstrapping can be used to label branches with a confidence value, as can be shown in Fig. 3.

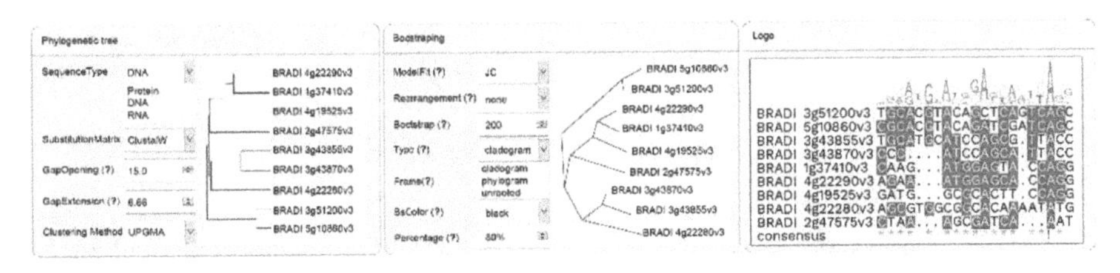

Fig. 3. Parameters to generate the phylogenetic tree, bootstrapping and logo.

3.3 Circos

A Circos diagram is used to show the orthology among a set of genes in a species when compared to some others species. The first step is to select the source genes and the target species as shown in Fig. 4.

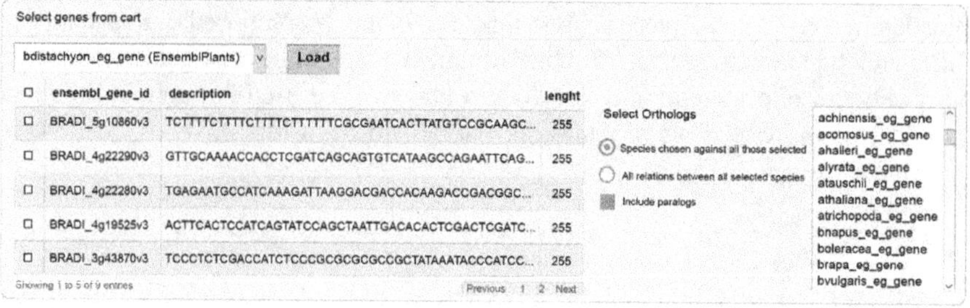

Select genes from cart

bdistachyon_eg_gene (EnsemblPlants) Load

	ensembl_gene_id	description	lenght
☐	BRADI_5g10860v3	TCTTTTCTTTTCTTTTCTTTTTTCGCGAATCACTTATGTCCGCAAGC...	255
☐	BRADI_4g22290v3	GTTGCAAAACCACCTCGATCAGCAGTGTCATAAGCCAGAATTCAG...	255
☐	BRADI_4g22280v3	TGAGAATGCCATCAAAGATTAAGGACGACCACAAGACCGACGGC...	255
☐	BRADI_4g19525v3	ACTTCACTCCATCAGTATCCAGCTAATTGACACACTCGACTCGATC...	255
☐	BRADI_3g43870v3	TCCCTCTCGACCATCTCCCGCGCGCGCCGCTATAAATACCCATCC...	255

Showing 1 to 5 of 9 entries Previous 1 2 Next

Select Orthologs

- Species chosen against all those selected
- All relations between all selected species
- Include paralogs

achinensis_eg_gene
acomosus_eg_gene
ahalleri_eg_gene
alyrata_eg_gene
atauschii_eg_gene
athaliana_eg_gene
atrichopoda_eg_gene
bnapus_eg_gene
boleracea_eg_gene
brapa_eg_gene
bvulgaris_eg_gene

Fig. 4. First step to select the data from orthologs and paralogs of the species.

User may select to show the orthologies from one species to many or among all the selected species. Paralogs can also be shown. The resulting circular diagram shows the involved chromosomes and their lengths and colours them by using a different colour per species.

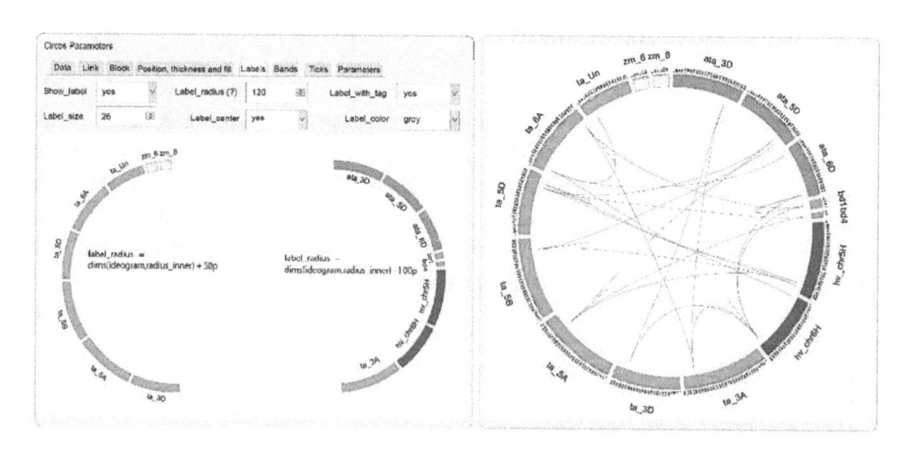

Fig. 5. Adjustable parameters to make the graph in Circos.

Figure 5 shows orthologs and paralogs of Triticum *aestivum* compared to Zea *mays*, Arabidopsis *thaliana*, Brachypodium *distachyon* and Hordeum *vulgare*. Advanced users may download the script automatically generated by EasyBio to include by his own any particular customization not available in the web platform. Orthologs and paralogs are shown by means of arcs, and chromosomes are shown as circular bands. The user may customize this diagram through EasyBio but only the main options are included to avoid over-whelming the user interface.

3.4 Gene Structure and Cis-Regulatory Elements

Cis are short DNA sequences found in the upstream of a gene that regulate its expression. Once the User Cart is fulfilled, the user has to select the length of the left flanking region where to search and the source of the Cis to look for: New PLACE, PlantCARE or user defined. Cis are shown as thin coloured vertical lines at the left of their corresponding gene structure. The way in which the structure of a gene is drawn can be customized too, as shown in Fig. 6, by selecting the colours and thickness of their components: UTR5', UTR3', introns and exons.

CIS Selected

ensembl_gene_id	cis_name	content	nro	lenght	strand	description
BRADI_3g51200v3	-10PEHVPSBD	TATTCT	171	6	-	barley (Hordeum vulgare)
BRADI_3g43855v3	-300CORE	TGTAAAG	665	7	-	maize (Zea mays); wheat (Tr...
BRADI_3g43855v3	XYLAT	ACAAAGAA	56	8	+	Arabidopsis thaliana
BRADI_4g19525v3	-300ELEMENT	TGTAAAAG	217	8	+	wheat (Triticum aestivum)
BRADI_4g22280v3	WUSATAg	TTAATGG	611	7	+	Oryza sativa (rice)

Fig. 6. Selection of the Cis to be displayed on the graph together with the adjustable parameters.

Due to the short length nature of Cis, the search may return a lot of results. Hence, before drawing them, the researcher should select only those s/he is interested in. After that, the whole diagram is drawn in interactive mode, as shown in Fig. 7. Cis colours can also be defined by the user.

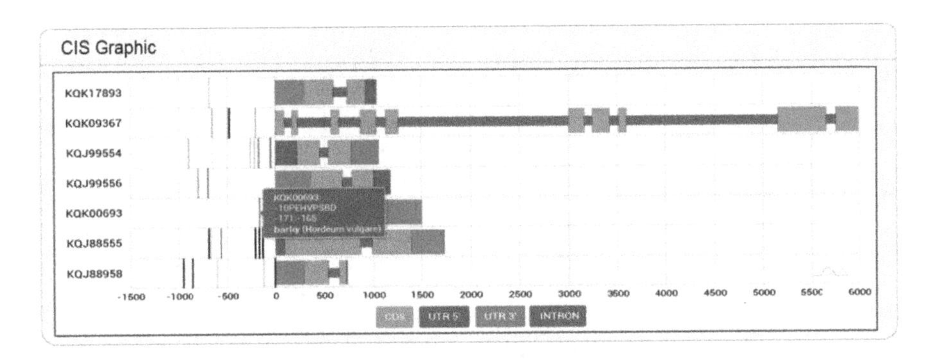

Fig. 7. Graph of Cis, UTR, Introns and Exons.

4 Conclusions

This work shows EasyBio, a web platform mainly oriented to genome-wide analysis of gene families and transcription factors that fosters the usability of bioinformatics tools using data from open databases. With this approach, geneticists can focus on the results of their analyses and they do not need to devote additional efforts to the installation, configuration or setup of the underlying tools. The user does not need to learn about complex procedures, purely technical, or to deal with involute command line consoles. Research among different teams and institutions is also encouraged because the results of different research lines can be easily compared.

In addition to be user-friendly, EasyBio is also free-to-use. Once we obtain a public domain for EasyBio, any researcher will be able to register and use it free of charge. This aspect of EasyBio will be even more important in the near future, where we are planning to incorporate some functionalities that consume computational resources.

In this sense, we are working on including new tools and capabilities into EasyBio, like access to the Phytozome datasets (phytozome.jgi.doe.gov) and dis-covering of gene regulatory networks through WGCNA (Weighted Gene Co-expression Network Analysis). Due to the importance of some crops in the human feeding, like wheat, we are also planning to interact with gene expression databases (wheat-expression.com) and Kallisto, a mapping tool for RNA-Seq, to generate gene expression heatmaps.

Acknowledgements. Funding support from projects P18-RT-992 from Junta de Andalucía (Andalusian Regional Government), Spain (Co-funded by FEDER), and by the Spanish Ministry of Science and Innovation project PID2019-108195GB-I00 are gratefully acknowledged.

References

1. Reuter, J.A., Spacek, D.V., Snyder, M.P.: High-throughput sequencing technologies. Mol. Cell **58**(4), 586–597 (2015)
2. Liu, H., Yang, L., Xin, M., Ma, F., Liu, J.: Gene-wide analysis of aquaporin gene family in Malus domestica and heterologous expression of the gene MpPIP2;1 confers drought and salinity tolerance in Arabidposis thaliana, Int. J. Mol. Sci. **20**, 3710 (2019)
3. To, V.T., et al.: Genome-wide analysis of the gras gene family in barley (Hordeum vulgare l.), Genes (Basel) **11**, 553 (2020)
4. Song, J., et al.: Genome-wide characterization and expression profiling of Squamosa Promoter Binding Protein-like (SBP) transcription factors in wheat (Triticum aestivum L.). Agronomy **9**, 527 (2019)
5. Mlecnik, B., Scheideler, M., Hackl, H., Hartler, J., Sanchez-Cabo, F., Trajanoski, Z.: PathwayExplorer: web service for visualizing high-throughput expression data on biological pathways. Nucleic Acids Res. **33**(SUPPL 2), W633–W637 (2005)
6. Kanehisa, M., Furumichi, M., Tanabe, M., Sato, Y., Morishima, K.: KEGG: New perspectives on genomes, pathways, diseases and drugs. Nucleic Acids Res. **45**(D1), D353–D361 (2017)
7. Nishimura, D.: BioCarta. Biotech Softw. Internet Rep. **2**(3), 117–120 (2001)
8. Salomonis, N., et al.: GenMAPP 2: new features and resources for pathway analysis. BMC Bioinform. **8**, 217 (2007)

9. Rocha, P.: Palmas, Plataforma Bioinformática. Palmas **29**(2), 19–28 (2008)
10. Esteban, F., Díaz, D., Hernández, P., Caballero, J., Dorado, G., Gálvez, S.: MC64: a web platform to test bioinformatics algorithms in a many-core architecture. In: 5th International Conference on Practical Applications of Computational Biology & Bioinformatics (PACBB 2011), vol. 93, pp. 9–16. Springer, Heidelberg (2011)
11. Afgan, E., et al.: The Galaxy platform for accessible, reproducible and collaborative biomedical analyses: 2018 update. Nucleic Acids Res. **46**, W537–W544 (2018)
12. Kinsella, R.J., et al.: Ensembl BioMarts: a hub for data retrieval across taxonomic space. Database **2011** (2011)
13. Goodstein, D.M., et al.: Phytozome: comparative platform for green plant genomics. Nucleic Acids Res. **40**(D1) (2012)
14. Monaco, M.K., et al.: Gramene 2013: Comparative plant genomics resources. Nucleic Acids Res. **42**(D1), D1193–D1199 (2014)
15. Lescot, M., et al.: PlantCARE, a database of plant cis-acting regulatory elements and a portal to tools for in silico analysis of promoter sequences. Nucleic Acids Res. **30**(1), 325–327 (2002)
16. Higo, K., Ugawa, Y., Iwamoto, M., Korenaga, T.: Plant cis-acting regulatory DNA elements (PLACE) database: 1999. Nucleic Acids Res **27**(1), 297–300 (1999)
17. Larkin, M., et al.: ClustalW and ClustalX version 2.0. Bioinformatics **23**, 2947–2948 (2007)
18. Díaz, D., et al.: MC64-ClustalWP2: a highly-parallel hybrid strategy to align multiple sequences in many-core architectures. PLoS ONE **9**(4), e94044 (2014)
19. Krzywinski, M., et al.: Circos: an information aesthetic for comparative genomics. Genome Res. **19** 1639–1645 (2009)

Development and Use of Dynamic Link Libraries Generated Under Various Calling Conventions

Cristian Gallardo[1(✉)], Andrey Pogrebnoy[1], and José Varela-Aldás[2]

[1] Tomsk Polytechnic University, 30 Lenin Avenue, Tomsk, Russia
kristianmaurisio1@tpu.ru, avpogrebnoy@gmail.com

[2] SISAu Research Group, Facultad de Ingeniería y Tecnologías de la Información y la Comunicación, Universidad Tecnológica Indoamérica, Quito, Ecuador
josevarela@uti.edu.ec

Abstract. Dynamic-link libraries are mainly used to extend the functionalities of the applications, decrease the use of memory, and disk space since it allows the reuse of code and procedures, sharing their access between multiple processes, applications, threads, or services. In this work, the coding of dynamic-link libraries and their generation under different calling conventions (stdcall, cdecl, and fastcall) are presented. The entire process is focused on 32-bits Windows operating systems, with TDM-GCC like the compiler and CodeLite like the IDE.Compared to using 64-bit libraries, there are many implementation and usage issues with 32-bit libraries, issues that this study address. An experimental result is presented in which a dynamic link library is created that is imported and used to give access to the data obtained from a RPLIDAR A1M8-R2 radar. The entire process of generating and using libraries in programming languages C, C++, Java, Python, C#, and VisualBasic is documented. The experimental result was carried out using Visual C++, Matlab 2015B, Unity 5.6.7f1, and Python.

Keywords: Calling conventions · Dynamic link libraries · Programming languages

1 Introduction

The creation and use of shared libraries were born along with the first basic operating systems in history, there are records of reuse of functions since the GM-NAA I/O (input/output) system of General Motors and North American Aviation, which It was an operating system created for the IBM 704. The appearance of shared libraries was due to the need to occupy the least amount of space and memory possible creating functions that were reused [1,2].

These container type files would be used by all subsequent operating systems, such as Windows, Linux, macOS, OS/2, Unix, and others. The use of libraries can be seen until today, especially in the management of interfaces and the use

Á. Rocha et al. (Eds.): ICITS 2021, AISC 1330, pp. 220–232, 2021.
https://doi.org/10.1007/978-3-030-68285-9_22

of hardware drivers, allowing all applications of the operating system to share their use, saving disk, memory, and processing resources; These characteristics are because the source code is unique and is physically separated from the applications, allowing each application to use said functions without having to copy the code within each one. At the response time level, it was appreciated that the use of libraries significantly increases the operating speed of operating systems, this is because libraries are instantiated once in operating memory and their functions can be used from any application without having to instantiate again, improving the performance not only of programs that use libraries but from the entire operating system [3–5].

Although the use of libraries focused on the interaction between the operating system and the hardware, over time it began to be implemented to increase the functionalities of the applications, functionalities such as: performing numerical calculations, sharing information, controlling processes, among others; giving way to the appearance of modularity features in applications, so the systems began to compact their common functionalities in libraries, concerning numerical calculations one of the preferred languages was Fortran [4,6], but their study is not limited only to the performance of functions, there are also studies on their behavior and how one could interact with them without knowing their content, perform interception of functions, in the same way, have been carried out studies of types of linking between executable files and libraries [3,7].

Although the use of libraries can be seen almost from the beginning of computer science, the presence of information on the subject in documents indexed in scientific databases is almost nil, and even more, information about the conduct of a study from its creation until its use in several applications, to this is added the confusion between calling convention. This problem is only found in 32-bit applications since in 64-bit architecture a consensus has been reached to use a single calling convention offering maximum compatibility between libraries and applications [8].

Currently, with the advancement of technology and computing capacity, the development and use of libraries has been neglected, but a study carried out in 2019 shows that even today, a notable improvement in performance can be seen when use libraries [9].

Due to this type of complex inconvenience to handle and the lack of information in scientific databases about the development and use of libraries, the following research is presented, in which the concepts, development, and use of libraries according to their calling convention are showed. Libraries will be created in C language and used in various programming languages (C, C++, C, Java, Python, Visual Basic). The use of libraries is not limited to the development, these libraries can also be used in applications such as the Microsoft Office suite that has support for the Visual Basic programming language, or commercial applications of mathematical calculations such as Matlab, thus showing the importance of this work not only for developers but also for users of commercial applications of specific use.

This document is divided into 5 sections including the introduction. Section 2 presents the study carried out. The experimental results are presented in Sect. 3. Finally, Sect. 4 shows the conclusions of the article.

2 Case Study

In the following proposal to create dynamic link libraries the C programming language is used, the project is compiled under the most common calling conventions; cdecl, stdcall, and fastcall. The entire process of coding, generating, and using dynamic link libraries is presented.

Internally the library can contain the number of functions required by the user. This section shows the process of sending and receiving variables, for which will be created 4 functions that work with values of type Integer, Floating, Double, and Chain of characters.

2.1 Caller, Callee and Calling Conventions

When talking about the interaction between applications and libraries two terms appear that identify the interaction, these are caller and callee, this can occur in the interaction of several files, for example between an executable application and a library; but this can also be given within the same application or library, specifically in the interaction between functions. The calling convention is the low-level implementation that determines the interaction between the Caller and the Callee when working with libraries, specifically determines the order and in which registers of the processor the parameters and the response values of the library will be stored, in the same way, the calling convention indicates which is responsible for cleaning the stack. This problem occurs only on 32-bit Windows systems since in 64-bit Windows architecture this was fixed by using a single calling convention called the Application Binary Interface (ABI) and thus achieving maximum compatibility [10].

In 32-bit Windows operating systems, there is a great compatibility problem at the level of calling conventions, since, in the beginning, each compiler proposed its way of handling the parameters and returns of functions with different procedures in handling the registry and cleaning the stack, therefore, there were several proposals for calling conventions for 32-bit Windows systems [11].

2.2 Creation of Libraries

There are important points to keep in mind when developing a library, for example, must be chosen between a static library and dynamic link library, this is important since there are applications that only support dynamic link libraries such as Matlab. For reasons of greater compatibility, dynamic link libraries are generated in this proposal. If static type libraries are required, just needed to change the compiler indicated in the project properties, the code does not require any modification. It is worth mentioning that the interaction between static and dynamic libraries is different.

Another problem to take into account is the size of the type of variables since different lengths are depending on the type of variable, the programming language, and the architecture for which it is compiled, for this reason, the type of variable and size to be used must be known. The developing propose is in C programming language, the corresponding type and size are: Variables of type Integer and Float have a size of 4 bytes, variables of type Double have a size of 8 bytes and character strings will depend on the initialization size taking into account that 1 byte is required for each character.

To specify the use of calling convention, each function must start with the __declspec modifier and must take dllexport as its argument. In this proposal we will create only a shared link library file, we will not create an import library that has a .lib extension and that offers features such as specifying export names of functions. In the source file, the different functions will be implemented and the header file will contain the function declarations including the different calling conventions which will be uncommented one by one to generate different libraries according to the calling convention. Three libraries are generated and renamed; testLibraryCdecl.dll, testLibraryStdCall.dll, and testLibraryFastCall.dll. The code to use is shown in (Fig. 1).

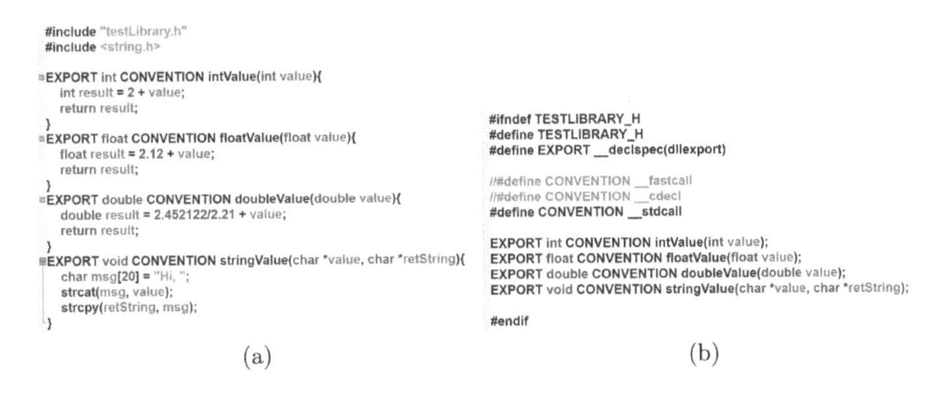

Fig. 1. Programming code: a) Source file .c; b) Header file .h

The libraries generated in the previous step can be used from many programming languages and applications, but will not be accessible from projects in the java programming language. For projects in Java programming language, it is necessary to generate a different library, the mechanism that allows us to interact with libraries written in C/C++ language from Java is called Java Native Interface (JNI).

It should be remembered that Java has a name dependency between the source file and its internal class, therefore it would be necessary to create a library for each Java project that is needed, as a solution to this problem, it is proposed to generate a file as part of a package that has the implementation of intermediate functions that interact with the declared JNI functions, since this way the library can be used in any java project only by creating a package and file with the same name as the source file used to generate the header file. Once the library is created it can be renamed, it is only required to specify the name of the library in the function parameter System.loadLibrary ("library_name");.

A .java file is created with the structure of the functions (Fig. 2c), this file is used by the javac tool to generate the .h header file (Fig. 2a), and based on this header file we create a source file and we implement the functions (Fig. 2b).

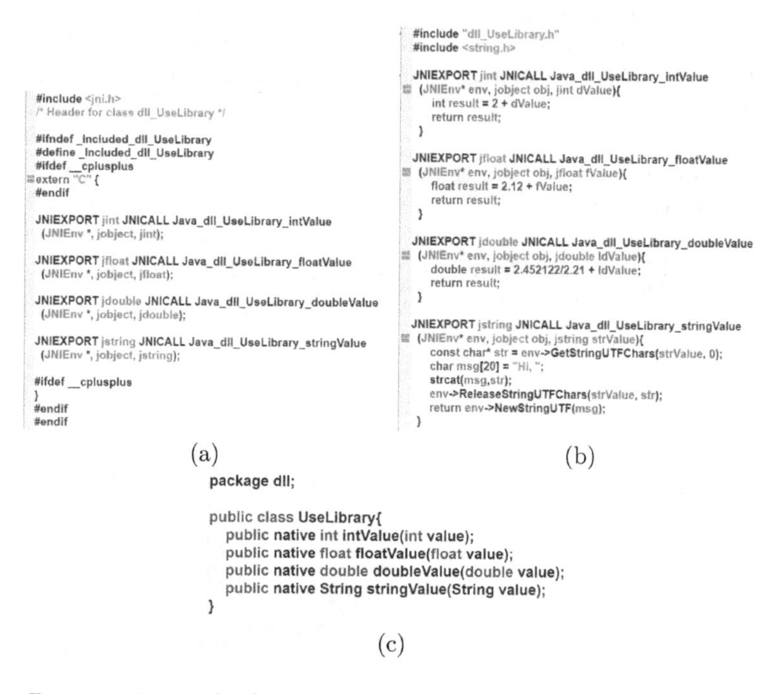

Fig. 2. Programming code for be using in Java projects: a) Generated file - dll_UseLibrary.h; b) Implementation of functions - UseLibrary.cpp; c) Base structure in Java - UseLibrary.java

2.3 Use of Libraries

The name of the functions within the export table may have been modified during the compilation process, adding decorators or symbolic names. To know the name that identifies each function within the export table, the DLL Export Viewer application was used [12].

In Fig. 3, the calling conventions of sections a, b, c are important, since it is necessary to know the exact name to call the functions, unlike the names exported by the library for Java projects. Java handles everything internally, in java the names of the functions although they change within the export table, when using them it is only necessary to refer to the name corresponding to the source code.

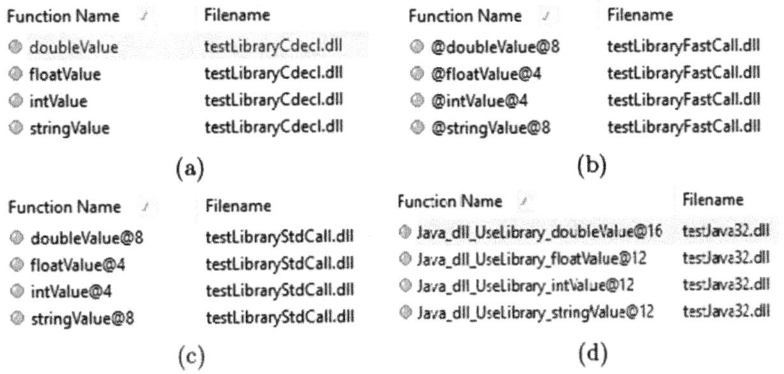

Function Name	Filename
doubleValue	testLibraryCdecl.dll
floatValue	testLibraryCdecl.dll
intValue	testLibraryCdecl.dll
stringValue	testLibraryCdecl.dll

(a)

Function Name	Filename
@doubleValue@8	testLibraryFastCall.dll
@floatValue@4	testLibraryFastCall.dll
@intValue@4	testLibraryFastCall.dll
@stringValue@8	testLibraryFastCall.dll

(b)

Function Name	Filename
doubleValue@8	testLibraryStdCall.dll
floatValue@4	testLibraryStdCall.dll
intValue@4	testLibraryStdCall.dll
stringValue@8	testLibraryStdCall.dll

(c)

Function Name	Filename
Java_dll_UseLibrary_doubleValue@16	testJava32.dll
Java_dll_UseLibrary_floatValue@12	testJava32.dll
Java_dll_UseLibrary_intValue@12	testJava32.dll
Java_dll_UseLibrary_stringValue@12	testJava32.dll

(d)

Fig. 3. Function names according to the Calling Convention: a) Cdecl; b) FastCall; c) StdCall; d) Java Projects

If you want to use a 32-bit library in a 64-bit Windows operating system, it is proposed to locate the libraries at C:\Windows\SysWOW64\ as it is the default location for file searching by a 64 bit Windows operating system running 32 bit applications. Keep in mind that on 64-bit Windows operating systems the C:\Windows\System32\ folder is directly accessible only by 64-bit applications. If a 32-bit library is located in the System32 directory, it will not be accessible from 32-bit applications by the apparently correct path that is in the System32 folder, to access it, the path C:\Windows\SysNative\ must be specified, which refers internally to the items in the System32 folder when 32Bits applications are calling, for more information check [13].

Use in C/C++ Project. The compiled library usage code is presented with the calling convention Cdecl, if you want to use a compiled library with Calling Convention FastCall we must change the creation code of new variable types adding the reserved word __fastcall and if you want to use a compiled library With calling convention stdcall the keyword WINAPI is added. The calling conventions Stdcall and Fastcall modify the variable names, so the information provided by the DLL Export Viewer application is used to know the names of the functions in the export table. The code is shown in (Fig. 4).

Use in C# Project. The library usage code compiled with the calling convention Cdecl is presented, if it is required to use a library compiled with Calling Convention StdCall must be added the EntryPoint parameter and indicate the

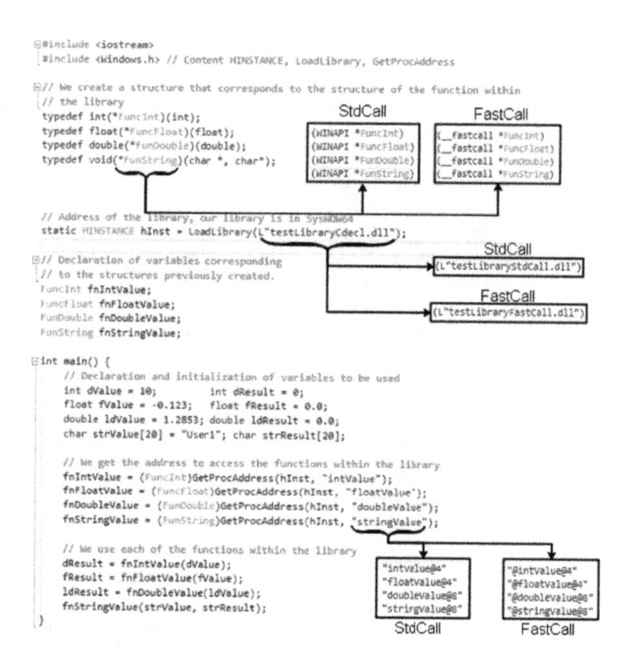

Fig. 4. Example code of library use in C++ project, the changes required for different Calling Conventions are presented.

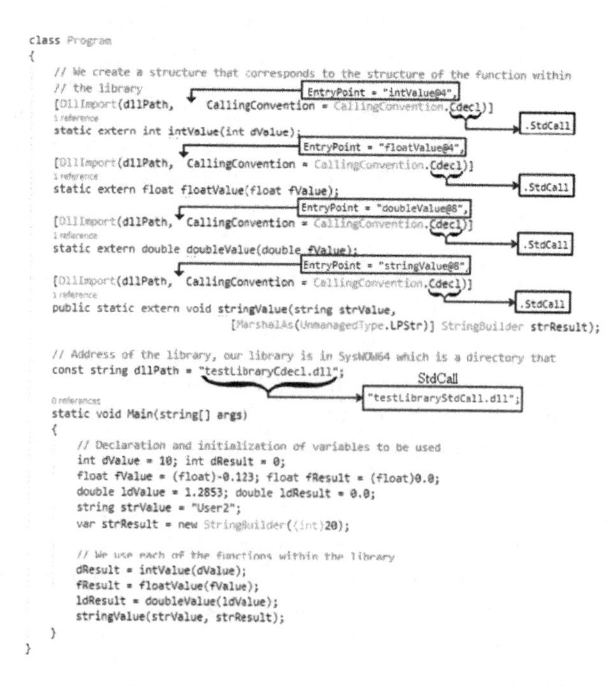

Fig. 5. Example code of library use in C# project, the changes required for Calling Convention StdCall are presented.

name of the function according to the export table, in the same way, it must be specified that the calling convention will be StdCall. Fastcall is not supported in 32-bit C#, the .Net framework only supports Cdecl, StdCall, ThisCall. Calling conventions like fastcall are declared in the framework enumerator but it is not implemented [14]. The code is shown in (Fig. 5).

Use in Python Project. A complete example of using libraries created under the calling convention Cdecl is presented, as well as the modifications to be made to use libraries compiled in calling convention StdCall. To use libraries generated in different calling conventions, you must know the Python support. Calling conventions like ThisCall and FastCall are not implemented in the ctypes library because they are not compatible with other compilers in any standard way [14]. The code is shown in (Fig. 6).

Fig. 6. Example code of library use in Python project, the changes required for Calling Convention StdCall are presented.

Use in Visual Basic Project. To use libraries generated from Cdecl, it is required to use the DllImport function found in System.Runtime.InteropServices. An example of using a library compiled with calling convention Cdecl and the necessary modifications for using libraries generated from calling convention StdCall are shown in Fig. 7.

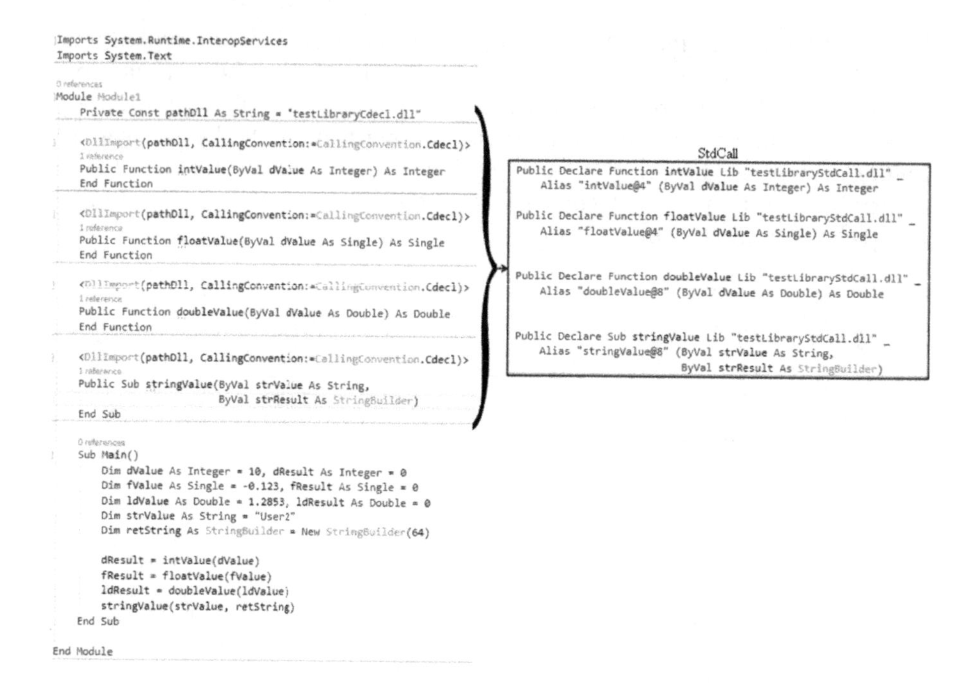

Fig. 7. Sample code for library use in Visual Basic project, the changes required for Calling Convention StdCall are presented.

Although Microsoft Excel handles the Visual Basic programming language, some code changes are required for it to work with dynamic link libraries. Microsoft Excel only supports calling convention StdCall [15]. For the parameters of the functions that use type Integer it must be declared of type Long both in sending and in returns of the functions, there is no type of variable StringBuilder so it is necessary to specify the type of variable as String, the variable that will contain the return of the stringValue function must be initialized as follows "strResult As String * 64". With these changes, the use of calling convention StdCall libraries in Excel projects with the use of Visual Basic would work.

Use in Java Project. Unlike the examples seen above, when using a library generated for Java projects, the function names are used without any modification, although decorators are added internally, as could be seen with the use of the DLL Export Viewer tool. Once a new java project has been created, a new package must be created that corresponds to the package name with which the library was created, the name "dll" was specified, after this, we can import the UseLibrary.java file used to create the header file, or create a new source file with the same name and copy into the file the entire content of the code. The code is shown in (Fig. 8).

```
package javaTest;

import dll.UseLibrary;

public class Main {

    public static void main(String[] args) {

        int dValue = 10; int dResult = 0;
        float fValue = (float) -0.123; float fResult = (float) 0.0;
        double ldValue = 1.2853; double ldResult = 0.0;
        String strValue = "User2"; String strResult;

        dResult = UseLibrary.useLibrary.intValue(dValue);
        fResult = UseLibrary.useLibrary.floatValue(fValue);
        ldResult = UseLibrary.useLibrary.doubleValue(ldValue);
        strResult = UseLibrary.useLibrary.stringValue(strValue);

    }
}
```

```
package dll;

public class UseLibrary{
    public native int intValue(int value);
    public native float floatValue(float value);
    public native double doubleValue(double value);
    public native String stringValue(String value);

    static {
        System.loadLibrary("testJava32");
    }

    public static UseLibrary useLibrary = new UseLibrary();
}
```

Fig. 8. Example code of library use in Java project.

3 Experimental Results

In this experiment, the following requirements are presented: Have access to the data of an A1M8-R2 radar from Matlab, Unity3D, and python. In Matlab, tests of a mathematical model on obstacle avoidance will be carried out, in Unity3D the visualization will be carried out and in Python, there is a WebSocket client that will send all the data to an external server. The highest possible read and write speed is required for data availability. It is not necessary to queue the data, only the most current set of values is required. The problem is that the device does not allow more than one process to read its data.

The use of Inter-Process Communication is proposed, specifically the use of named Shared Memories. The functions of creating, opening, reading, and writing on shared memories is a complex process to carry out in each programming language, so it is proposed to create a dynamic link library with the necessary functions. The IDE used is CodeLite and TDM-GCC-32 like a compiler.

3.1 Dynamic Link Library Creation

Figure 9 shows the header file inclusion process that contains the shared memory management functions, as well as a code fragment that shows the declaration and definition of functions understructure to generate libraries. The same structure shown in Fig. 1 is used.

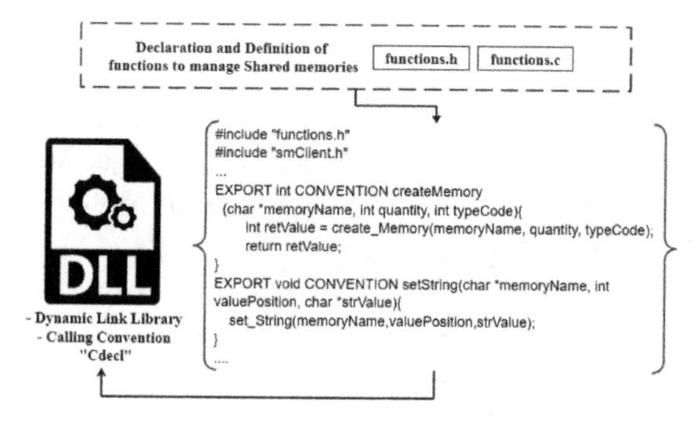

Fig. 9. Schema and source code snippet of the dynamic link library generation process

The experiment was carried out on a 32-bit Windows 8.1 operating system and the library was located in the folder C:\Windows\System32\.

3.2 Obtaining and Writing Values from Device

A console-type application is developed in C++ language using the Visual Studio 2012 IDE. The application uses the RPLIDAR A1M8-R2 SDK to access information such as SDK version, Serial Number, Firmware version, hardware revision number, health status, start data capture, stop data capture, get laser values. The functions to use are the last 3 mentioned.

When the application starts, the shared memory management DLL is instantiated and one shared memory is created that will store a string with the values obtained from the laser. A structure "angle-distance:angle-distance" is created with the values corresponding to a complete turn of the laser (approximately 360 pairs of values). The angle is given in degrees and the distance in millimeters (see Fig. 10).

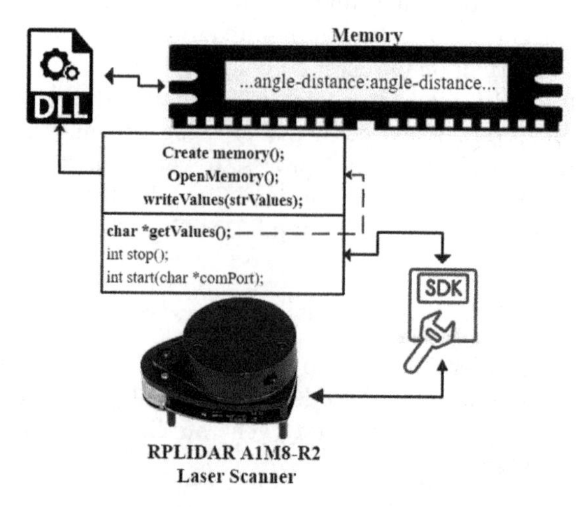

Fig. 10. Reading data from the radar and storage in shared memory

3.3 Reading Data from Applications

At this point, the information generated by the RPLIDAR A1M8-R2 is stored in a shared memory through the Dynamic Link Library. Then, the same library is instantiated within Unity 5.6.7f1, Matlab 2015B, and Application created with Python 3.8, all applications are 32 bits. The instantiation and use of libraries in Unity (C#) and python are detailed in Subsect. 2.3.

Figure 11 shows a code snippet of the instantiation and use of dynamic link libraries within Matlab. It should be noted that Matlab only supports the use of dynamic link libraries, it does not support static libraries.

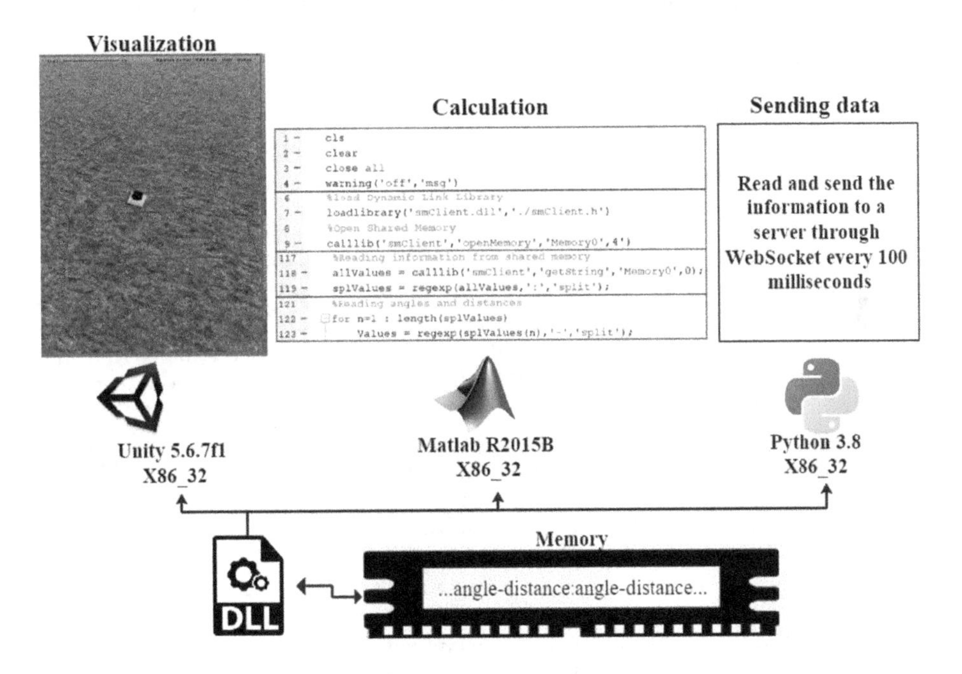

Fig. 11. Use of dynamic link library from various applications.

4 Conclusions

Once this study is concluded, it is possible to demonstrate that currently, the use of libraries shows to be a viable solution for storing functions that will be used by several applications concurrently, showing that it is a mechanism that helps to solve problems in an optimal way.

In the study carried out, the Cdecl calling convention shows great advantages both in compatibility and ease of use since it does not modify the names of the functions in the export table, unlike the other two analyzed calling conventions (StdCall, FastCall).

The only application found in this study that does not support the Cdecl calling convention is Visual Basic within Microsoft's office suite, this suite only supports the stdcall calling convention.

References

1. Hurd, C.C.: Early computers at IBM. Annal Hist. Comput. **3**(2), 163–182 (1981)
2. Ryckman, G.F.: 17. the IBM 701 computer at the general motors research laboratories. Annals Hist. Comput. **5**(2), 210–212 (1983)
3. Walsh, J.E., Redmon, W.: Method and system for dynamic-link library, December 1994. https://patentimages.storage.googleapis.com/5c/6c/86/c7f62b3fb9b181/US5375241.pdf
4. Corporation, M.: What is a DLL? December 2019. https://support.microsoft.com/en-us/help/815065/what-is-a-dll
5. Young, M.J.: Software Tools for OS/2: Creating Dynamic Link Libraries. Addison-Wesley Longman Publishing Co. Inc, Boston (1989)
6. David, R., Lemmon, J.L.S.: Creating and Using Dynamic-Link Libraries, pp. 191–218. Springer, New York (2005)
7. Hammond, R.P., Simsbury, C.: Method and system for dynamic interception of function calls to dynamic link libraries into a windowed operating system, August 2004. https://patentimages.storage.googleapis.com/e9/6b/88/5ab25990bc5463/US6779187.pdf
8. Cobertson, C., Leavitt, S., N.S.F.X: x64 calling convention, June 2020. https://docs.microsoft.com/en-us/cpp/build/x64-calling-convention?view=vs-2019
9. Ortega-Cabezas, P.M., Colmenar-Santos, A., Borge-Diez, D., Blanes-Peiró, J.J.: Application of rule-based expert systems in hardware-in-the-loop simulation case study: Software and performance validation of an engine electronic control unit. J. Softw. Evol. Process **32**(1), e2223 (2020). https://onlinelibrary.wiley.com/doi/abs/10.1002/smr.2223. e2223 JSME-18-0154.R2
10. Fog, A.: Calling convention for different CPP compilers and operating systems, December 2019. https://www.agner.org/optimize/calling_conventions.pdf
11. Cai, S., Mikeblome, ., C.R.M.J.G.J.M.S: Argument passing and naming conventions (2019). https://docs.microsoft.com/en-us/cpp/cpp/argument-passing-and-naming-conventions?view=vs-2019
12. Nirsoft: Dll export viewer. https://www.nirsoft.net/utils/dll_export_viewer.html
13. Batchelor, D., Satran, M.: File system redirector, May 2018. https://docs.microsoft.com/en-us/windows/win32/winprog64/file-system-redirector
14. Corporation, M.: Calling convention enum, October 2018. https://docs.microsoft.com/en-us/dotnet/api/system.runtime.interopservices.callingconvention?view=netframework-4.8
15. Caputo, L.: Working with DLLs, November 2014. https://docs.microsoft.com/en-us/office/client-developer/excel/working-with-dlls

IEC 61499 Based Control for Low-Cost Cyber-Physical Production Systems

Gustavo Caiza[1], Carlos A. Garcia[2], Mario Garcia-C.[2], Edmundo Llango[3], and Marcelo V. Garcia[2,4](✉)

[1] Universidad Politecnica Salesiana, UPS, 170146 Quito, Ecuador
gcaiza@ups.edu.ec
[2] Universidad Técnica de Ambato, UTA, 180103 Ambato, Ecuador
{ca.garcia,marioggarcia,mv.garcia}@uta.edu.ec
[3] Instituto Superior Tecnológico Cotopaxi, ISTX, 050102 Latacunga, Ecuador
mellangop@istx.edu.ec
[4] University of Basque Country, UPV/EHU, 48013 Bilbao, Spain
mgarcia294@ehu.eus

Abstract. Cyber-Physical Production Systems (CPPS) is the result of the combination of complex production processes with high communication capabilities. This joint work generates an improvement in the general system's performance and efficiency. CPPSs play an increasingly important role in the implementation of "Smart Factories" compatible with data exchange among all the conforming devices, under the concept of the Internet of Things (IoT). Current industry's needs demand the generation of modular and flexible production systems, so they can easily adapt to the changing market requests. To fulfill this need, the automation standard IEC-61499 is developed based on a distributed architecture and the use of control devices with high processing capabilities. The present work shows a low-cost embedded architecture designed using the IEC-61499 standard to control the operation of an analogic control process.

Keywords: IEC 61499 · Cyber Physical Production Systems (cpps) · Industry 4.0 · Embedded control

1 Introduction

Embedded systems have been driven by advanced semiconductor and communications technologies over the last decade, communicating through many means such as Wi-Fi, Bluetooth, and the network. Currently, there are large applications of Cyber-Physical Systems (CPS) in the field of industrial automation. One of the challenges of applying CPS in industrial automation is modeling the techniques for distributed automation software with integrated physical processes [2].

At present, factory automation systems need to address the different challenges such as very rapid technological advancement, global-competitive market,

Á. Rocha et al. (Eds.): ICITS 2021, AISC 1330, pp. 233–245, 2021.
https://doi.org/10.1007/978-3-030-68285-9_23

and product customization. These challenges lead to a new generation of automation systems based on the so-called Cyber-Physical Production Systems (CPPS) globally connected to form a Flexible System of Cyber-Physical Production Systems (SoCPPS). CPPS require the acquisition of data from the production system and intelligent data processing to extract information to improve the overall performance of the production system, to achieve this it is necessary to bridge the gap between the control systems and the upper layers of automation [5]. In addition, a set of embedded devices can be interconnected to serve a purpose in the system. This establishes a solid foundation for the Cyber-Physical (CPS) system that deals with the deep integration of virtual computing with physical processes [15].

Modern concepts such as the Internet of Things, Cyber-Physical Systems, and Industry 4.0 define the shape of future industrial automation systems. These concepts use state-of-the-art technologies such as low-power wireless communication, web services, and low-cost embedded devices, enabling for example enhanced monitoring and distributed control in factory automation applications [3]. Recent trends, such as the Internet of Things (IoT) or Cyber-Physical Systems (CPS), are based on the availability of reliable, high-speed networks as well as small, economical, and powerful hardware devices [6].

An automation standard widely accepted in real industries is the IEC 61131. This standard using the Programmable Logic Controller (PLC) as the main device to implement the control algorithms. Viewing this standard for the future does not address the requirements of complex smart factory systems. The new requirements are the portability, interoperability, increased reuse, and distribution of the automation control. For this reason, the International Electrotechnical Commission (IEC) developed the 61499 standard, which is a mature technology to enable intelligent automation [14].

The huge demands for robust techniques of industrial control for example decentralized or complex automation have accelerated the change to a distributed and reconfigurable engineering approaches. The IEC 61499 tends to design and implement distributed industrial process measurement and control systems, this is the standard for the future smart factories. IEC 61499 has the main goal of the implementation of an open, independent, and component-oriented framework to improve interoperability, reuse, portability, reconfigurability, and distribution of control software for complex distributed systems [11].

In the present work is proposing the development of CPPSs using low-cost devices to implement a closed-loop control system according to the IEC-6149 standard and the storage of variables in a database that will be used in the process is proposed.

The design of the document is as follows, Sect. 2 indicates the related work that has been used as a starting point for this research. Section 3 describes some concepts about IEC-64199 that can be used in the case study. Section 4 illustrates a case study where SIFBs designed under IEC 61499 is indicated. Section 5 shows the implementation proposal. Finally, some conclusions and work in progress are presented in Sect. 6.

2 Related Works

The aim of this section is to introduce existing technologies that can be adapted for CPPS automation with the IEC-61499 standard. In this sense, it gives an overview of the related works, paradigms, and implementation technologies, that is, oriented architectures to services and industrial systems.

In the last decade, the applicability of IEC 61499 in distributed control systems has been extensively studied in many projects such as intelligent networks [13], machining [12], manufacturing control [10] and process control [4]. These case studies have confirmed many advantages of IEC 61499 over PLC technology based on IEC 61131-3 in terms of design efficiency, interoperability, and code reuse [11].

Zoitl [16] investigates the use of the 4DIAC open-source environment compatible with IEC 61499 to develop modular and reusable control code. In addition, future improvements are identified to improve its usability for industrial automation applications. Concluding that the use of the open-source development methodology has a great potential to share risks and development efforts in order to establish new technologies for the mastery of industrial automation. Demonstrating the benefits of the 4DIAC open-source automation environment not only for the selected sample application but also for large industrial control applications.

Researches in [13], discuss the application of the IEC 61499 standard to design and implement applications based on events in Intelligent Networks. Showing that IEC 61499 is very suitable in the context of Smart Grids. In the case study, the example given is an energy exchange between several pairs of Intelligent Networks. The feasibility of IEC 61499 in the use case is evaluated on the basis of a prototypical implementation using the 4DIAC open-source framework. Concluding that IEC 61499 is a powerful tool for the engineering of applications based on Intelligent Networks events.

A useful and innovative approach in automation control of the FBs from the IEC 61499 standard is process planning. In this standard, the FBs are triggered by events to run internal algorithms in a controlled sequence. This feature opens up many new exciting application scenarios. The changes during the implementation and process plan execution using the IEC 61499 standard permits the improvement of the industrial systems using the distributed technologies.

Forms of integration have been proposed in works such as [7] for the tools and methodologies of IEC 61499 to be used with low-cost hardware and software platforms. Providing a set of SIFBs that encapsulate basic functions on low-cost hardware platforms so that they can run in low-cost runtimes according to IEC 61499. All this to facilitate the creation of complex distributed applications by combining these basic blocks with other blocks existing ones. Similarly, the study carried out in [6] has shown how the implementation of the IEC-61499 standard on low-cost platforms at the industrial level can be considered as an alternative in rapid measurements of real processes where the application of traditional methods is very expensive.

Adamson [1], shows i) the concept of manufacturing based on characteristics for the control of equipment; ii) the assignment of tasks in distributed and collaborative manufacturing environments is presented. Distributed control is performed by using function blocks of decision modules in intelligent networks that allow the performance of the production activities according to the actual manufacturing conditions.

Proposal of this paper is based on the starting point described in [6] to extend the scalability of the standard to more low-cost devices while adopting more advanced control techniques taking advantage of the devices' embedded system, proving that all the features accomplished because of the support of the different research groups in the previously mentioned studies, can be implemented in CPPS.

3 Related Technologies

3.1 IEC-61499

The automation standard IEC-61499 proposes a flexible and distributed control architecture. Its development base element is the Function Block (FB), which is an established software concept. FBs are created to contain the system operations inside their programming, having more robustness and turning them into reusable components. IEC-61499 guidelines establish that FBs can be programmed in any language specified in the previous IEC-61131 standard, but also in high-level programming languages such as Java, C, C++, Python, etc.

The generation of distributed applications compatible with this standard is done with the use of two software tools. The first one is a desktop engineering environment, able to provide an extensible development framework for modeling distributed control applications [8]. And the second tool is a runtime environment in charge of executing the generated functions under the operation scheme, both specified in the engineering environment. For the following work, 4DIAC-IDE^{TM} has been used as an engineering environment for the development of distributed control applications. This framework provides portability, interoperability as well as configuration capabilities. Similarly, the selected run-time is FORTE (4DIAC-RTE), which allows executing IEC 61499 applications on top of small embedded devices. FORTE provides portability to several operating systems including Windows and Linux.

An IEC-61499 distributed control system is composed of different applications interconnected using a communication bus. At the same, applications are the result of a FB's network logically interconnected by "wires", allowing the interchange of events and information among them. This type of operation is possible thanks to FB's physical distribution shown in Fig. 1, composed by i) a head in charge of the events flow, and ii) a body that manages the data flow.

In the IEC 61499 architecture, other models can be constructed based on FBs. For instance, the application model is set up by FBs networks, which is composed of basic FBs, Composed FBs, or Service Interface FBs (SIFBs) whose terminals are connected to other data or events. The second model is the resource

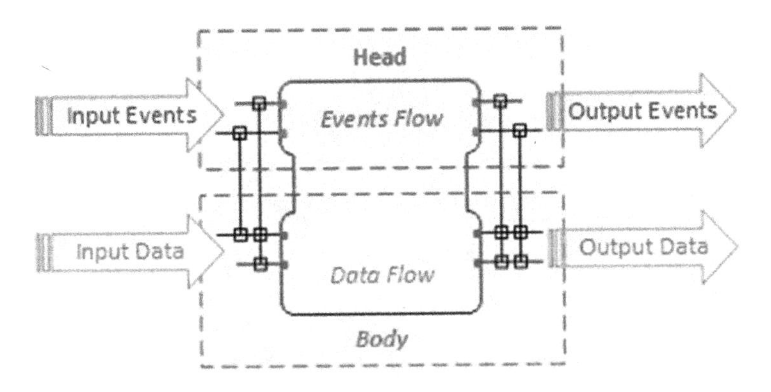

Fig. 1. FB Physical distribution

model which has some or one local application and interfaces for communications. Finally, the device model has one or more process interfaces terminals into one or more resources (See Fig. 2).

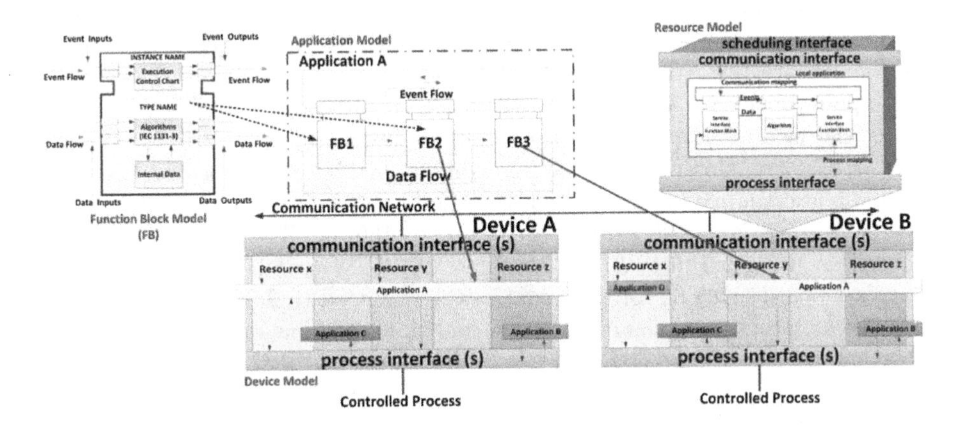

Fig. 2. IEC 61499 Architecture

4 Study Case

An industrial educative system is presented in the proposed case study, this system shows a scale factory automation process. The FESTO MPS-PA Compact Work-station has four systems to be controlled: level, flow rate, pressure, and temperature. This system allows the use of discontinuous or continuous controllers using PLC or low-cost devices as Raspberry PI (RPI), Odroids, or Arduinos boards. The close loops controls are depicted in Fig. 3.

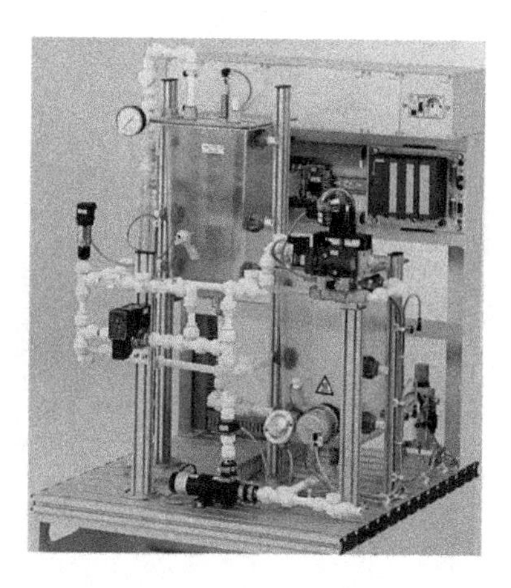

Fig. 3. FESTO^{TM} MPS-PA Compact Workstation

The close loops selected were the level and temperature loops (see Fig. 4) of this industrial system, i) Level Control Loop: The P&ID diagram shows an ultrasonic sensor to measure the process variable, which senses the liquid level in tank B102. The actuator of this loop is the pump P101. The tank B101 is used as a reservoir. The value should be kept on a certain level also if disturbances or setpoint changes occur. ii) Temperature Control Loop: The liquid is heating using the electric heater E104 this actuator is located at the bottom of the tank B101. The Cooling of the water is made by the cooler located below the profile plate; the maximum temperature for the system is 65 °C. The sensor of temperature is a PT100 calibrated from 0 to 100 °C. The heating actuator is controlled using a Pulse Width Modulation (PWM) that is generated by the low-cost board using as software control the standard IEC 61499.

The architecture of the hardware is comprised of a low-cost card such as the Beagle Bone Blackboard (BBB). This hardware receives and sends the sensor and actuators signals from the industrial system. The IEC 61499 control application was implemented using the 4DIAC framework into the BBB hardware. The FBs manipulate the General Purpose Input/Output ports (GPIO), control the industrial system, and store the values of the process tags using a transactional database.

5 The Proposed Implementation

5.1 Hardware Platform

A low-cost hardware platform is proposed to be the controller of this system. This kind of controller is known as a Single Board Computer (SBC), is a small computer built on a unique circuit board with all the standard components like microprocessor, RAM memory, GPIOs, and other functional hardware components baked-in. This SBC reduces the control system cost because has the

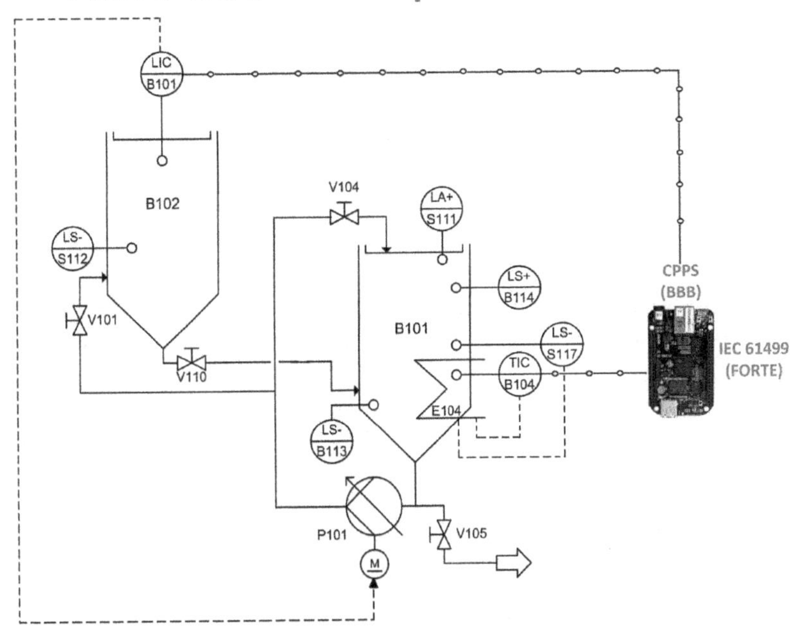

Fig. 4. PI&D of temperature and level process

same functions that an expensive PLC device. The BBB is selected as a control prototyping platform [9]. The BBB board is useful to implement Industry 4.0 applications because of their powerful embedded operating system that is great for implement a distributed system using IEC 61499 standard which cannot be implemented using traditional automation devices. The BBB board has a processor from Texas Instruments AM335x built-in by an Arm Cortex-A8. A 512 MB DDR3 RAM memory. A Hard Drive of 4 GB 8-bit eMMC onboard flash storage. Co-processors 2xPRU-ICSS and 46 pins from the GPIOs See Table 1.

Table 1. BBB electrical characteristics

Beagle Bone Black (BBB)	
SoC	OMAP3530
CPU	ARM Cortex-A8
RAM	512 MB DDR3
GPIOs	67xGPIO
USB	1 × USB 2.0
Networks	Ethernet 10/100

5.2 Function Blocks (FBs) Set for IEC-61499

The software tool used to design the FBs is the 4DIAC-IDE framework [5], a set of FBs is implemented to control the analog input/output GPIO port encapsulating the c++ libraries into the FBs. Furthermore, a FB encapsulating the stack communication to send and receive data from a transactional database is programmed. Two sets of FBs were developed, (1) FBs for GPIO analog manipulation and (2) FBs for Proportional-Integral-Derivative (PID) control implementation.

FBs for Inputs/Outputs Manipulation. This FBs set was designed to read and write the analog pins from the BBB board and use the Pulse-Width Modulation (PWM) output pins allowing them to connect with the sensors and actuators from MPS Compact Workstation controlling the level and temperature loops. The configuration of the tags associated with the pins is made using a configuration file in XML format. The XML file has (1) tags names to access the process variables, (2) pin's number to map the variables to GPIO port, (3) pins function, and (4) a small comment. The XML file is presented as XML Schema Definition (XSD) in Fig. 5.

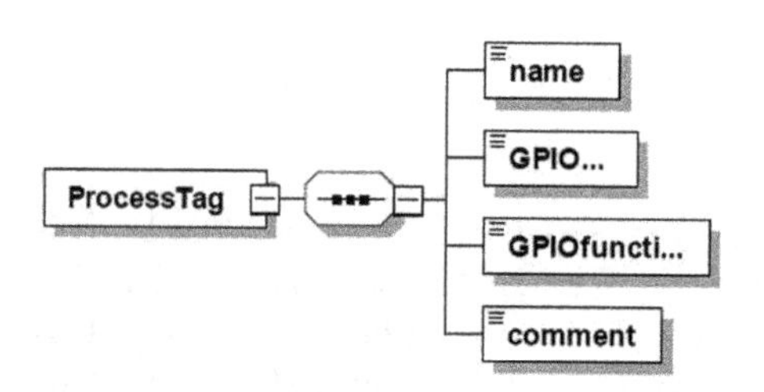

Fig. 5. XSD configuration file for process tags

1. **ANALOG_OUTPUT_R FB:** This FB is used to configure the EHRPWM output pins, the FB write an analog value within 4 to 20 mA, which is a common standard to control actuators. As all the FBs the inputs events are INIT and REQ to begin the writing procedure. The output events are the INITO and CNF. The terminals of this FB are explained in more detail below (see Fig. 6):
   - **NAME (Input, STRING):** The tag name of the port mapped to the pin on BBB board.
   - **OUT (Input, REAL):** Analog value written on the GPIO port.
2. **ANALOG_INPUT_R FB.** This FB read the analog value in the pin of the GPIO port. The FB has the previously events explained. These FBs work with the following Input/Output variables. These FBs work with the following Input/Output variables (see Fig. 7):

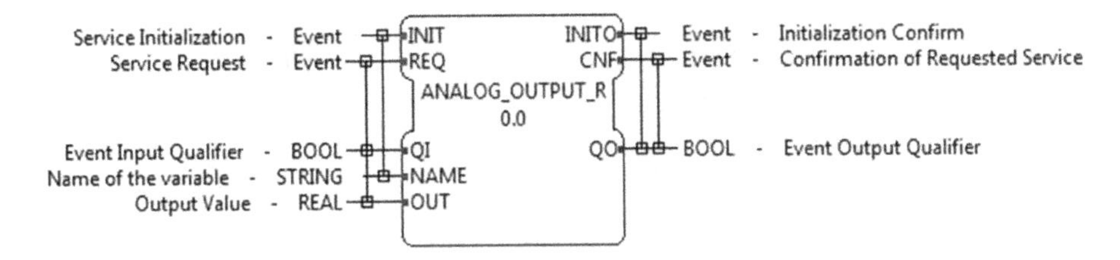

Fig. 6. ANALOG_OUTPUT_R FB

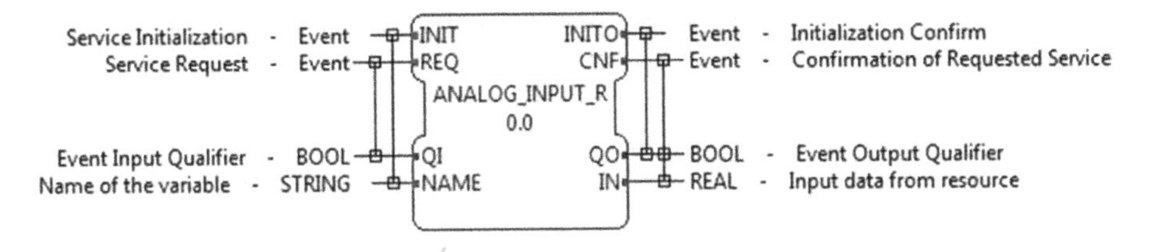

Fig. 7. ANALOG_INPUT_R FB

- **NAME (Input, STRING):** The tag name of the pin to be manipulated.
- **IN (Output, REAL):** The Analog value to be read on the inputs analog pins.

FBs for Analog Control. This kind of FBs has the PID control algorithms and algorithms and libraries to log data from the process with the aim of implement a historian's software characteristics under IEC-61499 to link the data acquisition and storage system with a relational database server. This FB grant secure and high-speed communication with databases giving rapid retrieval techniques increasing the amount of industrial data transactions.

1. **PID CONTROL FB** For the case study selected which represents a closed-loop control system, a Function Block was implemented containing the algorithm of a Proportional, Integral, and Derivative Control System as shown in Fig. 8.
   - **SETPOINT (Input, REAL):** SetPoint input value.
   - **SENSOR (Input, REAL):** Sensor input value.
   - **G_Kp (Input, REAL):** Proportional gain input value.
   - **G_Ki (Input, REAL):** Integral gain input value.
   - **G_Kd (Input, REAL):** Derivative gain input value.
2. **LOG DATA FB** This FB encloses the PostgreSQL library to send and receive messages from databases under IEC 61499. This library is an open-source object-relational database. An interesting feature of PostgreSQL is the multi-version concurrency control or MVCC. This method adds an image of the database state to each transaction. This allows the FBs to make transactions eventually consistent, offering great performance advantages. In Postgres, it is not required to use read locks when carrying out a transaction,

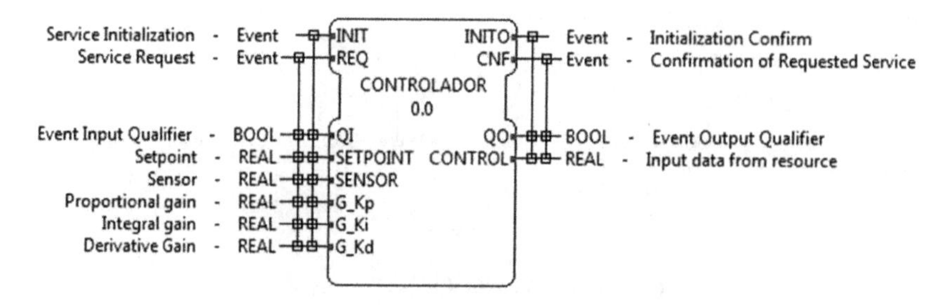

Fig. 8. PID CONTROL FB

which gives the FBs greater scalability. Also, PostgreSQL has Hot-Standby. This allows clients to search (read-only) servers while in recovery or standby mode. So FBs can do maintenance or recovery tasks without completely blocking the system (see Fig. 9). The terminals of this FB are:

- **USER (STRING):** the user name of the Postgres database for authentication.
- **PASSWORD (STRING):** The user password to authenticate to the database.
- **IP (STRING):** IP socket address to send and receive messages from the database.
- **TAG (STRING):** Process tags to log in database
- **VALUE (STRING):** Data value to be written in the database.

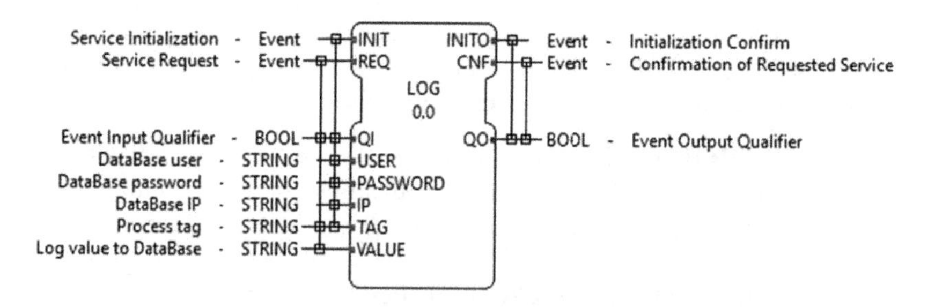

Fig. 9. LOG DATA FB

Functionality. As a final result of our work, based on the developed FB for the management of I/O of the BBB and for the closed-loop control of the Festo Cell. Finally, the respective networks of FBs have been generated, in order to achieve the correct operation of the plant process. The BBB controls the compact MPS-PA workstation and the Setpoint signal values and Sensor signal are stored in a database.

In these figures (see Fig. 10a and Fig. 10b) an execution control FB is used to periodically generate the trigger signals in charge of starting the activation of the REQ events sequence. With the REQ event activation, the I/O FBs start their continuous process consisting of obtaining the file location of the XML document generated for each board, relating the specified input or output name with the variables inside the file, returning the value of the physical magnitude at the selected GPIO pin (reading functions), and modifying the output value at the selected GPIO to match the logical input variable (writing functions). Similarly, the control FBs generate the respective control actions according to the received input values when the REQ event is triggered, in response to the algorithms developed based on the specific work of each module. Figure 10a) indicates the overall operation of the developed Function Blocks. While Fig. 10b) shows the response of the controller to the step of the reference or setpoint.

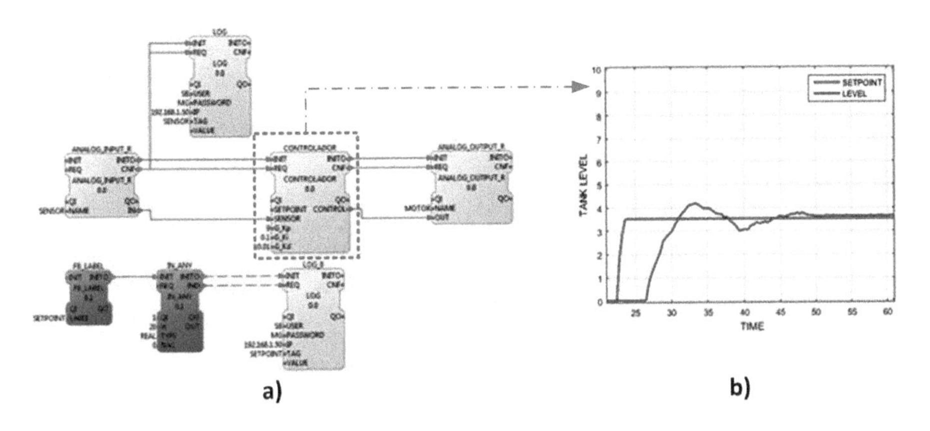

Fig. 10. FB Control Diagram. a) Distributed Control Application in 4DIAC-IDE. b) Control response to setpoint

6 Conclusions and Ongoing Work

The architecture proposed in this paper supports the creation of flexible manufacturing automation systems where sub-tasks are assigned to components in a way to separate concerns and support change. The developed analog control systems are implemented and tested on the distributing and sorting processes of the FESTO^{TM} FMS-200 industrial process, which allows having a notion of the behavior of the implemented algorithms in industrial environments

This work presents an approach for the control and access to field data in automation systems using FBs in low-cost CPPS architecture using applications of the IEC 61499 standard that are executed in the FORTE runtime. The time invested in the development of different control algorithms for controllers coming from non-related brands is drastically reduced thanks to the capability of using the same software development tool for this purpose. This limits the number of software tools that a person needs to the domain in order to generate a multi-brand distributed control system, focusing the learning time on mastering one specific tool.

By creating generic programs and eliminating the need to manually migrate the control algorithms from one brand's software tool to another, the new system presents enhanced configuration times by needing only to remap the application to another device in order to be able to download it in the new CPU.

These FBs implemented can be used for the construction of new applications with frames in the IEC 61499 standard for the development of closed-loop control systems of any type of the variable used in the industry. Also, the variables of the control processes can be stored in a database for their further analysis.

Future works focus on developing new functionalities such as integrating CPPS with Neural Networks.

Acknowledgment. This work was financed in part by Universidad Tecnica de Ambato (UTA) and their Research and Development Department (DIDE) under project CONIN-P-256-2019.

References

1. Adamson, G., Wang, L., Moore, P.: Feature-based control and information framework for adaptive and distributed manufacturing in cyber physical systems. J. Manuf. Syst. **43**, 305–315 (2017). https://doi.org/10.1016/j.jmsy.2016.12.003. http://www.sciencedirect.com/science/article/pii/S0278612516300905. High Performance Computing and Data Analytics for Cyber Manufacturing
2. Dai, W., Vyatkin, V., Chen, C., Guan, X.: Modeling distributed automation systems in cyber-physical view. In: 2015 IEEE 10th Conference on Industrial Electronics and Applications (ICIEA), pp. 984–989 (2015)
3. Derhamy, H., Drozdov, D., Patil, S., van Deventer, J., Eliasson, J., Vyatkin, V.: Orchestration of arrowhead services using IEC 61499: distributed automation case study. In: 2016 IEEE 21st International Conference on Emerging Technologies and Factory Automation (ETFA), pp. 1–5 (2016)
4. Drozdov, D., Patil, S., Dubinin, V., Vyatkin, V.: Formal verification of cyber-physical automation systems modelled with timed block diagrams. In: 2016 IEEE 25th International Symposium on Industrial Electronics (ISIE), pp. 316–321 (2016)
5. García, M.V., Irisarri, E., Pérez, F., Estévez, E., Marcos, M.: OPC-UA communications integration using a CPPs architecture. In: 2016 IEEE Ecuador Technical Chapters Meeting (ETCM), pp. 1–6 (2016)
6. García, M.V., Pérez, F., Calvo, I., Moran, G.: Developing CPPs within IEC-61499 based on low cost devices. In: 2015 IEEE World Conference on Factory Communication Systems (WFCS), pp. 1–4 (2015)
7. García, M.V., Pérez, F., Calvo, I., Morán, G.: Building industrial CPS with the IEC 61499 standard on low-cost hardware platforms. In: Proceedings of the 2014 IEEE Emerging Technology and Factory Automation (ETFA), pp. 1–4 (2014)
8. GmbH, D.C.P.: Framework for Distributed Industrial Automation and Control (4DIAC) (2010). http://www.fordiac.org

9. Guamán, Y., Ninahualpa, G., Salazar, G., Guarda, T.: Comparative performance analysis between MQTT and COAP protocols for IoT with Raspberry pi 3 in IEEE 802.11 environments. In: 2020 15th Iberian Conference on Information Systems and Technologies (CISTI), pp. 1–6 (2020)
10. Jovanović, M., Zupan, S., Prebil, I.: Holonic control approach for the "green"-tyre manufacturing system using IEC 61499 standard. J. Manuf. Syst. **40**, 119–136 (2016). https://doi.org/10.1016/j.jmsy.2016.06.008. http://www.sciencedirect.com/science/article/pii/S0278612516300279
11. Pang, C., Patil, S., Yang, C., Vyatkin, V., Shalyto, A.: A portability study of IEC 61499: semantics and tools. In: 2014 12th IEEE International Conference on Industrial Informatics (INDIN), pp. 440–445 (2014)
12. Querol, E., Romero, J.A., Estruch, A.M., Romero, F.: Norma IEC-61499 para el control distribuido aplicación al cnc. Jornadas de Automatica, pp. 3–5 (2014)
13. Rosenstatter, T., Wanger, R., Huber, S., Heistracher, T., Engel, D.: Applicability of IEC 61499 for event based smart grid applications. In: 2015 International Symposium on Smart Electric Distribution Systems and Technologies (EDST), pp. 278–283 (2015)
14. Thramboulidis, K.: IEC 61499 vs. 61131: a comparison based on misperceptions. J. Softw. Eng. Appl. **06**(08), 405–415 (2013). https://doi.org/10.4236/jsea.2013.68050. http://www.scirp.org/journal/doi.aspx?DOI=10.4236/jsea.2013.68050
15. Wang, S., Zhang, C., Jia, D.: Improvement of type declaration of the IEC 61499 basic function block for developing applications of cyber-physical system. Microprocess. Microsyst. **39**(8), 1255–1261 (2015). https://doi.org/10.1016/j.micpro.2015.07.004. http://www.sciencedirect.com/science/article/pii/S0141933115001039
16. Zoitl, A., Strasser, T., Ebenhofer, G.: Developing modular reusable IEC 61499 control applications with 4DIAC. In: 2013 11th IEEE International Conference on Industrial Informatics (INDIN), pp. 358–363 (2013)

Distributed Data Warehouse Resource Monitoring

Pedro Martins[1](✉), Filipe Sá[1], Filipe Caldeira[1], and Maryam Abbasi[2]

[1] CISeD – Research Centre in Digital Services, Polytechnic of Viseu, Viseu, Portugal
{pedromom,filipe.sa,caldeira}@estgv.ipv.pt

[2] Department of Computer Sciences, University of Coimbra, Coimbra, Portugal
maryam@dei.uc.pt

Abstract. In this paper, we investigate the problem of providing scalability (out and in) to Extraction, Transformation, Load (ETL) and Querying (Q) (ETL+Q) process of data warehouses. In general, data loading, transformation, and integration are heavy tasks that are performed only periodically, instead of row by row. Parallel architectures and mechanisms can optimize the ETL process by speeding up each part of the pipeline process as more performance is needed. We propose parallelization solutions for each part of the ETL+Q, which we integrate into a framework, that is, an approach that enables the automatic scalability and freshness of any data warehouse and ETL+Q process. Our results show that the proposed system algorithms can handle scalability to provide the desired processing speed in big-data and small-data scenarios.

Keywords: Scalability · Monitoring · Actuate · ETL · Data warehouse

1 Introduction

ETL tools are special-purpose software used to populate a data warehouse with up-to-date, clean records from one or more sources. The majority of current ETL tools organize such operations as a workflow. At the logical level, the E (extraction) can be considered as the capture of data flow from the sources, usually more than one with high-rate throughput. Then, we have T representing the transformation and cleansing of data. This corresponds to modifying data so that it will conform to an analysis schema. The L (load) represents loading the data into the data warehouse, where the data is stored to be queried and analyzed. When implementing these types of systems, besides the necessity to create all these steps, the user is required to be aware of scalability requirements that the ETL+Q (ETL and queries) might raise for this specific scenario.

When defining the ETL+Q, the user must have in mind the existence of data sources, where and how the data is extracted to be transformed (e.g., completed, cleaned, validated), the loading into the data warehouse, and finally the data

Á. Rocha et al. (Eds.): ICITS 2021, AISC 1330, pp. 246–255, 2021.
https://doi.org/10.1007/978-3-030-68285-9_24

warehouse schema, each of these steps requires different processing capacities, resources, and data treatment. However, in some applications scenarios, (e.g., near-real-time monitoring of telecom, energy distribution or stock market) ETL can be demanding in terms of performance. Most of the time, because the data volume is too large and one single, extraction, transform, loading or querying node is not sufficient. Thus, more nodes must be added to extract the data and extraction policies from the sources must be created (e.g., round-robin OR on-demand). The other phases, transformation, and load must also be scaled

In this paper, we study how to provide ETL+Q scalability with ingress high-data-rate in big and small data warehouses. We propose a set of mechanisms and algorithms, to parallelize and scale each part of the entire ETL+Q process, which is included in an auto-scale (in and out) ETL+Q framework. This framework is based on time bounds for the parts of the ETL+Q and/or the global ETL process, automatically scaling, to assure the desired time bounds.

The presented results prove that the proposed monitoring mechanisms and detection algorithms can scale-out when necessary.

In Sect. 2, we present relevant related work in the field. Section 3, we describe the architecture of the proposed system. Section 4 explains the main algorithms which allow to scale-out when necessary. Section 5 shows the experimental results. Finally, Sect. 6 concludes the paper and discusses future work.

2 Related Work

Works in the area of ETL scheduling include efforts towards the optimization of the entire ETL workflow [4] and individual operators in terms of algebraic optimization (e.g., joins or data sort operations). The work [2] deals with the problem of scheduling ETL workflow at the data level and in particular scheduling protocols and software architecture for an ETL engine to minimize the execution time and the allocated memory needed for a given ETL workflow. The second aspect in ETL execution that the authors address is how to schedule flow execution at the operations level (blocking, non-parallelizable operations may exist in the flow) and how we can improve this with pipeline parallelization [1].

The work [3] focuses on finding approaches for the automatic code generation of ETL processes which is aligning the modelling of ETL processes in the data warehouse with Model Driven Architecture (MDA) by formally defining a set of QVT (Query, View, Transformation) transformations.

Related problems studied in the past include the scheduling of concurrent updates and queries in real-time warehousing and the schedule of operators in data streams management systems. However, we argue that a fresher look is needed in the context of ETL technology. The issue is no longer the scalability cost/price, but rather the complexity it adds to the system. Previews presented recent works in the field do not address in detail how to scale each part of the ETL+Q and do not regard the automatic scalability to make ETL scalability easy and automatic. The authors focus on mechanisms to improve scheduling algorithms and optimizing workflow and memory usage. In our work, we assume that scalability in several machines and quantity of memory is not the issue. We focus on offering scalability for each part of the ETL pipeline process, without the nightmare of operators relocation and complex execution plans. Thus, in our work, we focus on scalability based on generic ETL process to provide the users desired performance with minimum complexity and implementations. We also support queries execution.

3 Architecture

In this section, we describe the main components of the proposed architecture for ETL+Q scalability.

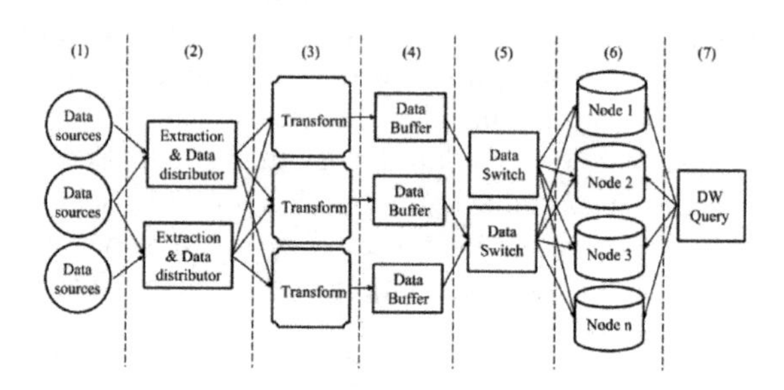

Fig. 1. Automatic ETL+Q scalability model

Figure 1 depicts the main processes needed to support total ETL+Q scalability with specific time bounds.

(1) Represents the data sources from where data is extracted from the system.
(2) The data distributor(s) is responsible for forwarding or replicating the raw data to the transformer nodes. The distribution algorithm to be used is configured and enforced at this stage. The data distributors (2) should also be parallelizable if needed, for scalability reasons.
(3) In the transformation nodes, the data is cleaned and transformed to be loaded into the data warehouse. This might involve data look-ups to in-memory or disk tables and further computation tasks. In Fig. 1 the transformation (3) is parallelized for scalability reasons.
(4) The data buffer can be in memory, disk file (batch files) or both. In periodically configured time frames/periods, data is distributed across the data warehouse nodes.
(5) The data switches are responsible for distributing (pop/extract) data from the "Data Buffers" and set it for load into the data warehouse, which can be a single-node or a parallel data warehouse depending on configured parameters (e.g., load time, query performance).

(6) The data warehouse can be in a single node or parallelized by many nodes. If parallelized, the "Data Switch" nodes will manage data placement according to configurations (e.g., replication and distribution). Each node of the data warehouse loads the data independently from batch files.
(7) Queries (7) are rewritten and submitted to the data warehouse nodes for computation. The results are then merged, computed and returned.

The main concepts, we propose are the individual ETL+Q scalability mechanisms of each part of the ETL+Q pipeline. By offering the solution to scale each part independently, we provide a solution to obtain configurable performance. Then, in future work based on user configuration parameters, a framework using these components scales the ETL+Q automatically when necessary.

4 Scaling Algorithms

In this section, we describe the algorithms which allow the framework to scale-in and scale-out each part of the ETL and Query process. For each part that we design for later, automatic scale in and out, we explain the scaling algorithms.

Extraction & Data Distributors - Scale Out: Depending on the number of existing sources and data generation rate and size, the nodes that process the extraction of the data from the sources might need to scale. The addition of more "extraction & data distributors" (2) depends on if the current number of nodes can extract and process the data with the correct period and inside the maximum extraction time (without delays). For instance, if the extraction period is specified as every 5 min and the extraction duration is 10 s, every 5 min the "Extraction & Data distributor" nodes cannot spend more than 10 s extracting data if so, a scale-out is needed. By scaling out the extraction, nodes will have fewer data to extract/process and more concurrent extraction, leading to performance improvement.

Extraction & Data Distributors - Scale In: To save resources when possible, the nodes that perform the data extraction from the sources can be set in standby or removed. This decision is made based on the last execution times. If previous execution times of at least two or more nodes are less than half of the configured maximum, one of the nodes is set on standby or removed, and the other one takes over.

Transform - Scale-Out: The transformation process is critical. If the transformation is running slow, data extraction at the refereed rate may not be possible, and information will not be available for loading and querying when necessary. The transformation step has an important queue, used to determine when to scale the transformation phase. If this queue reaches a limit size (by default 50%), then it is necessary to scale, because the actual transformer is not being able to process all data that is arriving. Another mechanism used to scale the transformation process is the user-configured maximum transformation execution time. If this time is exceeded then, the transformation must be scaled-out.

Transform - Scale In: The size of all queues is analyzed periodically. If this size at a specific moment is less than half of the limit size for at least two nodes and the average transformation time of at least two nodes is half of the specified then, one of those nodes is set on standby or removed, and another one of the low load nodes takes over.

Data buffer - Scale: The data buffer nodes scale-out based on the incoming memory queue size and the storage space available to hold data. Low data warehouse load frequency will require data buffers with storage space to keep the data until the scheduled load time. Thus, the data buffers scale dynamically as more storage space is necessary. Another scale-out situation is when the available incoming memory queue becomes above a certain threshold (by default 50%). This means that the data ingress rate is higher than the data swap speed; thus, nodes must scale-out not to lose data. By user request, the data buffers can also scale-in. In this case, the system will allow it if the data from any data buffer can be fitted inside other data buffer.

Data Switch - Scale: These nodes scale based on configured data rate limits. If after a data load process occurs the average limit extraction data rate is equal or above a certain threshold, then these nodes are set to scale. The data switches can also scale-in. In this case, the system will allow it if the average data rate from the previews load period is less than the maximum supported by each data switch.

Data Warehouse - Scale: The data warehouse scalability is detected after each load process. The loading process might include, among other operations: destroy indexes, load data, update materialized view, and rebuild indexes. The data warehouse load process has a limit time to be executed every time it starts. If that limit time is exceeded then, the data warehouse must scale.

The data warehouse scalability is not only based on the load & integration speed requirements, but also on the queries desired maximum execution time. The faster queries need to execute, more nodes will be necessary.

Because it is a computationally expensive operation, when an alarm is raised (the data warehouse needs to scale) the data warehouse nodes scale-in and scales-out, can only be triggered by user request and iff the average query execution time and the average load time respect the conditions 1 and 2 (where n represents the number of nodes):

$$\frac{(n-1) \times avgQueryTime}{n} \leq desiredQueryTime \tag{1}$$

and

$$\frac{(n-1) \times avgLoadTime}{n} \leq maxLoadTime \tag{2}$$

Every time the data warehouse scales-out or scales in the data inside the nodes needs to be re-balanced. The default re-balance process to scale-out is based on the phases:

- Replicate dimension tables;
- Extract information from nodes;
- Load the extracted information into the new nodes.

5 Experimental Setup and Results

In this section, we test the ability of the proposed auto-scale framework to automatic scale-out the ETL process when more performance is necessary to provide the desired results.

Considering that, we have data sources and extraction nodes to extract data. When the data flow is too high, a single data node can not handle all ingress data. In this section, we study how the extraction nodes scale to handle different data rates. The extraction process uses an on-demand approach to extract data, where an "automatic scaler" process orders the nodes to extract data from sources. There is a configured maximum allowed extraction time and a extraction frequency, represented by the Eqs. 3 and 4. If any of them is not respected, the system is set to scale-out.

$$\max_{extractionTime} < \max_{desiredExtractionTime} \tag{3}$$

$$\max_{extractionTime} < ExtractionFrequency \tag{4}$$

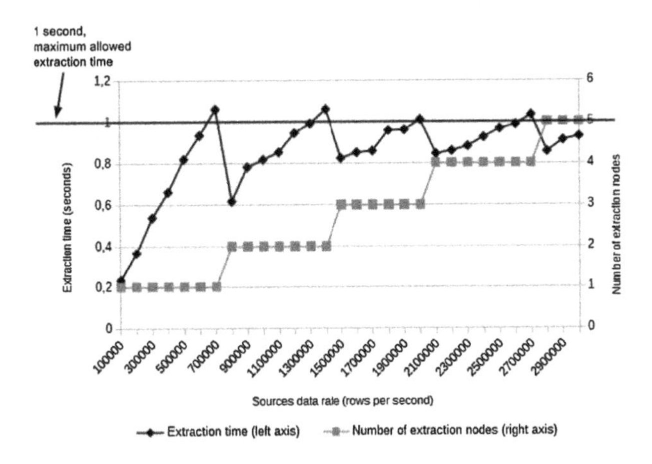

Fig. 2. Experimental results, extraction scalability

Figure 2 shows: In the left Y-axis is the average extraction time in seconds; In the right Y-axis is the number of nodes; The X-axis is the data-rate; the Blackline represents the extraction time; Grey line represents the number of nodes; The maximum allowed extraction time was set to 1 s maximum extraction time, with periodic extraction of 5 s.

As we can conclude from Fig. 2 experimental results, every time the maximum allowed extraction time has exceeded the system requested an additional extraction node to improve the extraction performance.

5.1 Transformation Scalability

During the ETL process, after data extraction, it is set for the transformation. In our tests, the transformation consists of converting the TPC-H dataset into SSB format. Because this process is computationally heavy, it is often required to scale the transformation nodes. Each transformation node has an entrance queue for ingress data, and an automatic scale monitors all queues. Once it detects that a queue is full above a certain (configured) threshold it starts the scaling process, this means that $Rate_{extract} \geq Rate_{transform}$.

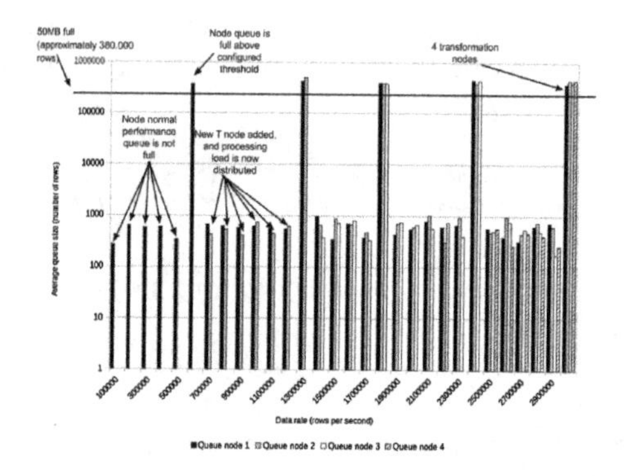

Fig. 3. Automatic transformation scalability for different data rates, 60 min processing per data rate. Transformation threshold set to 50 MB, approximately 380.000 rows

Figure 3 shows: Y-axis, average queue size in number of rows; X-axis, the data rate in rows per second; Each plotted bar represents a node queue size, up to 4 nodes; The limit queue size to trigger the scale mechanisms was set at 50 MB, approximately 380.000 rows; The maximum transformation time for each row was set to 1 s. Each measure is the average queue size of 60 s run.

During the experimental tests, as depicted in Fig. 3, the maximum allowed transformation time was never exceeded. However, the size of the queues increased while increasing the data rate, allowing to show that the proposed approach is efficient to scale the transformation nodes.

5.2 Data Buffer Nodes

These nodes hold the transformed data until it is loaded into the data warehouse. During all our tests, we used a single machine with 16 GB memory and 1 TB disk, all available to be used. If the available storage space becomes full, then, the automatic scale sets the system to scale the Data Buffer node (add one more node). However, during our tests, we never needed to do so since all transformed data could fit into memory until the next load (into the data warehouse) period.

5.3 Data Warehouse Scalability

In this section, we test the data warehouse scalability, which can be triggered or by the load process (because it is taking too long), or because of the queries time (they are taking more time than the desired execution time). If the maximum configured load time is exceeded, the data warehouse is set to scale.

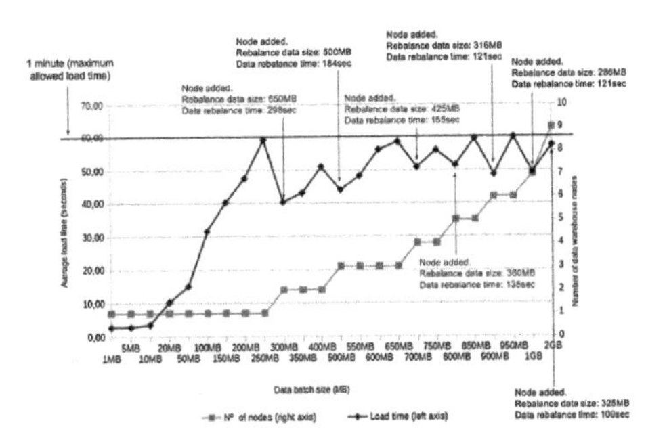

Fig. 4. Data warehouse load scalability

Experimental results from Fig. 4 show: Left Y axis, average load time in seconds; Right Y-axis, number of data warehouse nodes; X-axis, data batch size in MB; The maximum allowed load time, set to 60 s; Each time a data warehouse (scales) node is added, we show the data size that was moved into the new node and the required time in seconds (re-balance time).

Based on our experimental results, we conclude that the proposed method to scale the data warehouse when the bottleneck is related to the load time is efficient, improving the overall load performance. Note that, every time a new node was added the data warehouse required to be re-balanced (data distributed by the nodes evenly). This process requires three steps, first, extract (in parallel) the data from the existent nodes, second load the data into the new node, third load the new data (distributed and parallel) in batch and check if the load time is lower than the maximum allowed load time.

5.4 Query Scalability

When running queries, if the maximum desired query execution time (i.e. configured parameter) is exceeded then, the data warehouse is set to scale in order to offer more query execution performance. The following workloads were considered to test the proposed system:

- Workload 1;
  - 50 GB total size;
  - Execute Q1.1, Q2.1, Q3.1, Q4.1 randomly chosen;
  - Desired execution time per query: 5 min (300 s).
- Workload 2 (as workload 1 but with more sessions);
  - **1 to 8 sessions;**

Workload 1 studies how the proposed mechanisms scale the data warehouse when running queries. Workload 2 studies the scalability of the system when running queries and the number of simultaneous sessions (e.g., the number of simultaneous users) increases. Both workloads to deliver the configured execution time per query (300 s).

5.5 Query Scalability - Workload 1

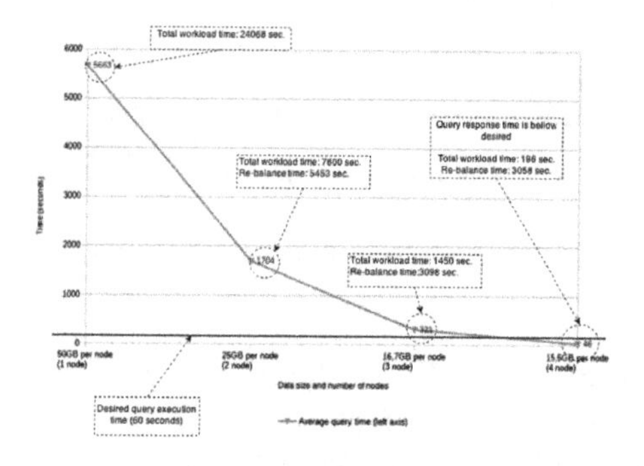

Fig. 5. Data warehouse scalability, workload 1

Figure 5 shows: The experimental results for workload 1; Y-axis, average execution time in seconds using a logarithmic scale; X-axis the data size per node and the current number of nodes; The horizontal line over 300 s represents the desired query execution time;

The results from Fig. 5, show that the proposed system can detect and scale the data warehouse nodes until the average query execution time is the desired.

5.6 Query Scalability - Workload 2

Figure 6 shows: The experimental results for workload 2; Y-axis, average execution time in seconds using a logarithmic scale; X-axis the number of sessions, the data size per node and the number of nodes; The horizontal line over 300 s represents the desired query execution time; The last result does not respect the desired execution time because of the limited resources for our tests, 12 nodes.

In Fig. 6, we show that while the number of simultaneous sessions increases the system scales the number of nodes to provide more performance; thus, the

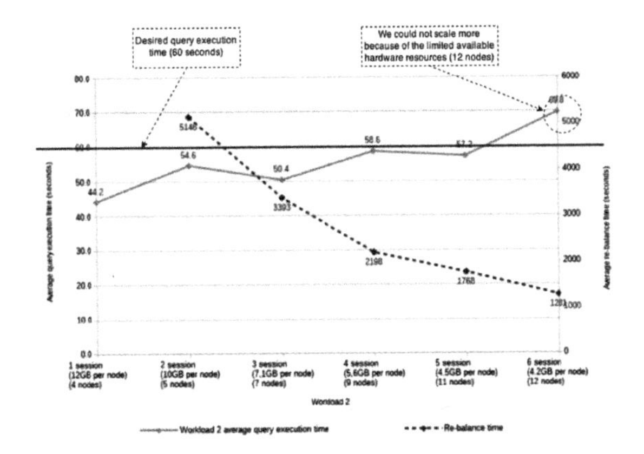

Fig. 6. Data warehouse scalability, workload 2

query average execution time follows the configured parameters. As our experimental results show the proposed system scales efficiently to provide the desired performance.

6 Conclusions and Future Work

In this work, we propose mechanisms and algorithms to achieve automatic scalability for complex ETL+Q, offering the possibility to the users to think solely in the conceptual ETL+Q models and implementations for a single server. The tests demonstrate that the proposed techniques can scale-out. Future work will investigate an auto-scale framework for scale-out and scale in any ETL+Q and, at the same time, providing data freshness and support for near-real-time data stream processing.

Acknowledgement. "This work is funded by National Funds through the FCT - Foundation for Science and Technology, IP, within the scope of the project Ref UIDB/05583/2020. Furthermore, we would like to thank the Research Centre in Digital Services (CISeD), the Polytechnic of Viseu for their support."

References

1. Halasipuram, R., Deshpande, P.M., Padmanabhan, S.: Determining essential statistics for cost based optimization of an ETL workflow. In: EDBT, pp. 307–318 (2014)
2. Karagiannis, A., Vassiliadis, P., Simitsis, A.: Scheduling strategies for efficient ETL execution. Inf. Syst. **38**(6), 927–945 (2013)
3. Muñoz, L., Mazón, J.-N., Trujillo, J.: Automatic generation of ETL processes from conceptual models. In: Proceedings of the ACM twelfth international workshop on Data warehousing and OLAP, pp. 33–40. ACM (2009)
4. Simitsis, A., Wilkinson, K., Dayal, U., Castellanos, M.: Optimizing ETL workflows for fault-tolerance. In: 2010 IEEE 26th International Conference on Data Engineering (ICDE), pp. 385–396. IEEE (2010)

Smart Mobility: A Systematic Literature Review of Mobility Assistants to Support Drivers in Smart Cities

Nelson Pacheco Rocha[1](✉), Ana Dias[2], Gonçalo Santinha[3], Mário Rodrigues[4], Carlos Rodrigues[3], and Alexandra Queirós[5]

[1] Medical Sciences Department and Institute of Electronics and Informatics Engineering of Aveiro, University of Aveiro, Aveiro, Portugal
npr@ua.pt

[2] Department of Economics, Industrial Engineering, Management and Tourism and GOVCOPP - Governance, Competitiveness and Public Policies, University of Aveiro, Aveiro, Portugal
anadias@ua.pt

[3] Department of Social, Political and Territorial Sciences and GOVCOPP - Governance, Competitiveness and Public Policies, University of Aveiro, Aveiro, Portugal
{g.santinha, cjose}@ua.pt

[4] Águeda School of Technology and Management and Institute of Electronics and Informatics Engineering of Aveiro, University of Aveiro, Aveiro, Portugal
mjfr@ua.pt

[5] Health Sciences School and Institute of Electronics and Informatics Engineering of Aveiro, University of Aveiro, Aveiro, Portugal
alexandra@ua.pt

Abstract. The objective of the study reported by this article was to identify information services using smart cities' infrastructures aiming to assist drivers. A systematic review was performed based on a search of the literature. From the systematic search of the literature 17 articles were retrieved and all of them aimed to contribute for the development of information services to assist drivers in smart cities. Considering the retrieved articles, it is possible to conclude that there is a lack of robust solutions and that the smart cities' infrastructures are underutilized.

Keywords: Smart cities · Smart mobility · Drivers assistance · Systematic review

1 Introduction

Within the smart city paradigm, smart mobility [1] is often seen as related to the use of information technologies to adequately orchestrate services designed to improve urban mobility [2]. In this respect a wide range of information services can be foreseen such as personal mobility assistants to help the surpassing of difficulties of the travelers

Á. Rocha et al. (Eds.): ICITS 2021, AISC 1330, pp. 256–266, 2021.
https://doi.org/10.1007/978-3-030-68285-9_25

facing multi-modal transit situations [3], algorithms to infer to mobility patterns [4, 5], or drivers' assistants, namely to advise drivers towards vacant parking spaces or to promote safe driving.

Intelligent transportation systems have been object of significative research [6–8] including aspects related to driver assistance [9, 10]. The systematic review reported by this article does not aim to analyze the advances in terms of the use of information services to assist drivers in general, but specifically to determine the current state of information services designed to use smart cities' infrastructures to assist drivers. This complements other studies (e.g., [11, 12]) and is useful to inform smart cities' stakeholders about state-of-the-art solutions and researchers about gaps of the current research.

2 Methods

The research study was informed by the following research questions:

- RQ1: What are the current research trends related to information services using smart cities' infrastructures to assist drivers?
- RQ2: What types of smart cities' data are being used?
- RQ3: What are the maturity levels of the solutions being reported?

Boolean queries were prepared to include all the articles that have in their titles, abstract or keywords one of the following expressions: 'Smart City', 'Smartcity', 'Smart-city', 'Smart Cities', 'Smartcities' or 'Smart-cities'. These broad search terms were selected to ensure that no pertinent studies were overlooked.

The resources considered to be searched were two general databases, Web of Science and Scopus, and one specific technological database, IEEE Xplore. The literature search was concluded in January 2020.

As inclusion criterion, the authors aimed to include all the articles that report the development of information services to assist drivers explicitly using smart cities' infrastructures.

Considering the exclusion criteria, the authors aimed to exclude all the articles not published in English, without abstract or without access to full text. Furthermore, the authors also aimed to exclude all the articles that report overviews, reviews, or solutions that do not require smart cities' infrastructures, as well as articles reporting studies not relevant for the specific objective of this systematic review.

The selection and classification of the articles were performed according the following steps:

- The authors removed the articles that were duplicated or without abstract.
- The authors assessed all titles and abstracts to remove all articles that were not published in English, or that are overviews or reviews, editorials, prefaces, and announcements of special issues, workshops, or books.
- The authors assessed all titles and abstracts for relevance against inclusion and criterion.

- The full texts of the eligible articles were retrieved and screened against inclusion and exclusion criteria.
- The full texts of the included articles were analyzed and classified.

In all these four steps the articles were analyzed by at least two authors and any disagreement was discussed and resolved by consensus.

Data from each included article were extracted using a standardized form including: article authors, title and year of publication; aim of the study being reported; details of the applications being reported; details of the applied research methods; methods applied for the assessment of the proposed applications; results; and authors' interpretations.

3 Results

From the search on Web of Science, Scopus, and IEEE Xplore, 21236 articles were retrieved.

The initial step of the screening phase yielded 21034 articles by removing the duplicates or the articles without abstracts (Fig. 1).

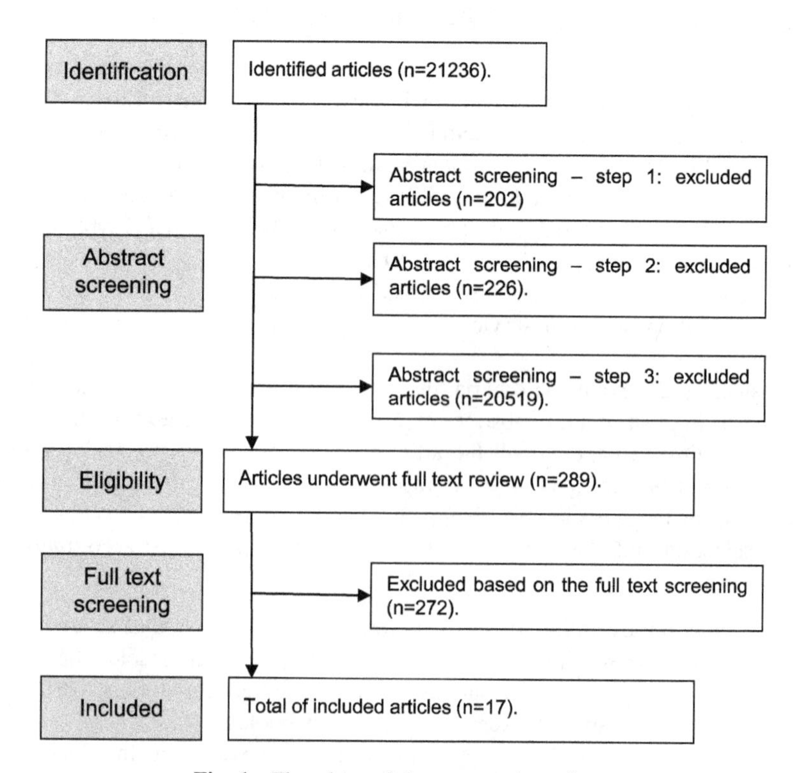

Fig. 1. Flowchart of the systematic review.

Based on titles abstracts and using semi-automatic techniques, 226 articles were removed since they were not published in English, or they are overviews or reviews, editorials, prefaces, and announcements of special issues, workshops, or books. Moreover, 20519 articles were removed because they do not target the development of information systems to assist drivers in the context of smart cities (Fig. 1).

Finally, the full texts of the remaining 289 articles were screened and 272 articles were excluded because they do not meet the inclusion criterion. Therefore, 17 articles were considered eligible for this systematic review and their aims are presented in Table 1.

Table 1. Aims of the articles considered eligible.

Article	Year	Aim
[13]	2013	To propose an information service to aggregate context data generated by the city and its citizens and data from internet of things devices to monitor and signalize the state of availability of each single parking space
[14]	2014	To propose an information service to detect if a vulnerable road user intends to cross a road in a risky zone, and to issue alerts to the vehicles nearby
[15]	2015	To propose an information service based on a social driving concept that allows to compare driving styles (e.g., fuel consumption) to provide rewards to motivate drivers to use their cars more efficiently
[16]	2015	To propose the architecture of a smart parking information service able to collect the occupancy state of parking spaces
[17]	2015	To propose an information service that warns drivers of potential dangers on the road using audio and visual alerts
[18] [19] [20]	2015 2015 2016	To propose the architecture of an information service aiming to reduce the probability of accidents involving vulnerable road users and vehicles
[21]	2016	To propose an architecture of an information service based on internet of things devices to promote efficient reservation of parking slots
[22]	2016	To propose a smart parking information service based on internet of things devices and cloud computing
[23]	2016	To propose a driver assistance based on the speed of vehicles for effective lane changing
[24]	2017	To propose a combination of internet of things and cloud computing technology to enable the drivers to reserve parking slots
[25]	2017	To propose a smart parking information service to allow the drivers to book a parking spot and to provide automatic cashless billing
[26]	2017	To propose the architecture of a crowdsourcing information system aiming to provide suitable parking options for drivers in a smart city
[27]	2017	To propose an information service based on internet of things to facilitate a safer driving by providing safety routes
[28]	2017	To propose an information service to assists drivers to perform safe and smooth crossings at the intersections
[29]	2019	To propose an index to measure the respect of drivers to pedestrian priority and an information service to measure this index aiming to influence positively the behavior of car drivers

3.1 Application Types

All the included articles intend to contribute for the development of information services to assist drivers using smart cities' infrastructures. Seven of them [13, 16, 21, 22, 24–26] propose smart parking services to advise drivers towards parking spaces, six [14, 17–20, 27] propose services to alert drivers of potential dangers, two [23, 28] present services to assist drivers in dangerous situations, and other two [15, 29] propose services to promote civilized behaviors.

Smart Parking. Since in most cities parking spaces for vehicles are a scare resource, the optimal management of vehicles parking areas represents a fundamental aspect of sustainable mobility within smart cities [13, 16]. In fact, it is estimated that 30% of the daily traffic congestion in an urban area is caused by vehicles cruising for parking spaces [16, 30]. This inefficient use of parking spaces leads to traffic congestion, waste of time and negative impacts for the natural environment.

Recent advances in low-cost, long-life battery powered internet of things devices bring the opportunity to develop new applications not only to advise drivers toward vacant parking spaces, but also to adequately monitor and manage the existing parking areas [16]. Consequently, various articles [13, 16, 21–25] suggest the development of smart parking services. The proposed solutions consist mainly in the on-site deployment of internet of things devices (e.g., infra-red or ultrasonic sensors) to monitor and signalize the state of availability of each single parking space. Moreover, article [26] propose a crowdsourcing system that aims to determine the most suitable parking options within a smart city.

The articles related to smart parking either present the architecture of the systems being proposed [16, 21, 26] or report the development of proof of concept prototypes [13, 22, 24, 25]. Additionally, article [26] also reports simulation related to the solution being proposed.

Alerts of Potential Dangers. Several applications are being designed to promote safe driving by alerting drivers of potential dangers [14, 17–20, 27], namely the proximity of vulnerable road users, such as pedestrians or cyclists [14, 18–20].

Vulnerable road users are among the users at higher risks of traffic accidents. Although the road fatalities reported in the European Union tend to decrease [18, 20, 31], the same does not happen with vulnerable road users. Moreover, accidents involving vulnerable road users occur frequently in facilities designed for them (e.g., pedestrian crossings or cycle lanes) that are close to the common traffic infrastructures [18, 20]. Therefore, several articles [14, 18–20] report studies aiming the development of applications to reduce the probability of accidents involving vulnerable road users and vehicles.

Three articles [18–20] report the same research study. The proposed information service aims both to provide more time to the drivers and vulnerable road users to take the appropriate maneuvers and avoid a possible collision by delivering warnings. For that, the vehicles are used as mobile sensors that share their positions, speed, and directions with the vulnerable road users to avoid collisions. The architecture is based on a centralized system that deploys specific wireless vehicular communications, mobile communications, and cloud computing that manages all the information obtained from the users (i.e., drivers and cyclists).

The articles present the architecture of the system and report a validation of the reliability of the messages delays. Moreover, the information service has been simulated in different scenarios to check how it performs in the worst conditions and has been deployed and tested in a real location. Based on the results, the authors concluded about the feasibility of the solution [18–20].

Article [14] propose a smart city information service to detect if a vulnerable road user intends to cross a road in a risky zone, and to issue alerts directed to the vehicles nearby. The service is responsible of performing the vulnerable road users' detection and classification. Therefore, low cost sensors (e.g., passive infrared sensors and non-contact temperature sensors) are used to determine if an object is in motion, the direction of its motion, and if it corresponds to a vulnerable road user or not. Preliminary results showed that the system was able to detect a pedestrian at risk under controlled scenarios using low computational complexity algorithms. As future work, the authors intend to build a prototype that will be able to simultaneously detect various vulnerable road users moving in different directions or moving at high speed.

Other two articles [17, 27] are not related to vulnerable road users, but also aim the provision of alerts of potential dangers situations.

Article [27] aims to promote safer driving by providing routes that include safety metrics, based on an internet of things architecture framework integrating smartphone sensing, in-vehicle sensing, wireless data communications, massive data gathering and processing, and cloud computing.

Moreover, the objective of the proof of concept prototype presented in [17] is to keep drivers informed of changing road and traffic conditions. For that, transmitter units are installed at fixed intervals along roads in the city and continuously broadcast information to receiver units located inside vehicles [17].

Assistance in Dangerous Situations. Concerning the assistance in dangerous situations, article [23] proposes a system to assist effective lane changing and article [28] describes a system aiming to assist drivers to perform safe and smooth crossings at the next intersection.

In article [23] the authors argued that in the previous works done in this field, the major studies have focused on using cameras followed by image processing for the lane change assistance and safety warning systems. In turn, according the approach presented, data from the Global Positioning System (GPS) were used to facilitate lane changing. Some tests were performed based on a prototype [23].

In article [28] wireless connectivity achieved by smart city provides the vehicles with data from the next intersection which are presented to the driver together with suggestions such as advisory speed to deal with dilemma zone, emissions and fuel consumption reduction or delay and stop mitigation [28]. The effectivity of the information service was evaluated from the perspective of the drivers using a driving simulator [28]. According the authors, experiment results showed that between 70% and 90% of subject drivers agree that the information service is a useful tool for improving the safety and mobility of their driving conditions [28].

Promotion of Civilized Behaviors. The main goal of two articles [15, 29] is the promotion of civilized behaviors.

Article [15] presents the proof-of concept of an information service-based transportation solution, called social driving, that combines the smart city concept with green transportation. According this concept, drivers can monitor different parameters of their cars and driving style, such as fuel consumption and CO_2 emission, and compare these values with the values of other drivers using similar cars and similar engine types. The solution creates different statistics and has a rewarding system to motivate drivers to use their cars more efficiently [15].

Finally, article [29] presents a social metric for the level of law compliance from drivers regarding pedestrians, at permitted crossings, the Index of Respect to Pedestrians Priority (IRPP). The IRPP calculation follows current regulations and includes factors that affect human security in pedestrian-driver interactions (e.g., ignoring pedestrian priority, speeding, improper parking or dangerous maneuvering), so that the index could be considered as a way of estimating the level of compliance of drivers and how this level evolves over time [29].

Moreover, an information service to measure the IRPP and to promote its increase over time was also presented. This service was tested on a university campus and it seems able to influence the behavior of car drivers in a positive way when interacting with pedestrians [29].

3.2 Data Sources

The included articles reported the integration of data from different types of sensors, as can be seen in Table 2. Moreover, two articles refer the acquisition of geo-tagged data [23, 26] and other three do not report data acquisition [17, 21, 29]. Only one article [13] reports the collection of city contextual information.

Table 2. Data sources.

Sensors' types	Articles
Vehicle detection sensors	[13, 16]
In vehicle sensors	[15, 27, 28]
Mobile sensors communicating the position of the vehicles	[18–20]
RFID	[16]
Infrared sensors	[22, 24, 25]
Passive infrared sensors	[22]
Ultrasonic sensors	[22]
Non-contact temperature sensors	[14]
Sensors to monitor environmental parameters	[16]

Analyzing Table 2 is possible to conclude that the types of data being collected and processed are very specific and represent just a small part of all types of data available in a smart city. Therefore, it should be questioned if the applications reported by the included articles could be deployed without smart cities' infrastructures. In fact, only one article [13] reports the collection of city contextual information (without details) and, considering the study reported by the articles [18–20], the authors referred smart cities as one of the possible scenarios. Furthermore, just five articles [13, 16, 18–20] tried to explain the alignment of the applications being reported and the smart city paradigm, and in remainder articles smart cities are mainly referred in the abstract and introduction.

3.3 Maturity of the Applications

More than half of the articles (i.e., the articles [13, 15–17, 21, 22, 24, 25, 27]) do not report the validation of the solutions being proposed.

Considering the remainder articles [14, 18–20, 23, 26, 28, 29], article [23] presents a performance evaluation of the technological solution while article [29] presents a preliminary validation of the developed application. Moreover, articles [26, 28] present and discuss simulation results of the solutions being proposed, and article [14] presents the validation results of an algorithm to detect pedestrians. Finally, [18–20] report results of simulations of the proposed service, as well as from its evaluation in real word conditions.

4 Discussion and Conclusion

Considering the current trends of research related to applications using smart cities' infrastructures to assist drivers (i.e., the first research question), smart parking was reported by seven articles [13, 16, 21, 22, 24–26] and six articles [14, 17–20, 27] (although three of them report the same study) propose information services to alert drivers of potential dangers. These two application types are complemented with two other types, each one with two articles: information services to assist drivers in danger situations [23, 28] and information services to promote civilized behaviors [15, 29].

In terms of data sources, one article (i.e., [13]) reports the collection of city contextual data (although it does not specify the contextual data). Moreover, the included articles report the gathering of several types of data: geo-tagged data [23, 26], data from vehicle detection sensors [13, 16], in vehicle sensors [15, 27, 28], mobile sensors communicating the position of vehicles [18–20], RFID [16], infrared sensors [22, 24, 25], ultrasonic sensors [22], non-contact temperature sensors [14] and sensors to monitor environmental parameters [16]. The articles do not refer the use of existing smart cities' data sources.

Regarding the maturity level of the proposed solutions (i.e., the third research question), it should be mentioned that most of the articles do not report the validation of the solutions being proposed. Moreover, other articles report prototypes that were developed to demonstrate the feasibility of the concepts. In fact, only three articles [18–20] reporting the same study refer the assessment by real users. Even so, ad-hoc questionnaires were used, the design of the experimental study and the measured outcomes were poorly described, and the number of participants was small.

This systematic review did not aim to analyze all the research being reported related to assistance to drivers. The focus was studies related to assistance to drivers that explicitly make use of smart cities' infrastructures. Therefore, it is possible to conclude that when considering smart cities' infrastructures, the articles reporting the development of information services to assist drivers are mainly focused on technology, with limited attention to value creation.

Although the authors tried, in methodological terms, to guarantee that the review selection and the data extraction of this systematic review were rigorous, it should be acknowledged that this study has limitations, namely the dependency on the keywords and the selected databases, or the fact that both grey literature and publications written in other languages than English were excluded. Nevertheless, after this systematic review is possible to state that, in the case of assistance to drivers in the smart city context, the researchers should focus on the development of added value solutions.

Acknowledgments. This work was financially supported by National Funds through FCT – Fundação para a Ciência e a Tecnologia, I.P., under the project UI IEETA: UID/CEC/001 27/2019.

References

1. Papa, E., Lauwers, D.: Smart mobility: opportunity or threat to innovate places and cities. In: 20th International Conference on Urban Planning and Regional Development in the Information Society (REAL CORP 2015), pp. 543–550. Real Corp (2015)
2. Benevolo, C., Dameri, R.P., D'Auria, B.: Smart mobility in smart city. In: Empowering Organizations, pp. 13–28. Springer, Cham (2016)
3. All, S.P., Klug, K.: Key to Innovation Integrated Solution Multimodal personal mobility. European Commission, Brussels (2013)
4. Tosi, D., Marzorati, S.: Big data from cellular networks: real mobility scenarios for future smart cities. In: 2nd International Conference on Big Data Computing Service and Applications (BigDataService), pp. 131–141. IEEE (2016)
5. Calabrese, F., Colonna, M., Lovisolo, P., Parata, D., Ratti, C.: Real-time urban monitoring using cell phones: a case study in Rome. IEEE Trans. Intell. Transp. Syst. **12**(1), 141–151 (2010)
6. Mangiaracina, R., Perego, A., Salvadori, G., Tumino, A.: A comprehensive view of intelligent transport systems for urban smart mobility. Int. J. Logist. Res. Appl. **20**(1), 39–52 (2017)
7. Sumalee, A., Ho, H.W.: Smarter and more connected: future intelligent transportation system. IATSS Res. **42**(2), 67–71 (2018)
8. Pauer, G.: Development potentials and strategic objectives of intelligent transport systems improving road safety. Transp. Telecommun. J. **18**(1), 15–24 (2017)
9. Reagan, I.J., Cicchino, J.B., Kerfoot, L.B., Weast, R.A.: Crash avoidance and driver assistance technologies-are they used? Transp. Res. Part F: Traffic Psychol. Behav. **52**, 176–190 (2018)

10. Yue, L., et al.: Assessment of the safety benefits of vehicles' advanced driver assistance, connectivity and low level automation systems. Accid. Anal. Prev. **117**, 55–64 (2018)
11. Rocha, N., et al.: Smart cities and public health: a systematic review. Procedia Comput. Sci. **164**, 516–523 (2019)
12. Rocha, N., et al.: A systematic review of smart cities' applications to support active ageing. Procedia Comput. Sci. **160**, 306–313 (2019)
13. Rico, J., et al.: Parking easier by using context information of a smart city: enabling fast search and management of parking resources. In: 27th International Conference on Advanced Information Networking and Applications Workshops, pp. 1380–1385. IEEE (2013)
14. Guayante, F., Díaz-Ramírez, A., Mejía-Alvarez, P.: Detection of vulnerable road users in smart cities. In: 8th International Conference on Next Generation Mobile Apps, Services and Technologies, pp. 307–312. IEEE (2014)
15. Ekler, P., Balogh, T., Ujj, T., Charaf, H., Lengyel, L.: Social driving in connected car environment. In: 21st European Wireless Conference, pp. 1–6. VDE (2015)
16. Mainetti, L., Patrono, L., Stefanizzi, M.L., Vergallo, R.: A Smart Parking System based on IoT protocols and emerging enabling technologies. In: 2nd World Forum on Internet of Things (WF-IoT), pp. 764–769. IEEE (2015)
17. Ksiksi, A., Al Shehhi, S., Ramzan, R.: Intelligent traffic alert system for smart cities. In: International Conference on Smart City/SocialCom/SustainCom (SmartCity), pp. 165–169. IEEE (2015)
18. De-la-Iglesia, I., Hernandez-Jayo, U., Perez, J.: CS4VRU: a centralized cooperative safety system for vulnerable road users using heterogeneous networks. In: 82nd Vehicular Technology Conference (VTC2015-Fall), pp. 1–5. IEEE (2015)
19. Hernandez-Jayo, U., De-la-Iglesia, I., Perez, J.: V-Alert: description and validation of a vulnerable road user alert system in the framework of a smart city. Sensors **15**(8), 18480–18505 (2015)
20. Hernandez-Jayo, U., Perez, J., De-la-Iglesia, I., Carballedo, R.: CS4VRU: remote monitoring and warning system for Vulnerable Road. In: 13th International Conference on Remote Engineering and Virtual Instrumentation (REV), pp. 153–158. IEEE (2016)
21. Taherkhani, M.A., Kawaguchi, R., Shirmohammad, N., Sato, M.: Blueparking: an IoT based parking reservation service for smart cities. In: 2nd International Conference on IoT in Urban Space, pp. 86–88. ACM (2016)
22. Khanna, A., Anand, R.: IoT based smart parking system. In: International Conference on Internet of Things and Applications (IOTA), pp. 266–270. IEEE (2016)
23. Joshi, J., Singh, A., Moitra, L.G., Deka, M.J.: DASITS: Driver assistance system in intelligent transport system. In: 30th International Conference on Advanced Information Networking and Applications Workshops (WAINA), pp. 545–550. IEEE (2016)
24. Mahendra, B.M., et al.: IoT based sensor enabled smart car parking for advanced driver assistance system. In: 2nd International Conference on Recent Trends in Electronics, Information & Communication Technology (RTEICT), pp. 2188–2193. IEEE (2017).
25. Hainalkar, G.N., Vanjale, M.S.: Smart parking system with pre & post reservation, billing and traffic app. In: 17th International Conference on Intelligent Computing and Control Systems (ICICCS), pp. 500–505. IEEE (2017)
26. Mitsopoulou, E., Kalogeraki, V.: Efficient parking allocation for smartcities. In: 10th International Conference on Pervasive Technologies Related to Assistive Environments, pp. 265–268. ACM (2017)

27. Taha, A.E.M.: Facilitating safe vehicle routing in smart cities. In: International Conference on Communications (ICC), pp. 1–5. IEEE (2017)
28. Lee, J., Gutesa, S.: Human factor evaluation of in-vehicle signal assistance system. In: SmartWorld, Ubiquitous Intelligence & Computing, Advanced & Trusted Computed, Scalable Computing & Communications, Cloud & Big Data Computing, Internet of People and Smart City Innovation (SmartWorld/SCALCOM/UIC/ATC/CBDCom/IOP/SCI), pp. 1–6. IEEE (2017)
29. Betancur, M.J., Restrepo, V., Pérez, J.J., Restrepo, J.A., Agudelo, A., Cuartas-Ramírez, D.: An approximation to the construction of pedestrian smart cities. In: 4th Colombian Conference on Automatic Control (CCAC), pp. 1–6. IEEE (2019)
30. Arnott, R., Rave, T., Schöb, R.: Alleviating Urban Traffic Congestion. MIT Press Books, Cambridge (2005)
31. EU Commission: Road safety in the European Union: Trends, statistics and main challenges. EU DG Mobility and Transport, Brussels (2015)

Evaluation and Modeling of Microprocessors' Numerical Precision Impact on 5G Enhanced Mobile Broadband Communications

Borja Bordel[1(✉)], Ramón Alcarria[1], Joaquin Chung[2,3], Rajkumar Kettimuthu[2,3], and Tomás Robles[1]

[1] Universidad Politécnica de Madrid, Madrid, España
{borja.bordel, ramon.alcarria, tomas.robles}@upm.es
[2] Argonne National Laboratory, Lemont, IL, USA
chungmiranda@anl.gov, kettimut@mcs.anl.gov
[3] Consortium for Advanced Science and Engineering, The University of Chicago, Chicago, USA

Abstract. Future 5G systems are envisioned to provide enhanced mobile broadband communications (eMBBC) service with bitrates in the order of 10Gbps. To support these large bitrates, the proposed schemes employ orthogonal frequency division multiplexing (OFDM) techniques to improve the spectral efficiency. These OFDM technologies are based on the calculation of Fast Fourier Transforms (FFT), whose result quality determines the orthogonal properties of the entire system. A low quality FFT calculation increases the spectral interferences and reduces the physical channel capacity, apart from the usual effects of numerical noise. Then, certain applications such as IoT solutions, where resource constrained devices with limited numerical precision are common, may experiment problems and practical limitations to be provided with eMBBC. It is essential to perform a scientific analysis of this situation to enrich and improve the future 5G standards. Therefore, in this paper we study from a theorical and simulation point of view the effective physical bitrate in 5G eMBBC scenarios including devices with limited numerical precision. We present a mathematical model considering both effects, the electromagnetic interferences and the numerical noise, and different numerical simulations analyzing relevant situations. As a conclusion, we present recommendations for future standards.

Keywords: 5G · Enhanced mobile broadband communications · Numerical models · Numeric noise · Orthogonal modulation

1 Introduction

Future 5G networks [1] are characterized by three basic innovative requirements [2]. Namely:

- Enhanced Mobile Broadband Communication (eMBBC) [3]: 5G networks should provide real bitrates above 100 Mbps at application level, with peaks around 20 Gbps.

Á. Rocha et al. (Eds.): ICITS 2021, AISC 1330, pp. 267–279, 2021.
https://doi.org/10.1007/978-3-030-68285-9_26

- Ultra-Reliable Low Latency Communications (URLLC) [4]: Network latency must be below 5 ms.
- Massive Machine Type Communications (mMTC) [5]: Future mobile networks must be able to operate in highly dense environments up to a million devices deployed per square kilometer.

Several different scientific proposals to deal with these innovative requirements have been reported [6], however (nowadays) 5G standards are only addressing eMBBC and URLLC uses cases, and only eMMBC has been standardized for standalone deployments [7]. Specifically, the new needed extremely high throughputs have caused the creation of new radio access technologies (RAT) that use enlarged wireless channels with higher physical capacities [8]. However, increased frequency channels are not enough to support the envisioned exponential growth in bitrates, and more efficient modulation schemes and physical layer control are also required. Then in 2019, the 3rd Generation Partnership Project (3GPP) proposed a global solution for RAT in 5G networks named as "5G New Radio (5GNR)", which was completely described in the Release 16 [9].

In this release, waveforms, frequency channels, operation frequencies, bandwidths and multiple access technologies are described. Specifically, in order to take as much advantage as possible from the available frequency spectrum, orthogonal frequency division multiplexing (OFDM) technologies [10] are implemented. This spectrum distribution (see Fig. 1) is the optimal manner to combine carriers in a multicarrier modulation architecture, leveraging the periodical zero values in sinusoidal signals in the frequency domain.

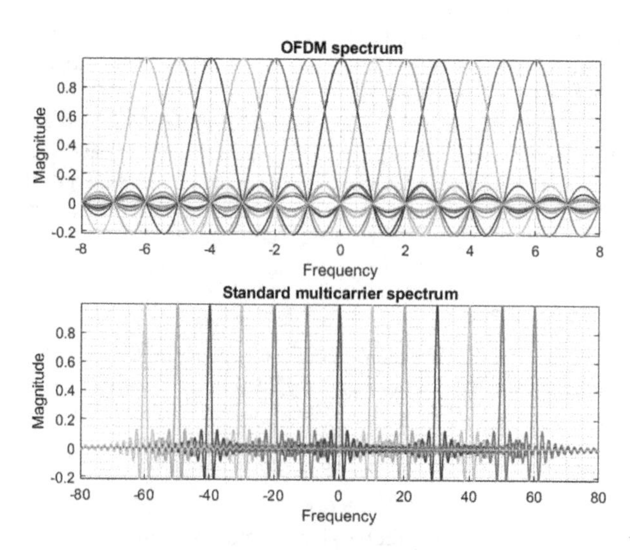

Fig. 1. Comparison between OFDM multicarrier architecture and a traditional multicarrier solution

In order to make OFDM schemes implementable at practical level, multicarrier signals are obtained through numerical procedures and using the Fast Fourier Transform (FFT) [11] instead of through costly transmissions chains. However, in this approach, the numerical quality of discrete signals obtained from FFT modules are critical, and they can highly affect the orthogonal properties of the global system. Specifically, if carriers are not correctly distributed, they tend to interfere one with each other, and then the Signal to Interference and Noise Ratio (SINR) tend to reduce. As a direct consequence, the physical bitrate in the wireless medium goes down.

5G networks are envisioned to provide high speed services to a very large catalogue of applications, including Cyber-Physical Systems [12] and Internet of Things (IoT) deployments [13], where most devices are resource constrained and are composed of microprocessors with a limited numerical precision. In these scenarios, numerical noise may be relevant, but, moreover, the quality of obtained FFT calculations might be affected. Rounding effects propagated through the large amount of operations involved in FFT calculations can generate a very unexpected impact. In this case, eMBBC could not be supported, but not because of the network, but because of limitations in the user devices. Regardless, the situation must be evaluated from a scientifical point of view, so future 5G standards can address the problem and include specific recommendations and solutions for these scenarios.

In this paper we analyze the impact of numerical precision from user microprocessors in the physical wireless channel capacity, specially in those cases where this precision is limited. We propose a mathematical model where both, the standard numerical noise and the interferences caused by error in the FFT calculation are considered. The signal processing and transmission laws, and the Shannon's law are employed to model this effect. Furthermore, we perform numerical simulations to determine the real impact of this effect and provide some recommendations for future standards.

The rest of the paper is organized as follows: Sect. 2 describes the proposed 5G New Radio scheme. Section 3 describes the main proposal including a mathematical model, and Sect. 4 includes some initial simulations and discussions about the proposed model. Finally, Sect. 5 presents our conclusions.

2 5G New Radio

Different carrier frequencies have been proposed for future 5G RAT [14], but distributed standards with Release 16 by 3GPP are only considering values below 52.6 GHz (however, frequencies up to 100 GHz are envisioned to be discussed in future releases). Currently, only two specific frequency ranges are being used to implement 5G NR:

- Frequency Range 1 (FR1) or Sub-6 GHz frequency bands. This range includes traditional frequency bands employed in 2G, 3G and 4G communications, as well as some new bands around 3.5 GHz or even 4.7 GHz

- Frequency Range 2 (FR2) or Millimeter waves (mmWave) in the range of 26.50 GHz to 43.50 GHz. This range includes traditional bands for satellite communications such as Ka-band or V-band

In all cases, 5G radio networks implement OFDM technologies, but with a variable sub-carrier spacing (SCS) (i.e., the distance between the peak value of two consecutive sub-carriers in the frequency domain is variable). This new OFDM technology is known as Multiple OFDM numerologies (contrary to 4G schemes where all sub-carriers are separated 15 KHz). OFDM numerology is represented by an integer parameter μ, so the SCS may be easily calculated (1).

$$SCS = \Delta f = 15 \cdot 2^{\mu} KHz \tag{1}$$

In 5G NR each user is provided with a frequency bandwidth known as Component Carrier (CC), whose maximum value is 400 MHz. Furthermore, in a frequency band up to 16 CC (carriers) can be created and introduced. Inside each CC, an OFDM multi-carrier structure is deployed, including 3300 subcarriers. Each group of 12 subcarriers is named as Physical Resource Block (PRB). OFDM in 5G NR may be supported by different constellations such as BPSK, QPSK (most commonly), 16-QAM, 64-QAM and 256-QAM. Figure 2 represents the expected spectrum structure in a 5G NR deployment.

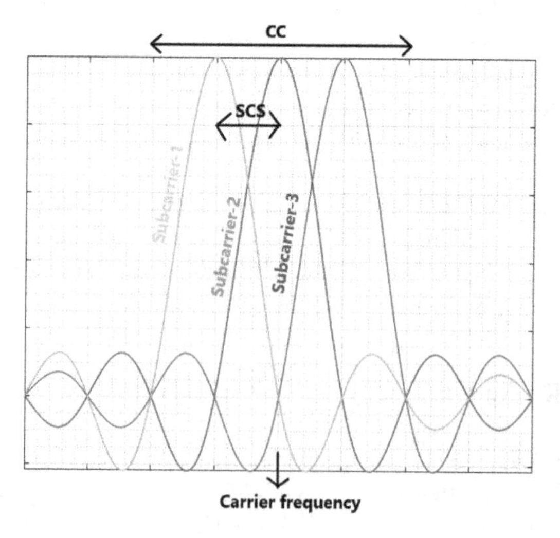

Fig. 2. Frequency spectrum in 5G NR

Each one of the described frequency ranges (see Table 1) is employed in a different scenario, and presents different spectral characteristics: modulation, sub-carrier spacing, CC, etc.

5G NR applies a TDMA (Time Division Multiple Access) scheme, where time is divided into frames with a duration of 10 ms. These frames are further divided into ten

Table 1. Frequency ranges and main characteristics in 5G NR

Parameter	FR1	FR2
Application	Macrocells, high user mobility	Small cells, low mobility
Frequency band	Sub 6 GHz	mmWave
SCS	15, 30, 60 kHz	60, 120 kHz
CC	5, 10, 15, 20, 25, 30, 40, 50, 60, 80, 100 MHz	50, 100, 200, 400 MHz
Frame/subframe duration	10 ms/1 ms	
Baseband modulation	BPSK, QPSK, 16QAM, 64QAM, 256QAM	
Max. Number of carriers/subcarriers	16/3300	
Spectrum organization	CP-OFDM, DFT-s-OFDM	

(10) subframes with a duration of 1 ms. Each subframe is, then, divided into different slots with a variable duration T_{slot} (2) depending on the numerology. Each slot contains a sequence of 14 different OFDM symbols sent in a row. If required, mini-slots containing only two, four or seven OFDM symbols could be defined, but in this paper we are not considering this possibility.

$$T_{slot} = \frac{1}{2^{\mu}}\, ms \tag{2}$$

In order to reduce as much as possible temporal interferences, standard OFDM scheme is enriched with a cyclic prefix separating each consecutive pair of symbols. This approach is known as CP-OFDM (Cyclic prefix OFDM). On the other hand, to allow user devices to manage 3300 subcarriers, OFDM scheme is supported by a signal processing algorithm (instead of by standard transmission chains) based on the Fast Fourier Transform or FFT (see Sect. 3). Specifically, a 4096 sample FFT must be calculated (the first power of two above 3300). This approach is known as DFT-s-OFDM or Direct Fourier Transform sparse OFDM. This low-cost OFDM implementation is, in fact, the base of our study and model.

3 eMBBC and Low Precision Microcontrollers: A Model

In this section, we analyze from a mathematical point of view the OFDM modulation process, so we can deduct the impact of a limited numerical precision in terms of electromagnetic interference and numerical noise.

3.1 OFDM Technologies

OFDM spectrum is organized in such a way that zero values of adjacent subcarriers are placed at the same point as maximum values of the central subcarrier (see Fig. 1). This condition is met only either when subcarriers are in quadrat (phase shift of $\frac{\pi}{2}$ radians) or when the relation between their angular frequencies is an integer number. In those situations, subcarriers are orthogonal if we consider subcarriers as vectors in the vector space $\mathcal{C}([-\infty, +\infty])$ of continuous functions in the real line $\mathbb{R}$, the traditional function scalar product and the Lebesgue integral.

Although OFDM may be used in baseband transmissions (such as in DSL-Digital Subscriber Line-technologies), in 5G networks the entire CC must be placed in the assigned frequency band, centered at f_c hertz. Thus, the temporal signal $s(t)$ corresponding to an OFDM spectrum can be easily deducted (3).

$$s(t) = \sum_{k=-N/2}^{\frac{N}{2}-1} d_k \cdot e^{\frac{j2\pi k}{T} \cdot t} \cdot e^{j2\pi f_c t} \tag{3}$$

Being, N the number of subcarriers (3300 in standard 5G networks), T the symbol period (calculated according to the numerology and the frame distribution), f_c the carrier frequency and, $\{d_k\}$ the symbols to be transmitted according to the employed baseband modulation (or constellation). Each subcarrier could be combined with a different constellation, but in this case, we are using the same constellation for all subcarriers: QPSK. Symbols are not represented with analog signals (sinusoidal signals) but with complex numbers, so the constellation is understood as the complex plane.

The second exponential term $e^{j2\pi f_c t}$ only contains information about the carrier frequency and can be extracted from the sum (4), while the first exponential term $e^{\frac{j2\pi k}{T} \cdot t}$ has, formally, the same mathematical expression as the inverse discrete Fourier Transform (IDFT). Only two differences can be seen: a scalar factor and the fact that IDFT works on discrete time, and the exponential term is defined in the continuous time. Then, a very efficient manner to obtain the analog temporal signal $x(t)$ represented by the sum (4) is through an IDFT calculation module in discrete time (numerical algorithm), followed for a Digital-to-Analog converter (DAC) (5). The obtained signal, then, is easily mixed with the carrier using standard communication techniques (see Fig. 3). As IDFT generates complex signals, both the real and the imaginary parts are combined in the radio signal using carriers in quadrature. Moreover, in order to increase as much as possible, IDFT is calculated through the most efficient algorithm: the Fast Fourier Transform (FFT). The reception process follows an equivalent mechanism.

$$s(t) = e^{j2\pi f_c t} \cdot \sum_{k=-N/2}^{\frac{N}{2}-1} d_k \cdot e^{\frac{j2\pi k}{T} \cdot t} \tag{4}$$

$$\begin{aligned} s(t) &= e^{j2\pi f_c t} \cdot [ADC(Re\{IDFT(\{d_k\})\}) + ADC(Im\{IDFT(\{d_k\})\})] \\ &= e^{j2\pi f_c t} \cdot x(t) \end{aligned} \tag{5}$$

Then, the radio signal in the wireless medium is affected by two numerical effects: the numerical noise introduced by ADC and the interferences in the frequency spectrum, due to impurities in the FFT calculation. These effects are, mainly, caused by the limited numerical precision of microprocessors and, at then, reduce the SINR in the aerial media. In conclusion, and because of the Shannon–Hartley theorem (6), the radio channel capacity (bitrate) C is going to decrease. In this initial work, we are assuming the SINR is frequency independent.

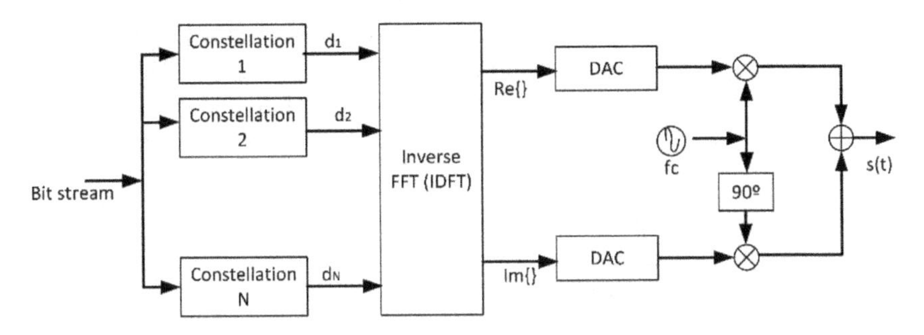

Fig. 3. OFDM transmitter

$$C = \int_0^{CC} \log_2(1 + SINR(f))df = CC \cdot \log_2(1 + SINR) \tag{6}$$

In particular, the SINR includes four relevant components: the signal power S, the interference power I caused by numerical errors in FFT calculation, the numerical noise N_n due to DAC conversion, and the combination of physical, electromagnetic, etc. noises and interferences N_0 (7).

$$SINR = \frac{S}{I + N_n + N_0} \tag{7}$$

The following subsections analyze these components in details.

3.2 Fast Fourier Transform

The FFT is an efficient manner to compute DFT. This algorithm decomposes DFT using a binary tree, so the global 2^m-sample DFT is obtained as the combination of two 2^{m-1}-sample DFT. This decomposition process is repeated until the original DFT is obtained as the combination of 2^{m-1} 2-sample DFT. This approach is only applicable if the DFT to calculate operates on 2^m points. Therefore, in 5G OFDM schemes a 4096-FFT is employed (as 4096 is the closest number to 3300 in the form 2^m). We are naming N_s the number of samples in the FFT.

In general, resource constrained microcontrollers operate with fixed point arithmetic and only real definitions. We are considering the point is placed at the left of the leftmost bit. In this context, the number of real additions a and real multiplications p required to calculate an FFT may be easily obtained (8). In this paper we are considering the Cooley–Tukey algorithm [15]. A microprocessor with a B bits (plus sign) architecture and signed operations, then, generates the following errors with each operation:

- A real binary addition generates an overflow. If this bit is null, no error is introduced. If this bit is not null, and we assume the result is shifted right to reduce as much as possible the numerical error, a distortion of 2^{-B} units is induced.
- A real multiplication of two B bits words has a length of $2B$ bits. The result must be, then, truncated to B bits another time. If all these bits are zero, no error is introduced. In any other case, an error between 2^{-2B} and 2^{-B} is generated.

Each one of these errors are characterized by an uniform probability distribution, whose extreme values are indicated and whose variances σ^2_{sum} and σ^2_{mul} can be calculated using the probability theory (9).

$$a = N_s \cdot \log_2 N_s$$

$$p = 2N_s \cdot \log_2 N_s \tag{8}$$

$$\sigma^2_{sum} = 2^{-(2B+1)}$$

$$\sigma^2_{mul} = \frac{2^{-2(B+1)}}{3} \tag{9}$$

These errors generate interference among subcarriers in the OFDM modulation. These interferences are caused by added power to frequencies were no power should appear. In this case, as said before, the same power I is added to all frequencies and may be calculated as the variance of the global error. If we assume all operations are physically independent in the FFT algorithm, the global variance may be obtained as the sum of all partial variances (10), and then the interference power is obtained.

$$I = \frac{1}{a+p}\left(a \cdot \sigma^2_{sum} + p \cdot \sigma^2_{mul}\right) \tag{10}$$

3.3 Numerical Noise

The numerical noise in the OFDM modulation is caused by the DAC, where each binary number is translated to a fixed analog voltage. As a result, no smooth continuity is actually generated. These steps in the generated analog signal cause an additive noise affecting the SINR.

In this paper, the DAC is characterized by the number of bits it manages B (equal to the number of bits in the microprocessor architecture, so errors are minimized), and the maximum voltage it can manage V_m. This voltage can be easily obtained from signal power if an unitary resistor is assumed to be the payload (11). In this context, 2^B steps can be defined, for an analog range of V_m volts. In the worst case, an error equal to a whole step would be generated, in the best case, no error is introduced. If all values have the same probability, the variance of the induced error may be easily obtained, and then, the numerical noise power (12).

$$V_m = \sqrt{S} \tag{11}$$

$$N_n = \frac{1}{12}\left(\frac{V_m}{2^B}\right)^2 \tag{12}$$

4 Experiments and Simulations

The 5G requirements must be met regardless the network configuration, even if the smallest CC is considered. Thus, in this section we are analyzing the behavior of bitrate (channel capacity) and SINR in that use case. Other network parameters are selected to represent a standard situation (see Table 2).

Table 2. Frequency ranges and main characteristics in 5G NR

Network parameter	Value
CC	5 MHz
S	25 dB
N_s	4096

The proposed model (7–11) in Sect. 3 is then employed to evaluate both relevant variables when microprocessors with different precision (number of bits B) are employed. Besides, different amounts of physical noise are considered.

Figure 4 shows the obtained results for the SINR parameter. As can be seen, for very small amounts of physical noise, the numerical effects determine the global behavior and values are independent from the N_0 variable.

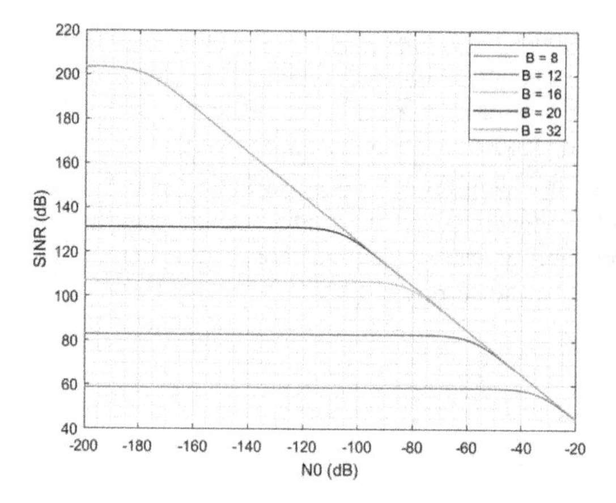

Fig. 4. Evolution in the SINR depending on the physical noise and the microcontroller precision

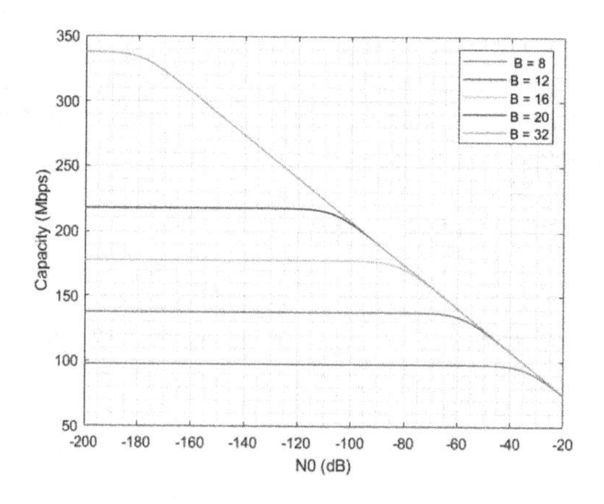

Fig. 5. Evolution in the bitrate depending on the physical noise and the microcontroller precision

In this area, the SINR value is related to the microcontroller precision, so new bit is added to the architecture, approximately, the SINR improves in 6 dB. Once a threshold value is passed, the physical noise is relevant and the SINR decreases linearly with N_0.

Figure 5 shows how this SINR evolution affects the effective channel capacity. Basically, the formal evolution is identical, although it is relevant to remark that for low-cost microcontrollers with a low-precision architecture (only eight bits), the 5G network cannot provide the user devices with real eMBBC.

SINR, and the bitrate, could be improved if higher signal powers are considered. However, errors in DAC also depend on this value in a direct way, so the SINR cannot increase easily using that approach. In those situations, we greatly recommend considering a subsampling/oversampling strategy. In that scheme, although a higher number of operations is involved, errors in the DAC can be reduced in M times (being M the oversampling ratio).

In order to show the potential effect of this subsampling/oversampling strategy, we are proposing a final scenario as example. In most future technological systems, such as Smart Grids, real-time applications only capture samples each few seconds. But nowadays even the cheapest microcontroller may obtain a new sample each few microseconds. This difference of almost six magnitude orders enable acquiring more samples than required and use them to reduce the numerical noise in the DAC. Using big numbers, the noise power reduces as many times (M) as the sampling speed is above the minimum. Figure 6 shows a possible example where $M = 100$. As can be seen, because of the reduction in the numerical noise, now eMBBC can be provided regardless the microprocessor precision. An improvement of 70 Mbps is obtained, although up to 100 Mbps could be reached in theory, because of errors in FFT which are not affected by the oversampling technique. As a disadvantage, the number of executed operations in the microprocessor grows up significantly, so the power con-

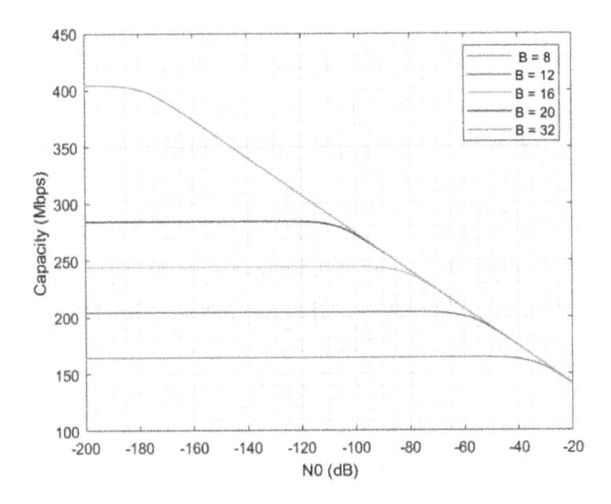

Fig. 6. Evolution in the bitrate depending on the physical noise and the microcontroller precision. Example for subsampling/oversampling strategy

sumption also goes up in a relevant way. This situation may be important in battery-based systems, where the battery could get drained faster.

5 Conclusions

In this paper we study from a theorical and simulation point of view the effective physical bitrate in 5G eMBBC scenarios including devices with limited numerical precision. It is provided a mathematical model considering both effects, the electromagnetic interferences and the numerical noise, and different numerical simulations analyzing relevant situations. The signal processing and transmission laws, and the Shannon's law are employed to model this effect.

Results show, under certain conditions, the effect of these numerical errors is relevant and then, some applications scenarios cannot be provided with eMBBC. A subsampling/oversampling scheme would help to address this problem.

In future works, we are analyzing the relation between the effective bitrate and the battery consumption is scenarios where the subsampling/oversampling scheme is implemented.

Acknowledgments. This work is supported by the Ministry of Science, Innovation and Universities through the COGNOS project.

References

1. Sánchez, B.B., Sánchez-Picot, Á., De Rivera, D.S.: Using 5G technologies in the Internet of Things handovers, problems and challenges. In 2015 9th International Conference on Innovative Mobile and Internet Services in Ubiquitous Computing, pp. 364–369. IEEE, July 2015
2. Bordel Sánchez, B., Alcarria, R., Robles, T., Jara, A.: Protecting physical communications in 5G C-RAN architectures through resonant mechanisms in optical media. Sensors **20**, 4104 (2020)
3. Gamage, H., Rajatheva, N., Latva-Aho, M.: Channel coding for enhanced mobile broadband communication in 5G systems. In Proceedings of the 2017 European Conference on Networks and Communications (EuCNC), Oulu, Finland, 12–15 June 2017, pp. 1–6 (2017)
4. Pocovi, G., Shariatmadari, H., Berardinelli, G., Pedersen, K., Steiner, J., Li, Z.: Achieving ultra-reliable low-latency communications: challenges and envisioned system enhancements. IEEE Netw. **32**, 8–15 (2018)
5. Bockelmann, C., Pratas, N.K., Nikopour, H., Au, K., Svensson, T., Stefanovic, C., Popovski, P., Dekorsy, A.: Massive machine-type communications in 5g: physical and MAC-layer solutions. IEEE Commun. Mag. **54**, 59–65 (2016)
6. Bordel, B., Alcarria, R., Robles, T., Sánchez-de-Rivera, D.: Service management in virtualization-based architectures for 5G systems with network slicing. Integrat. Comput.-Aided Eng. **27**(1), 77–99 (2020)
7. Ghosh, A., Maeder, A., Baker, M., Chandramouli, D.: 5G evolution: a view on 5G cellular technology beyond 3GPP release 15. IEEE Access **7**, 127639–127651 (2019)
8. Hoglund, A., Van, D.P., Tirronen, T., Liberg, O., Sui, Y., Yavuz, E.A.: 3GPP release 15 early data transmission. IEEE Commun. Stand. Mag. **2**(2), 90–96 (2018)
9. Sengupta, A., Alvarino, A.R., Catovic, A., Casaccia, L.: Cellular terrestrial broadcast–physical layer evolution from 3GPP release 9 to release 16. IEEE Trans. Broadcast. (2020)
10. Nee, R.V., Prasad, R.: OFDM for wireless multimedia communications. Artech House, Inc. (2000)
11. Nussbaumer, H.J.: The fast Fourier transform. In Fast Fourier Transform and Convolution Algorithms, pp. 80–111. Springer, Berlin (1981)
12. Bordel, B., Alcarria, R., Robles, T., Martín, D.: Cyber–physical systems: extending pervasive sensing from control theory to the Internet of Things. Pervas. Mob. Comput. **40**, 156–184 (2017)
13. Bordel, B., Alcarria, R., De Andrés, D.M., You, I.: Securing Internet-of-Things systems through implicit and explicit reputation models. IEEE Access **6**, 47472–47488 (2018)

14. Wang, Y., Li, J., Huang, L., Jing, Y., Georgakopoulos, A., Demestichas, P.: 5G mobile: spectrum broadening to higher-frequency bands to support high data rates. IEEE Veh. Technol. Mag. **9**(3), 39–46 (2014)
15. Mersereau, R., Speake, T.: A unified treatment of Cooley-Tukey algorithms for the evaluation of the multidimensional DFT. IEEE Trans. Acoust. Speech Signal Process. **29**(5), 1011–1018 (1981)

Design of a Fog Controller to Provide an IoT Middleware with Hierarchical Interaction Capability

Daniel S. do Prado(✉), Francisco L. de Caldas Filho, Lucas C. de Almeida, Lucas M. C. e. Martins, Fábio L. L. de Mendonça, and Rafael T. de Sousa Jr.

Faculty of Technology, Electrical Engineering Department, University of Brasília, Brasilia, DF 70910-900, Brazil
{daniel.prado,francisco.lopes,lucas.almeida,lucas.martins, fabio.mendonca}@redes.unb.br, desousa@unb.br

Abstract. With a large number of IoT devices currently and growing, it is of great importance to orchestrate them so that all exercises are fully functioning. This project focuses on the creation of low-cost IoT middleware capable of decentralizing responsibilities, processing, and decision making in a network. This work addresses the construction of an IoT middleware itself, showing the pillars based on its construction - using microservices in its composition - from the storage of information, dealing with particularities of the proposed architecture, up to its ability to react to certain scenarios. It also presents how to use fog concepts to decentralize middleware responsibilities at different points in a hierarchy in an IoT network. Finally, it is commented on the limitations of the following project and possible parts for improvement to be considered in future projects.

Keywords: Internet of Things · Middleware · Fog Computing · Microservices

1 Introduction

Large amounts of data are sent all the time on the Internet. When the scenario of the popularization of IoT technologies is added to the analysis, the projection is that there will be up to 3.9 billion IoT devices connected in the year 2022, an increase of 126% compared to 2021 [5]. With the increased demand for network resources, problems arise related to packet loss and the need for retransmission. Cordeiro et al. [6] explain the different classes of applications and how the consideration of losses and retransmissions are important depending on the nature of each communication process. Congested networks will generate packet queues up to the maximum buffer size on the routers, resulting in packet discard and requiring retransmission of messages do ages that operate using protocols such as TCP (Transmission Control Protocol).

Á. Rocha et al. (Eds.): ICITS 2021, AISC 1330, pp. 280–292, 2021.
https://doi.org/10.1007/978-3-030-68285-9_27

The disposal and retransmission of packets will directly impact the end time that the mediating component of the IoT infrastructure, called IoT middleware, has with the devices. In this context, the time it takes for data to leave final devices and reach a central processing unit are increased, which can take varied periods and, sometimes, these periods can get longer than expected for a specific application context. It is important to consider this delay, some applications need immediate responses. Although there is a universal tendency to use cloud computing in diverse contexts and varied applications. Considering an IoT network, this practice may not be efficient, since the variations in the time between the arrival and processing of information on the central servers of the distributed system can reach long periods depending on the demand for requests and the distance between the servers and the final devices.

Opportunely, Fog computing defines the replacement of the part, or all, of the responsibilities of the cloud with local processing, or as close as possible to the data source, as described by M. Aazam et al. [1]. In this way, it is the most suitable to overcome these obstacles, bringing more agility and allowing decision making independently of employees or human intervention [7]. When using fog computing concepts, much of the processing performed in robust data center units are exchanged for several low-cost devices with less processing capacity, such as computers on a single card (SBC).

According to Martins et al.[12], using microservices to build an IoT software is more versatile considering the modularity and low coupling of applications. Also, the less binding allows the inclusion or removal of modules without losing the functionality of the application. The user only has to define if it is necessary or not of that service for that application [12].

Our work focuses on the use of fog computing concepts to create, in SBC's, a middleware capable of working autonomously, making it possible to carry out actions to react to new environmental conditions. Its use will also be presented in the hierarchy, with each unit exercising its attributions within its contexts and, when necessary, seeking to know the state of adjacent applications for decision making.

This work is divided into five sections, this being the Introduction. Next, Sect. 2 reviews some concepts and published papers that are relevant to the composition of this work. Section 3 technically contextualizes the methodology used in the project and then Sect. 4 presents the measurements and verification made with the execution of this proposal. Finally, Sect. 5 comments on the difficulties, possibilities of extension, and future perspectives.

2 Related Works

The novelty of IoT technologies and the expectations related to their applications motivated the development of different protocols, with communications on different transmission frequencies and for an unprecedented variety of scenarios, with cloud integration and approaches in fog computing, which adds complexity to the work of integration, as can be concluded after careful reading by Al-Fuqaha et al. [2], Dizdarević et al. [8], and Noura et al. [14]. Therefore, the survey exposed in Rahman & Hussain [15] is a reference for the correct understanding of the taxonomy used in this work.

The works Datta et al. [7], Kum et al. [10], and Giang et al. [9] present approaches and concepts that clarify the choice and relevance of the use of fog computing in the context of IoT, whereas the works Menezes et al. [13], Martins et al. [11], and Caldas Filho et al. [3] are relevant examples of initiatives with similar and/or complementary objectives to the project in question, of integrating different technologies and abstraction of services within a platform. The main difference about the present work is related to the notion of rules implemented in a hierarchical chain of devices.

In Menezes et al. [13], it is possible to experimentally observe the advantages of using distributed architectures, however, even with the use of an execution module and tasks, the project still did not aim at the application of a complete ontology associated with the application of pre-established rules. In the proposition of Martins et al. [12], the concepts related to microservices are highlighted in the context of IoT, however, the problems related to the application of rules and the practical implementation of a distributed architecture for IoT customer data were not yet part of the objectives of the work. Therefore, the complete mechanic of rules that perform tasks is the most valuable addition of the project presented in this paper in relation of these two works.

In the work provided by Caldas Filho et al. [4], the level of integration achieved through the design and implementation of a semantic IoT gateway is extremely valuable, however, a microservice-oriented architecture with a more practical approach from the point of view of communications coming from the middleware itself for its network would prove to be more complex than in the present work, given the need to orchestrate different drivers for each technology.

There are still few studies that fully propose a set of hardware and software components that provide a complete and scalable infrastructure not only for data concentration, but also for modularization of services, external integrations and universal application of rules for logical action of actions, and these are the main points to be proposed in this work.

3 Proposed Architecture

The UIoT architecture described in [16] was used as the basis this work. It is composed by several modules and, for the purpose of this work, we highlight the DIMS and Gateway modules. The former is the service that has the function of storing data and the latter communicates with physical devices through different protocols. Together they form a more complex service, called UIoT Middleware. Like a structure divided into blocks, middleware can be assembled as required by the environment and the required processing power [12].

It is important to remember that an IoT middleware is usually built in datacenter environments, however, the UIoT middleware was developed in a minimalist interface, with the ability to be executed in SBC's, being able to be orchestrated by a central element, which controls the processing and distributed storage activities, forming a Fog Computing architecture. Remote middleware can perform the same tasks as cloud middleware, respecting the processing and storage limitations of the hardware that supports it.

As mentioned, our middleware is a composition of microservices that are independent of each other, which together make up a more complex set. In Fig. 1, the interior of a UIoT middleware is described, focusing only on the modules of interest for this project. The boxes with dotted borders highlight the elements that were developed for this work. These new additions into UIoT architecture bring new capabilities to the middleware, as the possibility to process data in a decentralized way.

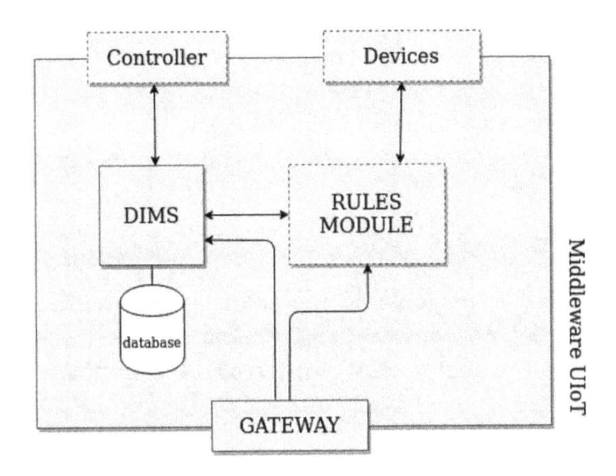

Fig. 1. Schematics of an UIoT Middleware.

3.1 Gateway

The Gateway instance is responsible for creating a communication bridge between the IoT devices in the environment with the UIoT Middleware. Using several communication protocols of different layers (such as HTTP, ZigBee, LoRA, among others), its function is to translate and standardize the received data and send it in JSON (JavaScript Object Notation) format to the API (Application Programming Interface) of the middleware data storage unit.

3.2 DIMS

The responsibility of the Database Interface Management System (DIMS) is for data persistence in the UIoT Middleware. NoSQL MongoDB technology was used as a database and the Python programming language for creating the API. It is up to DIMS to store data from the UIoT hierarchy in the following documents: client, service, data, and rule.

The client is an IoT device (hardware with embedded software) responsible for communicating with the gateway, sending information about its services, and receiving commands. The service is an instance of the IoT device where its function is to collect data from its attached sensor and send it to the client. Therefore, the data will be the smallest unit of information generated by the device, which is associated with a specific service that contains the value of the sensor, the collected time, and keys related to the service.

Developed for this work, the rule is the representation of a conditional interaction between the service and the instance of middleware output. In other words, it defines which services will be analyzed for a certain condition and which middleware output interface will be triggered when the condition becomes true. The first three documents in the database function in a hierarchy, where the client is the centralizer of the data that is in the services, and the rule connects services with interfaces.

After understanding how data comes out of a physical device and is stored in the UIoT Middleware, it is important to focus now on the central point of this work: The autonomy of middleware in the middle of a decentralized ecosystem. The function of the modules responsible for this work is described below.

3.3 Device Interface

The Device interface is responsible for creating communications between an UIoT Middleware with the devices on the local network, trying to understand what actions can be executed and send control commands to the hardware (client).

3.4 Rules Module

The rules module allows the node to have local processing capacity, analyzing data that reaches the middleware and executing actions previously established in the DIMS rule collection, essentially demanding no human interventions. Starting with a data entry event, it is checked if there are rules related to that service, if any, the module is in charge of verifying its activation condition, that if it is satisfied, interaction is made with the described middleware exit interfaces.

3.5 Controller

The controller is able to create asynchronous communication between different UIoT Middlewares, making it possible to request or send information between different network nodes. A single point on the network can request data from other adjacent locations for analysis, as well as send control instructions to other middleware.

As shown in Fig. 2, the controller has 6 major classes: Sender, Receiver, Core, Notify, Getter, and Setter. Each of them has a specific function within the controller, and will be explained below: The **Receiver** is responsible for receiving data from outside middlewares and forwarding it to the controller core. The **Sender** has the function of receiving requests from the Core and sending them

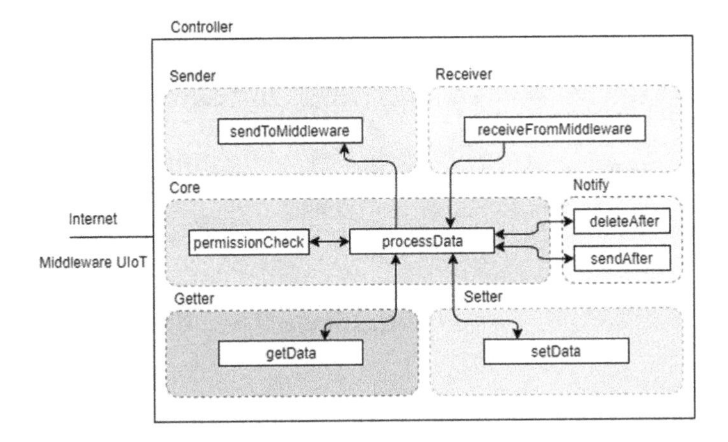

Fig. 2. Architecture of controller.

to other middleware. The instance **Getter** and **Setter** are similar, being responsible for requesting data from the database or including new data, respectively. **Core** is the centerpiece of this application. It is responsible for orchestrating requests and handling them. When any of the high-end applications (Getter, Setter, Sender, and Receiver) interface with this part of the application, the permission is checked, and finally, happen the request redirection.

Finally, the class **Notify** has the function of automating this module to either send data to other middlewares every period or delete data from the local database after some time. That is very useful to trigger interactions of automatic data sending.

For a request between different network points, it is necessary to inform parameters in the body of the request, such as the destination address, the purpose of the request, the content to be sent, among others.

It is already possible to create an autonomy of the nodes with reactive interactions for events only with the rules module. By adding the Controller, the possibility of visualization and interaction between different points of the network is added.

As shown in Fig. 3, a hierarchical UIoT topology was created using three middlewares to test the proposed architecture: one central to the network that represents a college, and two others (with associated Gateways) that represent the departments of that college. IoT devices periodically send their data to the gateways, subsequently sending them to the associated middleware, as already exemplified in Menezes et al. (2019) [13]. The central node, on the other hand, has the function of gathering data from its associated middleware. This proposal can be analyzed in two different contexts. When considering departments, you can only see the IoT devices connected to them. When analyzing the central middleware, you have a broad and complete view of the network because it can use subordinate node's data.

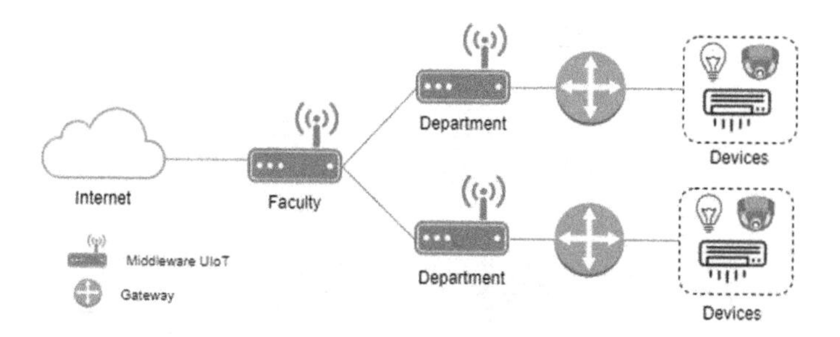

Fig. 3. Proposed topology.

4 Tests and Results

For the tests, two different scenarios were considered for the applicability of the solution: For the first scenario, the interactions between middlewares and IoT devices were evaluated using the rule as the focus. For the second scenario, the interaction between different UIoT Middleware was tested, focused on the exchange of information between nodes in the proposed hierarchy.

4.1 Scenario 01

A virtual machine with the Ubuntu 18.04 LTS operating system was used for the simulations. The modules DIMS, Gateway, Rules Module, and Device Interface were used in the middleware assembly to create a middleware with rule execution capabilities. We simulate IoT devices via software to assure they have a repeatable behavior to minimize the risk of a physical-world scenario of total or partial uncontrolled variables.

In this test scenario, 3 IoT devices were configured to send their data to the local UIoT middleware (sensors) and 2 IoT devices were able to receive data from the middleware (actuators). The sensor devices are a security camera, a temperature sensor, and a lighting sensor. The IoT camera uses a security camera and computer vision to detect the number of people in the environment. The temperature sensor reports an integer value of the room's ambient temperature. The lighting sensor in the room informs the state of the room: illuminated or not illuminated. The actuator devices are an air conditioning controller and a lighting controller. The air conditioning control can command the device to turn on, off, or set a temperature. Lastly, the lighting controller can command the lamps to change their state on or off.

With these devices already connected to the middleware, we could test rule execution feature. It is important to highlight that the rules can be conveniently shaped according to the situation. Considering our registered devices and our simulated scenario, it was decided to register five rules, as listed below:

1. Rule 01: If the IoT camera informs that there are no people in the room, the middleware should ask to turn off the air conditioner and to turn off the lights;

2. Rule 02: If the IoT camera reports that there is exactly 1 (one) person in the room, then the middleware should ask to turn on the lights;
3. Rule 03: Case the IoT camera reports that there are more than 1 (one) person in the room and the temperature sensor informs that the temperature is above 24°, then the middleware should ask to the air conditioner controller to adjust its temperature to 22°;
4. Rule 04: Case the IoT lighting sensor reports that the room is lit, then the middleware should ask to the lighting controller to turn lights off;
5. Rule 05: Case the IoT temperature sensor informs that the environment is below 22°, the middleware should ask to turn off the air conditioning.

For the first test, changes were made to the values of the devices individually, excluding data between one test and another. For the second test, the sensors were changed to emulate a situation more similar to the real, with the mixture of sending information and without excluding data.

Test 1: For unit tests, the rules for the IoT camera, temperature controller, and lighting controller were individually checked. The values and results are in the following Tables 1, 2, and 3.

Table 1. Camera results

Received value	Result
0	Rule 01
1	Rule 02
2	None

Table 2. Temperature results

Received value	Result
20	Rule 05
22	None
24	None
26	None

Table 3. Lighting results

Received value	Result
1	Rule 04
0	None

As noted in Table 1, when the camera device detected 0 or 1 person, the respective rules had only one sensor as an input and therefore were directly activated. When the sensor reported that there were 2 people, the rule in question depended on the value of the camera and the value of the temperature device, which was not being sent, so the rule was not activated. For the temperature sensor, Table 2 shows that the conditions of 20, 22, 24, and 26° of temperature were tested. Similar to the one observed in the results of the camera, the rules were only activated if its registration was done exclusively involving only the device. So only rule 05 fulfilled the condition. Finally, considering the lighting device, only one rule was exclusively dependent on the value of the device, and as expected, it was performed, as shown in Table 3.

Test 2: In this test, the sensors send information in a more similar way to a real situation in which multiple devices simultaneously send data to the middleware.

Table 4. Random data sending

Time	Interaction	Activated rule	Environment
1	Camera = 0	01	Air conditioning = Off, Light = Off
2	Camera = 1	02	Air conditioning = Off, Light = On
3	Temperature = 25	None	Air conditioning = Off, Light = On
4	Lighting = 1	04	Air conditioning = Off, Light = Off
5	Camera = 2	03	Air conditioning = 22, Light = Off

Table 4 shows this test results, with emphasis on the Time column that describes the order in which information was received.

As shown in the row of time 4 in Table 4, when the lighting sensor identifies that the room is lit, an action is taken to turn off the lights in the room. In the daytime, this interaction makes sense, but at night there is a problem: if a person reaches the environment, by rule 02 the lights are activated, but if they are activated, the lighting device will recognize that the environment is lit and will process rule 04 by turning off the lights in the room. This leads us to question the condition of occurrence of events for the execution of the rules since the rule fulfilled exactly what was proposed to be done. As noted, care must be taken when registering a rule with only one service at its entry, as this device only needs to reach the activation condition for the event to occur. The most appropriate is to further filter the scenarios of interest, including other IoT devices at the entrance so that the rule's execution takes place in a more specific scenario. In the case of rule 04, it would be more interesting to include, in addition to the lighting device, the absence of people in the environment so that, in a night situation, this rule does not run while there are people in the place.

Scenario 02: For this scenario, three UIoT middlewares were used at different hierarchical levels: one representing a faculty and the other two representing departments (as shown in Fig. 3). For the department's middleware, the one used in tests 2 of scenario 01 was reused. The first possible action to be taken from the faculty's point of view is to request data from the department's middleware. Using the Control module, a message is sent asking the departments to send their information. After the requests were made, it was possible to observe that the database was populated with the desired information. A label called "middleware id" is an indication that informs us of the origin of this data. So, we concluded that the faculty has then, in its local database, the combination of everything that happens in the departments. Applying filters, it is possible to select the view by device, by the time of data reception, or by analyzed middleware. Analyzing the ambient lighting IoT device, it is possible to notice the activation of rule 4 whenever it informed the middleware that the environment was illuminated, according to Fig. 4.

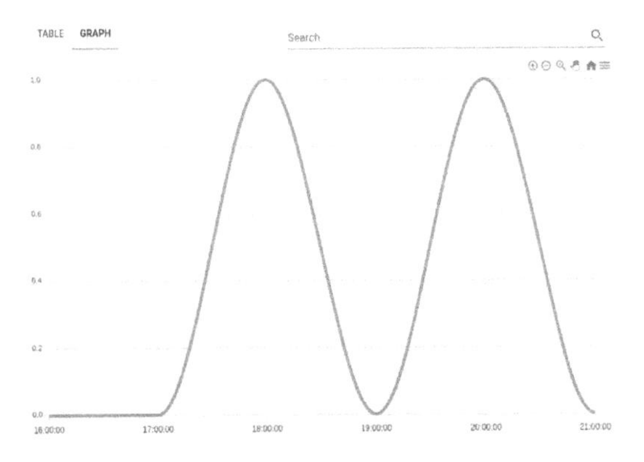

Fig. 4. Illumination chart.

Now it is possible to interfere in rule 04 of this middleware, excluding or modifying it. In this case, we can change the rule to the light sensor not only consider the amount of light but the number of people. Imagining that for security reasons, it is important for the faculty to keep the lights on its premises turned on while there are people present. As the faculty node is the only one in this topology that knows what happens to all departments, it can periodically analyze in its database if there are people present on site. If there are, it can register rules at the lower nodes to keep the lights on, while when there are no more people at the faculty node, register rules already know they have to turn off the lights. It is important to note that as the faculty node is an information centralizer, analyzing its database is the same as analyzing all the databases of the associated nodes at once. If we included in the topology a middleware superior to the existing one, called a campus, and assume that it centralizes data from several faculties, visualizing its data would be equivalent to analyzing the set of faculties at once.

It is important to highlight that, for this paper, data analysis must be done manually, but that it is possible to implement another service capable of executing the actions described here but in an automated way. Also, it is important to point out that the tests were carried out in a local environment, there were no problems with the range of the network between the middlewares. In a common situation with this implementation, it is necessary to take note of this problem and find suitable solutions, since Middleware will usually be on different networks. Only the agent with a broad view of the network that made changes to registered rules, but each department is also allowed to change its own rules according to its reality.

5 Conclusion and Future Works

The fog computing architecture helps IoT solutions to handle two major issues: the high volume of data sent over the network and the high latency in communications. Nonetheless, implementing an IoT middleware in a fog computing architecture can become a quite challenging task, especially when there are large amounts of sensors, actuators, and even middlewares hierarchically below a single network node. For instance, IoT middlewares communication should handle relationships in 1:N, N:1, and N:N modes.

Taking advantage of the microservices architecture used by UIoT middleware, it was possible to include a new service, called Controller, which added to the middleware the capability to communicate coordinately with other middleware's instances. Besides, it was also necessary to make it communicates efficiently with each internal module of the IoT middleware in order to avoid loosing time by processing several suboptimal operations.

Even with these challenges, we showed that it was possible to verify the benefits of the adopted strategies: local processing and the possibility of distribution of responsibilities. The use of microservices was a differential to reduce complexity in development and the composition of the UIoT middlewares proved to be light and efficient, even in contexts of low processing power, in addition to successfully decentralizing data and decisions on the network. Still, using the same component, it was possible to arrange different hierarchies, with different levels and contexts, and to operate in environments with little infrastructure.

Despite all the benefits that can be extracted from this proposal, one issue arose from our analysis: is it worthwhile to allow a subordinate middleware to ask the central middleware instead of just waiting for central middleware initiatives?

Bearing in mind, however, the challenges mentioned, the first proposal for extending the work is straightforward: it is necessary to objectively evaluate parameters for the choice of the hardware responsible for supporting the UIoT middlewares and the maximum complexity level (network size) that must be associated with every middleware. Also, investigating which contexts and quantities of end devices validate the application of this architecture, considering that the adoption of fog computing, even though it required less processing power, requires more complex projects and more distributed devices.

Also, it is extremely important to implement components and methodologies that take advantage of the proposed architecture to work with big data tools and techniques, as well as assess whether storage technologies can become a bottleneck for certain types of services, such as operations of high-precision mechanical machine environments in time and space.

Finally, considering the distribution of middleware instances over different networks and the Internet, it is necessary to provide them with mechanisms that improves its security. We want to explore a registration protocol between IoT middlewares, specially those with blockchain-based schemes.

Acknowledgment. The authors would like to thank the support of the Brasilian research, development and innovations agencies CNpq (Projects INCT Seg-Ciber 465741/2014-2, PQ-2 312180/2019-5 and LargEWiN BRICS2017-591), CAPES (PROJECTS FORTE 23038.007604/2014-69 and PROBRAL 88887.144009/2017-00) and FAPDF (UIoT Projects 0193.001366/2016 and SSDDC 0193. 001365/2016), as well as the support of the LATITUDE/UnB Laboratory (SDN Project 23106. 099441/2016-43), cooperation with the Ministry of the Economy (TEDs DIPLA 005/2016 and ENAP

083/2016), the Office of Institutional Security of the Presidency of the Republic (TED 002/2017), the Attorney General's Office (TED 697,935/2019) and the Administrative Council for Economic Defense (TED 08700.000047/2019-14).

References

1. Aazam, M., Zeadally, S., Harras, K.A.: Fog computing architecture, evaluation, and future research directions. IEEE Commun. Mag. **56**(5), 46–52 (2018). https://doi.org/10.1109/MCOM.2018.1700707
2. Al-Fuqaha, A., Khreishah, A., Guizani, M., Rayes, A., Mohammadi, M.: Toward better horizontal integration among IoT services. IEEE Commun. Mag. **53**(9), 72–79 (2015)
3. Caldas Filho, F.L.D.: Proposta de um Gateway IoT em Computação Fog com Técnicas de Aceleração WAN. Master's thesis, Universidade de Brasília, Brasília, DF, Brazil (2019)
4. Caldas Filho, F.L.D., Martins, L.M.C.E., Araújo, I.P., Mendonça, F.L.L.D., da Costa, J.P.C.L., de Sousa Júnior, R.T.: Design and evaluation of a semantic gateway prototype for IoT networks. In: Companion Proceedings of the10th International Conference on Utility and Cloud Computing, UCC 2017 Companion, pp. 195–201. ACM, Austin, TX, USA (2017). https://doi.org/10.1145/3147234.3148091
5. Cisco System: Cisco visual networking index: Global mobile data traffic forecast update. 2017–2022. Technical report, Cisco System (2018)
6. Cordeiro, C.d.M., Sadok, D.H., Kelner, J., Pinto, P.d.F.: Establishing a trade-off between unicast and multicast retransmission modes for reliable multicast protocols. In: Proceedings 8th International Symposium on Modeling, Analysis and Simulation of Computer and Telecommunication Systems (Cat. No. PR00728), pp. 85–91, San Francisco, CA, USA (2000). https://doi.org/10.1109/MASCOT.2000.876432
7. Datta, S.K., Bonnet, C., Haerri, J.: Fog computing architecture to enable consumer centric internet of things services. In: 2015 International Symposium on Consumer Electronics (ISCE), pp. 1–2. IEEE (2015). https://doi.org/10.1109/ISCE.2015.7177778
8. Dizdarević, J., Carpio, F., Jukan, A., Masip-Bruin, X.: A survey of communication protocols for Internet of Things and related challenges of fog and cloud computing integration. ACM Comput. Surv. **51**(6) (2019). https://doi.org/10.1145/3292674
9. Giang, N.K., Blackstock, M., Lea, R., Leung, V.C.M.: Developing IoT applications in the fog: a distributed dataflow approach. In: 2015 5th International Conference on the Internet of Things (IOT), Seoul, South Korea, pp. 155–162 (2015). https://doi.org/10.1109/IoT.2015.7356560
10. Kum, S.W., Moon, J., Lim, T.B.: Design of fog computing based IoT application architecture. In: 2017 IEEE 7th International Conference on Consumer Electronics - Berlin (ICCE-Berlin), Berlin, Germany, pp. 88–89 (2017). https://doi.org/10.1109/ICCE-Berlin.2017.8210598
11. Martins, L.M.C.E., Caldas Filho, F.L.D., de Sousa Júnior, R.T., Giozza, W.F., da Costa, J.P.C.L.: Increasing the Dependability of IoT Middleware with Cloud Computing and Microservices. In: Companion Proceedings of the10th International Conference on Utility and Cloud Computing, UCC 2017 Companion, Austin, TX, USA, pp. 203–208. ACM (2017). https://doi.org/10.1145/3147234.3148092

12. Martins, L.M.C.E., Caldas Filho, F.L.D., de Sousa Júnior, R.T., Giozza, W.F., Costa, J.P.C.L.d.: Microservices Adoption Proposal in IoT [Proposta de Adoção de Microsserviços em IoT]. In: Atas das Conferências IADIS Ibero-Americanas WWW/Internet 2017 e Computação Aplicada 2017, pp. 63–70. IADIS Press, Vilamoura, Algarve, Portugal (2017)
13. Menezes, J.T.M.D., da Costa, P.H.L., da Cunha, D.F., de Caldas Filho, F.L., Martins, L.M.C.E., de Mendonça, F.L.L.: Development of hierarchical model of middlewares with application of fog computing for iot networks [Desenvolvimento de Modelo Hierárquico de Middlewares com Aplicação de Fog Computing para Redes IoT]. In: Atas das Conferências Ibero-Americanas WWW/Internet 2019 e Computação Aplicada 2019, Lisbon, Portugal, pp. 155–162. IADIS Press (2019)
14. Noura, M., Atiquzzaman, M., Gaedke, M.: Interoperability in Internet of Things: taxonomies and open challenges. Mob. Networks Appl. **24**(3), 796–809 (2019). https://doi.org/10.1007/s11036-018-1089-9
15. Rahman, H., Hussain, M.I.: A comprehensive survey on semantic interoperability for Internet of Things: state-of-the-art and research challenges. Trans. Emerging Telecommun. Technol. **n/a**(n/a), e3902 (2020). https://doi.org/10.1002/ett.3902
16. Silva, C.C.d.M., Ferreira, H.G.C., de Sousa Júnior, R.T., Buiati, F., Villalba, L.J.G.: Design and Evaluation of a Services Interface for the Internet of Things. Wireless Personal Commun. (2016). https://doi.org/10.1007/s11277-015-3168-6

Intelligent Jacket for Monitoring Mobility of People with Reduced Disabilities

Andrés Pérez[1], Ximena Acaro[1], Maria Molina[1], Juan Yturralde[1], Lidice Haz[2,3](✉), and Teresa Guarda[2,3]

[1] Universidad de Guayaquil, Guayaquil, Ecuador
andres.perezf@ug.edu.ec, tguarda@gmail.com
[2] Universidad Estatal Península de Santa Elena, La Libertad, Ecuador
victoria.haz@hotmail.com
[3] CIST – Centro de Investigación en Sistemas y Telecomunicaciones, Universidad Estatal Península de Santa Elena, La Libertad, Ecuador

Abstract. IoT represents the expansion of technology in all aspects of everyday life. There are various initiatives according to the needs of the people. In this sense, visual impairment constitutes a greater difficulty for mobilization in unfamiliar environments. The obstacles present in the environments increase the risks and damage to the physical integrity of the person suffering from said disability. The use of wearable IoT technology allows the integration of different services that provide a practical and reliable solution. This work presents the design of a smart jacket that improves the mobility of people with visual disabilities. The design consists of two components of hardware and software. The first integrates an electronic design with ultrasonic sensors implemented in a smart jacket whose purpose is to determine the obstacles present on the road, and the second a mobile application that allows you to send a message or make a call to a pre-registered contact in case of emergency. In addition, the results allowed to validate the functional and non functional requirements defined for the project. Finally, this work contributes to improving the quality of life of this group of people in relation to their care and mobility independently.

Keywords: Smart jacket · Wearable · IoT · GSM · Geolocation · Mobility · Visual disability

1 Introduction

The Internet of things is the interconnection of the objects of the physical world through the Internet. These objects are equipped with sensors, actuators and communication technology that allows to control and monitor the activities for which they have been programmed. The technology is oriented to facilitate the development of new applications and the improvement of applications in different areas such as industry, education, health, energy, among others [1].

Today, technology helps improve the lives of people with disabilities. It promotes the use of different electronic devices and computer applications that facilitate the

Á. Rocha et al. (Eds.): ICITS 2021, AISC 1330, pp. 293–302, 2021.
https://doi.org/10.1007/978-3-030-68285-9_28

execution of daily tasks, as well as helping to control and monitor possible risk situations. [2].

Assistance and rehabilitation technologies increase, maintain or improve the functional capabilities of people with disabilities. They promote the independence and empowerment of the person to access or perform a task that was unable to fulfill, or that had great difficulty in executing. The term "rehabilitation technology" is used to refer to devices that help people recover psychomotor functions after illness or injury [3, 4].

The design of smart clothes is becoming more frequent. Companies from various industries of fashion, sports, medicine, among others, invest resources to integrate technology with clothing [5]. For example, Google, in collaboration with Levis clothing brand, created the Jacquard project a smart jacket, capable of communicating with a smartphone through a hand gesture such as sliding and touching the cuff of the sleeves. This jacket allows you to communicate with other people, use GPS navigation or control applications without using the mobile phone [6].

Following this concept and framed in the use of IoT, it is possible to design an intelligent device to improve the mobility of people with visual disabilities. This work presents the design of a jacket that identifies and warns to the blind people the type of obstacle it faces (wood, metal, amorphous). This detection is carried out through motors that are located at the back of the jacket, and that emit vibrations. The device has an alert system that sends an SMS or makes a call to pre-registered and stored contacts on a chip. It means, it works like a panic button, which if pressed would alert a family member to go to the scene of the incident according to the geolocation generated.

The document is organized as follows, the second section offers a description of the applied technologies to people with visual impairment; the third section describes the design of the wearable IoT device; the results of the proposal are presented in the fourth section; and finally the conclusions and recommendations.

2 Applied Technologies for People with Visual Impairment

Visual impairment is defined as a condition that directly affects the functioning of the eye. It can range from moderate difficulty in the perception of light signals to zero ability to see or total blindness. This results in the person having a limited scope of the information that is perceived from the surrounding environment. Which significantly compromises the physical integrity of the sufferer.

The degeneration of this sense at a high level makes the individual more susceptible. Therefore, it is necessary to explore new options that allow you to better understand the environment in which you interact and move [7].

This population group generates a social interest that promotes the concept of accessibility that these people require in urban areas, public transportation, parking, customer service, among others. Thus, it is important to integrate the use of information and communication technologies to generate new products that help improve the care and comfort of this group of people [8].

In the market there are mechanical prototypes designed for people with visual disabilities. Technology has facilitated the creation of devices that convert mechanical objects into intelligent entities through the implementation of artificial intelligence. For example, vehicles with intelligent systems capable of stopping their gear by witnessing an amorphous object (person) on their way [9].

There are other technological proposals aimed at detecting objects along the way and thereby alerting blind people, for example, the design of an electronic walking stick for blind people [10]. The use of artificial vision systems with ability to recognize characters, which allow access to information for people with visual impairment. An example of this implementation is the development of a ring equipped with an artificial vision system for the recognition of artificial intelligence characters and techniques, which allows not only to recognize the text present in any document and read it aloud, but also to inform by vibrating signals the end, the jump and the line discontinuity [11].

Other work corresponds to the development of a system to alert students with visual disabilities about potential dangers within the institution's facilities. The implementation consists of a system based on radio frequency identification (RFID) with sensors installed in strategic points of the educational center, and a mobile application. This application communicates to the user through a voice synthesizer on the device about the variants that are presented in the infrastructure of the environment in which it moves, being able to anticipate and avoid risk situations [12].

Another example is Teubica, a mobile application developed to improve mobility and monitoring of blind people. The application indicates the location in real time to the user with disabilities and to a monitor or guardian in charge of their care. This system informs the exact position in real time, preventing the case of loss due to the disorientation of the disabled person [13].

In conclusion, there are several electronic devices and computer applications that facilitate the mobility and location of people with visual impairments, improving their care and performance in their daily activities. Most of these applications integrate artificial intelligence techniques that allow to detect amorphous objects within a context in order to describe and alert the blind to their presence.

3 IoT Smart Jacket

The IoT smart jacket proposed in this work, facilitates the mobility of blind people. It implements ultrasonic sensors that determine the proximity of solid and amorphous objects. These detections are alerted to the person through vibrating motors located in different areas of the waistcoat. The jacket design is small and is designed to support the circuit. The system uses sensors which are programmed at a detection distance of maximum 1 m. The vibrations begin in a proximity range of 20 cm minimum distance with obstacles. All components are powered by independent sources included in the back box. The panic button is integrated in the GSM module. This button works as a contingency in cases of accidents or emergencies of the person. The generation of the alert is by sending an SMS message or a call for help to pre-registered contact in the system. Next, the design and implementation phases of the smart jacket prototype are described.

3.1 Smart Jacket Design

The waistcoat or smart jacket was made with foamix EVA material, which allows flexible, lightweight and easily made molds due to its thermoplastic. In addition, it has characteristics such as impermeability, high temperature resistant, recyclable and non-toxic. The functional requirements of the jacket are described below:

- Issue an object detection alert from a proximity of 20 cm.
- Motors should emit vibration levels in ascending mode when they detect a distance closer to the obstacle.
- Send a text message and/or make a call through the cellular network to pre-registered contacts when the panic button is pressed.
- The jacket must have an approximate autonomy of 6 h.
- Lightweight design, easy to use, and containing a reflective tape as a warning for pedestrians and vehicles.

Figures 1, 2 and 3 show the 3D design of the prototype. In this process the dimensions and weight of each of the electronic components were analyzed. The prototype was designed in a standard size considering the average weight and height of a person in the experimental group (15 cm wide and 35 cm high). The design allows an adjustment to the sides of the jacket. The sensors are located on the front of the jacket; and in the central part the panic button, as shown in Fig. 4. On the back, a 17 × 17 × 8.5 cm box containing the arduino microcontroller, GPS module and batteries is integrated.

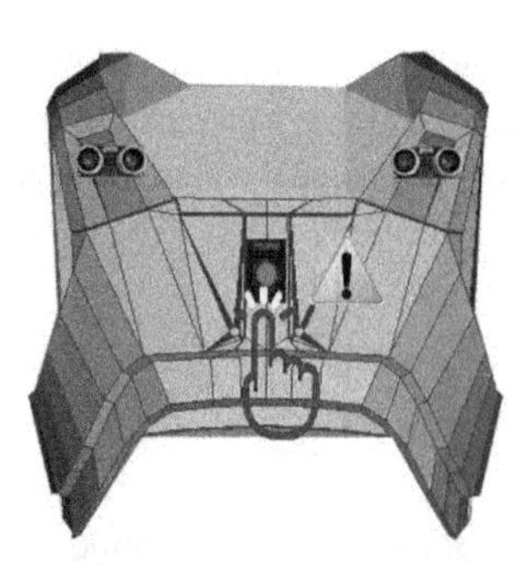

Fig. 1. Frontal design of the intelligent jacket

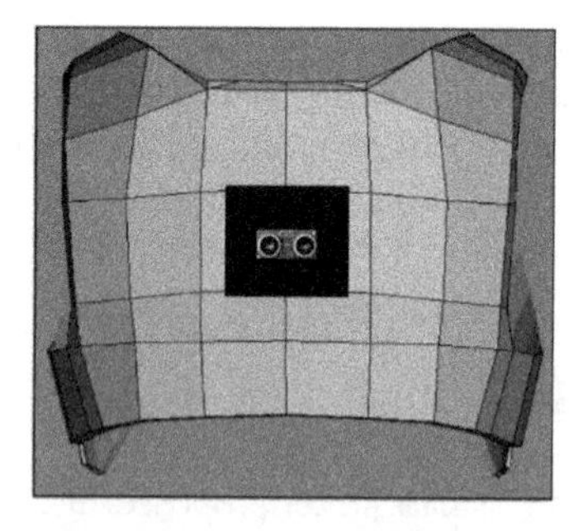

Fig. 2. Back design of the smart jacket

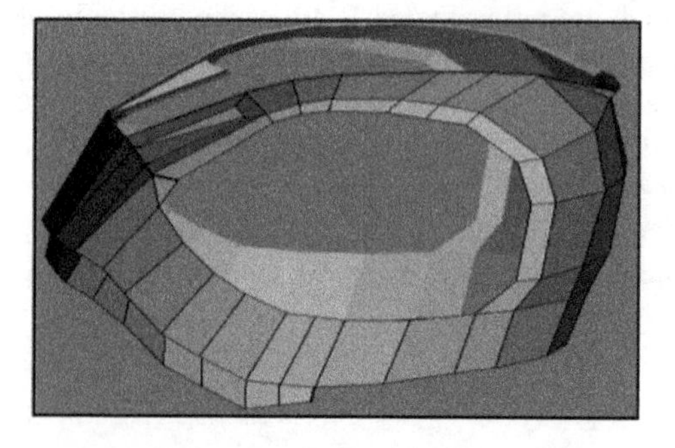

Fig. 3. Side sight of the intelligent jacket

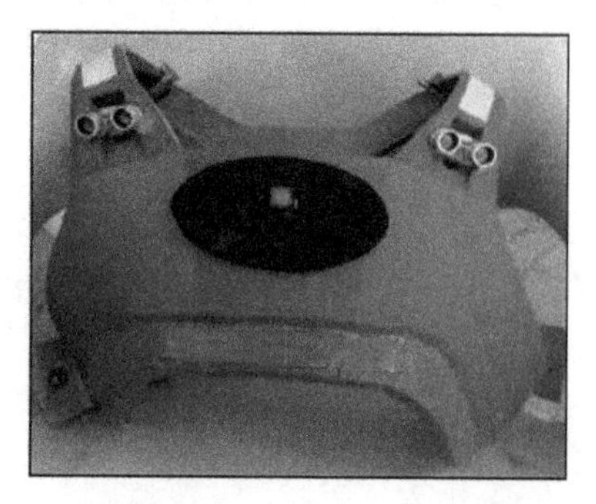

Fig. 4. Jacket with ultrasonic sensors

3.2 Electronic Components

The jacket contains three HC-SR04 ultrasonic sensors, 3 vibrating motors, a GPS module and an Arduino microcontroller, these components are detailed in Table 1 and their technical characteristics in Table 2. The sensors detect obstacles at a maximum distance of 1.5 m. The GSM module in conjunction with the panic button allows you to send an SMS text message with the exact location to a pre-registered contact.

Table 1. Electronic components of the jacket

Hardware	Type	Model
Sensor	Ultrasonic	HC-SR04
Motor	Vibrator	Mini 1020
Battery	Lithium	4800 mAh
Module	GPS	SIM 900
Microcontroller	Arduino Uno	Geo Neo 6M v2

Table 2. Technical characteristics

Description	Module HC-SR04	Módulo GSM	Arduino UNO-R3
Working Voltage	DC 5 V	3.4–4.5 V	1.8–5.5 V
Working Current	15 mA	-	0.2 mA
Max Range	4 m	-	-
Min Range	2 cm	-	-
GPRS class 10	-	Max 85.6 kbps (downlink)	-
Memory	-	-	32 kbytes

Figure 5 shows the connectivity of electronic components and their operation. The Arduino microcontroller handles the reception of analog signals emitted by ultrasonic sensors. These will be analyzed through the analog-digital converter (ADC) and converted into measured values in meters. Vibration motors then determine the approach distance to the object or person. The GPS module allows you to determine the location coordinates of the blind person when you require help or are in a danger situation. These coordinates are sent along with the SMS alert issued by the SIM module by pressing the panic button on the jacket.

The sensor functionality tests were carried out according to the parameters exterior and interior environment, opening angle, directions and fixed distances. Sensors detect objects in outdoor and indoor environments. In the external environments a variation is taken in relation to the crossing of the obstacles by the different points of the sensors. The calibration of the sensors is optimized in the programming code. The engine vibration reception time based on the signal received by the sensors is immediate 0.01 s, in either of the sensors and in both test environments.

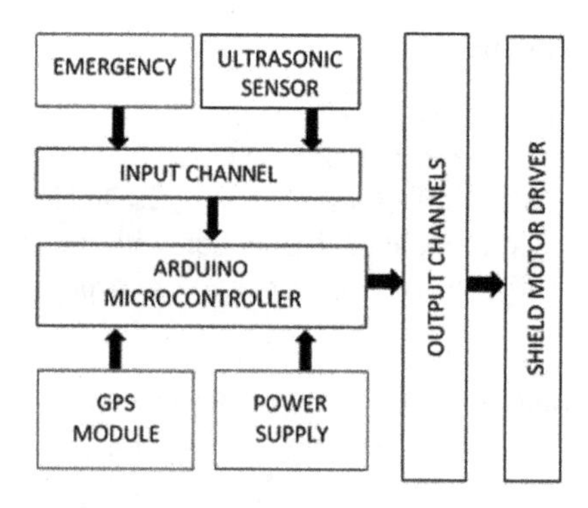

Fig. 5. Layout of the system

3.3 Detection of Obstacles

The levels of these vibrating motors will be managed by the power they emit when encountering an object detected by the sensors, as shown in the tests carried out in Table 3. Engine calibration will depend on the proximity of the object with the sensor, and these values will be weighted with figures from 1 to 5, with 5 being the highest vibration level, according to the following conditions:

- If the detected object is a distance of 0.15 m, the motor will vibration of greater force.
- If the object is between 0.16 m and 0.5 m the motor will send less frequent and force vibration.
- If the object is more than 0.5 m away the motor vibration will be almost imperceptible.

Table 3. Vibrator motor power

Distance [m]	0.1	0.25	0.5	0.75	1
Inner right arm	5	5	5	4.5	4
Inner left arm	5	5	5	4.5	4
Inner back	5	4	4	3	3
Outer right arm	5	5	4.5	4	3
Outer left arm	5	5	4.5	4	3
Outer back	5	4	4	3	3

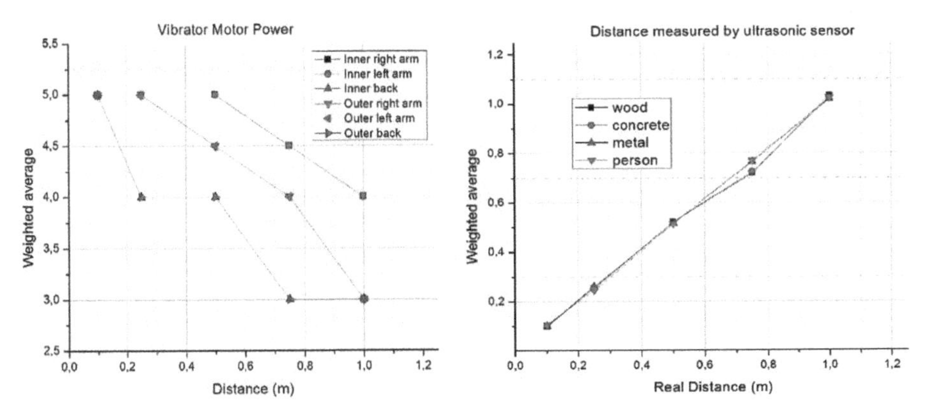

Fig. 6. Power vibrator motor

Fig. 7. Distance measured by ultrasonic sensor

In the Fig. 6 demonstrate the records received in the realized tests of 6 motive vibrators integrated in the prototype, which answer at the force levels depending on the distances towards some object, for example, at major distance the vibration force diminishes and on having diminished the distance with the object, the vibration increases; this way there will help the person not clairvoyant to react of immediate form and to stop or to take precaution (Table 4).

Table 4. Distance measured by ultrasonic sensor

Real Distance [m]	Distance measured by ultrasonic sensor in wood [m]	Distance measured by ultrasonic sensor in concrete [m]	Distance measured by ultrasonic sensor in metal [m]	Distance measured by ultrasonic sensor over person [m]
0.1	0.1	0.103	0.1	0.107
0.25	0.253	0.256	0.26	0.244
0.5	0.52	0.52	0.515	0.513
0.75	0.72	0.725	0.765	0.77
1	1.032	1.024	1.018	1.023

Figure 7 shows the detection tests performed with several objects with the ultrasonic sensors, such as wood, concrete and metal made with blind people of average height 1.70 m; checking that the results are similar, therefore the detection of objects is reliable and differs against any object that will be in its displacement when the prototype is used; making it a reliable, safe and easy prototype to maneuver.

3.4 SIM Module

The SIM module acts as a transmitter of the alert messages, that is, when the emergency button is pressed, an SMS alert message and an automatic call to a pre-registered contact are generated. This module is located on the back of the jacket. Figure 8 show the sending of the SMS. The operating conditions are as follows: send SMS, make voice calls, send GPS coordinates, through the data package, GPRS.

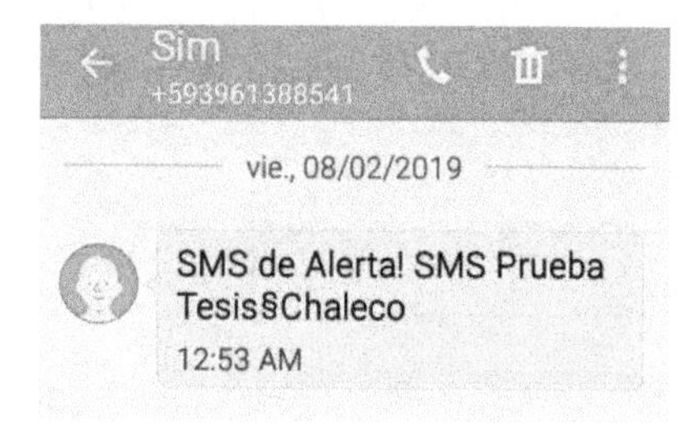

Fig. 8. SMS message alert

4 Experiments

During the tests realized with the experimental group, showed in the Figs. 9 and 10; who used the jacket for 5 daily hours for 7 days; it was demonstrated that the displacement and the evasion of the obstacles of the blind persons improved in 70% with regard to its initial situation.

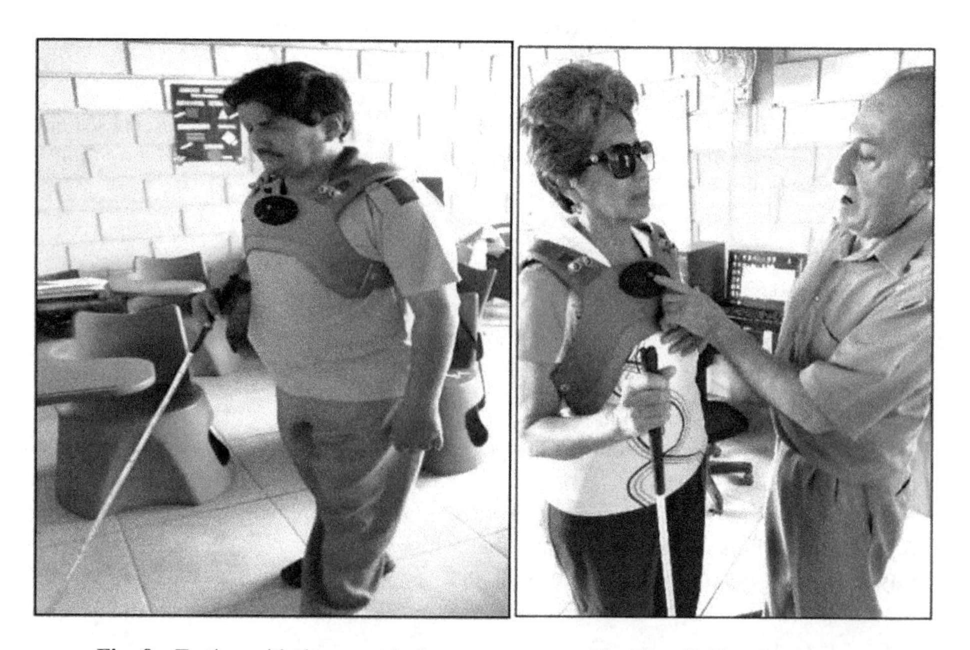

Fig. 9. Testing with the smart jacket

Fig. 10. SMS message alert

Testing of the alert system was 99.9% successful; that is to say, the alerts were successfully delivered in all tests performed. Figure 11 shows the sms message to alert you to an emergency that includes geographic coordinates and custom text. The alert is generated with a delay time of 5 s after pressing the panic button.

SMS de Alerta! SMS Prueba
Tesis_Chaleco
https://maps.google.com/maps
?q=0.000000+0.00000

Fig. 11. Emergency message

To improve the prototype, it is convenient to use an embedded board that allows developing a device with greater communication possibilities. This will reduce the number of existing cables. In rainy climates or with higher humidity, it is recommended not to use the device, its aesthetic design allows protection in this environment, although the sensors may not work properly. During the battery recharging time, it is not feasible to manipulate the jacket, because it could generate a bad maneuver and stop working.

5 Conclusions

The pilot test of the smart jacket showed favorable results for the orientation of people with visual disabilities in internal and external environments. It generated an increase in their autonomy, security and tranquility during their displacement. This significantly improves their quality of life thanks to the perception of independence and freedom of these people.

The cost of the prototype is feasible due to the selection of the electronic components and the manufacturing material, its small sizes, lighter weights and its availability, make it more attractive and competitive in the market. In addition, the aesthetic design of the device allows greater mobility to the user thanks to the coupling of the elements. This design can also vary according to the user's requirement.

The use of sensors anticipates obstacles according to the determined range. It is considered a distance of 1 m maximum and a pussy of 15 cm in internal or external environments covering the 4 cardinal points of the person. The duration of the batteries for vibrator motors, the arduino board and the GSM module, meets the requirements established for the prototype design.

The variation of the vibrating motor located on the back of the jacket has a variation, due to the characterization of the sensor that detects amorphous objects. This is parameterized in the sensor programming according to what is established in the design. The message or call that is generated when pressing the panic button is executed with 99.9% effectiveness. The favorable use of the GSM module and the panic button are forced to work with the monetary balance saved on the SIM card, therefore, it is necessary to verify that there is an available balance and carry out periodic recharges.

In future works, the development and implementation of a computer platform that integrates the smart jacket with a web application that functions as a control and monitoring center is proposed; and that allows the notification area to be sectorized more efficiently, enabling the generation of statistical reports. It is possible to add more services to the smart jacket such as voice transmission during the emergency.

References

1. Madakam, S., Ramaswamy, R., Tripathi, S.: Internet of Things (IoT): a literature review. J. Comput. Commun. **03**(05), 164–173 (2015)
2. Sánchez, A.J., et al.: Geolocation Applied to Emergency Care Systems for Priority Groups. International Conference on Information Theoretic Security. Springer, Cham (2018)
3. Hand Function. Springer International Publishing (2019)
4. DIS Comprehensive design in the development of assistive technology, https://disjournal.ibero.mx/index.php/DISJournal/article/view/38. Accessed 17 Feb 2020
5. Piwek, L., Ellis, D.A., Andrews, S., Joinson, A.: The rise of consumer health wearables: promises and barriers. PLoS Med. **13**(2) (2016)
6. Kumar, A., Rudnicki, C., Chatterjee, R., Mina, K., Onyeije, O., Starner, T.: JacquardToolkit: allowing interactions with Levi's Jacquard jacket, En Actas del 23º Simposio internacional sobre computadoras portátiles, pp. 106–108, septiembre 2019
7. Beltrán, J., Zepeda, J., Maciel, M., Larios, V., Espinoza, J., Martínez, J.: Technologies to support the transfer and access to information for people with visual disabilities. Inventum **14**(26), 70–78 (2019)
8. Koon, R.A., Vega, M.E.D.L.: The technological impact on people with disabilities (2014)
9. Arriola Oregui, I.: Object detection based on Deep Learning and applied to autonomous vehicles (2018)
10. Lizárraga González,C.R.: Proposal for the design of an electronic baton for the blind to improve the quality of their daily commute (2018)
11. Shilkrot, R., Huber, J., Liu, C., Maes, P., Nanayakkara, S.C.: FingerReader: a wearable device to support text reading on the go. In: CHI 2014 One of a CHInd, pp. 2359–2364 (2014)
12. Cruz Felipe, Md., Pinargote Ortega, J.M., Demera Ureta, G.P., Vera Zambrano, E.M., Mosquera Alcívar, R.A.: Alert system for visually impaired students at UTM. Revista Científica**1**(31), 85–95 (2018)
13. de Tristán, G., Arcia, A., Pérez, R., Montes, H.: Mobile application for monitoring people with visual disabilities. Tecnología y accesibilidad, vol. 1, pp. 93–100 (2016)

Smart Home Control System Using Echo Dot

José Varela-Aldás[1,2](✉), Jorge Buele[1], and Myriam Cumbajin[1]

[1] SISAu Research Group, Facultad de Ingeniería y Tecnologías de La Información y La Comunicación, Universidad Tecnológica Indoamérica, Ambato, Ecuador
{josevarela, jorgebuele, myriamcumbajin}@uti.edu.ec
[2] Department of Electronic Engineering and Communications, University of Zaragoza, Zaragoza, Spain

Abstract. In the past, home automation in residential environments was a luxury, but today it has become a human need. Although costs have decreased, there are households that do not have sufficient infrastructure for this type of technology. For this reason, this document describes the implementation of a low-cost intelligent system that makes it easy to control the facilities of a department. The actions are coordinated using the Amazon's Alexa digital voice assistant, included in the Echo Dot smart device. In the hardware part, conventional household elements are used, coupled to an accessible electronic system whose central processing unit is the ESP8266 embedded card. The link between the stations is made using a free server that works as a bridge to transfer the information. The tests carried out validate this proposal and with it, an initial prototype is presented that can be installed in low-income housing and educational institutions.

Keywords: Smart homes · Internet of Things · Digital voice assistants · Web server

1 Introduction

Home automation allows technology to be used to automate processes in different environments [1–3]. Home automation is routinely applied in homes and buildings, but it should be noted that its use in industrial areas is also feasible [4]. Since its inception in the 70s, it has been an innovative proposal, but it has not been available to everyone, due to economic and infrastructure issues. [5]. Subsequently, with the appearance of the Internet of Things (IoT), it has been possible to give greater enhancement and provide greater benefits to users [6]. Although the IoT can be used in various fields such as education, industry, commerce, tourism, entertainment, etc., its use at home has a great influence [7, 8].

The application of the IoT can occur in conjunction with other technological tools such as artificial intelligence (AI) [9]. In this way, it seeks to have a greater interaction with the human being, knowing their needs, and trying to anticipate. A clear example of this are the virtual assistants currently on offer, such as: Google's Assistant, Apple's Siri, Microsoft's Cortana and Amazon's Alexa [10, 11]. These voice assistants

Á. Rocha et al. (Eds.): ICITS 2021, AISC 1330, pp. 303–312, 2021.
https://doi.org/10.1007/978-3-030-68285-9_29

(VAs) are now included in smart devices that have a built-in microphone and speaker, which allows communication with these. These popular devices since 2017, most of these do not have a screen and highlight Facebook Portal, Google Home Hub, Amazon Echo Dot and Echo Spot. All these devices allow the user to organize their agendas and personal information, train, entertain and even automate their home, with an increasingly reduced investment [9, 12].

VAs have become more popular today, since the use in various applications has demonstrated the correct interaction with the human being. In [13] se implementa Amazon Echo Dot Devices en los hogares de adultos mayores, donde se evalúan sus interacciones sociales y se categoriza ontológicamente a los asistentes de voz. Amazon Echo Dot Devices is implemented in the homes of older adults, where their social interactions are evaluated and VAs are categorized ontologically. This work provides recommendations for the design of future technologies that provide accompaniment to the elderly. Voice activation and control now have a close relationship with service and support robotics. For this reason, [14] describes a system that includes a humanoid robot torso that is controlled by an IoT device, and simulates an assisted workload process. A review of several papers is presented in [11], in which users with a variety of disabilities are using the Amazon Echo as a tool for speech therapy and support for caregivers. Becoming a benchmark for possible use in school and university classrooms.

The interaction of these devices is not only carried out with the human being, but with other elements that could be connected to the same wireless network. This has made it important to analyze the level of privacy of the data that these devices provide to the user, as can be seen in [15]. This document proposes to carry out 4 attacks on the system and see the behavior, at the same time that they describe a programming framework to increase security in programming. In [16] it is mentioned that while the user interacts verbally with the VA, content that is recorded could be filtered without consent. With this proposal, the VA is continuously blocked, preventing it from recording the user's voice, unless a specific voice command is issued. This improves the security level of the system, which could be being compromised without any kind of control.

Applications with other Virtual Assistants should also be reviewed, to get a better perspective. Therefore, in [17] is describe a system for the control of certain domestic devices using Google Home and Raspberry Pi. Tests have shown that this prototype works, although they are very scarce. For its part, in [18] a research is presented that analyzes the satisfaction levels of the assistants and establishes a new improved method. Experimental results indicate that this model based on specific consultation questions has significant statistical improvements at several baselines in terms of common rating evaluation metrics. Finally, [19] describes the development of a proprietary system intended to replace a conventional assistant, using multimodal dialogue systems. In this way, the user is given a system that allows the recognition of gestures, images and obviously voice.

As seen in [20], conventional environments can improve with the application of AI and IoT, allowing to have intelligent work environments (Smart Office) and also homes with more functionalities. This represents the acquisition of smart sensors that are coupled to these systems, which is why this document proposes a low-cost smart system with high performance. Using the ESP8266 embedded card and a reduced electronic design, it is possible to couple the conventional elements of a home. While the control of the necessary actions to automate the environment is done using the Echo Dot device and the Alexa voice assistant.

This document is composed of 4 sections, including the introduction and related works in Sect. 1. The materials and methods that have been used in hardware and software are described in Sect. 2. The preliminary results obtained are described in Sect. 3, and the conclusions and discussions in Sect. 4.

2 Materials and Methods

2.1 Proposal Design

The proposed system starts from the need to voice control conventional electrical devices in the home, so the scheme in Fig. 1 is proposed. The user interacts with the smart speakers located in different places in the home, and a virtual assistant installed in the speakers it recognizes and sends the orders dictated by voice. To control the outputs, a server is used as a container, i.e., the virtual assistant writes the desired state of the output to the server and a controller performs the reading to carry out the action requested by the user. The controller requires an IoT platform with wireless connection to the internet, it controls a relay module through digital outputs, to supply power to the output devices.

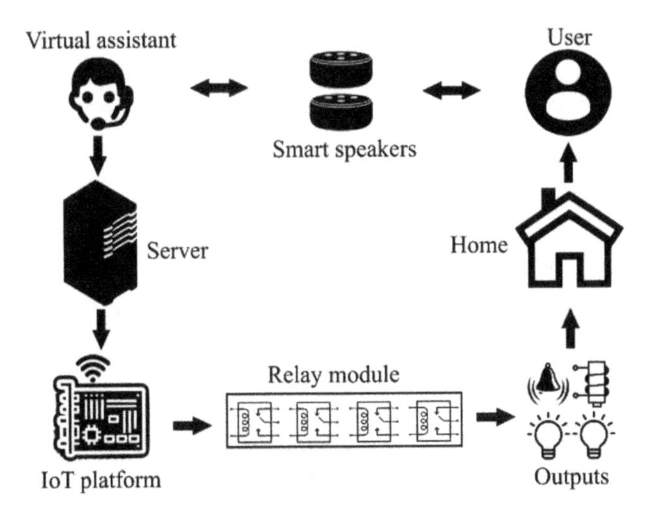

Fig. 1. General scheme of the proposal.

The system architecture has two main components. The hardware including the distribution of the household devices and the electronic circuit, the household elements are the smart speakers and the output devices, and the electronic circuit is connected to the network and the output switches. On the other hand, the software consists of the virtual assistant and the server, Amazon's artificial intelligence allows interaction with the virtual assistant, and the server known as Sinric contains the states of the outputs in the cloud.

2.2 Hardware

The elements are installed in an apartment of 75 m^2, it has a main room, a bathroom, a kitchen, a dining room, a living room and another room. On the other hand, the output devices are two lamps, an electric lock and a bell. The physical components are distributed as shown in Fig. 2, one lamp is located in the main room and the other illuminates the kitchen, dining room and living room, the electric lock secures the main entrance door and the bell is located in the kitchen. The smart speakers to use are the third-generation Echo Dot from the Amazon brand, with an approximate cost of 40 USD (US dollars), it has a Wi-Fi connection, 4 far-field microphones and comes integrated the virtual assistant Alexa. One of the speakers is located in the main room and the other is in the living room, with a distance of 12 m between these. The electronic circuit is installed in the main room, from where it is wired to all the output devices located within the department.

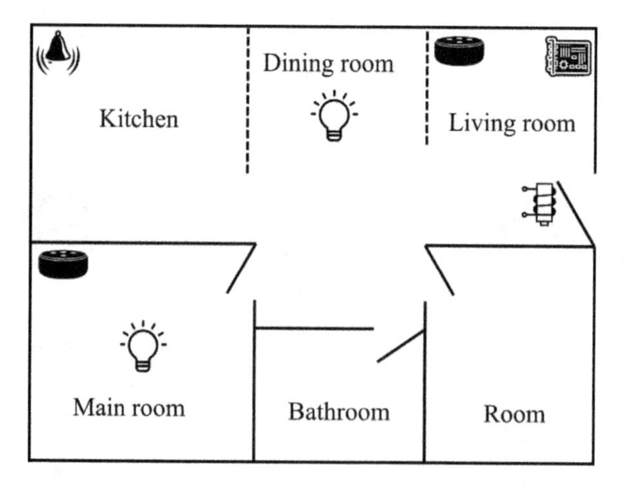

Fig. 2. Location of the physical components in the department.

The electronic circuit is based on the NodeMCU card, this is an IoT platform built with the ESP8266 chip, ESP-12 module, with direct connection to the Wifi network and digital input and output pins; The connections made are presented in Fig. 3. The circuit is designed for direct connection to the alternating current source of the domestic electrical system, for this the HKL-PM01 power module is required, which includes an AC/DC converter and a voltage regulator (5 V) to power the ESP8266 and the relay module. The digital pins D1–D4 are used to control the activation of the relays in the following order: D1.- Bell, D2.- Lock, D3.- Lamp 2, D4.- Lamp 1. In the same way, the

AC source supplies the outputs through the normally open contacts of the relays. The implementation cost of this electronic circuit is 90 USD including all the elements.

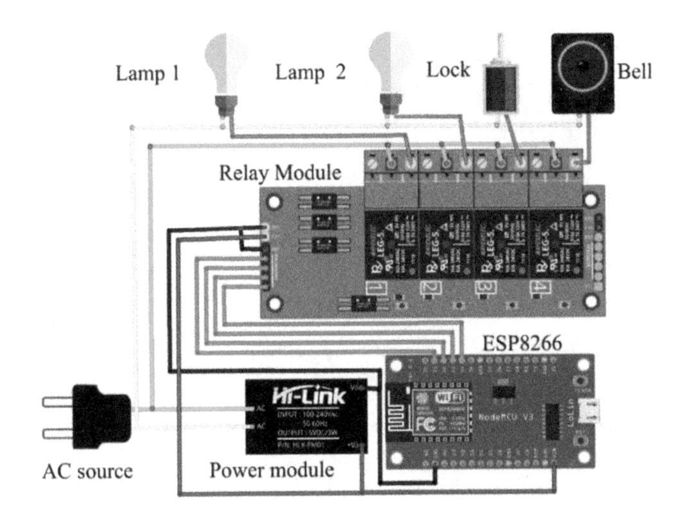

Fig. 3. Electronic circuit design

2.3 Software

The purpose of the server is to contain the virtual devices of the outputs to be controlled, this server is used by the Alexa virtual assistant and the IoT NodeMCU platform. The tool used is Sinric, this is a server developed for home automation applications with compatibility for Alexa and Google Home, this service is free, to enter it is required to create an account and install the Skill Sinric in the Amazon Alexa mobile application. Each account generates a unique API key that allows access to the server for reading and writing, and requires the creation of Smart home devices, in the case of this proposal Switch-type devices are created to control the 4 outputs of the home. Similarly, each device generates a unique identification to relate to the digital outputs of the IoT platform.

The program installed on the ESP8266 microcontroller uses the WebSocket communication protocol to access Sinric. Figure 4 shows the flow chart of the developed code.

At the beginning of the program, the components to be used are configured: pins D1 to D4 are established as digital outputs, it is connected to the Wi-Fi network by entering the network identification and password. In addition, the connection via WebSocket to the address "iot.sinric.com" is begined using port 80, the event function for WebSocket protocol interruptions is set, and authorization is requested from the server using the API key. Within the loop, validation conditions are followed until the status of the output device is defined. Before proceeding, the server connection is reviewed, and the information is read. The data received in String format are: the device identification, the action and the value. If the action indicates "setPowerState", it

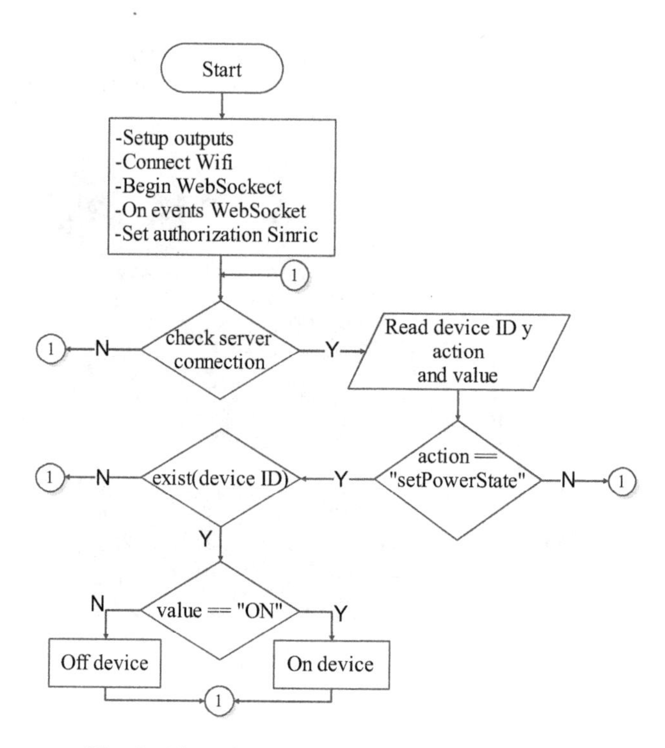

Fig. 4. Flow chart of the main program developed.

proceeds to identify which device to update, otherwise the loop is restarted. Finally, the value indicates the status of the output device, "ON" to turn on and "OFF" to turn off, this data is compared and the digital output is activated or deactivated as appropriate, to continue the infinite repetition of the loop. The application used to implement the code is IDE Arduino, requiring the libraries: Arduino.h, ESP8266WiFi.h, WebSocketsClient.h, ArduinoJson.h and StreamString.h.

3 Results

The system is implemented and evaluated in the apartment where three adults reside, two occupy the main room and one the extra room, in a period of 30 days. The objective has been to analyze the performance of the system in a real and everyday environment, observing the successes and errors of the system that are registered by the Amazon Alexa application and by users. The virtual devices created on the server have the following names: Foco1, Foco2, Puerta y Timbre (names in Spanish), corresponding to the lamp 1, lamp 2, lock and bell devices, respectively. The Echo Dot located in their design positions are shown in Fig. 5, the Echo Dot 1 (main room) in black and the Echo Dot 2 (living room) in white.

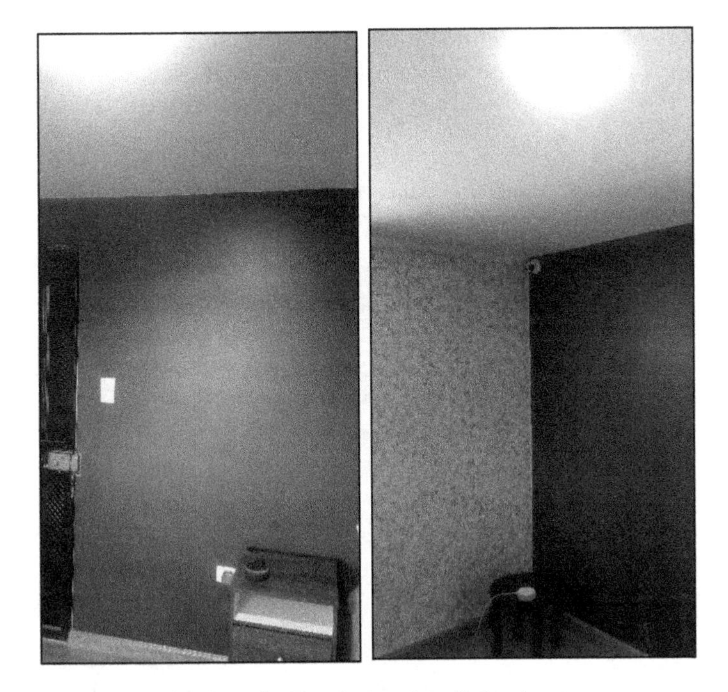

Fig. 5. Echo Dot devices installed at home.

When a command is generated by voice (Alexa turn on "Foco 1") there are three possibilities for the system to respond:

Success. - Alexa understands the order and manages to send the information to Sinric. The NodeMCU receives the command and executes it by turning the output device on or off.

Error 1. - It is a local error and it happens when the problem appears in the Echo Dot due to factors that prevent executing the order. The reasons could be that the order is not understood by the virtual assistant or that it is difficult to connect to the server. In this case, Alexa notifies that the order cannot be executed.

Error 2. - It is a remote error and it happens when there is a problem in the IoT card, i.e. the order is sent correctly to the server, but the ESP8266 cannot access the updated information. This usually happens due to server crashes or connection problems. In this case Alexa mistakenly confirms the execution of the order. This failure can be attributed to the technology used, due to the server being disconnected or the electronic card not responding correctly.

Regarding the treatment of errors, in any of the error cases, the user should not retry the order in order not to generate noise in the data collection, so that the manual switches are kept to activate the devices with a second option.

Table 1 presents the results of the system operation, observing the number of hits, errors 1 and errors 2 with their respective percentages, classified by output device, and according to the receiving speaker. In total 355 orders are registered, of which 27 generate a type 1 error and 21 a type 2 error, in general there are more errors for local issues (7.6%) than for errors in the remote circuit (5.9%), this It is due to the human factor that does not adequately inform what it requires. These results are acceptable for the proposal because in no case are there an excessive number of errors.

Table 1. Home automation control system operating results.

Device	Orders	Errors 1	Errors 2	% Errors 1	% Errors 2
Echo Dot 1					
"Foco 1"	91	6	4	6.6%	4.4%
"Foco 2"	24	4	2	16.7%	8.3%
"Puerta"	13	3	2	23.1%	15.4%
"Timbre"	5	0	1	0%	20%
Echo Dot 2					
"Foco 1"	35	3	3	8.6%	8.6%
"Foco 2"	127	8	5	6.3%	3.9%
"Puerta"	51	2	3	3.9%	5.9%
"Timbre"	9	1	1	11.1%	11.1%
Total	355	27	21	7.6%	5.9%

In addition to the features detailed in this proposal, it is possible to control the output devices using a mobile device with the Amazon Alexa application. It has a switches user interface to activate or deactivate the devices, or it also supports dictate by voice orders to the mobile device instead of the Echo Dot. This option allows to operate the home automation system from outside the home making remote use of the installed system.

4 Conclusions and Discussions

In this work, a home automation control system is implemented using the Alexa virtual assistant that is incorporated into the Echo Dot smart speaker and implementing an electronic circuit that receives the information and executes it. As a linking method, a server is available as an online container. The controlled devices in the home are 2 lamps, a lock and a bell. The results have been generated in a real environment for 30 days and in everyday conditions, observing the number of hits and the number of errors, verifying a correct operation of the system and a number of acceptable failures for domestic use. The main advantage of the proposal is the low cost, since it requires 170 USD for the implementation (2 Echo Dot and the electronic circuit), all the components are easily accessible and there is online support for all the tools used.

The main contribution of this work is to analyze the performance of the proposal in a real environment, with middle-aged users who interact on a daily basis with the system. Based on this, the utility can be questioned, observing considerable failure percentages that must be improved to optimize the system response. On the other hand, the importance of these proposals for human-machine interaction has been demonstrated, contributing to the emotional state [13]. This work makes available a quick and easy implementation model to improve the living conditions of people with difficulties to use basic household devices. However, the lack of security that these systems can present must still be considered [15]. The evidence of results is scarce in [17], besides that Google Home presents a greater difficulty to understand the user's voice commands. On the other hand, both [18] and [19] are interesting proposals with little application that will be added for future work, in order to expand the benefits offered to the user.

In future work, it is planned to improve the characteristics of this prototype system, including more input and output devices, increasing the benefits of the virtual assistant, and improving the system response. In addition, it is planned to implement the proposal in an educational institution to analyze the usability of the system.

References

1. Toschi, G.M., Campos, L.B., Cugnasca, C.E.: Home automation networks: a survey. Comput. Stand. Interfaces. **50**, 42–54 (2017). https://doi.org/10.1016/j.csi.2016.08.008
2. Sovacool, B.K., Furszyfer Del Rio, D.D.: Smart home technologies in Europe: A critical review of concepts, benefits, risks and policies (2020). https://doi.org/10.1016/j.rser.2019.109663
3. Saá, F., Varela-Aldás, J., Latorre, F., Ruales, B.: Automation of the feeding system for washing vehicles using low cost devices. In: Advances in Intelligent Systems and Computing, pp. 131–141 (2020). https://doi.org/10.1007/978-3-030-32033-1_13.
4. Marikyan, D., Papagiannidis, S., Alamanos, E.: A systematic review of the smart home literature: a user perspective. Technol. Forecast. Soc. Change. **138**, 139–154 (2019). https://doi.org/10.1016/j.techfore.2018.08.015
5. Beaudin, M., Zareipour, H.: Home energy management systems: a review of modelling and complexity (2015). https://doi.org/10.1016/j.rser.2015.01.046
6. Risteska Stojkoska, B.L., Trivodaliev, K.V.: A review of Internet of Things for smart home: challenges and solutions (2017). https://doi.org/10.1016/j.jclepro.2016.10.006
7. Alaa, M., Zaidan, A.A., Zaidan, B.B., Talal, M., Kiah, M.L.M.: A review of smart home applications based on Internet of Things (2017). https://doi.org/10.1016/j.jnca.2017.08.017
8. Liu, L., Stroulia, E., Nikolaidis, I., Miguel-Cruz, A., Rios Rincon, A.: Smart homes and home health monitoring technologies for older adults: a systematic review, (2016). https://doi.org/10.1016/j.ijmedinf.2016.04.007
9. Kaye, J.J., Fischer, J., Hong, J., Bentley, F.R., Munteanu, C., Hiniker, A., Tsai, J.Y., Ammari, T.: Panel: voice assistants, ux design and research. In: Conference on Human Factors in Computing Systems - Proceedings (2018). https://doi.org/10.1145/3170427.3186323
10. Hoy, M.B.: Alexa, siri, cortana, and more: an introduction to voice assistants. Med. Ref. Serv. Q. **37**, 81–88 (2018). https://doi.org/10.1080/02763869.2018.1404391
11. Terzopoulos, G., Satratzemi, M.: Voice assistants and artificial intelligence in education. In: ACM International Conference Proceeding Series (2019). https://doi.org/10.1145/3351556.3351588.

12. Ammari, T., Kaye, J., Tsai, J.Y., Bentley, F.: Music, Search, and IoT: How people (really) use voice assistants. ACM Trans. Comput. Interact. 26 (2019). https://doi.org/10.1145/3311956
13. Pradhan, A., Findlater, L., Lazar, A.: "Phantom friend" or "just a box with information": personification and ontological categorization of smart speaker-based voice assistants by older adults. In: Proceedings of the ACM on Human-Computer Interaction (2019). https://doi.org/10.1145/3359316
14. Erol, B.A., Wallace, C., Benavidez, P., Jamshidi, M.: Voice activation and control to improve human robot interactions with IoT perspectives. In: World Automation Congress Proceedings, pp. 322–327 (2018). https://doi.org/10.23919/WAC.2018.8430412
15. Fernandes, E., Jung, J., Prakash, A.: Security analysis of emerging smart home applications. In: Proceedings - 2016 IEEE Symposium on Security and Privacy, SP 2016. pp. 636–654 (2016). https://doi.org/10.1109/SP.2016.44
16. Gao, C., Chandrasekaran, V., Fawaz, K., Banerjee, S.: Traversing the quagmire that is privacy in your smart home. In: IoT S and P 2018 - Proceedings of the 2018 Workshop on IoT Security and Privacy, Part of SIGCOMM 2018. pp. 22–28 (2018). https://doi.org/https://doi.org/10.1145/3229565.3229573.
17. Peng, C.Y., Chen, R.C.: Voice recognition by Google Home and Raspberry Pi for smart socket control. In: Proceedings - 2018 10th International Conference on Advanced Computational Intelligence, ICACI 2018. pp. 324–329 (2018). https://doi.org/https://doi.org/10.1109/ICACI.2018.8377477.
18. Hashemi, S.H., Williams, K., Kholy, A. El, Zitouni, I., Crook, P.A.: Measuring user satisfaction on smart speaker intelligent assistants using intent sensitive query embeddings. In: International Conference on Information and Knowledge Management, Proceedings, pp. 1183–1192 (2018). https://doi.org/10.1145/3269206.3271802
19. Kepuska, V., Bohouta, G.: Next-generation of virtual personal assistants (Microsoft Cortana, Apple Siri, Amazon Alexa and Google Home). In: 2018 IEEE 8th Annual Computing and Communication Workshop and Conference, CCWC 2018. pp. 99–103 (2018). https://doi.org/10.1109/CCWC.2018.8301638
20. Horch, A., Kubach, M., Robnagel, H., Laufs, U.: Why should only your home be smart?-A vision for the office of tomorrow. In: Proceedings - 2nd IEEE International Conference on Smart Cloud, SmartCloud, pp. 52–59 (2017). https://doi.org/10.1109/SmartCloud.2017.15

Impact of the Multiplatform Mobile Applications and Their Technological Acceptance Model in Tourist Georeferenced Management

Hernán Naranjo-Ávalos[1], Jorge Buele[2,3](✉), Franklin Castillo[2], Bryan Torres[1], and Franklin W. Salazar[1]

[1] Universidad Técnica de Ambato, Ambato 180103, Ecuador
{hf.naranjo,btorres8559,fw.salazar}@uta.edu.ec
[2] SISAu Research Group, Universidad Tecnológica Indoamérica, Ambato 180212, Ecuador
{jorgebuele,franklincastillo}@uti.edu.ec
[3] Universidad Internacional de La Rioja, Logroño 26006, Spain

Abstract. Georeferenced tools have made it possible to use the information collected through electronic collaboration environments in a better way. However, no investigation on the quantification of the influence of georeferenced tourist-oriented software tools has been found. The document aims is to validate the technological acceptance of tourist georeferencing applications using a multiplatform prototype that allows the registration of sites of interest. A web and mobile prototype are implemented to georeference data provided by personnel. The application prototype was validated by a group of 125 university students and people linked to tourism using the Technology Acceptance Model. The results were interpreted with the help of the Kendall Tau-b correlation analysis where highly significant positive correlation values were obtained.

Keywords: Georeferencing · Tourism · Technological Acceptance Model · Mobile applications · Digital transformation

1 Introduction

Tourism management requires diversifying its management model to promote agile and dynamic processes [1, 2]. Currently, the economies of various countries benefit from foreign exchange that comes from tourism activities [3–5]. In Ecuador, according to the Tourism report of the year 2019, this sector contributes 2.2% of the GDP (gross domestic product). In Ecuador until 2019 with an arrival of foreigners of around 1,471,968 and an estimated foreign exchange inflow of 2,287.5 USD million for the country [6, 7]. Although the relevance of tourism in the economy of the countries is quite evident, the need to disseminate and promote it using digital environments is an issue that involves the participation of the government, business sector and the community [8, 9]. The use of geo-referenced data has become a fundamental element of information management [10, 11].

Á. Rocha et al. (Eds.): ICITS 2021, AISC 1330, pp. 313–322, 2021.
https://doi.org/10.1007/978-3-030-68285-9_30

The digital transformation has generated several changes in the tourism industry, as can be detected in [12, 13]. The results included the long-term economic, political and social consequences of digital transformation that can lead to digital colonialism in the tourist destination. An analysis of the scientific literature of current works was carried out to find the impact of georeferenced tourism management in multiplatform mobile applications based on a model of technological acceptance from various research edges. As can be seen in [14–17], technologies such as augmented reality, artificial intelligence, digital platforms and web ontologies are linked with activities related to tourism.

The use of technology makes it easier for people of all ages to go on tourism. In [18] a review is described that shows how older adults use ICT with greater confidence and independence to travel. The MEAN framework (acronym for: MongoDB, ExpressJS, AngularJS, NodeJS), stands out mainly for its platform-level independence. These benefits lie in the transversal use of the JavaScript language from the client-side to the server-side, while JSON (JavaScript Object Notation) allows information to flow between structural components natively and flexibly [19].

In the presented literature, no references have been found for quantitative usability studies in which the influence of this type of application is determined from a collaborative perspective. The objective of this study is to show the results of a quantitative evaluation of attitude towards use, intention to use, perceived utility and ease of use in digital tourism management tools within the area of electronic citizen collaboration. This research is of an experimental type, starting with the development of a multiplatform application. Additionally, verifying the benefits of the new development frameworks and their technical feasibility of implementation in the context of the current pandemic. For the evaluation, the Technological Acceptance Model (TAM) was used with the participation of 125 people.

This document is made up of 4 sections, including the introduction and review of applications related to tourism management and the contribution of multiplatform frameworks in Sect. 1. Work methodology used in the development of this proposal is shown in Sect. 2. In Sect. 3 the results obtained in the investigation are described, analyzing the impact of the perceived utility and ease of use from the users' feedback. Conclusions and future works of the research are described in Sect. 4.

2 Methodology

In the first phase of the project, a multiplatform application was developed using STACK MEAN with a multilayer architecture; logically dividing the functionalities into presentation, services, data and persistence layers as shown in Fig. 1. The system uses the pattern Architectural MVVM (Model-View-ViewModel) and uses the type REST services layer for communication with the data layer. Regarding the authentication process, the Firebase cloud service is used. CRUD (Create, read, update, delete) operations are managed in the data layer through the Mongoose Object Data Modeling (ODM) library. Data is stored in the MongoDB document-oriented database.

The application information schema consists of six collections, which can be seen in Fig. 2. The documents are in BSON (binary representation of JSON) format helping HTTP requests to have a short response time. These types of models benefit aspects such as speed and response times in transactional operations.

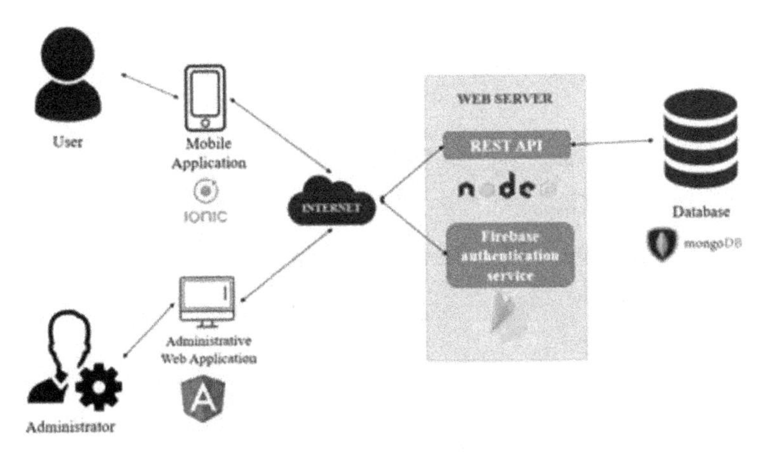

Fig. 1. Application architecture.

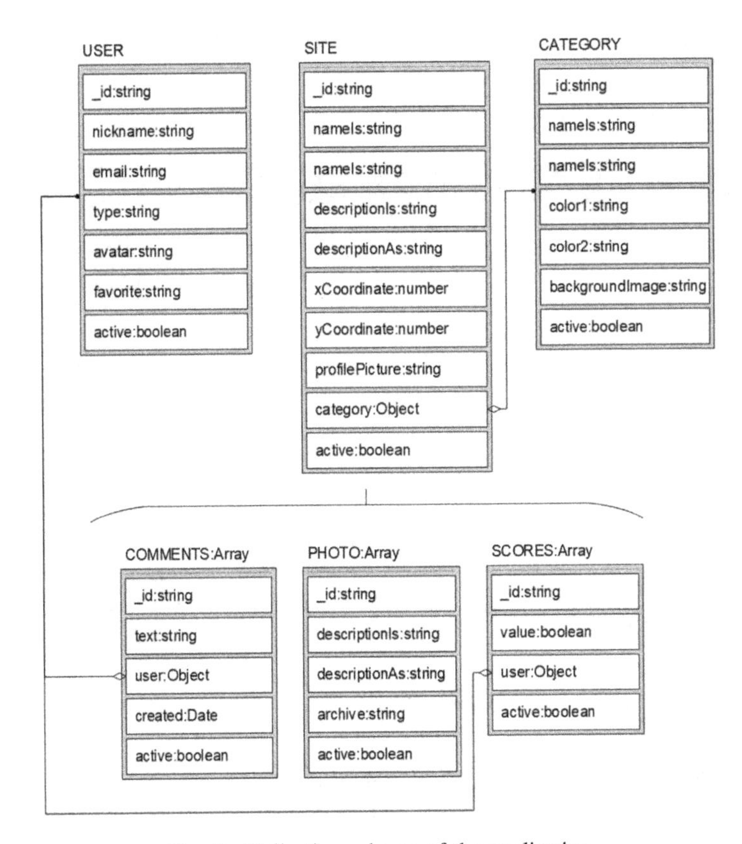

Fig. 2. Collection scheme of the application.

For its part, the georeferenced information of the application was managed using the Google Maps API. Data recovered from the business layer is used through JavaScript following the process as shown in Fig. 3.

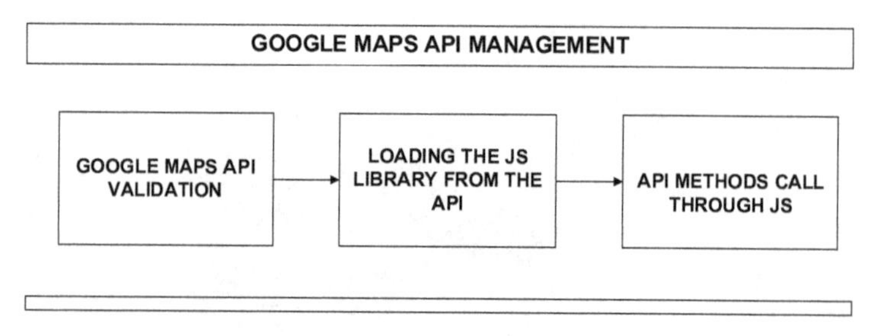

Fig. 3. Application's architecture

The API provides functions that help the manipulation of data obtained from the database servers and, after interpreting it, it can be shown in the map component of the application. The information first group focuses on the description of sites of interest, pointing out variables of the typology of the place, description, image gallery and feedback comments as illustrated in Fig. 4. The second group belongs to the georeferenced data which can be deployed into two or three dimensions using the Google Maps API.

#	Name	Location	Coordinates	Cover Photo	Actions
1	Cevallos Park	Sucre, Joaquín Lalama, Luis A. Martínez & Pedro Fermín Cevallos Street	**Latitude:** -1.2412560286061165 **Longitude:** -78.62692466785312		View Details Delete
2	Montalvo Park	Simón Bolívar, Mariano Castillo, Antonio José de Sucre & Juan Montalvo Street	**Latitude:** -1.2422732096149502 **Longitude:** -78.62933071143766		View Details Delete
3	The Cathedral	Bolívar y Montalvo Street	**Latitude:** -1.2417304558712203 **Longitude:** -78.6289944636128		View Details Delete
4	Montalvo's House	Montalvo N° 03-50 y Bolívar Street	**Latitude:** -1.2416636286694 **Longitude:** -78.62967008849827		View Details Delete

Fig. 4. Description of tourist places in the application.

Through an authentication process, users can log in to the application to visualize the whole tourism-interest sites registered. Information is sort by different criteria, taking care of a logic order in the finding process. As part of the functionality of the prototype, users can list relevant places that afterward, they can be shared with the community. Eventually, users have the ability to qualify and provide feedback showing their experiences about the journey in a certain locality.

As it is a multiplatform application, the prototype has the ability to fulfill the same functionalities of the web platform from the mobile environment as illustrated in Fig. 5. In this collaborative application, users can include comments, and scores and additionally, view the gallery such as it is shown in Fig. 6. Plus, the application constitutes a contact platform in which entities and businesses can add personalized content and links to their websites. The map component displays clean content in two and three dimensions.

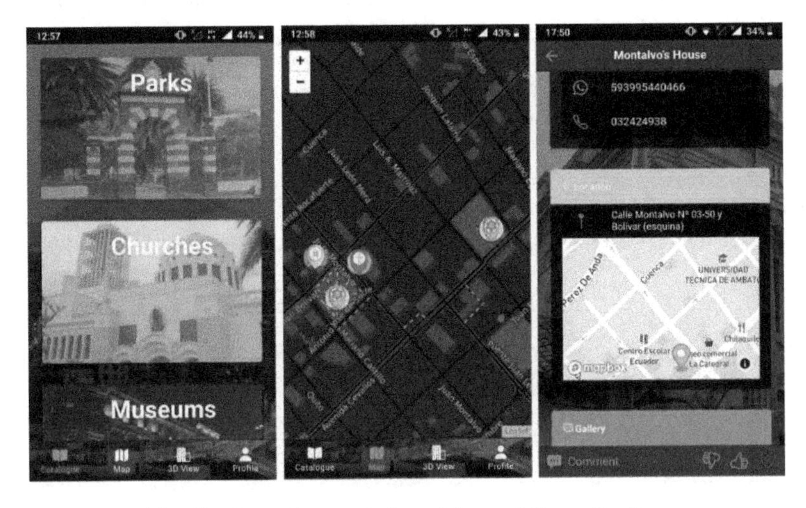

Fig. 5. General presentation of the mobile application.

Fig. 6. Manual entry of new places in the mobile application.

3 Tests and Experimental Results

3.1 Design of the Experiment

The experiment was made in two main groups. The first sample group was set-up by 25 entity managers and organizers related to tourist attraction places from the city of Ambato. The second group was made up of a sample of 100 university students linked to Tourism and Engineering professional degrees. The importance of working with these groups is to link aspects such as the usage of Information and Communication Technologies (ICT) and keep a clear view of the positive impact of tourism.

The experiment process was made in four phases, as illustrated in Fig. 7. Phases such as training, touristic sites registration, information review, and TAM poll, as well as execution times, are described in Table 1 [20].

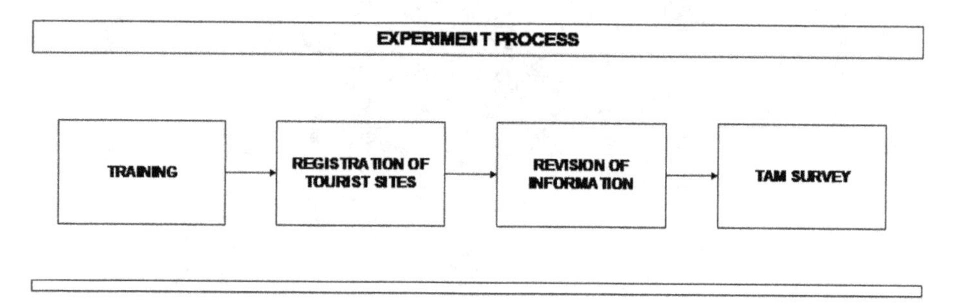

Fig. 7. Experiment process.

Table 1. Training phases.

Phase	Actions	Average time
Training	In this stage, the participants were trained in the usage of the application's functionality	20 min
Touristic site registration	In this cycle, the participants had access to the functionality of the application, in which they could register information on tourist sites in the city of Ambato	2 weeks
Information review	With the information recorded collaboratively, participants were asked to review the recorded information (textual, graphic, and geo-referenced). In this phase, the distribution of the points of greatest interest on the map component can be seen	40 min
TAM Poll	In the last experiment phase, a TAM-based poll was made up to the participants with the aim of evaluating the impact of the app from the technological acceptance criteria	10 min

3.2 Experimental Results

From the TAM quantitative analysis of the participants, the following results are deployed: 63% of the participants were women; the 57% of the surveyed people were under 25 years old, the 20% is made up from 26 to 40 years old and the 23% are people older than 40 years old. 52% of the surveyed ones are related to Tourism career, 41% with engineering careers and the remaining 7% were regarding administrative character careers. 20% is related to entities that offer tourism-related services. 73% of the surveyed people go for at least one touristic activity once a year. The 84% did not have prior experiences with collaborative tools that allow interaction and feedback with touristic experiences from other people.

For the result analysis of the research, it has been used the SPSS software tool. The validity and reliability of the questionnaire items were determined through of the Cronbach Alfa analysis using a Likert scale with 100% truly data that owns a 0.821 value.

- The ease of use perceived when using the tool impacts positively in the attitude of the application's usage.
- The application's ease of use promotes its implementation as part of the information managing for an efficient transport service.
- The feedback of the information is strongly correlated with the time optimization of users regarding the location of sites of interest
- The ease of use of the map allows intuitive access to the georeferenced information.
- The interactive map is a fundamental component of the proposal and promotes the future usage of the tool in the analyzed segment.

Table 2 values belong to the criteria of the mean and standard deviation values (based on a Likert scale with values from 1 to 5). It confirms that the software built for this study got the highest acceptance levels.

Table 2. Statistic values of TAM criteria.

TAM criteria	Mean	Standard deviation
Usage attempt	4.67	0.29
Perceived utility	4.35	0.48
Perceived ease of use	4.52	0.31
Attitude to the use	4.33	0.51

Additionally, it was demonstrated the technological acceptance criteria results do not follow a normal distribution when applying the Kolmogorov-Smirnov test, with a reference value of $p = 0.00$. Thus, for the data analysis, it was used the Tau-b from Kendall with the aim to establish a non-parametrical correlation. The resultant correlation between the elements was significantly positive as shown in Table 3.

Table 3. Tau-b from Kendall Results.

N = 125		P1	P2	P3	P4
Do you consider yourself satisfied whether using the application you can recover information regarding important touristic sites?	CC	,548**	,205**	,190**	,261**
	BS	0,000	0,000	0,000	0,000
When analyzing the proposed software proto-type, do you consider that in short time you can become an expert in this application?	CC	0,070	,227**	,124*	,174**
	BS	0,200	0,000	0,023	0,003
When analyzing the proposed software prototype, do you believe that its design makes it easier to use?	CC	,233**	0,082	,387**	,276**
	BS	0,000	0,104	0,000	0,000
Is the information access through the map easy to use for you?	CC	,212**	0,089	,278**	,307**
	BS	0,000	0,094	0,000	0,000
Does the map component highlight the information value that is managed in the application?	CC	,261**	,207**	,314**	1,000
	BS	0,000	0,000	0,000	
Does the generated reports highlight the information value that is managed in the application?	CC	,242**	0,075	,431**	,130*
	BS	0,000	0,150	0,000	0,015
Is significantly high the map usage?	CC	,205**	,156**	,303**	,298**
	BS	0,000	0,003	0,000	0,000
Would you suggest the use of the application for tourist information management?	CC	,156**	,424**	,302**	,367**
	BS	0,004	0,000	0,000	0,000
Would you use the application for better feedback and information management of the touristic places of your location?	CC	,267**	0,076	,611**	,255**
	BS	0,000	0,142	0,000	0,000

Abbreviations: P1: The software prototype finds this to be easy to use. P2: The prototype does not require much effort to use. P3: Easy access to information. P4: Map component helps understand the information. CC: Correlation Coefficient. BS Bilateral Significance.

4 Conclusions and Future Works

This research gets into the applicative type ones, thus the scientist foundation cannot be clearly evidenced. The functionality of the software prototype in this research is relevant since it allows users to contribute, consume, rate, and promote touristic information based on personal experiences. As the project has been developed from a multiplatform perspective and with open-source technologies, allows it to broaden its spectrum of use while simplifying the maintenance processes of the application. The study shows quantitatively that users (personnel associated with tourist activities and students) agreed with the use of this type of application. The results of the TAM poll let the authors confirm that the ease of use, usage attempt, and utility are highly correlated with the handling of the tool.

The map component becomes a relevant element in the proposal. Taking into account the current post-pandemic policies and strategies for economic reactivation of the autonomous governments, it is considered that this tool can contribute significantly to the tourism promotion in the national context and easily replicated in any location. The applicative inclusion as a touristic tool not only can benefit this sector, but it can be used as a very helpful tool inside education knowledge.

The main limitation in this research, which at the same time became a threaten to the trustful results, is the social confinement that does not allow touristic activities with normal parameters. Thus, it is highly recommended that in the close future, when health guarantees offer a more normal social environment, the evaluation can be managed again with a similar group of participants.

References

1. Li, Y., Hu, C., Huang, C., Duan, L.: The concept of smart tourism in the context of tourism information services. Tour. Manag. **58**, 293–300 (2017). https://doi.org/10.1016/j.tourman.2016.03.014
2. Varela-Aldás, J., Palacios-Navarro, G., García-Magariño, I.: Immersive virtual reality app for mild cognitive impairment. RISTI - Rev. Iber. Sist. e Tecnol. Inf. **2019**, 278–290 (2019)
3. Irawan, H., Akmalia, G., Masrury, R.A.: Mining tourist's perception toward Indonesia tourism destination using sentiment analysis and topic modelling. In: ACM International Conference Proceeding Service, pp. 7–12 (2019) https://doi.org/10.1145/3361821.3361829
4. Buele, J., Espinoza, J., Pilatásig, M., Silva, F., Chuquitarco, A., Tigse, J., Espinosa, J., Guerrero, L.: Interactive system for monitoring and control of a flow station using labVIEW. In: Advances in Intelligent Systems and Computing. pp. 583–592 (2018) https://doi.org/10.1007/978-3-319-73450-7_55
5. Ukpabi, D.C., Karjaluoto, H.: Consumers' acceptance of information and communications technology in tourism: A review. Telematics Inf. **34**(5), 618–6444 (2017). https://doi.org/10.1016/j.tele.2016.12.002
6. Rivera, M.A.: The synergies between human development, economic growth, and tourism within a developing country: An empirical model for ecuador. J. Destin. Mark. Manag. **6**, 221–232 (2017). https://doi.org/10.1016/j.jdmm.2016.04.002
7. Rodríguez-Fernández, M.M., Sánchez-Amboage, E., Martínez-Fernández, V.A.: The emergent nature of wine tourism in Ecuador and the role of the Social medium Facebook in optimising its positioning. Espacios. 38, (2017)
8. Heripracoyo, S., Adi, S.: Implementation of Tourism Business Web. In: Proceedings of 2019 International Conference Information Management Technology ICIMTech 2019. vol. 1, pp. 30–35 (2019). https://doi.org/10.1109/ICIMTech.2019.8843716
9. Zaman, K., el Moemen, M.A., Islam, T.: Dynamic linkages between tourism transportation expenditures, carbon dioxide emission, energy consumption and growth factors: evidence from the transition economies. Curr. Issues Tour. **20**, 1720–1735 (2017). https://doi.org/10.1080/13683500.2015.1135107
10. Bura, D., Singh, M., Nandal, P.: Predicting secure and safe route for women using google maps. In: Proceedings of International Conference Machine Learning Big Data, Cloud Parallel Computer Trends, Prespectives Prospect. Com. 2019. pp. 103–108 (2019) https://doi.org/10.1109/COMITCon.2019.8862173

11. Luthfi, A.M., Karna, N., Mayasari, R.: Google maps API implementation on IOT platform for tracking an object using GPS. In: Proceedings of - 2019 IEEE Asia Pacific Conference Wireless Mobile, APWiMob 2019. pp. 126–131 (2019) https://doi.org/10.1109/APWiMob48441.2019.8964139
12. Abbasian Fereidouni, M., Kawa, A.: Dark side of digital transformation in tourism. In: Lecture Notes in Computer Science (including subseries Lecture Notes in Artificial Intelligence and Lecture Notes in Bioinformatics). pp. 510–518 (2019) https://doi.org/10.1007/978-3-030-14802-7_44
13. De Bernardi, P., Bertello, A., Shams, S.M.R.: Logics hindering digital transformation in cultural heritage strategic management: An exploratory case study. Tour. Anal. **24**, 315–327 (2019). https://doi.org/10.3727/108354219X15511864843876
14. Ma, X., Sun, J., Zhang, G., Ma, M., Gong, J.: Enhanced expression and interaction of paper tourism maps based on augmented reality for emergency response. In: ACM International Conference Proceeding Series. pp. 105–109 (2018) https://doi.org/10.1145/3289430.3289442
15. Topal, I., Uçar, M.K.: In tourism, using artificial intelligence forecasting with tripadvisor data: year of Turkey in China. In: 2018 International Conference on Artificial Intelligence and Data Processing, IDAP 2018. Institute of Electrical and Electronics Engineers Inc. (2019) https://doi.org/10.1109/IDAP.2018.8620874
16. Wardhana, H., Mustofa, K., Sari, A.K.: Utilization of semantic web rule language for tourism ontology. In: Proceedings of the 3rd International Conference on Informatics and Computing, ICIC 2018 (2018). https://doi.org/10.1109/IAC.2018.8780474
17. Buele, J., Varela-Aldás, J., Castellanos, E.X., Jadán-Guerrero, J., Barberán, J.: 3D Object Reconstruction Using Concatenated Matrices with MS Kinect: A Contribution to Interiors Architecture. (2020) https://doi.org/10.1007/978-3-030-58817-5_49
18. Klimova, B.: Senior tourism and information and communication technologies. In: Lecture Notes in Electrical Engineering. pp. 440–445. Springer Verlag (2017) https://doi.org/10.1007/978-981-10-5041-1_71
19. Bharath, M., Reddy, K.V., Dey, R.: Implementation of IoT architecture for intruder alert system using MQTT protocol and MEAN stack. In: 2018 4th International Conference on Computing Communication and Automation, ICCCA 2018 (2018) https://doi.org/10.1109/CCAA.2018.8777526
20. Salazar, F.W., Naranjo-Ávalos, H., Buele, J., Pintag, M.J., Buenaño, É.R., Reinoso, C., Urrutia-Urrutia, P., Varela-Aldás, J.: Prototype System of Geolocation Educational Public Transport Through Google Maps API. (2020) https://doi.org/10.1007/978-3-030-58817-5_28

A Systematic Review of Context-Aware Technologies Applied to Buildings Comfort

Ana Isabel Martins[1], Ana Carolina Oliveira Lima[2], Paulo Bartolomeu[2], Lucilene Ferreira Mouzinho[2], Joaquim Ferreira[3], and Nelson Pacheco Rocha[4](✉)

[1] Institute of Electronics and Informatics Engineering of Aveiro, Department of Electronics, Telecommunications and Informatics, University of Aveiro, Aveiro, Portugal
anaisabelmartins@ua.pt

[2] Institute of Telecommunications, Department of Electronic, Telecommunications and Informatics, University of Aveiro, Aveiro, Portugal
{ana.carolina.lima,bartolomeu,lucileneferreira}@ua.pt

[3] Institute of Telecommunications, Higher School of Technology and Management of Águeda, Águeda, Portugal
jjcf@ua.pt

[4] Institute of Electronics and Informatics Engineering of Aveiro, Department of Medical Sciences, University of Aveiro, Aveiro, Portugal
npr@ua.pt

Abstract. Context-aware technologies can support automation and control of systems promoting inhabitants' comfort while safeguarding energy efficiency. In this study, a systematic review of the literature was performed to analyze how context-aware systems with impact in the comfort of buildings' inhabitants are being developed and how their translation is being addressed. Seventeen studies were included in this systematic review. As a main conclusion, the included articles report comfort solutions based on context-aware technologies still far from being consolidated, mainly because there is a lack of robust evaluation trials.

Keywords: Systematic literature review · Buildings' comfort · Energy efficiency · Context-aware

1 Introduction

Since people spend most of their time indoors [1] the quality of the indoor environment (e.g., air quality, ventilation, thermal conditions, illumination, and acoustics) have a huge impact in the comfort of buildings' inhabitants and consequently in their health and well-being conditions [2, 3].

The environmental control of buildings has been considered a canonical application of context-aware technologies in assisting decision making by using contextual information [4]. Context-aware technologies have been able to detect, sense and interpret environmental variables and characteristics of the inhabitants and to control

Á. Rocha et al. (Eds.): ICITS 2021, AISC 1330, pp. 323–332, 2021.
https://doi.org/10.1007/978-3-030-68285-9_31

the environment, generally for the purposes of being energy efficient and guaranteeing inhabitants' comfort [4].

Although there are articles systematizing evidence related to buildings technologies (e.g., [5–13]), to the best of the author's knowledge, there are no systematic reviews of the literature related to the use of context-aware technologies to promote the inhabitants' comfort. Therefore, the study reported by this article aimed to analyze how context-aware systems with impact in the comfort of buildings' inhabitants are being developed and how their translation (i.e., the processes required to move research findings from labs to practical use) is being addressed.

2 Methods

Considering the aforementioned objective, the authors performed a systematic review of the literature considering the following research questions:

- RQ1: What are the aims and outcomes of the studies?
- RQ2: What is the level of maturity of the systems being reported?
- RQ3: How is being addressed the translation of the systems being reported?

Boolean queries were prepared to include all the articles published until February 2020 that have their titles, abstract or keywords conform with the following expression: ('intelligent' OR 'smart' OR 'wise' OR 'cognitive') AND ('building' OR 'domestic' OR 'home' OR 'dwelling' OR 'house') AND ('efficient' OR 'green' OR 'eco' OR 'ecological' OR 'environment-friendly' OR 'environmental') AND ('comfort' OR 'well-being').

The searched resources included two general databases, Web of Science and Scopus, and one specific technological database, IEEE Xplore.

As inclusion criterion, the authors aimed to include all the articles published in scientific journals or in conference proceedings that have as main purpose the explicit use of context-aware technologies in modern buildings to promote the inhabitants' comfort while safeguarding energy efficiency.

Considering the exclusion criteria, the authors aimed to exclude all the articles not published in English, without abstracts or without access to the full text. Furthermore, the authors also aimed to exclude all articles that: do not report primary research results, such as reviews, surveys, or editorials; report solutions that are not intended to be used inside buildings (e.g., streets lighting); do not report results of the application of information technologies, but other technologies (e.g., innovative construction materials); do not report implementations of systems aiming the inhabitants' comfort, but rather partial solutions such as algorithms or support studies and tools; and do not report the use of context-aware technologies.

After the removal of duplicates and articles without abstracts, the analysis of the remaining articles was performed according to the following three steps: first, the authors assessed all titles and abstracts according to the outlined inclusion and exclusion criteria and those clearly out of the scope of the this study were removed; then, the authors assessed the full text of the remainder articles according to the outlined inclusion and exclusion criteria; finally the retrieved articles were classified.

In these steps, all the articles were analyzed by at least two authors. Any disagreement was discussed and resolved by consensus.

In order to analyze the translation of the solutions being reported, the authors distinguished the following development stages to classify the maturity level of the studies: i) Initial - the article describes the requirements elicitation which could include, in some cases, forms of co-design involving potential end users; ii) Design - the article describes the general overview of the system architecture or some of the respective components; iii) Technical testing - the article presents results of the performance evaluation of the system or some of its components (e.g., the performance of a specific algorithm); iv) Prototype testing - the article describes a laboratory evaluation involving end users (e.g., a usability test) of a minimally working version of the system being developed; v) Pilot testing - the article describes a real-world evaluation by end users in their daily context during a certain period; vi) Mature - the article describes a system that has been tested by end users, amended in some way and that is ready for deployment.

These stages are related to the different development phases, each one with a different maturity level of the development life cycle of information systems: requirements, analysis, design and implementation, testing, and evolution or maintenance.

3 Results

A total of 2144 articles were retrieved from the initial search on the Web of Science, Scopus, and IEEE Xplore.

Based on the title and abstract screening, 1826 articles were removed due to the following reasons: articles that were duplicated; articles without abstract; articles reporting overview or reviews; editorials, prefaces, and announcements of special issues, workshops, or books; articles reporting solutions intended for outdoor spaces; articles out of context for the present study since they do not report the application of information technologies (e.g., architecture, education, certification, building techniques or new materials); or articles that, although reporting systems based on information technologies, their focus was not the use of context-aware technologies to promote the comfort of the inhabitants.

Afterward, the full texts of the remaining 318 articles were analyzed and 301 articles were excluded due to the following reasons: the full texts were not in English; did not report the implementation of systems, but rather partial solutions such as algorithms or support studies and tools; reported studies where the main purpose was energy efficiency and not comfort; or were not focused on the use of context-aware technologies.

Therefore, 17 articles were considered eligible for this systematic review of the literature and their aims are presented in Table 1. To determine the maturity level of the systems, the 17 included articles were classified according to the development stages previously defined. The results of this classification are presented in Table 2. None of the studies were classified as initial or mature.

Table 1. Aims of the articles considered eligible.

Article	Year	Aim
[14]	2011	To propose a system able to predict inhabitants' behavior patterns, including the estimation of total number of occupants and occupancy duration, to optimize thermal comfort
[15]	2011	To propose an information system using cameras to analyze occupancy and human activity to optimize visual comfort
[16]	2013	To propose an automatic thermostat control system based on the mobility prediction of inhabitants, using contextual information obtained by mobile phones
[17]	2015	To present an automated lighting control framework which dynamically learns inhabitants' lighting preferences, models human visual comfort and controls light dimming to promote visual comfort
[18]	2015	To present a natural light reproduction system based on wireless sensor networks to restructure lighting conditions of windowless spaces
[19]	2015	To present a system to automatically control luminance and temperature of university classrooms, which was based on a platform that enables rapid development of Internet of Things and context-aware technologies for energy efficiency in smart buildings
[20]	2016	To propose an intelligent lighting control that recognizes the locations and behaviors of the teacher and students in a classroom to promote visual comfort
[21]	2016	To propose a system aiming to manage electrical load while attaining the building comfort
[22]	2016	To present a smart ambience system that analyzes the inhabitants' lighting habits considering different environmental context variables and inhabitants' needs to optimize visual comfort
[23]	2017	To propose a machine-to-machine technology infrastructure aiming to reduce power consumption without compromising the inhabitants' comfort
[24]	2017	To propose a smart heating, ventilation, and air conditioning (HVAC) system aiming to achieve a trade-off between energy costs and thermal comfort
[25]	2018	To propose a decision engine to use the environment contextual information to promote visual comfort
[26]	2018	To present the ongoing work of creating a smart Internet of Things desk
[27]	2019	To propose a context modeling and a reasoning architecture for observing the environment to improve thermal and visual comfort
[28]	2019	To propose a semantic framework to enable the representation of appliances' energy consumption to optimize comfort
[29]	2019	To propose an ontology to organize contextual information regarding the inhabitants, their needs, and the devices deployed inside the environment, to optimize comfort
[30]	2020	To propose smart lighting system to adjust the illumination conditions of classrooms according to specific learning context

Table 2. Classification of the articles according to the development stage of the systems being reported.

Development Stage	Articles
Design	[19, 21, 26–28]
Technical Testing	[14, 15, 18, 20, 22, 29]
Prototype Testing	[16, 17, 23–25]
Pilot Testing	[30]

3.1 Design

The studies reported by five [19, 21, 26–28] of the 17 articles were classified as in the design stage, since in general the articles present the systems' architectures or some of the respective components. Furthermore, all the five articles are focused on multiple perspectives comfort.

Article [27] introduces a framework that comprises various types of sensors (i.e., temperature, relative humidity, illumination, and levels of carbon dioxide, oxygen, and benzene) and a context modelling and reasoning module for observing the environment, aiming to improve thermal and visual comfort.

Article [19] presents a context manager that automatically controls the illumination and the temperature, while displaying power consumption. The data collected by sensors include the environmental temperature and the presence and location of the inhabitants (i.e., presence and electric power sensors).

The system presented in [21] was built upon an infrastructure aiming to manage electrical load while attaining buildings comfort using context-aware technologies. The authors do not refer sensors to collect environmental data but propose the use of fingerprint and passive infrared (PIR) sensors to identify and locate the occupants.

Article [26] proposes a smart Internet of Things desk that can personalize the environment around the occupants and can act as a support system to drive their behaviors towards better environmental settings (i.e., thermal and visual comfort) and improve posture and ergonomics. To determine the occupants' identity and location, motion and radio-frequency identification (RFID) sensors are proposed. In turn, environmental data are collected by volatile organic compounds and particle matter sensors.

Finally, article [28] presents a semantic framework aiming at providing inhabitants with the possibility of having personalized indoor comfort in their living environments and at helping them in scheduling their daily activities requiring appliances. The activities of the inhabitants are estimated using electric power sensors. The sensing network is further complemented with luminance, acoustic and relative humidity sensors.

3.2 Technical Testing

The studies reported by six articles [14, 15, 18, 20, 22, 29] have been considered in the technical testing development stage. One of these six articles is focused on multi perspective comfort [29], one is focused on thermal comfort [14], while the remainder four are focused on visual comfort [15, 18, 20, 22].

Article [29] uses an ontology to organize contextual information (i.e., location of the inhabitants using RFID) to leverage knowledge regarding the inhabitants, their needs, and the devices deployed inside the environment. Several simulation scenarios are presented, and the authors conclude about the possibility to reduce energy consumption.

Article [14] proposes a solution to optimize thermal comfort by using knowledge of the inhabitants, more precisely by estimating the number of inhabitants and the duration of their activities by deploying a large-scale sensor network. The inhabitants' activities are estimated by using cameras while the environmental data are collected by luminance, acoustic, temperature, relative humidity, and carbon dioxide sensors. The article also presents simulation results according to which there could be 19% in energy saving with 88% thermal comfort satisfaction using an inhabitant behavior-based control approach [14].

Article [15] reports on the use of cameras and a distributed processing method to estimate inhabitants' location and activity in a room to feed light control algorithms in order to optimize inhabitants' visual comfort and lightning energy costs. The article also reports the assessment of the activity analysis methods' performance based on hidden Markov models and on conditional random fields, and a simulation of the light control based on the designed occupancy reasoning algorithms [15].

Article [20] presents a control system aiming to create the appropriate illumination environment of a classroom by learning schedules and by recognizing the locations of the teacher and students by means of sensors (i.e., presence sensors and electric power sensors, respectively, to determine the location of the occupants and their activities). The environmental data are collected using luminance and correlated color temperature sensors. The performance of the suggested system was assessed using a classroom lighting scenario.

Article [18] implements a natural light reproduction system able to adapt the illumination environments to the changing situation of the inhabitants. Considering this aim, luminance and correlated color temperature sensors are used to gather the illumination conditions, and an aggregation of infrared, ultrasonic, microwave and electric power sensors turn possible the determination of the location of the inhabitants. The performance of the proposed system was evaluated considering the conformity with natural light of the light controlling units [18].

Finally, the system presented by [22] analyses the inhabitants' illumination habits, to automate the ambiance dynamically considering the data gathered by luminance, temperature, relative humidity, and pressure sensors. The applied learning models were compared by creating a visualization of predicted light across the different types of models.

3.3 Prototype Testing

Five of the 17 studies were included in the prototype testing development stage [16, 17, 23–25]. One of these five articles is focused on multi perspective comfort [23], two are focused on thermal comfort [16, 24], and two are focused on visual comfort [17, 25].

The system presented in [23] was built upon a machine-to-machine technologies infrastructure to reduce home power consumption to the minimum without compromising inhabitant comfort (i.e., thermal, visual and air quality) in a single or multiuser environment. The environmental data are collected by luminance sensors, while touch screens, microphones and electric power sensors are used for the location of the inhabitants and the inference of their activities. The authors compare the accuracy among various activity inference algorithms in diverse circumstances using a dataset collected from scratch in a smart home lab [23].

Article [17] presents a system to continuously analyze ambient information, using luminance sensors, and derive dynamic models of the inhabitants' visual comfort, safeguarding their preferences under different control scenarios. The identification and location of the inhabitants is based on the use of cameras, mobile phones and motion sensors. In terms of assessment, an experiment was conducted involving two subjects and focused on visual comfort preferences. The results indicated that it is possible more than 10% energy savings retaining comfort levels above 90% or more than 35% savings retaining comfort levels above 75% [17].

The system presented by [25] is able to analyze the inhabitants' illumination habits, considering different environmental context variables (i.e., luminance, temperature, relative humidity and carbon dioxide) and inhabitants' needs (i.e., using a mobile phone for identification and location of the inhabitants) in order to automatically learn about their preferences and automate the ambience dynamically. A prototype of the system was installed in a smart room. During a 2-week period, users interacted with the system using a mobile application to control devices and visualize real-time and historic data. The authors concluded that the main functions of the system can be executed regularly without any major delays and not causing any discomfort for the inhabitant [25].

Although different strategies were considered, articles [16, 24] propose solutions to optimize thermal comfort by using knowledge of the inhabitants: article [16] describes a control system based on the mobility prediction of the inhabitants, using contextual information obtained by mobile phones; and article [24] describes a network of sensors and actuators to support scheduling mechanisms according to which HVAC working times can be selected aiming to achieve a trade-off between energy costs and thermal comfort of inhabitants. Article [16] does not refer sensors to collect environmental variables but propose the use of mobile phones to identify and locate the inhabitants. In turn, article [24] proposes the use of temperature sensors to monitor the environment, as well as PIR sensors to locate the inhabitants.

In both studies [16, 24] the assessment of the systems involved end users. The results of testing the network of sensors and actuators presented in [24] indicate that the system ensures the inhabitants' thermal comfort with traditional approaches. In turn, the experiment reported by [16] involved 21 participants and allowed the authors to conclude that their approach can decrease energy consumption by more than 25% and predict at least 70% of the transit cases.

3.4 Pilot Testing

Finally, in the pilot testing development stage one article [30] was included. Article [30] reports a system to improve the illumination conditions of classrooms by adjusting brightness, correlated color temperature, and luminance distribution to specific learning contexts. The location of the occupants is determined using PIR sensors. In turn, various types of sensors are used to determine the illumination conditions (i.e., luminance and correlated color temperature sensors) and the air quality (i.e., formaldehyde density and particle matter sensors).

A user acceptance evaluation in various real classrooms is mentioned [30], but no details are presented about this evaluation.

4 Discussion and Conclusion

Concerning the first research question (i.e., what are the aims and outcomes of the studies?), we might conclude that the retrieved articles present different approaches to integrate context-aware technologies in systems aiming to improve the inhabitants comfort, namely multiple perspective comfort (seven articles [19, 21, 23, 26–29]); visual comfort (seven articles [15, 17, 18, 20, 22, 25, 30]); and thermal comfort (three articles [14, 16, 24]).

Except two articles [22, 27], all the remainder articles refer the use of sensors to collect data related to the inhabitants. The main concern is the inhabitants' positioning to provide personalized comfort zones across different spaces, by inferring occupancy or the number of inhabitants [14–21, 23–26, 29, 30]. Due to the limitations of the sensors some articles report the use of multiple sensing techniques supported by different types of sensors with the same purpose (e.g., [20, 24]).

Moreover, the retrieved articles also consider a great diversity of sensors to monitor the environmental conditions, namely luminance [14, 17, 18, 20, 22, 23, 25, 27–30], correlated color temperature [18, 20, 29, 30], temperature [14, 19, 22, 24, 25, 28], relative humidity [14, 22, 25, 28], pressure [22] and air quality [14, 25–30].

Looking for the maturity level of the systems being reported (i.e., the second research question), prototypes were developed to evaluate the performance of the technologies being used [14, 15, 18, 20, 22, 29] or to allow the assessment of their performance in a laboratory environment involving users [16, 17, 23–25]. Just one article, [30], reports an evaluation considering a real-world space with real end users, but the details of this evaluation were not presented.

Therefore, this systematic review of the literature shows that the retrieved articles report solutions for buildings still far from being consolidated for translation (i.e., the third research question). Most of them do not report evidence about the evaluation of the systems in real environments with real end users. It is evident the lack of robust evaluation trials focusing scaling aspects (e.g., number of services being provided, amount of context being used and long-term usage characteristics), issues that are essential to translate the solutions being developed. This is one of the major barriers to the translation of the solutions being proposed and developed.

Acknowledgments. The present study was developed in the scope of the Smart Green Homes Project [POCI-01-0247 FEDER-007678], a co-promotion between Bosch Termotecnologia S.A. and the University of Aveiro. It is financed by Portugal 2020, under the Competitiveness and Internationalization Operational Program, and by the European Regional Development Fund.

The publication was financially supported by National Funds through FCT – Fundação para a Ciência e a Tecnologia, I.P., under the project UI IEETA: UID/CEC/00127/2019.

References

1. Klepeis, N.E., et al.: The national human activity pattern survey (nhaps): a resource for assessing exposure to environmental pollutants. J. Exp. Sci. Envir. Epidemiol. **11**(3), 231–252 (2001)
2. Allen, J.G., et al.: Green buildings and health. Current Envir. Health Reports **2**(3), 250–258 (2015)
3. Clausen, G., et al.: The combined effects of many different indoor environmental factors on acceptability and office work performance. HVAC&R Res. **14**(1), 103–113 (2008)
4. Schilit, B., Adams, N., Want, R.: Context-aware computing applications. In: First Workshop on Mobile Computing Systems and Applications, pp. 85–90. IEEE (1994)
5. Darko, A., Zhang, C., Chan, A.P.: Drivers for green building: a review of empirical studies. Habitat Int. **60**, 34–49 (2017)
6. Zhao, X., Zuo, J., Wu, G., Huang, C.: A bibliometric review of green building research 2000-2016. Arch. Sci. Rev. **62**(1), 74–88 (2019)
7. Soares, N., et al.: A review on current advances in the energy and environmental performance of buildings towards a more sustainable built environment. Renew. Sustain. Energy Rev. **77**, 845–860 (2017)
8. Jung, W., Jazizadeh, F.: Human-in-the-loop HVAC operations: A quantitative review on occupancy, comfort, and energy-efficiency dimensions. Appl. Energy **239**, 1471–1508 (2019)
9. Marikyan, D., Papagiannidis, S., Alamanos, E.: A systematic review of the smart home literature: a user perspective. Technol. Forecast. Soc. Change **138**, 139–154 (2019)
10. Petroşanu, D.M., Căruţaşu, G., Căruţaşu, N.L., Pîrjan, A.: A review of the recent developments in integrating machine learning models with sensor devices in the smart buildings sector with a view to attaining enhanced sensing, energy efficiency, and optimal building management. Energies **12**(24), 4745 (2019)
11. Bavaresco, M.V., D'Oca, S., Ghisi, E., Lamberts, R.: Technological innovations to assess and include the human dimension in the building-performance loop: A review. Energy Build. **202**, 109365 (2019)
12. Panchalingam, R., Chan, K.C.: A state-of-the-art review on artificial intelligence for smart buildings. Intell. Build. Int. 1–24 (2019)
13. Martins, A.I., et al.: Enabling green building's comfort using information and communication technologies: a systematic review of the literature. In: 2020 World Conference on Information Systems and Technologies, pp. 197–208. Springer, Cham (2020)
14. Dong, B., Lam, K.P.: Building energy and comfort management through occupant behaviour pattern detection based on a large-scale environmental sensor network. J. Build. Perform. Simul. **4**(4), 359–369 (2011)
15. Lee, H., Wu, C., Aghajan, H.: Vision-based user-centric light control for smart environments. Pervasive and Mobile Comput. **7**(2), 223–240 (2011)

16. Lee, S., Chon, Y., Kim, Y., Ha, R., Cha, H.: Occupancy prediction algorithms for thermostat control systems using mobile devices. IEEE Trans. Smart Grid **4**(3), 1332–1340 (2013)
17. Malavazos, C., Papanikolaou, A., Tsatsakis, K., Hatzoplaki, E.: Combined visual comfort and energy efficiency through true personalization of automated lighting control. In: 2015 International Conference on Smart Cities and Green ICT Systems (SMARTGREENS), pp. 1–7. IEEE (2015)
18. Kim, Y.J., Kwon, S.Y., Lim, J.H.: Design and implementation of the natural light reproduction system based on context awareness in WSN. Int. J. Distr. Sensor Netwk. **11**(9), 781584 (2015)
19. Kamienski, C., et al.: Context-aware energy efficiency management for smart buildings. In: IEEE 2nd World Forum on Internet of Things (WF-IoT), pp. 699–704. IEEE (2015)
20. Lee, H.S., Kwon, S.Y., Lim, J.H.: A development of a lighting control system based on context-awareness for the improvement of learning efficiency in classroom. Wireless Pers. Commun. **86**(1), 165–181 (2016)
21. Putri, G.A.A., Nugroho, L.E.: Context modeling for intelligent building energy aware. In: International Conference on Smart Green Technology in Electrical and Information Systems (ICSGTEIS), pp. 161–166. IEEE (2016)
22. Yin, X., Keoh, S.L.: Personalized ambience: an integration of learning model and intelligent lighting control. In: IEEE 3rd World Forum on Internet of Things (WF-IoT), pp. 666–671. IEEE (2016)
23. Lu, C.H., Wu, C.L., Weng, M.Y., Chen, W.C., Fu, L.C.: Context-aware energy saving system with multiple comfort-constrained optimization in M2M-based home environment. IEEE Trans. Autom. Sci. Eng. **14**(3), 1400–1414 (2015)
24. Marche, C., Nitti, M., Pilloni, V.: Energy efficiency in smart building: a comfort aware approach based on Social Internet of Things. In: 2017 Global Internet of Things Summit (GIoTS), pp. 1–6. IEEE (2017)
25. Meurer, R.S., Fröhlich, A.A., Hübner, J.F.: Ambient intelligence for the internet of things through context-awareness. In: International Symposium on Rapid System Prototyping (RSP), pp. 83–89. IEEE (2018)
26. Aryal, A., Anselmo, F., Becerik-Gerber, B. Smart IoT desk for personalizing indoor environmental conditions. In: 8th International Conference on the Internet of Things, pp. 1–6. ACM (2018)
27. Mughal, S., Razaque, F., Malani, M., Hassan, M. R., Hussain, S., Nazir, A.: Context-aware indoor environment monitoring and plant prediction using wireless sensor network. In: International Conference for Emerging Technologies in Computing, pp. 149–163. Springer, Cham. (2019)
28. Spoladore, D., Mahroo, A., Trombetta, A., Sacco, M.: ComfOnt: a semantic framework for indoor comfort and energy saving in smart homes. Electronics **8**(12), 1449 (2019)
29. Degha, H.E., Laallam, F.Z., Said, B.: Intelligent context-awareness system for energy efficiency in smart building based on ontology. Sustain. Comput.: Inf. Syst. **21**, 212–233 (2019)
30. Sun, B., Zhang, Q., Cao, S.: Development and implementation of a self-optimizable smart lighting system based on learning context in classroom. Int. J. Envir. Res. Public Health **17**(4), 1217 (2020)

Multi-agents Cooperation Supporting Smart Hydroponic Crop

Manuel J. Ibarra[1(✉)], Yonatan Mamani-Coaquira[1], Olivia Tapia[1], Edgar Alcarraz[1], Vladimiro Ibañez[2], and Yalmar Ponce[3]

[1] Universidad Nacional Micaela Bastidas de Apurímac, Abancay, Peru
manuelibarra@gmail.com, ymamanic@gmail.com, oly_ana@hotmail.com, alcarraz.e@hotmail.com
[2] Universidad Nacional del Altiplano Puno, Puno, Peru
vibanez@unap.edu.pe
[3] Universidad Nacional Jose María Arguedas, Andahuaylas, Peru
yalmar@unajma.edu.pe

Abstract. This article presents smart hydroponic farming which provides a flexible and intelligent environment improving the quality of vegetable production. The main objective of this research is to develop a multi-agent system and test the functionality and interaction of the sub-systems. The tests were performed with lettuce plantation for 30 days. The model integrates six agents that cooperate in real-time for optimal production of vegetables; for this purpose, each agent helps the system using the knowledge acquired during the process.

Keywords: Monitoring · Cooperation · Hydroponic · Farm · Smart · Agent · Crop · Lettuce

1 Introduction

Currently, around 84.1% of South-America's population live in urban areas, reducing the agricultural production rate [1]; this has increased the possibilities to make urban agriculture. Also, it is possible to carry out production on the roofs of houses to save space. As the world population grows, it also increases the demand for food products. Due to this growing demand, maybe a food crisis in the coming years [2] and with urban agriculture, part of the problem of food shortages in the markets can be solved. China is one of the most densely populated countries with megacities and will continue to expand in number and size, which will require increasing food production. The total roof space in China is approximately one million hectares, which can be used for growing vegetables [3].

Nowadays, there are several continuing challenges in the hydroponic crop domain, mainly when they are supported by the Internet of Things (IoT), because it allows automating monitoring permanently each of the production variables, using sensors and communication networks during the production process [4].

Á. Rocha et al. (Eds.): ICITS 2021, AISC 1330, pp. 333–339, 2021.
https://doi.org/10.1007/978-3-030-68285-9_32

Also, BigData could help to analyze data, predict situations, improve many activities in real-time, process the collected data sometimes, and it is necessary to have supercomputers. However, currently, vast sets of data can be analyzed on powerful but affordable, notebooks, desktop computers or Raspberry Pi devices with standard open software tools [5].

2 Related Works

Gonzales-Briones et al. [6], in his paper 'A multi-agent system framework for autonomous crop irrigation', proposes the design and implementation of the system divided into sub-systems that collect data using sensors through wireless communication for autonomous crop irrigation. The multi-agent system improves the efficiency and effectiveness of agricultural production. First, analyzing the data collected by sensors, then, extracting some patterns, and finally predict events. The main idea is improving the crop and anticipates such events giving localized responses with automated actuators. It has been tested in an agricultural environment to optimize irrigation in a corn crop. The soil quality and its climatic conditions were analyzed, extracting information about the needs of cultivated corn and making efficient irrigation decisions based on those needs, reducing water consumption by 17.16% compared to traditional automotive irrigation.

Kim and Yeo [7] present the research titled *Self-healing Multi-Agent Prototyping System for Crop Production*. They propose the design and architecture for a proactive self-healing system that solve internal problems using self-awareness as contextual information for crop production monitoring. For this purpose, the system is divided into Agents that analyze the log context to perform self-healing and self-diagnosis. The result shows that the system is useful, and practical experiments confirmed it.

3 The Proposed Multi-agent System

Multi-Agent system is a system divided into two or more systems, and have a common objective to solve a problem. In last decades different multi-agent models and intelligent systems were developed, for example, smart home control [8], and visual design [9]

The Multi-Agent agents that cooperate in this system are:

A) Energy Control Agent

This agent is in charge of supplying the electrical energy by time slices, to save the consumption of electrical power, the "submersible water pump" runs 10 min every 2 h. In general, plants require water and oxygenation from time to time and not permanently. It has been tested with the cultivation of certain vegetables, and their growth is similar when the pump is permanently operating. The energy helps to run the motor that raises the water approximately 1 m high and then falls by gravity and recirculates in the system.

B) Smart Crop Agent

This agent contains the vegetables and needs the nutrients as input. The vegetable containers do not have an angle of inclination.

Researchers mention that containers must have inclination angle between 5%–15%, it allows for flowing the water by gravity [10, 11]; however, the problem is that water pumps must work all the time, and increases electrical energy consumption.

C) Sensors Agent

This agent measures the most critical variables, such as pH, electrical conductivity, water temperature, water volume, light intensity, plant growth, among others. Sensors need to be well-calibrated to give the right measurements at the right time. The sensors send their acquired values to the database of the monitoring center.

D) Monitoring Center Agent

This agent collects all the data obtained by the sensors; the agent uses data-mining and big-data techniques to process the information, and then sends to the farmer. It is also responsible for generating alerts to users, for example, the pH for the optimal lettuce growing must be between 5.5 and 6.5, so when the sensor measures a value outside this range, then the system sends a warning message through email, social network or text.

E) Nutrient Mixing Agent

This agent controls the quantity and quality of nutrients. Plants need some micro-nutrients (Iron chelate, Manganese sulfate, Boric acid, Zinc sulfate, Copper sulfate) and macro-nutrients (Potassium nitrate, Ammonium nitrate, Triple calcium superphosphate, Calcium nitrate) for growing. There are some technologies to make Nutrient Mixing. Mixing nutrients using fuzzy logic rules-based inference [12, 13]; another approach is to use neural networks [14].

F) Farmer Agent

The Farmer agent continuously updates the data obtained by the sensors, and the farmer has information for decision-making from anywhere and anytime. The main advantage is that the farmer can change the system configuration values from his computer or mobile device.

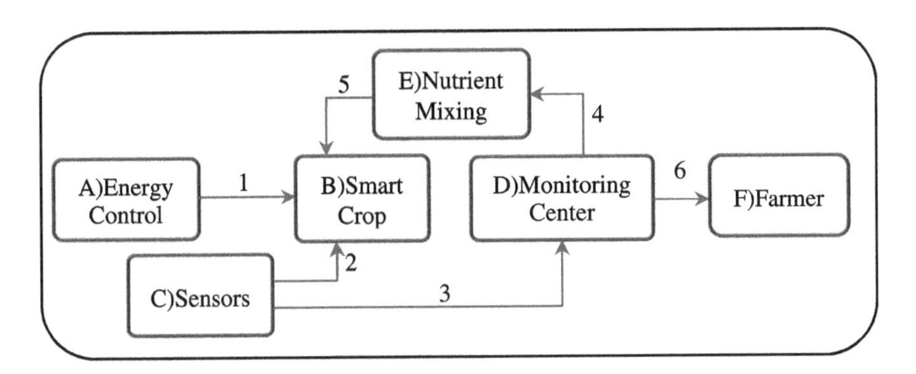

Fig. 1. Multi-agents interaction

Figure 1 shows the Multi-agents interaction with each system agents. The *Energy Control Agent* sends the electric energy by time slice to the *Smart Crop Agent.* The *Sensors Agent* reads the values of variables (pH, Temperature, and others), then sends these values to the *Monitoring Center*. If the *Monitoring Center* detects that nutrient solution is not into the correct parameters, then the *Nutrient Mixing Agent* mixes the new nutrient solution and adds to the *Smart Crop Agent.* Finally, the *Farmer Agent* receives the notification and has a control panel to see the evolution of the production.

4 Functionality and Evaluation of the Proposed System

For the evaluation of the multi-agent system, we planted lettuce; the planting began with germination on February/15/2020, and the harvest was on March/16/2020, 30 days in the hydroponics system. It is necessary to clarify that, before starting the plantation, the germination lasted 27 days.

As an example, we can see in Fig. 2 pH measuring, the y-axis shows the pH value and the x-axis shows the number of measures (ten times a day), during the 30 days of the growth in the hydroponic multi-agent system. During the process, the system sent six alerts to the user (farmer) because the pH value was reaching the lower and upper limits.

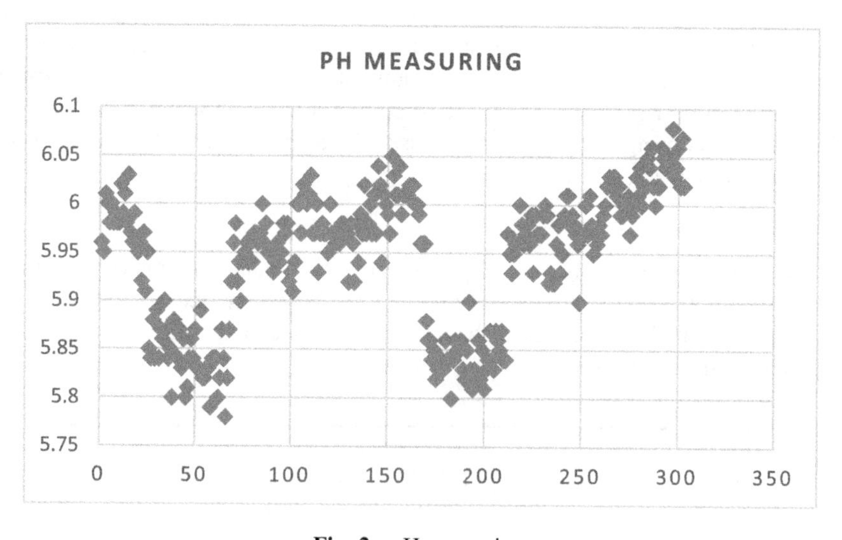

Fig. 2. pH measuring

Figure 3 shows the picture of lettuce growth in the multi-agent system; in this case, the communication between sensors and Arduino mainboard was using wired cables. The communication between raspberry pi (the mini-server device [15]) and the user agent was using HTTP protocol and Wireless Network communication.

Fig. 3. Multi-agent system in operation

The Electronic devices used in this research are shown in Table 1.

Table 1. Electronic Devices for the multi-agent.

Device	Description
Electric timer	This device automates the turning on and off of the submersible water pump. It is used to manage the turning on and off of an electronic device
Arduino ONE	Open source microcontroller board based on ATmega328P microchip; this board has been developed by Arduino
Temperature Sensor	A submersible probe in water to measure Temperature (DS18B20)
pH Sensor	An electronic device used to measure the pH of the water containing the solution to feed the plant (pH Electrode)
Raspberry Pi	A mini-computer called Raspberry Pi was used, which fulfils the function of receiving the data read by the sensors and saving it in a database

5 Conclusion and Future Work

This research work has developed a system with cooperative work in real-time among 6 agents or sub-systems called multi-agents. The tests were made with lettuce plantation during January and February 2020, in which the communication among the agents has worked properly.

As future work, we will use a 3D printer to encapsulate the electronic components, and thus more safety and ergonomics in the parts. We will also carry out exhaustive tests of the system with other types of plantations such as tomato or onion. Finally, as future work, the intention is to add a prediction agent that uses artificial intelligence and machine learning to determine, through image processing, the growth, weight and colouration of the planted plant.

Acknowledgements. Thanks to the Micaela Bastidas National University of Apurimac for financial support.

References

1. Nations, U.: Department of Economic and Social Affairs Population Dynamics https://population.un.org/wup/Download/
2. Gashgari, R., Alharbi, K., Mughrbil, K., Jan, A., Glolam, A.: Comparison between growing plants in hydroponic system and soil based system. In: Proceedings of 4th World Congress Mechanical Chemical Material Engineering (2018) https://doi.org/10.11159/icmie18.131
3. Liu, T., Yang, M., Han, Z., Ow, D.W.: Rooftop production of leafy vegetables can be profitable and less contaminated than farm-grown vegetables. Agron. Sustain. Dev. **36**, 41 (2016)
4. Tzounis, A., Katsoulas, N., Bartzanas, T., Kittas, C.: Internet of things in agriculture, recent advances and future challenges. Biosyst. Eng. **164**, 31–48 (2017)
5. Senthilvadivu, S., Kiran, S.V., Devi, S.P., Manivannan, S.: Big data analysis on geographical segmentations and resource constrained scheduling of production of agricultural commodities for better yield. Procedia Comput. Sci. **87**, 80–85 (2016) https://doi.org/10.1016/j.procs.2016.05.130
6. González-Briones, A., Castellanos-Garzón, J.A., Mezquita-Martín, Y., Prieto, J., Corchado, J.M.: A multi-agent system framework for autonomous crop irrigation. In: 2nd International Conference Computer Application Information Security ICCAIS (2019) https://doi.org/10.1109/CAIS.2019.8769456
7. Kim, H.K., Yeo, H.: Self-healing multi agent prototyping system for crop production. In: Advances in Intelligent Systems and Computing. pp. 31–43. Springer Verlag (2014) https://doi.org/10.1007/978-3-319-07596-9_40
8. Hawdziejuk, K., Grabska, E.: Cooperation of agents in the agent system supporting smart home control. In: Luo, Y. (ed.) Cooperative Design, Visualization, and Engineering, pp. 57–64. Springer International Publishing, Cham (2017)
9. Palacz, W., Ślusarczyk, G., Strug, B., Grabska, E.: Visual design using case-based reasoning and multi-agent systems. In: Luo, Y. (ed.) Cooperative Design, Visualization, and Engineering, pp. 48–56. Springer International Publishing, Cham (2017)

10. Soares, H.R., Silva, Ê.F. de F., Silva, G.F. da, Pedrosa, E.M.R., Rolim, M.M., Santos, A.N.: Lettuce growth and water consumption in NFT hydroponic system using brackish water. Rev. Bras. Eng. Agr{\'\i}cola e Ambient. **19**, 636–642 (2015)
11. Rubio Mena, C.: Automatización de un cultivo hidropónico NFT para el control de temperatura, riego y mezcla de solución nutritiv, ubicada en la zona urbana de Quito (2017)
12. Yolanda, D., Hindersah, H., Hadiatna, F., Triawan, M.A.: Implementation of real-time fuzzy logic control for NFT-based hydroponic system on Internet of Things environment. In: Proceedings of 2016 6th International Conference System Engineering Technology ICSET 2016. pp. 153–159 (2017) https://doi.org/10.1109/FIT.2016.7857556
13. Chang, C.L., Hong, G.F., Fu, W.L.: Design and implementation of a knowledge-based nutrient solution irrigation system for hydroponic applications. Trans. ASABE. **61**, 369–379 (2018) https://doi.org/10.13031/trans.11564
14. Cho, W.-J., Kim, H.-J., Jung, D.-H., Han, H.-J., Cho, Y.-Y.: Hybrid signal-processing method based on neural network for prediction of NO3, K, Ca, and Mg ions in hydroponic solutions using an array of ion-selective electrodes. Sensors (Switzerland). 19, (2019) https://doi.org/10.3390/s19245508
15. Ibarra, M.J., Huaraca, C., Soto, W., Palomino, C.: MLMS: Mini learning management system for schools without internet connection. In: Twelfth Latin American Conference on Learning Technologies (LACLO), . pp. 1–7 (2017)

Agent-Oriented Approaches for Model-Based Software Testing: A Mapping Study

Jose Ramírez-Méndez(✉), Christian Quesada-López, Alexandra Martínez, and Marcelo Jenkins

Universidad de Costa Rica, San José, Costa Rica
{jose.ramirez16,cristian.quesadalopez,alexandra.martinez, marcelo.jenkins}@ucr.ac.cr

Abstract. Automated software testing reduces manual work, increases test coverage, and improves error detection. Model-Based Testing (MBT) is a testing approach that automatically executes test cases generated from a model representing the system behavior. The parallelization of MBT process stages, such as model creation and exploration, or test case generation and execution, could improve its scalability to handle complex systems. Agent-Oriented Software Testing (AOST) refers to the use of intelligent agents focusing on the automation of complex testing tasks. AOST could improve the testing process by providing a high level of decomposition, independence, parallel activation, intelligence, autonomy, sociality, mobility, and adaptation. In this work, we conducted a systematic mapping study of the existing AOST approaches for MBT. We identified 36 primary studies over the period 2002–2020. We classified agent approaches according to the MBT process stages, and tasks and roles covered as part of their implementation. We found 25 implementations of AOST approaches in the test case generation stage, 20 in the test execution, 10 in the model construction, and 3 in the test criteria selection. Studies reported the test generator role 25 times, test executor role 20 times, and the monitor-coordinator of activities 12 times. Additional studies to understand the benefits of agent-oriented approaches for model-based testing are required.

Keywords: Model-based testing · Agent-oriented testing · Systematic mapping review

1 Introduction

Software testing is the process of dynamically verifying that a program provides the expected behaviors given a finite set of test cases [5]. Testing involves executing each test case with a different set of input values to find as many bugs as possible before releasing software product [21]. Automating software-testing activities offers several benefits, notably better test coverage, reduced human effort, increased error detection, and confidence in the system quality [7].

Á. Rocha et al. (Eds.): ICITS 2021, AISC 1330, pp. 340–349, 2021.
https://doi.org/10.1007/978-3-030-68285-9_33

Model-Based Testing (MBT) is an approach that automatically generates test cases from a model representing the system behavior, and executes these cases to validate system requirements and consistency [5]. MBT has four stages [19]: model design, test selection criteria, test case generation, and test case execution. MBT has shown to reduce the overall cost, time, and effort of the testing process, and aid with requirements traceability and evolution [21]. Although tools that automate the MBT process exist, some activities still require manual work [22]. Tasks such as test case generation, prioritization, minimization, and execution are often time-consuming and strenuous. One alternative to address these challenges seems to be agent-oriented approaches [3].

Agent-Oriented Software Testing (AOST) applies agent entities to support the automation of complex software testing activities. AOST conducts software-testing activities by using agent-based paradigms, artificial intelligence techniques, and software engineering practices [10]. MBT challenges can be addressed by taking advantage of agent-oriented approaches. An agent is a software entity capable of executing tasks and interacting with other entities (software, hardware, humans) with a high-level of decomposition, independence, and parallel activation [13,16,17]. AOST could help MBT become more cost effective and efficient [3].

This works presents a systematic mapping study of existing AOST approaches for automating the MBT stages. We classified these approaches according to the MBT stages where agents were used, and the tasks and roles those agents perform. The rest of the paper is structured as follows: Sect. 2 presents the related work. Section 3 describes the mapping study design. Section 4 presents and discusses the results, and Sect. 5 describes the conclusions.

2 Related Work

This section presents secondary studies conducted in agent-oriented software testing (AOST) and model-based testing (MBT). Arora and Bhatia [3] reported a systematic review of test case generation approaches for regression testing using AOST. They identified approaches, platforms, and methodologies in 115 studies, where 56 were agent-based approaches for software testing. They classified testing categories such as regression, web, and object-oriented testing for model-based, structural-based, fuzzy-based, and performance-based testing.

Kumaresen et al. [10] stated a definition for the Agent-Based Software Testing research area. They conducted a systematic mapping analyzing 41 studies related to AOST. They identified the agent design architectures, testing levels (such as acceptance and integration), and testing areas (functional, non-functional, white-box, and regression testing). Villalobos et al. [21] conducted a tertiary study in which 12 literature surveys and 10 systematic literature reviews on MBT approaches were selected. They reported on a hierarchical taxonomy scheme to classify MBT areas, tools, and challenges.

To our knowledge, no prior mapping studies have classified agent-oriented testing approaches implemented to automate model-based testing stages.

3 Methodology

We designed and reported our study based on the guidelines by Petersen et al. [14].

3.1 Goal and Research Questions

The goal of this mapping study was to analyze Agent-Oriented Software Testing (AOST) approaches in terms of the agents' roles and tasks, the MBT process stages in which agents were implemented, and the coverage criteria used in the evaluations. Our research questions (RQ) were as follows:

- **RQ1**: What MBT process stages have been implemented by agents?
- **RQ2**: What roles and tasks have been assigned to the agents during the MBT process?
- **RQ3**: What test coverage criteria have been used in agent-oriented MBT evaluation?

The first research question (RQ1) identifies AOST approaches that attempt to automate the MBT process stages. We classify the approaches into one of four stages [19]: (1) model construction, (2) test criteria selection, (3) test case generation, and (4) test case execution. The second research question (RQ2) characterizes the approaches in terms of the agent implementation methodology. We also reported the roles and tasks attributed to each agent entity. Finally, the third research question (RQ3) determines what coverage metrics have been reported for MBT evaluations in the context of AOST.

In the following section the search protocol is presented, which has been first developed by the first author, and later been reviewed by the other authors. The construction, refinement, and validation of the mapping design were based on 7 control studies [2,4,6,8,11,12,23]. We identified the control studies with an exploratory search on agent-oriented approaches for MBT. We used control studies to refine the search string and selection process of relevant studies.

3.2 Search Process and Study Selection

A PICO (Population, Intervention, Comparison, and Outcomes) model was developed to identify keywords and formulate the search string. In our study, the population is model-based testing, and the intervention is AOST approaches. We do not consider comparison and outcomes to achieve the highest coverage. After several refinement searches, we defined the following search string:

("model-based test" OR "model based*test" OR "test case generat*" OR "model driven*test" OR "software test") AND ("multi-agent" OR "multi agent" OR "agent-based" OR "agent based" OR "agent-oriented" OR "agent oriented" OR "AOSE" OR "ABSE" OR "ABST" OR "AOST")*

Inclusion and Exclusion Criteria. During the research process, we defined the inclusion and exclusion criteria for determining relevant studies. We applied the criteria in studies based on the title, keywords, and abstract. We use three databases: IEEE, Scopus, and Web of Science. The set of inclusion criteria was (I1) studies that reported Model-Based Testing approaches following the characterization presented in [19], (I2) studies that reported agent-oriented approaches for software testing, (I3) only studies that were written in English, and (I4) type of study is primary. We established an exclusion criterion (E1) for studies not available in full text. In the first stage, we extracted a collection of 487 studies. After removing duplicates, we conserved 418 articles. Next, 12 relevant studies were identified after the inclusion and exclusion criteria. The remaining collection of articles were examined using paper screening by reading the introduction and conclusion sections. After that, 24 articles were classified as relevant and then added up to the prior 12 for a total of 36 relevant studies for detailed reading and data extraction. Figure 1 summarizes the search and selection process.

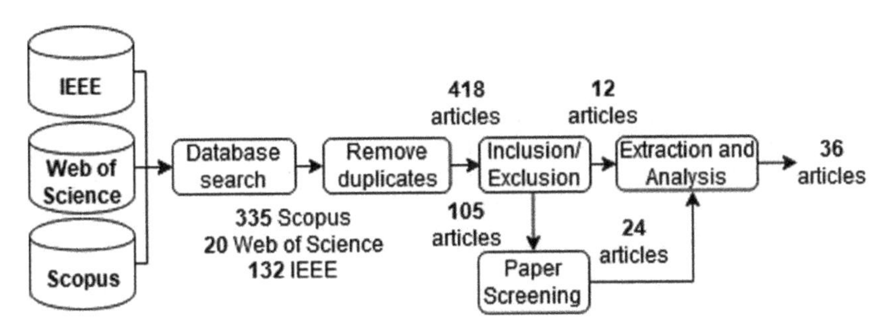

Fig. 1. Search and selection process

Data Extraction. For each research question, we defined a set of aspects to be extracted from relevant studies in Table 1. Firstly, we extracted general information from studies, and then the components for each research question. We classified the studies following the recommendations for constructing classification schemes for mapping studies [14]. All studies' information was tabulated into the data extraction form. To answer the research questions, we obtained groups from the classification, trends, and identified patterns. In addition, we prepared a descriptive analysis of the results.

Threats to Validity. The following threats to validity should be considered for this mapping study. The analysis was done only considering primary studies. Furthermore, the search string was calibrated by several testing runs. Nevertheless, relevant studies could not be identified by automated searches due to the absence of selected keywords. For the searching process, we used renowned databases in computer science and software engineering. The inclusion and exclusion, data extraction, and classification processes were done by the primary author of this

Table 1. Extracted items

RQ	Components
General	Year, Authors, Study Type, Forum, Forum Type
RQ1	MBT process stages improved agent-oriented approaches (model contruction, test criteria selection, test case generation, and test case execution)
RQ2	Reported agent-oriented methodology and implementation details (agent roles and their duties/tasks)
RQ3	Coverage criteria

mapping study; then, it can represent a bias for the studies' selection and classification. Lastly, the mapping study results can only be generalized for the set of selected literature.

4 Results

Details of the selected studies and the extraction form are available at http://tiny.cc/zb81tz. Studies are identified with the prefix S1-S36 that is used for reporting results. The first author of this study did the extraction and classification processes.

4.1 Stages of the MBT Process Implemented by Agents

MBT process consists of 4 main stages: model construction, test criteria selection, test case generation, and test case execution [19]. In 10, studies agent-oriented approaches proposed agents related to model construction tasks. Agents were in charge of analyzing web pages and node-link graph generation, serving as intermediaries with model construction testing tools, in GUI application for state capturing, in generating model representations from source code, and in searching over models executing errors and traces tracking.

In test criteria selection, three studies addressed this stage. They focused on impact analysis for criteria selection in regression testing, and coverage criteria selection for instrumented code in fault detection testing.

Regarding the test case generation stage, 25 studies reported agent-oriented approaches, carrying out tasks such as: generating performance tests in web components, creating optimized test plans, generation of test scenarios based on WSDL specifications, generation of test case from trace-logs files, test case generation from weighted paths, database test case generation, and test case generation triggered by source code changes.

Finally, for the test case execution stage, 20 studies used agent-oriented approaches for web-based testing, GUI action testing, test scripts execution, and multi-agent systems testing using mock agents. Table 2 lists each stage and their related studies.

Table 2. Stages of the MBT process

MBT Stage	Count	Studies
Model construction	10	S2, S4, S5, S6, S12, S16, S21, S24, S27, S32
Test criteria selection	3	S27, S34, S35
Test case generation	25	S1, S2, S4, S6, S7, S8, S11, S12, S13, S14, S16, S18, S19, S20, S22, S23, S24, S25, S28, S29, S31, S32, S33, S35, S36
Test case execution	20	S1, S2, S3, S4, S6, S7, S8, S9, S10, S11, S13, S14, S15, S17, S18, S23, S24, S25, S26, S30

4.2 Methodologies, Roles and Tasks Assigned to the Agents During the MBT Process

Agent-Oriented Software Engineering (AOSE) methodologies aim to structure the development process of multi-agent systems.

Multi-agent Systems Engineering (MaSE) consists of an agent-oriented methodology for developing heterogeneous multi-agent systems. It supports a set of models for describing agent types, goals, behaviors, and communication interfaces. Two main phases are part of MaSE: the analysis phase for specifying what the systems should do, and the design phase, determining how the systems will achieve its goals [17]. Prometheus methodology provides means for designing and implementing intelligent agents, supported by three phases: system specification phase, architectural design phase, and detailed design phase. This methodology has been integrated into a multi-agent system platform called JACK [17]. Finally, the Belief-Desired-Intention (BDI) model consists of three main elements: (1) Beliefs representing the information that an agent recognized about the world and they are updated by the environment perception. (2) Desires, also known as agents's goals and tasks, for achieving a specific belief and test situations. (3) Intentions that lead to a set of actions that an agent has committed to performing [9].

In Table 3 we present the identified methodologies among implementations. Five studies used a BDI approach, one used MaSE, and one used Prometheus.

Table 3. Agent-Oriented Software Engineering (AOSE) methodologies

Methodology	Count	Studies
BDI	5	S3, S10, S17, S30, S33
MaSE	1	S13
Prometheus	1	S16
Others	29	S1, S2, S4-S9, S11, S12, S14, S15, S18-S29, S31, S34, S36

Moreover, 29 studies were classified as others in with authors provide their own methodology.

An agent role is a capability that the agent exposes for doing a set of actions. Roles are designed for carrying out tasks to achieve specific goals or sub goals [18]. We extracted eight generic roles based on the original implementation. The roles and tasks are shown in Table 4. Model Builder, Test Criteria Selector, Test Generator, and Test Executor carry out tasks related to the MBT stages.

Additionally, other roles were extracted and classified in the extraction form as "own solution roles" for MBT non-generic activities. Model Builder was held in 10 studies. This role involves constructing the model and supporting data retrieval for different sources, and capturing the system's execution traces and actions during execution. Test criteria selector was found in 3 studies for test case selection, classification, and analysis of the test suite's impact. Twenty-five studies comprised Test Generator role in several related activities: test data generation, test cases, plans, and scenarios generation. A test oracle aims at verifying whether a test case succeeds or fails [20].

Test Oracle's role was only in three studies for test result validation according to expected model behaviors. The Test Executor role was described in 20 studies concerning test case execution, test result reporting, and communication with testing tools for test script execution. Lastly, we identified a set of roles that aims at supporting the agent-based approach, in particular, Monitor-Coordinator, Knowledge Manager, and User Assistant. The Monitor-Coordinator was recognized in 12 studies and Knowledge Manager for managing model, plans, and test repositories in ten studies. The User Assistant role developed human user guidance, instruction data input, and communication interface with other agents.

Table 4. Roles and tasks

Role	Count	Studies
Model Builder	10	S2, S4, S5, S6, S12, S16, S21, S24, S27, S32
Test Criteria Selector	3	S27, S34, S35
Test Generator	25	S1, S2, S4, S6, S7, S8, S11, S12, S13, S14, S16, S18, S19, S20, S22, S23, S24, S25, S28, S29, S31, S32, S33, S35, S36
Test Oracle	4	S2, S4, S13, S18
Test Executor	20	S1, S2, S3, S4, S6, S7, S8, S9, S10, S11, S13, S14, S15, S17, S18, S23, S24, S25, S26, S30
Monitor-Coordinator	12	S2, S4, S6, S7, S11, S12, S13, S14, S15, S16, S18, S32
Knowledge Base Manager	10	S2, S4, S6, S7, S8, S11, S13, S18, S26, S31
User Assistant	7	S2, S4, S6, S13, S16, S18, S32

4.3 Test Coverage Criteria

Test coverage criteria measure how do the generated test cases cover all the model's represented behavior. Table 5 summarizes coverage criteria in agent-oriented approaches for MBT. Model coverage can be measured over states, transitions, transition pairs, actions, and paths, among other types. Model coverage represents the degree, expressed as a percentage, to which the model elements are planned or exercised by a given test suite [19].

Thereby, model coverage depends on the paradigm adopted for representing the system behavior, such as state-based, transition-based, activity-based and stochastic notations [20]. Five studies presented model coverage with several paradigm notations, in particular, an actionable knowledge model using state-based, agent state machines, activity graphs, and robotic finite state machines in transition-based, and UML diagrams for activity-based notations.

Path coverage typically involves covering all the business scenarios in activity-based or covering cyclic or not cyclic paths in state-based and transition-based notations [15]. Path coverage was reported in 6 studies for node-link coverage in web testing, database statement coverage of relevant nodes in the control flow graphs, input variables in paths and their domain, and optimal path coverage from a source node to a destination node. Branch coverage, also known as decision coverage criteria, consists of the decisions that modify the control flow during test suite execution [19]. Branch coverage was found in 3 studies in control-flow graphs for database testing, in syntax tree code representations, and transition table scenarios. Plan coverage involves agent-based approaches for testing multi-agent systems (MAS) regarding the plans to achieve the desired goals and tasks. Three studies reported plan coverage in software testing for MAS under the BDI model methodology.

Code and model coverage metrics are complementary for MBT [1]. Code and statement coverage criteria comprise source code statements covered by the test suite. This criterion was reported in 3 and 2 articles respectively. Scenario coverage was found in 2 studies that describe test cases using formal abstract notations [20]. Other six coverage criteria were complementary and reported one time concerning the study goals and agent-based implementation. The aspect-oriented constraint coverage focuses on testing functional and non-functional

Table 5. Coverage criteria

Coverage	Count	Studies
Path coverage	6	S2, S12, S13, S20, S29, S36
Model coverage	5	S6, S9, S16, S32, S33
Branch coverage	3	S12, S27, S29
Plan coverage	3	S10, S14, S17
Statement coverage	3	S3, S12, S27
Code coverage	2	S5, S35
Scenario coverage	2	S7, S11

aspects of software components [S1]. The agent unit coverage involves the BDI agent as a unit for testing, including its states, plans, and goals [S30]. Control flow [S8], error-guessing list [S25], link and node [S2], and scenario coverages [S3] were described among the studies as well.

5 Conclusions

In this mapping study, we reported the results of 36 primary studies published between 2002 and 2020 on agent-oriented approaches for Model-Based Testing (MBT). We classified these approaches based on the MBT process stages and the agent entities' roles and tasks. Agent-oriented software testing (AOST) approaches have been implemented in the test case generation, test execution, model construction, and test criteria selection stages.

Belief-Desired-Intention (BDI), Multi-agent Systems Engineering (MaSE), and Prometheus models were reported among the studies. Eight generic roles were identified: model builder, test generator, test executor, test criteria, selector, test oracle, monitor-coordinator, user assistant, and knowledge base manager. Finally, path and model coverage criteria were the most criteria reported among agent-oriented MBT evaluations.

AOST approaches could be used to tackle MBT tasks such as test case generation, test prioritization, test minimization, and test execution in a effective and efficient way. Further work should be address to evaluate the benefits of agent-oriented approaches for model-based testing.

Acknowledgments. This work was partially supported by the University of Costa Rica No. 834-B8-A27. We thank the Empirical Software Engineering Group at UCR for the valuable feedback and help.

References

1. Chapter 4 - selecting your tests. In: Utting, M., Legear, B., (eds.) Practical Model-Based Testing, pp. 107 – 137. Morgan Kaufmann (2007). https://doi.org/10.1016/B978-012372501-1/50005-3
2. Arora, P., Bhatia, R.: Agent based regression testing framework. In: 2014 International Conference on Signal Propagation and Computer Technology (ICSPCT 2014), pp. 411–414 (2014)
3. Arora, P., Bhatia, R.: A systematic review of agent-based test case generation for regression testing. Arab. J. Sci. Eng. **43**, 1–24 (2017). https://doi.org/10.1007/s13369-017-2796-4
4. Arora, P., Bhatia, R.: Agent-based regression test case generation using class diagram, use cases and activity diagram. Procedia Comput. Sci. **125**, 747–753 (2018)
5. Bourque, P., Fairley, R.E., Society, I.C.: Guide to the Software Engineering Body of Knowledge (SWEBOK(R)): Version 3.0, 3rd edn. IEEE Computer Society Press, Washington, DC, USA (2014)
6. Devasena, M.G., Valarmathi, M.: Multi agent based framework for structural and model based test case generation. Procedia Eng. **38**, 3840 – 3845 (2012) https://doi.org/10.1016/j.proeng.2012.06.440.

7. Dudekula, M.R., Katam, R., Kiran, M., Petersen, K., Mäntylä, M.V.: Benefits and limitations of automated software testing: systematic literature review and practitioner survey. In: 2012 7th International Workshop on Automation of Software Test (AST), pp. 36–42 (2012)
8. Gardikiotis, S.K., Lazarou, V.S., Malevris, N.: Employing agents towards database applications testing. In: 19th IEEE International Conference on Tools with Artificial Intelligence(ICTAI 2007), vol. 1, pp. 173–180 (2007)
9. Guerra-Hernández, A., Seghrouchni, A., Soldano, H.: Learning in bdi multi-agent systems. In: Computational Logic in Multi-Agent Systems, pp. 39–44 (2004). https://doi.org/10.1007/978-3-540-30200-1_12
10. Kumaresen, P., Frasheri, M., Enoiu, E.: Agent-based software testing: A definition and systematic mapping study. ArXiv **abs/2007.10224** (2020)
11. Mahali, P., Acharya, A.A., Mohapatra, D.: Model Based Test Case Generation and Optimization Using Intelligent Optimization Agent, vol. 339, pp. 479–488. Springer India (2015)
12. Malz, C., Göhner, P.: Agent-based test case prioritization. In: 2011 IEEE Fourth International Conference on Software Testing, Verification and Validation Workshops, pp. 149–152 (2011)
13. Padmanaban, R., Thirumaran, M., Suganya, K., Priya, R.: Aose methodologies and comparison of object oriented and agent oriented software testing. In: Proceedings of the International Conference on Informatics and Analytics (2016)
14. Petersen, K., Vakkalanka, S., Kuzniarz, L.: Guidelines for conducting systematic mapping studies in software engineering: an update. Inf. Softw. Technol. **64**, 1–18 (2015)
15. Christmann, A., Kramer, B.L.A.M.N.M.T.M., Schulz", S.: Foundation level syllabus – model-based tester. International Software Testing Qualifications Board (2015)
16. Salima, T., Askarunisha, A., Ramaraj, N.: Enhancing the efficiency of regression testing through intelligent agents. In: International Conference on Computational Intelligence and Multimedia Applications (ICCIMA 2007), vol. 1, pp. 103–108 (2007)
17. Shehory, O., Sturm, A. (eds.): Agent-Oriented Software Engineering: Reflections on Architectures, Methodologies, Languages, and Frameworks. Springer, Berlin (2014)
18. Sivakumar, N., Vivekanandan, K.: Agent oriented software testing – role oriented approach. Int. J. Adv. Comput. Sci. Appl. **3**(12) (2012)
19. Utting, M., Legeard, B., Bouquet, F., Fourneret, E., Peureux, F., Vernotte, A.: Chapter two - recent advances in model-based testing. In: Memon, A., (ed.) Advances in Computers, vol. 101, pp. 53 – 120. Elsevier (2016)
20. Utting, M., Pretschner, A., Legeard, B.: A taxonomy of model-based testing. Software Testing, Verification and Reliability **22** (2012). https://doi.org/10.1002/stvr.456
21. Villalobos, L., Quesada-López, C., Martinez, A.: A tertiary study on model-based testing areas, tools and challenges: Preliminary results (2018)
22. Villalobos-Arias, L., Quesada-López, C., Martínez, A., Jenkins, M.: Evaluation of a model-based testing platform for java applications. IET Software **14**(2), 115–128 (2020)
23. Yu, S., Ai, J.: Software test data generation based on multi-agent. In: Advances in Software Engineering, vol. 4 (2010). https://doi.org/10.1007/978-3-642-10619-4_23

Fifth-Generation Networks and Vehicle-to-Everything Communications

Edgar E. González(✉), Flavio D. Morales, Rosario Coral, and Renato M. Toasa

Universidad Tecnológica Israel, E4-142, Marieta De Veintimilla y Fco. Pizarro, Quito, Ecuador
{eegonzalez,fmorales,rcoral,rtoasa}@uisrael.edu.ec

Abstract. The development of the fifth generation (5G) networks means an improved network offering optimised key performance indicators for new services. One of these services is aimed at the development of Intelligent Transport Systems (ITS). Therefore, this paper offers an overview of technologies for vehicular communications through the 5G cellular network, use cases derived from these technologies and an analysis of their potential requirements. Thus, an application for vehicular communications is presented. After the study, it is concluded that the development of the new 5G networks is adapted to the requirements of the use cases for vehicular communications, aimed at reducing the number of accidents on the roads, decrease travel times and reduce the emission of pollutant gases into the environment.

Keywords: 5G · Vehicle-to-everything (V2X) · ITS

1 Introduction

With the incorporation of new wireless communications technologies, new use cases and application scenarios have been identified. The application scenarios for 5G networks are [1]: Enhanced Mobile Broadband (eMBB), Massive Machine Type Communications (mMTC) and Ultra Reliable Low Latency Communication (uRLLC). As an application of these new technologies, their use has been identified in the development of new intelligent transport systems that allow, among other aspects, an efficient management of traffic on the roads, and thus reduce the number of accidents, decrease travel time, reduce fuel consumption and thus reduce pollution to the environment. In addition, these new systems make it possible to incorporate new use cases ranging from new services based on vehicle communications to complex systems, for example, autonomous vehicles. Such systems, known as ITS (Intelligent Transport System), can be based on different technologies and therefore support different use cases and requirements. 5G New Radio is a technology based on 3GPP mobile networks that support new

Á. Rocha et al. (Eds.): ICITS 2021, AISC 1330, pp. 350–360, 2021.
https://doi.org/10.1007/978-3-030-68285-9_34

requirements and use cases. NR C-V2X is the technology defined for vehicular communications over 5G networks. It is important to point out that, due to the performance and better KPIs that are expected to be achieved with the fifth generation networks, certain use cases are exclusive to this type of networks, since they meet the latency, bandwidth and network reliability requirements.

The forecasts for the incorporation of new V2X technologies in the automotive market are promising. Thus, projections indicate that by 2025, at a global level, C-V2X technology incorporated in vehicles from different manufacturers will represent a market of more than 30 billion dollars [2]. Since the incorporation of these new vehicle technologies is of great interest to different sectors, this paper analyses the defined use cases and the potential requirements that allow appropriate operation in different usage scenarios.

This paper is organized as follows. Section 2 discusses the importance of 5G networks in incorporating services based on vehicular communications. Next, in Sect. 3, a heterogeneous environment for vehicular communication services is proposed. In addition, different QoS parameters that should guarantee the networks for such services are analyzed through simulations.

2 5G NR and V2X

In the 3GPP technical specifications and reports, four types of vehicle communications (V2X) have been defined (see Fig. 1): vehicle-to-vehicle communications (V2V- vehicle-to-vehicle), vehicle-to-infrastructure (V2I- vehicle-to-infrastructure), vehicle-to-pedestrian (V2P- vehicle-to-pedestrian) and vehicle-to-network (V2N) [3]. These four types of vehicle communications are intended to provide intelligent transport services in order to improve road safety, better traffic management, reduction of pollution, etc.

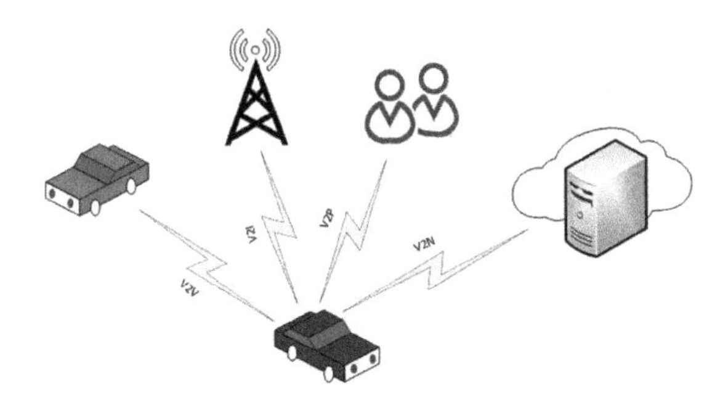

Fig. 1. V2X communications representation.

2.1 NR C-V2X Use Cases

Services that can be implemented over 5G networks can be advanced services with strict quality of service requirements. However, basic services, some of which can be offered with Long Term Evolution (LTE) networks, are possible. Table 1 shows some basic use cases defined for NR C-V2X, where it can be noted that the requirements of latency, message packet size and speed support, depends on the specific use case. Latency is a primary parameter for V2X services, since a delay in the transmission or reception of messages can affect the correct performance of the services offered and, in the worst case, can lead to counterproductive results. With respect to the size of the packets to be transmitted, it depends on the type of message being sent, whether it be by means of the V2V, V2I, V2P or V2N service. Additionally, it is essential that the network guarantees an adequate connection to vehicles circulating at high speeds.

Table 1. Potential requirements for basic NR C-V2X services. C-V2X [4].

Use case	Latency (ms)	Packet size (Bytes)	UE speed (km/h)
Forward collision warning	100	50–300	–
Emergency vehicle warning	100	50–300, max. 1200	280
Emergency stop use	100	1200	160
Queue warning	100	1200	160
Automated parking system	100	50–400	160
Pre-crash sensing warning	20	50–300	160
V2N traffic flow optimisation	1000	50–300	160
Road safety services	100	300	160

Other basic use cases defined for NR C-V2X include: road safety services, speed warning messages in curves, collision warnings to pedestrians or cyclists.

3GPP Release 15 introduces some improvements to 4G LTE networks aimed at the development of 5G NR. The added improvements, related to V2X service, consist in providing high transmission rates and lower latencies for V2N communications. Additionally, improvements are added in direct communications over PC5, transmission diversity and use of higher order modulations (64 QAM). Thus, the 3GPP in [4] has identified 25 new use cases for V2X services. These use cases can be grouped into 5 areas: Vehicle platooning, advanced driving, extended sensors, remote driving, and general purpose use cases. The first implementations of the NR-based C-V2X technology are expected to take place in 2023 and will be backwards compatible [5].

Next, some defined use cases for 5G C-V2X [4] are exposed:

V2X Support for Vehicle Platooning. Vehicle platooning consists of a group of vehicles that travel relatively close together with the aim of reducing fuel consumption, decrease pollution, and reduce the number of drivers in the platoon.

To maintain an adequate distance between vehicles in the platoon, it is necessary to share information on distance between vehicles, speed, acceleration, braking intention, etc.

V2X Support for Remote Driving. This use case refers to the control of driving a vehicle remotely, either by a human operator or by applying cloud computing. It should be noted that, unlike the case of an autonomous vehicle, remote driving requires fewer sensors installed in the vehicle. Furthermore, sophisticated artificial intelligence algorithms are not required. For the case of remote driving controlled by a human operator, a video camera can be placed on board the vehicle, from which the operator can send control signals remotely, based on the video information received. Another case of use can be the driving of buses remotely, which normally must follow a predefined route. For this case it should also be taken into account that not only is video transmission from the outside of the bus required, but also video information from inside the bus is required, so that the remote operator can react to actions such as when a passenger boards or gets off the bus. In the case of remote driving using cloud computing (without human operators), coordination between vehicles can be established and controlled by the cloud. Thus, the cloud can coordinate the route each vehicle can take and thereby reduce congestion, total travel time, fuel savings, etc. For this use case the network must support transmission rates of up to 1 Mbit/s in Downlink (DL) and 20 Mbit/s (2 H.265/HEVC HD 10 Mbit/s videos) in UL for V2X services for speeds up to 250 km/h. In addition, network reliability for UL and DL must be guaranteed to be greater than 99.999% and network latency (end to end) must not exceed 5 ms.

Collective Perception of Environment (CPE). Through this use case vehicles that are within a certain range of coverage exchange information from the sensors installed. This type of information exchanged is known as collective environmental perception (CPE), and can be used by vehicles to receive or send alert messages, and thus avoid accidents.

Cooperative Collision Avoidance (CoCA) of Connected Automated Vehicles. Unlike LTE V2X use cases, where the exchange of Cooperative Awareness Messages (CAM) is proposed, for NR V2X the exchange of sensor information and a list of commands for actions such as braking and acceleration, as well as lateral and longitudinal control information are specified. Key Performance Indicators (KPI) defined for this use case are Transmission rates up to 10 Mbit/s for CoCA message exchange, message size up to 2 kByte, latencies less than 10 ms and more than 99.99% network reliability.

Tethering via Vehicle. This use case allows a vehicle to offer Internet access to the vehicle occupants or user in the vicinity. The idea is to take advantage of the possibility of installing antennas and transmitters in a vehicle without having the same restrictions of space and resources compared to a user equipment. Compared to other modes of Internet access available to users, this use case allows for a reduction in the battery consumption of the devices and increases data rates.

Teleoperated Support (TeSo). Through this use case there is the possibility that a remote operator can take control, for a period of time, of a connected autonomous vehicle. For this case there are specific requirements such as: network latency of maximum 20 ms, support for transmission rates of up to 25 Mbit/s in UL and 1 Mbit/s in DL, and a network reliability of 99.999%. UL requires more data transfer due to the video information that will be sent to the remote operator; DL only requires the sending of data information related to the V2X service.

3D Video Composition for V2X Scenario. Through this use case UEs have a camera that captures video from the environment and sends this information to a server. The server can be remote or a server physically located near the UEs. The server can process and combine the video information from all users and create a 3D video of the environment. The 3D videos can be used, for example, for accident evaluation by the police. The requirements for this use case imply that the server can synchronize the video data from different UEs, considering UE speeds up to 250 km/h. In addition, network must support traffic transfer per user, in UL, of up to 10 Mbit/S (4K/UHD video quality).

Information Sharing for Partial Automated Driving. Due to the good projection of 5G networks in different services that require strict requirements, one of the applications that is being strongly promoted is in the use of autonomous vehicles [6]. For this use case, aspects of cooperative perception and cooperative manoeuvres are considered. The first aspect requires the transmission and exchange of high resolution data between vehicles circulating in a given area. Such information can be acquired by means of high resolution cameras or Light Detection and Ranging (LIDAR). In the second aspect, the vehicles need to exchange their detailed trajectory. Up to 50 Mbit/s per link are required for cooperative perception, and up to 3 Mbit/s for cooperative maneuvers. Other KPIs that the network must guarantee for this use case are low latency, high reliability, wide communication range and support for high density of connected devices

2.2 What Role Do 5G Networks Play in V2X Communications?

5G networks play a fundamental role in the generation of new services and applications, as they offer better Key Performance Indicators (KPI) than other predecessor networks. For 5G NR networks 3 usage scenarios can be defined: mMBB, mMTC, and uRLLC [1].

eMBB (enhanced Mobile BroadBand). This use case refers to services that require certain bandwidth parameters to operate efficiently. It is possible to mention, for example, applications that require the transmission of multimedia information in high definition, so the network must offer sufficient bandwidth. It is important to indicate that the bandwidth requirements are relative to the environment or area where the service will be offered (in dense urban environments the bandwidth requirements are higher than in open or rural areas). Features for 5G networks compared to IMT-A networks are: increase of peak transmission rates (Gbits/s) by up to 20 times, up to 10 times improved traffic capacity per area (Mbit/s/m2), improvements in spectral efficiency (bits/s/Hz) by up to 3 times, transmission rates per user (Mbit/s) by up to 10 times.

mMTC (massive Machine Type Communications). This use case is defined to allow the connection of a high density of devices to the network. A feature of this use case, in addition to supporting the connection of a large number of devices, is to maintain low power consumption of the devices. For this use case, 5G networks will allow up to 10 times more devices to be connected, and will improve the energy efficiency of the network by up to 100 times, compared to IMT-A networks. It is important to note that the need to reduce energy consumption is oriented from the network side as well as the user side.

uRLLC (ultra-Reliable Low Latency Communication). In this use case are applications that require strict latency parameters and network reliability to function properly. Thus, the performance of 5G networks compared to IMT-A networks is a greater mobility support (km/h) for users and reduction of latency (ms) by up to 10 times.

Table 2 shows the performance of the 5G networks planned as part of the IMT-2020 project.

Table 2. KPIs for IMT-2020 networks [1].

KPI	Projected value
Peak data rate	20 Gbit/s
User experienced data rate	100 Mbit/s
Mobility	500 km/h
Latency	1 ms
Connectivity density	1000000 devices/km^2
Area traffic capacity	10 Mbit/s/m^2

Requirements for different applications may vary depending on the service. In applications for handheld terminals, latency requirements are less than 10 ms, network availability of 99.9999% and transmission rates in the order of kbit/s and Mbit/s. For applications in the robotics industry the latency requirements are less than 1 ms, network availability 99.9999% and transmission rates in the order of kbit/s and Mbit/s. For applications involving the use of sensors the latency requirements are less than 100 ms, network availability of 99.99% and transmission rates in the order of kbps; in the latter, the energy savings of the devices is fundamental [1]. When analyzing the performance of 5G networks, it should be borne in mind that some services simply could not function properly with other networks. Regarding the performance of the planned 5G networks, they will guarantee low latencies (sub-millisecond), transmission rates in the order of Gbit/s, network reliability of up to 99.999%. The 3 usage scenarios reviewed above are directly related to the V2X service, however, it depends on the specific application in which it will be used. For example, network latency and reliability requirements for the autonomous vehicle application are more demanding than for CPE service.

3 QoS Simulations and Usage Scenario

In order to provide a specific analysis of the application of 5G and V2X communications in different scenarios, a usage scenario is proposed below. A general diagram of the use of V2X communications to improve safety and traffic on the roads is shown in Fig. 2. This diagram can be part of an overall ecosystem known as "Smart Cities".

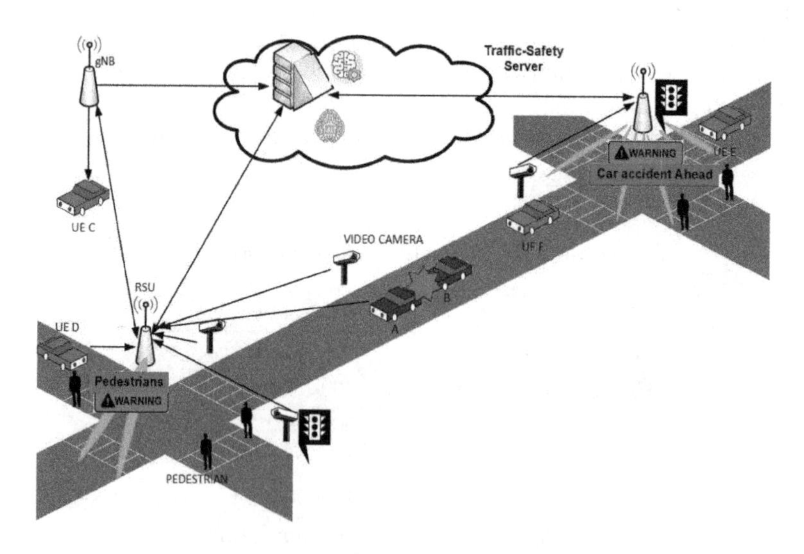

Fig. 2. 5G V2X usage scenario.

First, in Fig. 2 it is assumed that there is an accident involving vehicles A and B. It is assumed that vehicle B does not have V2X service incorporated, then it was not possible to establish prior communication to alert drivers of the risk situation. After the accident, the surrounding vehicles that have the V2X service can receive accident alert messages. For example, vehicle E may receive an alert indicating that there is an accident ahead. In the case of vehicle F, it is essential that it receives the alert message in a timely manner, so that the driver can react in a timely manner. Thus, these communication latency levels are offered by 5G networks (below 10 ms). Assuming that the vehicle F incorporates V2X service, it is possible to establish a direct link with vehicle A, through the PC5 interface. In addition, since the information is being stored and processed on a server, the surrounding vehicles can receive alternative routes to minimize the congestion due to the accident. The server, which can be an Edge Server, can incorporate emerging technologies of AI, Cloud Computing, Big Data, Machine Learning, etc. that allow an adequate processing of the information in real time. With respect to the RSU, they can be connected to the server by means of a 5G gNB, or they can also be infrastructures connected directly to the traffic server.

The use cases for V2X communications and 5G networks also include support for self-driving cars, in which case the latency, bandwidth and network reliability requirements are more demanding. On the other hand, pedestrians who have their terminals with V2X support, can transmit and receive alerts either from surrounding vehicles or 5G Network. For example, vehicles D and E in Fig. 2, can receive warnings that there are pedestrians crossing the street. Video cameras can transmit information in real time to the cloud. This information can be processed by complex artificial intelligence algorithms to precede the evolution of traffic on the roads, in order to determine alternate routes. The stored video information can also be used to create a 3D map of the environment, which can be used by the police to evaluate an accident or event that occurred on the street. In this case, it is necessary that the network supports a high transmission of information, especially in the uplink, due to the high-definition information sent by vehicles to the cloud. In addition, in the traffic server the information transmitted must be properly synchronized and filtered.

In order to have a stronger idea of the importance of an adequate response time in communications, simulations were carried out to determine the maximum reaction time as a function of the separation between vehicles on the roads. Such simulations have been performed using different functions and scripts implemented in the MATLAB tool. The theoretical basis of the simulations has been based on [7]. The scripts can be easily reconfigured to change the input parameters of the simulations. In Fig. 3, the results of the simulations are shown after configuring two vehicles, V1 and V2, which travel at average speeds of 30 km/h and 120 km/h, respectively.

The curves shown in Fig. 3 clearly indicate that the reaction time is shorter at higher speeds. On the other hand, a smaller gap between vehicles requires a quick reaction to avoid a possible collision between the vehicles. As shown in Fig. 3, the reaction time for separations between vehicles below 10 m, is below 100 ms. Therefore, a robust network is required to ensure this requirement, i.e., 5G NR.

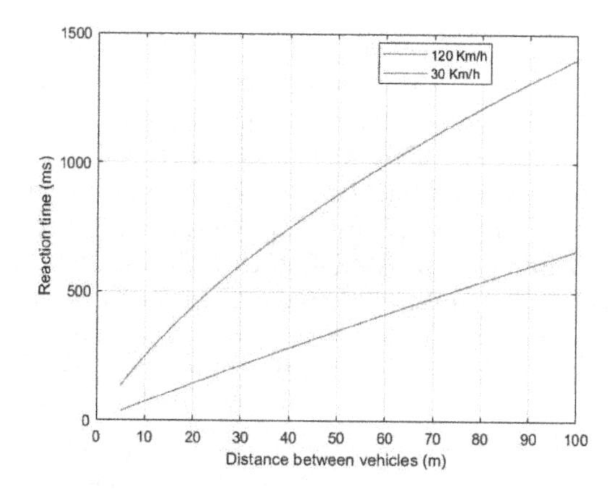

Fig. 3. Maximum reaction time as a function of the separation distance between cars.

Next, simulations were performed to verify the probability of collision considering a vehicle system that only transmits warning messages (message transmission) and another one that only transmits alerts (alert transmission). In the first case, it is the driver who must react to the messages received, while in the second case, the vehicle applies the brakes automatically. As expected, Fig. 4 shows how the probability of a collision between vehicles decreases significantly as the distance between vehicles increases. However, the probability of collision is lower when the braking action is performed automatically by the intelligent system incorporated into the vehicle. In this case, the simulations have been carried out considering: probability of not delivering a package correctly of 10%

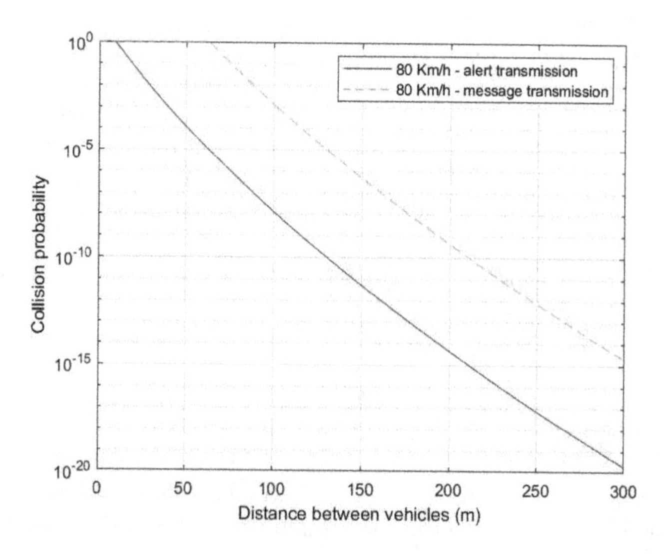

Fig. 4. Collision probability as a function of the separation distance between cars.

for average speeds of 80 km/h, transmission frequency of 10 messages per second and average driver reaction time of 0.5 s. Additionally, in the simulations performed it can be verified that, if the average driver reaction time is higher, a greater separation between vehicles is required to decrease the probability of collision.

As discussed above, the reliability of communications that the network must ensure is crucial in certain applications. Thus, simulations were carried out in order to study the necessary retransmissions that guarantee a certain probability that a packet will be properly delivered. Figure 5 shows the packet retransmission rate (pps) required for a certain probability of not delivering packets correctly (pv2v) as a function of the separation between vehicles. The results indicate that by increasing the spacing between vehicles, packet retransmissions can be decreased. However, as the separation distance between vehicles decreases, the packet transmission rate necessary to ensure a certain probability of lost packets increases significantly. As can be assumed, by increasing the probability of lost packets, it is necessary to increase the packet transmission rate.

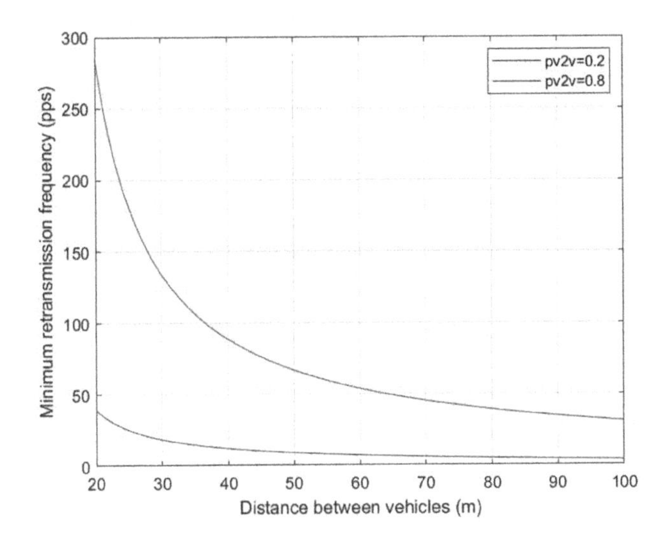

Fig. 5. Minimum packet retransmission rate vs. separation distance between vehicles.

4 Conclusions and Future Work

Since different services based on vehicular communications demand different potential requirements, it is necessary to characterize the performance of the networks in different scenarios. Thus, this work concludes that the key performance indicators of 5G networks enable advanced services in intelligent transport systems. Therefore, the rise of new V2X services will be directly linked to the development of 5G networks.

As future works derived from this research, architectures will be proposed for different specific C-V2X services that have been defined in the standard. In addition, End-To-End link simulations will be carried out to study the latency, QoS and bandwidth requirements of different applications.

References

1. ITU-R: IMT Vision – Framework and overall objectives of the future development of IMT for 2020 and beyond. Itu-R M.2083-0, vol. 0 (2015). https://www.itu.int/dms_pubrec/itu-r/rec/m/R-REC-M.2083-0-201509-I!!PDF-E.pdf
2. Bhutani, A., Wadhwani, P.: Cellular Vehicle-to-Everything Market, 2025 Forecast Report, June 2019. https://www.gminsights.com/industry-analysis/cellular-vehicle-to-everything-c-v2x-market. Accessed 13 Apr 2020
3. 3GPP: Study on LTE support for Vehicle-to-Everything (V2X) services. 3rd Generation Partnership Project (3GPP), Technical Report (TR) 22.885, version 14.0.0, December 2015
4. 3GPP: Study on enhancement of 3GPP Support for 5G V2X Services. 3rd Generation Partnership Project (3GPP), Technical report (TR) 22.886, version 16.2.0, December 2018
5. Hetzer, D., Muehleisen, M., Kousaridas, A., Alonso-Zarate, J.: 5G connected and automated driving: use cases and technologies in cross-border environments. In: 2019 European Conference on Networks and Communications (EuCNC), Valencia, Spain, pp. 78–82 (2019)
6. Yamanaka, N., Yamamoto, G., Okamoto, S., Muranaka, T., Fumagalli, A.: Autonomous driving vehicle controlling network using dynamic migrated edge computer function. In: 2019 21st International Conference on Transparent Optical Networks (ICTON), Angers, France, pp. 1–4 (2019). https://doi.org/10.1109/ICTON.2019.8840520
7. Nekovee, M.: Quantifying performance requirements of vehicle-to-vehicle communication protocols for rear-end collision avoidance. In: VTC Spring 2009 - IEEE 69th Vehicular Technology Conference, Barcelona, pp. 1–5 (2009). https://doi.org/10.1109/VETECS.2009.5073822

A Mobile Application for Improving the Delivery Process of Notifications

Heriberto Ureña-Madrigal(✉), Gustavo López, Ignacio Díaz-Oreiro, and Luis Quesada

University of Costa Rica, San José, Costa Rica
{heriberto.urenamadrigal,gustavo.lopez_h,ignacio.diazoreiro, luis.quesada}@ucr.ac.cr

Abstract. At present, there are systems in charge of classifying and sending notifications to smart devices at different times. However, there are not many studies that demonstrate the effectiveness of these systems in real world settings. We propose a method that classifies and prioritizes notifications by analyzing only the content of the notification and the sender of the message. We also developed a system implementing this method. User diaries were used to analyze the behavior of the system in real world situations, and the results showed that the implemented system significantly reduces interruptions to users. Additionally, the user experience of the system was evaluated through the standardized questionnaire UEQ (User Experience Questionnaire). The results obtained were positive in most of the scales of this instrument, above the average according to UEQ benchmarks. However, aspects such as stimulation and creativity can be improved in the future to motivate users to use the system.

Keywords: Mobile application · Android · Notification

1 Introduction

The possession of mobile and smart devices has increased considerably in the last years, becoming the favorite devices for users to access information and interact with other people [1]. This increase brings new ways of interrupting users [2] due to the amount of data that need to be processed by the user at the same time [3].

Smart devices have a built-in notification delivery functionality [4]. In most cases, notifications are delivered immediately, causing distractions and reducing people's attention to the different tasks they perform [5]. In the last years, smart devices can detect breakpoints for delivering notifications managing to minimize interruptions in the development of people's tasks [6].

There is research showing that the interaction between people and smart devices is associated with poor performance while performing tasks [7]. Therefore, before this research, a literature review about the delivery process of mobile

Á. Rocha et al. (Eds.): ICITS 2021, AISC 1330, pp. 361–371, 2021.
https://doi.org/10.1007/978-3-030-68285-9_35

notifications was conducted to learn about delivering push notifications to the user. Although there are many investigations related to the detection of opportune moments to present the notifications to the users by reducing interruptions, only a few studies show the effectiveness of these systems to reduce interruptions [8], as well as the behavior of these systems in real-world situations [9]. After performing the literature review, we developed the following research question: How can interruptions caused by mobile notifications to the user be reduced?

The rest of the paper is organized as follows: Section 2 briefly explains the concepts of human attention, interruption, notification, and breakpoint. Section 3 highlights the related work; Sect. 4 describes the methodology and Sect. 5 describes the obtained results. In Sect. 6 the conclusions of the proposed work are presented.

2 Background

In this section, we explain four key concepts required to understand the work proposed in this research. These concepts are human attention, interruptions, breakpoint, and notifications. Additionally, the concepts of user diary and sensors necessary to understand the stages of analysis and review of results will be explained.

First, human attention is the voluntary application of a person's mental activity or senses to a certain stimulus or object [10]. In the same way, interruption is a concept closely related to human attention and is defined as stopping or impeding a person's attention to a specific activity [10].

This research aims to improve the way in which notifications are delivered to the user, so it is important to define the meaning of breakpoint, which consists of the limit between two activities of the user that defines a good moment to carry out actions and reduce the impact on interruptions of the person [13].

Due to the different tools and frameworks that Android provides to manipulate the notifications received, this research will be developed in that operating system. Google, on its official site for Android mobile developers, defines notification as a message that Android displays to users when they are browsing outside the application to provide reminders, messages and other valuable information. Additionally, Google defines sensors as devices capable of providing raw data with high precision and accuracy, useful for monitoring the movement or three-dimensional positioning of a device, and for monitoring changes in the environment near a device [11].

Another important concept to clarify is the user diary, which is a tool used to collect information from participants about the use or evaluation of a system over a period of time. The collection occurs through repetitive recording of relevant information from the system. User diaries are normally applied to collect specific information regarding attitudes, behaviors, or feelings [12].

3 Related Work

There is research related to the loss of attention in people during the reception of notifications from mobile or smart devices. Such is the case of the article: "Silence Your Phones": Smartphone Notifications Increase Inattention and Hyperactivity Symptoms [14]. In this paper, the authors mention how interruptions derived from notifications can cause attention deficit in people who have not been diagnosed with this disorder.

Some studies were focused to reduce the lack of attention in people caused by the use of mobile devices; one of them consists of application-based interventions [2]. They inform the user how long they have been using certain applications. Likewise, studies were found that suggest delivering notifications at different breakpoints between people's activities to avoid constant interruptions caused by notifications [15].

Other research projects have studied the manipulation of mobile notifications with the aim of displaying alerts in a timely manner. Such is the case of Attelia, presented in 2016, in which the authors develop a machine learning system to recognize breakpoints during user activities, thereby allowing other applications to deliver notifications in a less intrusive way [3]. The system we proposed aims to not only inform when a breakpoint was detected but also classify the received notifications. On the other hand, InterrupMe, a system developed in 2014, uses parameters such as the activity that the user is carrying out, emotions, time and sensors to determine the appropriate times to deliver notifications [16]. The system that we propose differs from the ones already mentioned in that it will not use external sensors or parameters such as location, but only the text and the receiver of the notification will be analyzed.

As shown in the literature review, all the found studies were made for words of the English language. No studies were found regarding classification and prioritization systems for words of the Spanish language. The following section explains the different steps of the methodology carried out to implement the proposed method. Starting from the extraction of the different words in Spanish for the categories (important and urgent) to the final stage of the implementation of the system.

4 Methodology

4.1 Determining Keywords for Classifying Notifications

There are many algorithms and frameworks that process and classify text written in English language in order to reduce interruptions when delivering notifications [3,16]. Due to the lack of available studies related to this specific language, we focused on developing a method that prioritizes notifications written in Spanish to contrast the available tools developed for the English language.

To determine the words that users consider key to classify a notification as urgent or important, we ran a survey with 60 users of Android and iOS cellphones, including male and female participants between 18 and 50 years old. The gathered keywords were used to implement and develop the algorithm on which this research is based. In this survey, we also asked users for the two applications that present the highest number of notifications, we found that WhatsApp and Gmail are the applications that have the biggest flow of notifications. Based on these results, we developed a system that classifies and prioritizes notifications for WhatsApp and Gmail applications in Android devices and delivers them to the user at appropriate moments. The rest of the notifications are shown to the user normally by the operating system and are not processed by our method.

4.2 Identifying Opportune Moments to Deliver Notifications

To identify opportune moments to deliver notifications, we developed a system that analyzes their content to determine their importance and urgency. When a notification is urgent, it means that it requires immediate attention, and important means that a contribution in the long term is foreseeable. Table 1 shows the priority classifications of a notification according to its urgency and importance levels [17].

Table 1. Priority classification of notification

Denomination	Priority levels	Behavior
Exigency	Urgent and important	Not only should be delivered immediately but should also avoid any other distraction until the notification is acknowledged
Warning	Urgent but not important	Should be delivered as soon as possible and requires acknowledgment. These should be attended, or they can cause problems later
Routine	Important but not urgent	Are delivered as soon as possible assuring low-intrusion levels. This type of notifications is relevant for the user but can be sometimes deferred
Informative	Somehow important but not urgent	Usually, are queued to be delivered in batch, but timing is relevant for some users. These are usually information services that the user subscribes
Irrelevant	Neither important nor urgent	Should be delivered in a batch and timing is not essential. An offer to unsubscribe should be available. These are usually not subscribed by the user

Using the content and the recipient of the notification, we developed an algorithm that classifies and prioritizes every notification in one of the five categories mentioned in Table 1. According to the behavior of each category, we will deliver the notification at the right moment. For the proposed system, we are defining four static breakpoints to deliver batched notifications (mid-morning, lunch,

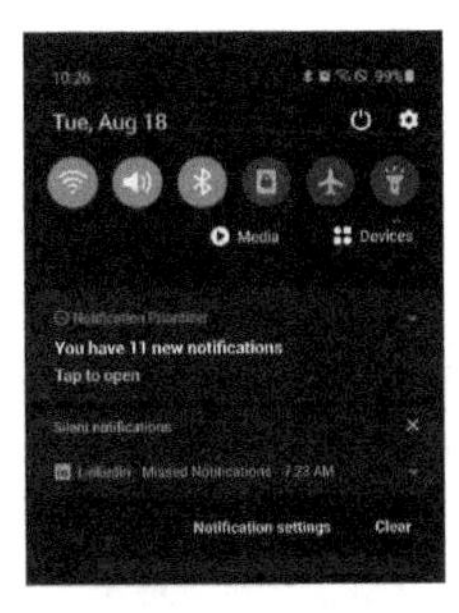

Fig. 1. Triggered notification from batched notifications

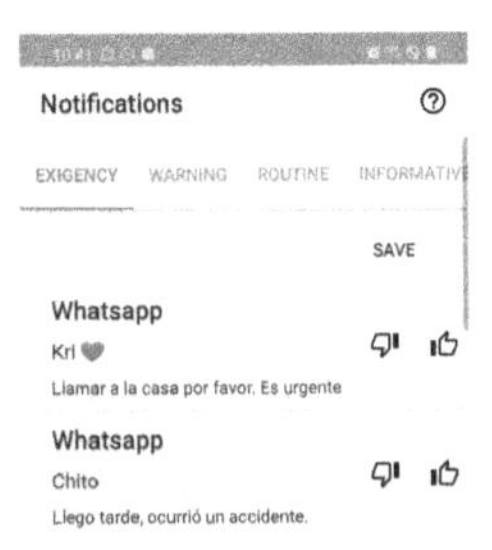

Fig. 2. System classification screen

mid-afternoon, dinner), however it could be improved in future work with the data collected by the scored notifications.

When the app receives a notification that must be delivered immediately or detected a breakpoint for delivering a set of previously classified notifications, it sends a notification to inform the user. Figure 1 shows the application delivering a set of notifications to the user at a given breakpoint. After pressing the received notification, the user is redirected to the main screen of the application where the notifications classified in the different categories are shown. This screen allows the user to score the different notifications, this score will be used in future work for improving the algorithm. Figure 2 shows the notifications classified in the category of exigency for a specific user.

Proposed Solution: Figure 3 shows a flow chart that explains the functionality developed in this method from the moment the user receives the notification until the notification is processed by the prioritization algorithm and classified in each of the proposed categories. This figure also explains the process for the notifications that do not belong to WhatsApp or Gmail.

4.3 Evaluating the Notification Classifier System

In the conducted literature review, we found that there are many studies related to the detection of breakpoints to deliver notifications [3]; however, there is not

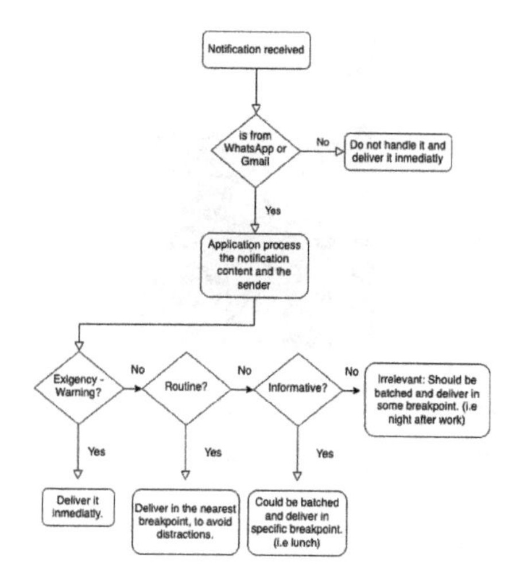

Fig. 3. Flow chart of the proposed method

enough research that shows the behavior of such systems in real-world situations. For evaluating this behavior, we developed a user diary and a user experience questionnaire.

User Diary: In order to gather valuable information while using the application, we created a user diary that collects specific information from eight users while using the system in real-world situations. The diary consists of an online questionnaire with five questions that the user should complete every day during seven days that allowed to extract information related to the time the application was used, the user's feelings, whether another activity was being carried out, the percentage of notifications that were classified correctly, and the name of the user.

User Experience Questionnaire: In order to evaluate the user experience of the system created in this research, we developed a one-minute-long video showing the main functionality of the application. The video will be evaluated in conjunction with a user experience questionnaire (UEQ). The results obtained from this UEQ allowed us to gather information related to the user experience of the system that will be contrasted with data obtained from the user diaries.

5 Results

One of the main results that we want to extract is the percentage of notifications that was handled correctly. For doing that, we developed a section in the system that allows users to score the classification of the notification (good or bad). After processing the results from the user diaries, we found that 65% of the participants

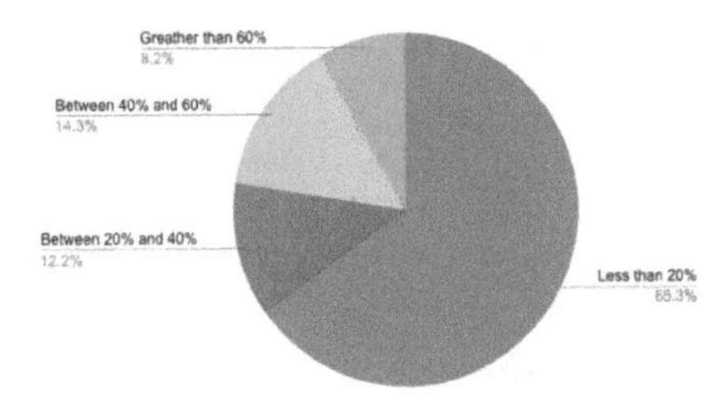

Fig. 4. Poorly classified notifications

consider that only 20% or less of the notifications were poorly classified, and only 8% of the participants consider that most of the 60% of the notifications were poorly classified. The results are significantly positive since only 8% are related to a negative evaluation. Figure 4 shows the rest of the gathered results for the classification of the notifications.

Likewise, the results gathered from the user diaries indicate that 68% of the notifications were correctly classified, and at least 32% of the notifications could not be classified.

Users said that the number of notifications was too high and scoring them one by one resulted frustrating. Users' feelings while using the system were also gathered by the user diary.

The processed data show that almost 34% of the users were frustrated while evaluating the application, and this aspect is highly related to the number of unrated notifications by the users. The rest of the results show the following percentages: 4.1% felt surprised, 8.2% felt happy while using the application and 53.1% felt confident.

To process the results of this UEQ analysis, we used the tools proposed by [18]. This UEQ was answered by 92 users that were not part of the process for real time evaluation to avoid information bias. From the results of the UEQ, we found that the highest mean dimension (1.992) is perspicuity. This result means that the application is easy to get familiar with and to learn how to use. On the other hand, the lowest mean dimension is stimulation (1.280). Table 2 shows the mean values for each dimension.

The UEQ is based on three aspects: attractiveness, pragmatic quality and hedonic quality. The pragmatic quality refers to the perception of the technical focus on achieving goals while the hedonic quality refers to aspects related to the emotions of the user [19]. The obtained results for each one of the aspects show that hedonic quality has the lowest mean value (1.41). Yet, we still consider this a positive evaluation since this low mean value is related to the data gathered from the user diaries, where some users felt frustrated or not motivated when using the application. On the other hand, attractiveness has the highest mean value (1.79).

Figure 5 shows the results of the analysis of each dimension of UEQ. Values between −0.8 and 0.8 represent a neutral evaluation, values greater than

Table 2. Mean for UEQ scale. Confidence intervals (p = 0.05) per scale

Scale	Mean	Std. Dev.	N	Confidence	Confidence	Interval
Attractiveness	1.786	0.784	91	0.161	1.625	1.947
Perspicuity	1.992	0.958	91	0.197	1.795	2.189
Efficiency	1.648	0.870	91	0.179	1.470	1.827
Dependability	1.407	0.803	91	0.165	1.242	1.572
Stimulation	1.280	1.000	91	0.205	1.075	1.486
Novelty	1.538	1.055	91	0.217	1.322	1.755

0.8 represent a positive evaluation, and values lower than −0.8 represent a negative evaluation of the corresponding scale. Based on the results in Fig. 5, all dimensions are in the range of the positive evaluation. The lowest evaluation (stimulation) is related to the data we obtained from the user diaries, where the user feels frustrated to use and score every notification. Another related concept is the mistrust generated by having to analyze the contents of each notification.

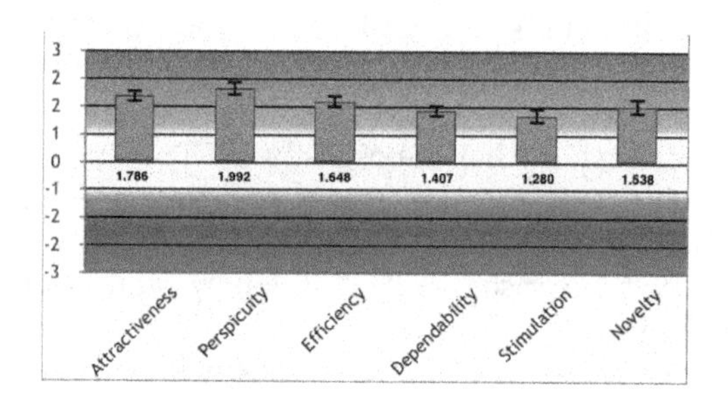

Fig. 5. Graphic of UEQ scale results

There is a benchmark tool from [18] that can be used to determine if the application is good or not. This benchmark has five categories to evaluate the product. These categories are excellent, good, above average, below average, and bad.

Figure 6 shows the result of the system proposed in this research with the benchmark data. Attractiveness, perspicuity, efficiency and novelty are the highest categories achieved by the system, all of them in the good category (10% of the benchmark results are better than the evaluated product, 75% of the results are worse.). This means that this application is liked by the users, is easy to get familiar with, responds quickly and is considered creative. The rest of the categories, dependability and stimulation, are categorized in the above average category (25% of the benchmark results are better than the evaluated product, 50% of the results are worse). They mean that the system can also be fun and secure for the users.

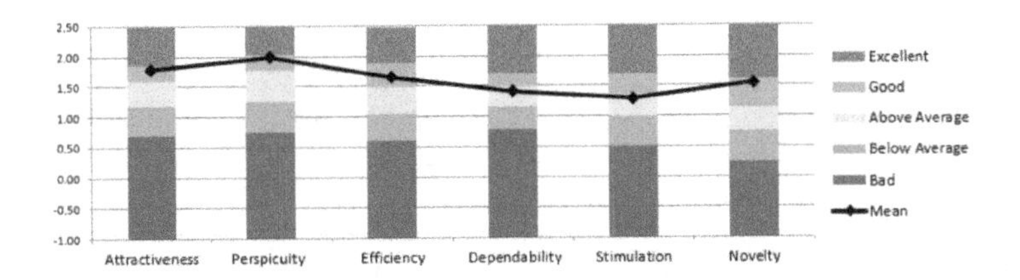

Fig. 6. Notification classifier benchmark

6 Conclusions and Future Work

In summary, based on the work carried out, it was possible to obtain key words in Spanish to classify a notification according to its urgency and importance. It was also possible to create a system capable of processing and classifying WhatsApp and Gmail notifications for the Android operating system. In addition, a comparison was made between the results obtained by the user diaries and the UEQ.

Evaluating the user experience aspects of the developed application is important for comparing system performance in real-world situations because there is not enough research to demonstrate the effectiveness of this type of system. Although very positive results were obtained in most evaluations, there are aspects that can be improved, such as the stimulation and the novelty (hedonic quality).

The results obtained by the UEQ show that the application has a lot of potential with the results referring to the user experience. However, this data is contrasted with the results obtained from user diaries where, in real life situations, people felt frustrated and in some cases distrustful because their notifications were being analyzed by third-party tools.

The real-time system evaluation shows that the developed system considerably reduces interruptions to users, particularly due to the grouped delivery of routine and irrelevant notifications, which prevents users from being disturbed by events that do not require immediate attention.

For future work, the classification algorithm can be improved to make a more accurate classification and prioritization of notifications. In order to achieve that improvement, two options that can be considered are integrating device sensors and analyzing previous notifications that provide more context during the classification process. However, all these types of improvements can have a negative impact on hedonic quality results, i.e it can increase the distrust of the users.

References

1. Teh, P.S., Zhang, N., Tan, S., et al.: Strengthen user authentication on mobile devices by using user's touch dynamics pattern. J. Ambient Intell. Human. Comput. **11**, 4019–4039 (2019)
2. Carthaigh, S.: The effectiveness of interventions to reduce excessive mobile device usage among adolescents: a systematic review. Neurol. Psychiatry Brain Res. **35**, 29–37 (2020)
3. Okoshi, T., Nozaki, H., Nakazawa, J., Tokuda, H., Ramos, J., Dey, A.: Towards attention-aware adaptive notification on smart phones. Pervasive Mob. Comput. **26**, 17–34 (2016)
4. Turner, L., Allen, S., Whitaker, R.: Reachable but not receptive: enhancing smartphone interruptibility prediction by modelling the extent of user engagement with notifications. Pervasive Mob. Comput. **40**, 480–494 (2017)
5. Weber, D., Void, A., Exler, A., Schroder, S., Bohmer, M.: Intelligent notification and attention management on mobile devices. In: 16th International Conference on Mobile and Ubiquitous Multimedia, pp. 561–565, November 2017
6. Visuri, A., Berkel, N., Okoshi, T., Goncalves, J., Kostakos, V.: Understanding smartphone notifications' user interactions and content importance. Int. J. Hum.-Comput. Stud. **128**, 72–85 (2019)
7. Stothart, C., Mitchum, A., Yehnert, C.: The attentional cost of receiving a cell phone notification. J. Exp. Psychol. Hum. Percept. Perform. **41**, 893–897 (2015)
8. Kunzler, F., Kramer, J., Kowatsch, T.: Efficacy of mobile context-aware notification management systems: a systematic literature review and meta-analysis. In: 2017 IEEE 13th International Conference on Wireless and Mobile Computing, Networking and Communications (WiMob), pp. 131–138 (2017)
9. Okoshi, T., Tsubouchi, K., Tokuda, H.: Real-world large-scale study on adaptive notification scheduling on smartphones. Pervasive Mob. Comput. **50**, 1–24 (2018)
10. Visuri, A., Berkel, N.: Attention computing: overview of mobile sensing applied to measuring attention. In: 2019 ACM International Symposium on Wearable Computers, pp. 1079–1082, September 2018
11. Notifications Overview — Android Developers. Developer.android.com. https://developer.android.com/guide/topics/ui/notifiers/notifications. Accessed 2 June 2020
12. DeLongis A., Hemphill K.J., Lehman D.R.: A structured diary methodology for the study of daily events. In: Bryant, F.B., et al. (eds.) Methodological Issues in Applied Social Psychology. Social Psychological Applications to Social Issues, vol. 2. Springer, Boston (1992)
13. Okoshi, T., Tsubouchi, K., Taji, M., Ichikawa, T., Tokuda, H.: Attention and engagement-awareness in the wild: a large-scale study with adaptive notifications. In: 2017 IEEE International Conference on Pervasive Computing and Communications (PerCom), pp. 100–110 (2017)
14. Kushlev, K., Dunn, E., Proulx, J.: "Silence your phones": smartphone notifications increase inattention and hyperactivity symptoms, May 2016
15. Turner, L., Allen, S., Whitaker, R.: Interruptibility prediction for ubiquitous systems: conventions and new directions from a growing field. In: 2015 ACM International Joint Conference on Pervasive and Ubiquitous Computing, pp. 801–812, September 2015
16. Pejovic, V., Musolesi, M.: InterruptMe: designing intelligent prompting mechanisms for pervasive applications. In: 2014 ACM International Joint Conference on Pervasive and Ubiquitous Computing, pp. 897–908, September 2014

17. López, G., Guerrero, L.: A conceptual framework for smart device-based notifications. J. Ambient Intell. Human. Comput. (2020)
18. Hinderks, A., Schrepp, M., Thomaschewski, J.: User Experience Quitionnaire. http://www.ueq-online.org/. Accessed 4 Aug 2020
19. Schrepp, M., Hinderks, A., Thomaschewski, J.: Applying the user experience questionnaire (UEQ) in different evaluation scenarios. In: Design, User Experience, and Usability. Theories, Methods, and Tools for Designing the User Experience, pp. 383–392. Springer, Cham (2014)

Software and Systems Modeling

Solving Errors Detected in Feature Modeling Languages: A Proposal

Samuel Sepúlveda[1,2](✉), Alonso Bobadilla[1], Manuel Espinoza[1], and Victor Esparza[1]

[1] Departamento de Ciencias de la Computación e Informática, Universidad de La Frontera, Temuco, Chile
[2] Centro de Estudios en Ingeniería de Software, Universidad de La Frontera, Temuco, Chile
samuel.sepulveda@ufrontera.cl, {a.bobadilla02,m.espinoza11,v.esparza02}@ufromail.cl

Abstract. It is common to describe *Software Product Lines* and manage its variability with the aid of a *feature model (FM)*. In this light, it shows that there are ambiguity issues concerning FM, which result in redundancy problems, anomalies, inconsistency, and mainly semantics issues. We propose a study regarding the feature modeling that considers the common aspects and the deficiencies in syntax, semantics, and semiotic clarity detected in the use of these modeling languages and the tools implemented from these. The initial results from this proposal show that the corrections of errors using feature modeling languages are feasible.

Keywords: Feature model · Syntactic · Semantics · Semiotic · Errors

1 Introduction

The software requirements are the descriptions of the services and constraints to be considered. These requirements are elicited, analyzed, verified, and documented. In the elicitation stage, there is a person responsible for writing in natural language the specifications of the requirements, complementing their descriptions with various models, e.g., description of organizational processes [1] or goal-oriented [25], among others modeling languages. Then, from the requirements representations, making the abstractions of the structural elements of the software system that expects to solve the problem understudy. Several authors review these aspects [29].

From what the software industry has experienced, attention has long been to different ways to develop software, including Software Product Lines (SPL). These are considered an intensive set in software systems that share a set of features to satisfy the needs of a specific market segment. An SPL is developed from a common set of fundamental elements and in a pre-established manner [10]. One detail of the elements that comprise an SPL is mentioned by [8], who regard such elements as the SPL architecture, a set of software components, and

Á. Rocha et al. (Eds.): ICITS 2021, AISC 1330, pp. 375–385, 2021.
https://doi.org/10.1007/978-3-030-68285-9_36

a set of products. The SPL as a discipline at both the academic and industrial levels is fully active and full of challenges, such as developing different types of software systems and new technologies, among others [5].

The motivation arises from the idea of avoiding the incorrect use of FM. In terms of modeling features, this proposal considers the common aspects and the deficiencies in syntax, semantics, and semiotics detected in these modeling languages and the tools implemented from these. The ambition of this proposal is to contribute to the SPL community with a solution for well-used feature modeling languages and with no presence of errors. It is possible to help the researchers and practitioners, to create FM avoiding the error identified. At the industry level, the results of this proposal could be applied to build algorithms and develop tools to support feature modeling.

The remainder of this paper is organized as follows. Section 2 shows the main concepts related to the proposal. Section 3 explains the methodology. Section 5 provides a discussion about our main findings and results. Section 6 outlines the main related work. Finally, Sect. 7 presents the conclusions and further lines of research.

2 Background

2.1 Software Product Lines

SPL are developed in two stages: *domain engineering and application engineering* [3]. In the domain engineering stage, the common and variant elements are described. The application engineering stage is where the individual products of the SPL are built by reusing domain devices and exploiting the variability of the SPL.

A main concept in SPL development is *variability*, which gives SPL the flexibility required to diversify and differentiate products [16]. For example, a software product must be able to adapt to the needs of each client or allow options for some specific configuration, so the products can reach different market segments [30]. As for domain engineering, it is common to describe SPL and manage its variability with the aid of FM [4].

2.2 Feature Modeling

FM appeared for the first time as part of the FODA method [22]. This model is still present, but with slight variations and adaptions for some SPL methods based on visual representations for the features of the product. The structure of an FM is a type of tree where its root node represents the product family, and the features are organized throughout the tree. These features can be assembled to give rise to particular software products [11]. The feature modeling has been the most relevant topic for SPL in recent years, having the best evolution behavior in terms of the number of published papers and references [20].

An example of FM can be seen in Fig. 1, which shows that a mobile phone must be able to call and have a screen, but only some have GPS. Another issue to bear in mind is that some of these features can depend on others for their inclusion, whereas the presence of certain features can prevent the selection of others. For example, if a mobile phone has GPS, it cannot have a basic screen.

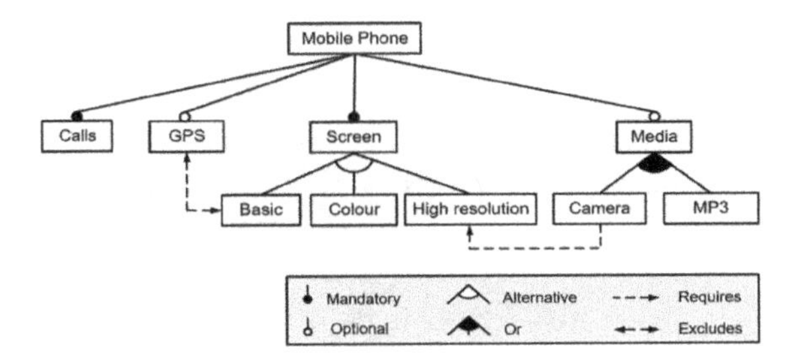

Fig. 1. Example of a feature model extracted from [7].

2.3 Formal Models

In the context of SE, the traditional mechanism for specifying syntactic aspects and well-constructed expressions using a modeling language is through the definition of a meta-model [32]. The author defines a meta-model as a specification model of a class or system type, where each system in that class is, in itself, a valid model expressed in a certain modeling language. Then, a meta-model makes statements about what can be expressed for valid models of a particular modeling language.

The semantics of modeling languages in scientific communities has been a topic from Philosophy of Science and it is an area that has arrived at useful frameworks and formalization. In particular, Bunge's proposal is considered. It includes the existence of modeling (representational) languages and discusses semantics applied to such representations [9].

The Principle of Semiotic Clarity establishes that it should be a 1:1 correspondence between semantic constructs and graphical symbols [28]. When this 1:1 correspondence is not accomplished, one or more anomalies can occur. Semiotic clarity allows eliminating the symbol deficit, symbol overload, and the symbol redundancy of visual notations [19].

3 Methodology

The methodology that guides the proposal is based on an adaptation of *Design Science*, a results-based computer systems research methodology [37]. It seeks to extend the limits of human and organizational skills by solving practical problems through the creation of new and innovative devices [26]. To solve this problem, Wieringa suggests using a workflow called the regulative cycle, which is divided into 4 phases. Additionally, the workflow is adapted by adding 2 of the most widely used methods in SE to phases 1 (conceptual analysis) and phase 4 (proof of concept) [17]. The need for a solution to correct the errors in the FM languages is considered the main practical problem. The knowledge problem is associated with the identification and formalization of the errors detected in FM. The complete workflow appears in Fig. 2 and, each of the 4 phases is detailed below.

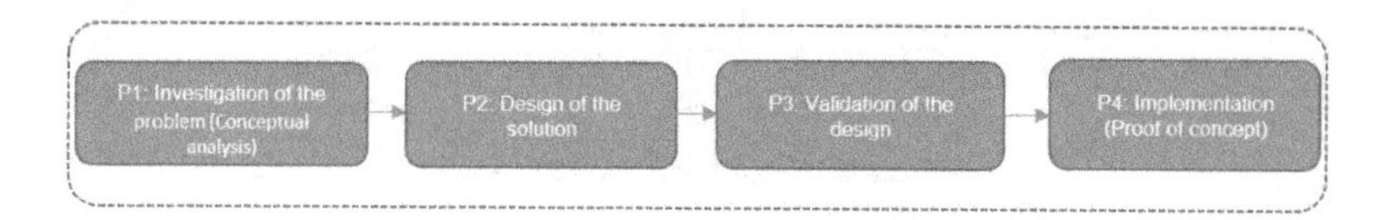

Fig. 2. Phases in design science.

Phase 1 (P1): This phase aims to describe the problem under study and eventually predict what could happen if nothing is done about it to solve it. The conceptual analysis will be used to understand the problem and to know the different proposals that could currently exist to solve them.

Phase 2 (P2): This phase consists of the process of specifying a possible solution to the main problem. For this proposal, the design consists of formalizing the errors detected for the FM.

Phase 3 (P3): This phase consists of the process of evaluating whether the design of the solution will bring stakeholders closer to the established goals.

Phase 4 (P4): This phase consists of the process of implementing the goals established and guided by the previously made design. The Design Science methodology is fully based on delivering a new and innovative artifact as a result and the fundamental questions within this methodology are: What is the use of the new artifact? What shows its usefulness? The answer to these questions depends on the presented evidence. The proof of concept will be used to focus efforts on what is strictly essential for the development of the proposal, the fulfillment of the objectives, and the solution to the problem.

The methodologies considered in each phase are consistent with what was suggested by [18]. Therefore, we may state that this proposal is rooted in the epistemology of the discipline to propose two large methodological stages: (i) investigation of the problem domain, and (ii) validation of the design and its implementation. It would make it possible to ensure that the feasibility is suggested in the hypotheses for this proposal.

4 Solution Proposal

The motivation that guides this proposal arises from the idea of avoiding the incorrect use of modeling languages for FM in SPL. This proposal focuses on Domain Engineering, in particular in the sub-stage Domain Requirements Engineering [30].

In terms of modeling features, this proposal considers the common aspects and the deficiencies in syntax, semantics and semiotics detected in these modeling languages and the tools implemented from these. This proposal pretends to contribute to the SPL community with a solution for well-used feature modeling languages and avoiding the presence of errors. It is possible to help the analysts (researchers and practitioners), to construct FM while avoiding the errors identified. At the industry level, the results of this proposal could be applied to build algorithms and develop tools to support the correct feature modeling. The main idea consists of creating a *Framework for Solving Errors in Feature Modeling Languages (FraSE-FML)*, that includes a set of formal models and algorithms that allow classify and solve the set of errors detected in the use of feature modeling languages.

To identify a set of errors in the use of FM languages, this proposal will endeavor to establish the state of the art around feature modeling errors, for which a systematic literature review will be performed [23]. This systematic review will be carried out according the protocol designed [36]. In addition, an initial survey will be applied to a group of experts from academia and industry to collect their personal impressions regarding the problem.

To formalize the identified errors, the proposal considers the use of meta models to formalize the syntactic aspects [32]. In order to formalize the semantic aspects, the concepts associated with Bunge's modeling languages [9] will be used. To formalize the semiotic aspects, the Moody's semiotic clarity principle [28] will be used.

To correct the identified errors, we consider doing this. At syntactic level, a set of algorithms and OCL restrictions will be added. At semantic level, a set of algorithms to correct semantic anomalies such as ambiguity and polysemy among others will be added. Finally, at semiotic level, a set of algorithms will correct the symbol deficit, symbol overload, and the symbol redundancy.

Finally, to validate the proposal a proof of concept will be carried out. Also, a survey will be applied to experts from the academia and industry and a controlled experiment. The survey will be administered to a group of experts regarding the main characteristics of the proposal and how it solves the problem of errors using FM. On the other hand, the controlled experiment will include two groups of people, ideally experts from academia and industry, otherwise with postgraduate and undergraduate students. The first group will receive training to carry out the steps defined in the proposal. The second group will not receive any training. The results of both groups will be compared according to a set of parameters to be defined.

5 Initial Results and Discussion

5.1 Results

To clarify the concept of *error* used in this proposal, we present an example. The examples are at the syntactic, semantic and, semiotic levels.

Syntactic error: this can be observed considering some relations between features. For example, the relation *"excludes"*, that means a feature can prohibit selecting other feature. This situation is shown in Fig. 3. If the FM language allows that the *daughter feature F2* can exclude the *parent feature F1*, this is an error, because no daughter can prohibit the selection of a parent.

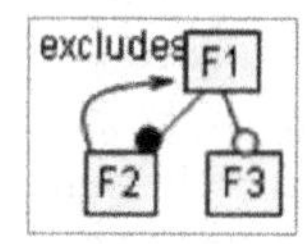

Fig. 3. Syntactic error for *excludes* relation.

An initial solution for this kind of error is to propose a meta-model (see Fig. 4), and a set of OCL constraints (see Table 1) for preventing the occurrence of the ill-formed sentences in an FM. This meta-model is explained in [35].

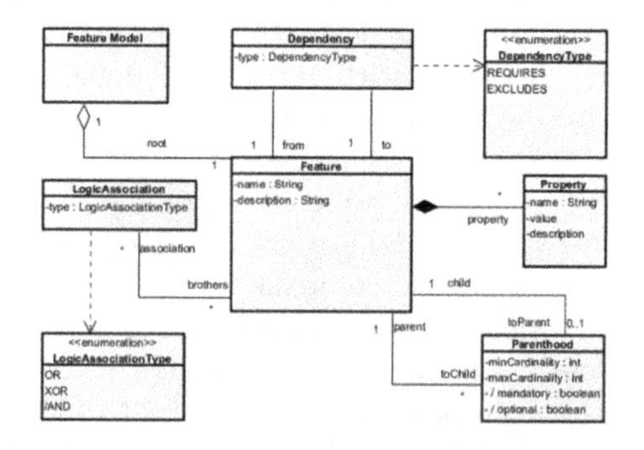

Fig. 4. Meta model proposed for FM [35].

A prototype tool called FMxx[1], that is based on the meta-model and implementing the definitions and OCL constraints showed in Fig. 4 and Table 1 is presented in [12–14].

Semantic error: this can be observed in the concept of *"feature"*. A feature is traditionally represented by a rectangle with its name inside. But, this feature could represent a functional or non-functional requirement and, in a broader sense, it could represent an *"asset"*. We can observe three different meanings for the same concept. Then, the key question is which concept to choose at any given moment. This situation is presented in Fig. 5.

[1] http://estocolmo.ceisufro.cl.

Table 1. Summary of Defs. & OCL constraints [35].

Definition	Context	Derive
DEF01. Attribute mandatory.	Parenthood:: mandatory: Boolean	minCardinality>0
....	....	...
Constraint	**Context**	**Invariant**
CST01. Cardinality constraint.	Parenthood	minCardinality≥0 and minCardinlaity≤maxCardinality
....	....	...

Concept
Graphic Representation
Meaning
feature
Feature
- Functional requirement
- Non-functional requirement
- Asset
?

Fig. 5. Semantic error for the concept *feature.*

An initial solution for this kind of error is to formalize and solving the semantic aspects of FM language. In particular the Bunge's proposal was used [9]. For solving these errors, we define four categories of semenatic anomalies (ambiguity, polisemy, idealism, and symbolic limitation), then a theoretical solution is proposed. This complete proposal can be reviewed in [33, 34].

Semiotic error: this can be observed in the concept of *"feature attribute"*. This concept is considered in different FM proposals, but there is not a general explanation of what a *"feature attribute"*, i.e. there is not a formal mapping from the modeling language to the FM universe of discourse. That is, we acknowledge the concept and its meaning for the FM community, but several FM languages do not describe a symbol for a *"feature attribute"*. So, this presents a symbol deficit of visual notation. This situation is presented in Fig. 6.

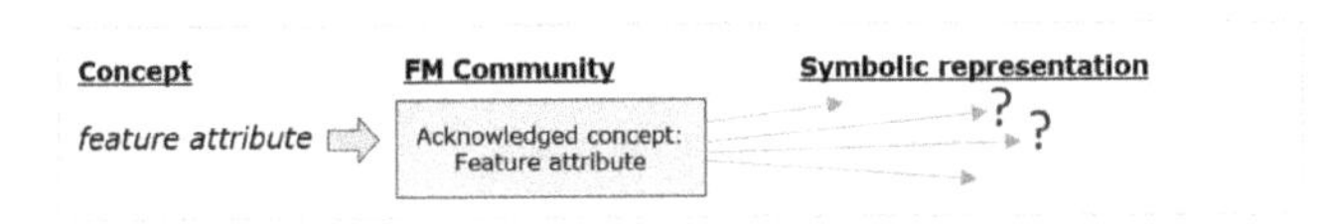

Fig. 6. Semiotic error for the concept *feature attribute.*

We do not have yet a formal solution to solve the semiotic errors. This work is under development, and we planned to use Moody's semiotic clarity formalization proposal.

5.2 Discussion

FM languages lack a proper conceptual foundation. This causes significant differences in the set of concepts included in each FM method. The existence of different dialects, each one with its own nuances, making difficult for the novice practitioner to make an informed decision about which one to use.

It must be considered that the use of FM involves a set of challenges, including in particular: (i) FM languages lacking their own conceptual base, cause differences in the set of concepts included in each method that adopts FM, and (ii) current tool support and its maturity level are quite limited. In this light, it may be noted that nowadays there are ambiguity issues with regard to FM, which result in redundancy problems, anomalies, inconsistency and mainly semantics issues.

These problems could be initially explained by two factors (F1, F2).
**F1:* There is no standard meta model accepted by the community, which gives rise to diverse and usually incompatible models.
**F2:* The tools not being based on a single meta model accepted by the community results in tools that, although they allow the creation an analysis of FM, do not reduce the possibility of creating incorrect diagrams.

These two factors allow the creation of incorrect models (syntax), and incorrect use (and interpretation) of the elements of the model are up to the user and their knowledge of the domain (semantics).

Thus, semantics of FM has been questioned and several analyses and approaches have been generated in order to support a useful modeling language for a wide Software Engineering community. We sustain that most of these proposals do not point to the real semantic problem. In order to show that we have used the Bunge's semantic framework and we have instantiated it to show why neither formalization nor meta models are limited perspectives to deal with FM semantics [33,34].

The semantic problem of FMs deals with more than a coherent and useful proposal, it is a social issue which requires not only a wide participation but that traditional scholars of SPL community be involved, i.e. at least traditional an recognized members of SPL research community.

The semiotic problem of FMs deals with the fact that there are acknowledged concepts (and their meaning) for the FM community, but many FM languages do not describe a symbol to represent these concepts. Then, this presents a symbol deficit of visual notation.

6 Related Work

In the first decade of the 2000s, a group of researchers posed a set of challenges to consider when using FM [4,6,15]. These challenges consider: lack of a proper conceptual foundation, proposals are focused on functional features, current tool support is very limited and, most of proposals have not proved their viability in real-world environments, among others.

More recently a considerable number of proposals are addressing the errors associated with FM. Among these we can consider the slicing technique that allows SPL practitioners to find semantically meaningful decomposition of FMs [2]. An ontological approach that measures the semantic similarity is suggested by [21]. How the defect prediction models can be used to identify defective features is presented in [31]. A generic algorithm for explaining different anomalies in FM is proposed in [24]. An overview on FeatureIDE's support to ensure the quality of FM and configurations is given by [27].

Our proposal shares with the aforementioned studies the concern for the correct use of the FM. On the other hand, we consider the joint use of syntactic, semantic and semiotic aspects to solve the errors detected in the FM.

7 Conclusions and Future Work

This work presented the main concepts and initial findings about the errors identified in the FM.

From the evidence presented, it is noted that the issues and challenges considered for this proposal are completely conceivable and are an open line of research within the SPL community. Thus it is possible to verify fundamental research topics: variability & requirements management, and feature modeling as relevant lines of inquiry within the Software Engineering and Software Product Lines community.

As future work, we plan to carry out a systematic literature review to establish the state of the art of feature modeling errors, then to develop a framework to formalize, classify and solve the identified errors for FM.

Acknowledgments. Samuel Sepúlveda thanks to Vicerrectoría de Investigación y Postgrado, Universidad de La Frontera, Proyecto DI20-2015.

References

1. van der Aa, H., Leopold, H., del Río-Ortega, A., Resinas, M., Reijers, H.A.: Transforming unstructured natural language descriptions into measurable process performance indicators using hidden markov models. Inf. Syst. **71**, 27–39 (2017)
2. Acher, M., Collet, P., Lahire, P., France, R.B.: Slicing feature models. In: 2011 26th IEEE/ACM International Conference on Automated Software Engineering (ASE 2011), pp. 424–427. IEEE (2011)
3. Apel, S., Batory, D., Kästner, C., Saake, G.: Software product lines. In: Feature-Oriented Software Product Lines, pp. 3–15. Springer (2016)
4. Asikainen, T., Mannisto, T., Soininen, T.: A unified conceptual foundation for feature modelling. In: 10th International Software Product Line Conference (SPLC'06), pp. 31–40. IEEE (2006)
5. Bashroush, R., Garba, M., Rabiser, R., Groher, I., Botterweck, G.: Case tool support for variability management in software product lines. ACM Comput. Surv. (CSUR) **50**(1), 1–45 (2017)

6. Batory, D.: Feature models, grammars, and propositional formulas. In: International Conference on Software Product Lines, pp. 7–20. Springer (2005)
7. Benavides, D., Trinidad, P., Ruiz-Cortés, A.: Automated reasoning on feature models. In: International Conference on Advanced Information Systems Engineering, pp. 491–503. Springer (2005)
8. Bosch, J., Florijn, G., Greefhorst, D., Kuusela, J., Obbink, J.H., Pohl, K.: Variability issues in software product lines. In: International Workshop on Software Product-Family Engineering, pp. 13–21. Springer (2001)
9. Bunge, M.: Semántica I. Sentido y referencia. Editorial Gedisa (2011)
10. Clements, P., Northrop, L.: Software product lines, course notes of product line systems program. Carnegie Mellon University. Pittsburgh, PA, USA, Software Engineering Institute (2003)
11. Czarnecki, K., Wasowski, A.: Feature diagrams and logics: there and back again. In: 11th International Software Product Line Conference (SPLC 2007), pp. 23–34. IEEE (2007)
12. Esperguel, M., Sepúlveda, S.: From UML/OCL to ADOxx specifications: how to do it. In: 2018 IEEE International Conference on Automation/XXIII Congress of the Chilean Association of Automatic Control (ICA-ACCA), pp. 1–6 (2018). https://doi.org/10.1109/ICA-ACCA.2018.8609724
13. Esperguel, M., Sepúlveda, S., Monsalve, E.: FMxx: a proposal for the creation, management and review of feature models in software product lines. In: 2017 36th International Conference of the Chilean Computer Science Society (SCCC), pp. 1–7 (2017). https://doi.org/10.1109/SCCC.2017.8405152
14. Esperguel, M., Sepúlveda, S.: Feature modeling tool: a proposal using ADOxx technology. In: 2016 XLII Latin American Computing Conference (CLEI), pp. 1–9. IEEE (2016)
15. Etxeberria, L., Mendieta, G.S., Belategi, L.: Modelling variation in quality attributes. VaMoS **7**, 51–59 (2007)
16. Galster, M., Weyns, D., Tofan, D., Michalik, B., Avgeriou, P.: Variability in software systems—a systematic literature review. IEEE Trans. Softw. Eng. **40**(3), 282–306 (2013)
17. Glass, R.L., Ramesh, V., Vessey, I.: An analysis of research in computing disciplines. Commun. ACM **47**(6), 89–94 (2004)
18. Glass, R.L., Vessey, I., Ramesh, V.: Research in software engineering: an analysis of the literature. Inf. Softw. Technol. **44**(8), 491–506 (2002)
19. Goncalves, E., Castro, J., Araujo, J., Heineck, T.: A systematic literature review of iStar extensions. J. Syst. Softw. **137**, 1–33 (2018)
20. Heradio, R., Perez-Morago, H., Fernandez-Amoros, D., Cabrerizo, F.J., Herrera-Viedma, E.: A bibliometric analysis of 20 years of research on software product lines. Inf. Softw. Technol. **72**, 1–15 (2016)
21. Itzik, N., Reinhartz-Berger, I.: Generating feature models from requirements: structural vs. functional perspectives. In: Proceedings of the 18th International Software Product Line Conference: Companion Volume for Workshops, Demonstrations and Tools-Volume 2, pp. 44–51 (2014)
22. Kang, K.C., Cohen, S.G., Hess, J.A., Novak, W.E., Peterson, A.S.: Feature-oriented domain analysis (foda) feasibility study. Carnegie-Mellon Univ Pittsburgh Pa Software Engineering Inst, Technical report (1990)
23. Kitchenham, B.: Procedures for performing systematic reviews. Keele, UK, Keele University **33**(2004), 1–26 (2004)
24. Kowal, M., Ananieva, S., Thüm, T.: Explaining anomalies in feature models. ACM SIGPLAN Not. **52**(3), 132–143 (2016)

25. Loucopoulos, P., Kavakli, E.: Capability oriented enterprise knowledge modeling: the CODEK approach. In: Domain-Specific Conceptual Modeling, pp. 197–215. Springer (2016)
26. von der Maßen, T., Lichter, H.: Deficiencies in feature models. In: Workshop on Software Variability Management for Product Derivation-Towards Tool Support, vol. 44, p. 21 (2004)
27. Meinicke, J., ThŘm, T., Schr, R., Benduhn, F., Leich, T., Saake, G., et al.: Quality assurance for feature models and configurations. In: Mastering Software Variability with FeatureIDE, pp. 81–94. Springer (2017)
28. Moody, D.: The "physics" of notations: toward a scientific basis for constructing visual notations in software engineering. IEEE Trans. Softw. Eng. **35**(6), 756–779 (2009)
29. Ncube, C., Lim, S.L.: On systems of systems engineering: a requirements engineering perspective and research agenda. In: 2018 IEEE 26th International Requirements Engineering Conference (RE), pp. 112–123. IEEE (2018)
30. Pohl, K., Böckle, G., van Der Linden, F.J.: Software Product Line Engineering: Foundations Principles and Techniques. Springer Science & Business Media, Heidelberg (2005)
31. Queiroz, R., Berger, T., Czarnecki, K.: Towards predicting feature defects in software product lines. In: Proceedings of the 7th International Workshop on Feature-Oriented Software Development, pp. 58–62 (2016)
32. Seidewitz, E.: What models mean. IEEE Softw. **20**(5), 26–32 (2003)
33. Sepúlveda, S., Cares, C., Cachero, C.: A theoretical approach to solve ambiguity and polysemy in feature models. In: Proceedings of 6th International Workshop on Advanced Software Engineering, IWASE, vol. 14 (2014)
34. Sepúlveda, S., Cares, C., Cachero, C.: Feature modeling languages: denotations and semantic differences. In: 7th Iberian Conference on Information Systems and Technologies (CISTI 2012), pp. 1–6. IEEE (2012)
35. Sepúlveda, S., Cares, C., Cachero, C.: Towards a unified feature metamodel: a systematic comparison of feature languages. In: 7th Iberian Conference on Information Systems and Technologies (CISTI 2012), pp. 1–7. IEEE (2012)
36. Sepúlveda, S., Díaz, J., Esperguel, M.: Systematic literature review protocol identification and classification of feature modeling errors. arXiv preprint arXiv:2010.15545 (2020)
37. Wieringa, R.: Design science as nested problem solving. In: Proceedings of the 4th International Conference on Design Science Research in Information Systems and Technology, pp. 1–12 (2009)

A Review of Learning-Based Traffic Accident Prediction Models and Their Opportunities to Improve Information Security

Pablo Marcillo(✉), Lorena Isabel Barona López,
Ángel Leonardo Valdivieso Caraguay, and Myriam Hernández-Álvarez

Departamento de Informática y Ciencias de la Computación,
Escuela Politécnica Nacional, Ladrón de Guevara E11-25 y Andalucía,
Edificio de Sistemas, 170525 Quito, Ecuador
{pablo.marcillo,lorena.barona,angel.valdivieso,
myriam.hernandez}@epn.edu.ec

Abstract. The World Health Organization (WHO) affirms that the first cause of death for people 5 to 29 years old is traffic accidents. Their victims are not only drivers but also passengers and road users. Therefore, reducing the number of accidents by providing technological advances is fundamental. This study presents the states of the art of traffic accident prediction models based on machine learning and security for connected vehicles. On one hand, it was determined that most of the models are seen as classification problems and are solved using deep learning and neural network algorithms. Also, some issues to be solved such as the integration of spatial heterogeneity or the solution of high dimensionality were identified. On the other hand, recent research has been mainly focused on resolving the security of communication links and identity and liability threats. The authentication into a VANET via digital signatures is one example of the direction toward where the research points.

Keywords: Machine learning · Traffic accident prediction · Heterogeneous data sources · Security · VANET

1 Introduction

The rapid urbanization and the need for mobilization have required the incorporation of means of transportation. It has also involved the rapid growth of the number of accidents, which has become an enormous issue for societies. According to WHO [1], the number of road traffic deaths worldwide reached the number of 1.35 million in 2016. However, more dangerous is that traffic accidents are considered the first cause of death for children and young people aged 5 to 29 years. Although the rate of deaths in comparison to the population has stabilized, it remains high and beyond expectations.

Á. Rocha et al. (Eds.): ICITS 2021, AISC 1330, pp. 386–395, 2021.
https://doi.org/10.1007/978-3-030-68285-9_37

Reducing the number of accidents is not only a matter of implementing strict traffic policies and technology infrastructure but also conducting more research about traffic accident prediction. It is known that a small percentage of traffic accidents are caused by external factors [2]. Considering that traffic accidents are also caused by random human factors [3], the inclusion of information about the driver, such as medical or physiological conditions, in prediction models is fundamental. However, not only the driver's data must be taken into account, also pedestrian characteristics and their mobility. WHO affirms that more than half of fatal casualties in traffic accidents globally are pedestrians, cyclists, and motorcyclists [1]. It is also known that reckless pedestrians cause traffic accidents especially in urban areas.

The occurrence of accidents depends on many factors in the majority of cases. It is complex to determine the causes of traffic accidents because their causes can vary during the time, from one place to another, or one driver to the next [4]. Without forgetting that many other factors such as mechanical problems, potholes in the road, or poor visibility due to fog, snowfall, or rainfall can also cause accidents. Indeed, traffic accidents are also caused by fault of other vehicles, in which case, it would be considered as a multi-vehicle accident. In that way, exchanging certain information among vehicles to avoid traffic accidents would be worthy.

Modern vehicles come equipped with a considerable amount of electronic components for control and communication, and also different types of sensors and actuators. Those components communicate among themselves or with others through four types of communication: in-vehicle network, Vehicular Adhoc Network (VANET), Wireless Personal Area Network (WPAN), and mobile networks. The need for establishing security mechanisms or models and countermeasures for components and their communications has become imperative in the industry and academia.

The purpose of this study is to present the states of the art of traffic accident prediction models based on machine learning and security for connected vehicles. For this purpose, a score of articles from journals and conferences have been classified and analyzed. Consequently, it was possible to establish the most common algorithms, learning techniques, data sources, solutions for security threats, shortcomings, and challenges of each one of the solutions.

This article is organized as follows. Section 2 presents the related work about traffic accident prediction and security for connected vehicles. Then, Sect. 3 introduces some classification schemes, followed by Sect. 4 that introduces a summary of proposals organized by performance evaluation metrics. Sect. 5 presents the new directions and research challenges. Finally, the conclusions of this study are presented in Sect. 6.

2 Related Work

This work has been conceived as an introductory study. It presents the states of the art of traffic accident prediction models based on machine learning and security for connected vehicles. The citation databases used in this work were Science Direct, Scopus, IEEExplore, and Google Scholar. To accomplish the article selection process some inclusion criteria have been defined. Thus, the articles that have been selected are from journals and conferences, published five years ago, written in English, and also which directly answered the following question: *What are the most relevant and recent works about traffic accident prediction models and security for connected cars?*. A summary of the proposals found is presented as follows.

2.1 Traffic Accident Prediction Models

- Ren et al. [5] propose a prediction model based on a Recurrent Neural Network (RNN) (**PA01**).
- Chen et al. [6] present a prediction model that uses the Stack denoise Autoencoder (SdAE) neural network (**PA02**).
- Park et al. [7] propose a prediction model that uses a framework based on MapReduce algorithm. It is based on k-Means as a clustering algorithm and Linear Regression as a classification algorithm (**PA03**).
- Yuan et al. [4] present a prediction model based on a Convolutional Long-Short Term Memory (LSTM) Neural Network (**PA04**).
- Bao et al. [8] propose a LSTM model with spatial features (**PA05**).
- Lin et al. [9] present a prediction model based on a Bayesian Network and a new Frequent Pattern (FP) Tree variable selection method (**PA06**).
- Dogru et al. [10] propose a detection model based on the Random Forest classifier (**PA07**).
- Sharma et al. [11] present a prediction model based on the Support Vector Machine (SVM) with Gaussian kernel (**PA08**).
- Huan et al. [12] propose a prediction framework based on a Hierarchical Fusion Network named Deep Dynamic Fusion Network (DFN) (**PA09**).
- Zhou et al. [3] present a deep learning framework based on a variant of Convolutional Neural Network (CNN) that fuses urban data (**PA10**).

2.2 Security for Connected Vehicles

- Agarwal et al. [13] propose an authentication protocol for VANETs called ESPA, which is based on cross-cryptography; and a resilient control schema for connected vehicles, which is based on the CACC algorithm and includes an approximation algorithm (**PB01**).
- Yadav et al. [14] present a security framework based on IDS for authentication of vehicles within VANETs through digital signatures based on HMAC and CRL (**PB02**).
- Sharma et al. [15] propose a conceptual model of an Intrusion Detection and Prevention System for connected cars using Artificial Neural Networks (ANN) (**PB03**).
- Li et al. [16] present a search system based on ciphertext retrieval that exploit RSUs as super peers for Connected Vehicular Cloud Computing (**PB04**).

- Contreras et al. [17] propose an architecture model for IoV based on seven layers that includes support for inter vehicular communication protocols in VANETs (**PB05**).

3 Classification

For the first state of the art, two classification schemes were designed. The first schema is a three-level taxonomy based on the structures proposed by [18,19]. The first level corresponds to the *learning type*, the next one to the *learning technique*, and the third one corresponds to the *function type*. The second schema has only one level, and this corresponds to the *data source*. The data sources defined for this scheme are *vehicle data*, *driver's data*, *weather conditions*, *traffic conditions*, *external factors or events*, and *pedestrian mobility*. Table 1 and Table 2 present the classification of proposals using these schemes.

A great number of authors have conceived the traffic accident prediction as a classification problem. The most common function type and algorithm among solutions are Deep Learning and Long Short-Term Memory (LSTM) Neural Network respectively. Almost all solutions use weather and traffic conditions as data sources. Some of the most common variables between these data sources are traffic accidents and the road network infrastructure. Also, very few solutions have used driver's data and pedestrian mobility as data sources, and only one has incorporated alcohol and drug use as a cause of traffic accidents.

Table 1. Classification by learning type

Proposals	Algorithms	Function types	Learning techniques
PA01	Long Short-Term Memory (LSTM) Neural Network	Deep Learning	Classification
PA02	Stack Denoise Autoencoder	Deep Learning	Classification
PA03	k-Means/Logistic Regression	Clustering/Regression	Clustering/Regression
PA04	Convolutional Long Short-Term Memory (LSTM) Neural Network	Deep Learning	Regression
PA05	Convolutional Long Short-Term Memory (LSTM) Neural Network	Deep Learning	Classification
PA06	k-Nearest Neighbor/Bayesian Network	Instance Based/Bayesian	Classification
PA07	Random Forest	Ensemble	Classification
PA08	Support Vector Machine		Classification
PA09	Multilayer Perceptron (MLP)	Neural Networks	Classification
PA10	Convolutional Neural Network	Deep Learning	Classification

Likewise, one classification scheme was designed for the second state of the art. It uses the taxonomy to classify security threats for connected vehicles proposed by [20], which is based on the security aspects shown as follows.

Table 2. Classification by data source

Proposals	Data source					
	Vehicle data	Driver's data	Weather conditions	Traffic conditions	External factors/events	Pedestrian mobility
PA01			description of daily weather and air quality	traffic accidents and traffic flow (GPS records)		
PA02				traffic accidents		human mobility (GPS records)
PA03			description of weather	speeds of cars, traffic volume, and traffic accidents		
PA04			hourly precipitation amount, temperature, and wind speed	traffic accidents, road network information, and traffic volume		
PA05			description of weather	traffic accidents, taxi trip (GPS records), road network information, land use, and population data		
PA06			description of weather and visibility	traffic accidents, traffic volume, speeds of cars, and occupancy		
PA07	traveling speed and location (GPS records)					
PA08	traveling speed	alcohol/drug use				
PA09				traffic accidents and points of interest data	city event reports	
PA10			meteorological data	road network infrastructure and population data	event calendar data	

- Security of communication links (**A1**): It includes the security of four types of communications: in-vehicle network, VANET, WPAN, and mobile networks.
- Data validity (**A2**): It refers to the reliability of all the information exchanged between ECUs, OBUs, RSUs, SPs, and SCs.
- Security of devices (**A3**): It refers to the security of electronic devices, both hardware and software (firmware).
- Identity and liability (**A4**): Identity refers to the ability to be who one says and liability to prove that an entity (owner, car, driver, etc.) is responsible for an event or specific information.

Table 3 presents the solutions analyzed in this study and Table 4 introduces the classification of solutions using the proposed schema.

Table 3. Solutions grouped by proposals

Proposals	Solutions
PB01	**S1**: Cross Cryptography Protocol (ESPA)
	S2: Cooperative Adaptive Cruise Control Algorithm (CACC)
	S3: Identity Based Security Schema
PB02	**S4**: Security Framework Based on Intrusion Detection System (IDS)
PB03	**S5**: Conceptual Approach for Intrusion Detection and Prevention System (IDPS)
PB04	**S6**: Cypher Text Based Search System for RSU
PB05	**S7**: An IoV Model based on seven layers

4 Performance of the Prediction Models

The performance of the different solutions has been established according to some measurement values. The authors have presented their performance metrics based on the learning technique applied. For classification problems metrics such as Prediction Accuracy Rate (PAR), False Positive Rate Or Fall-Out (FPR), True Positive Rate or Sensitivity (TPR), and F1 Score. For regression problems metrics such as Mean Absolute Error (MAE), Mean Squared Error (MSE), Mean Relative Error (MRE), Root Mean-Square Error (RMSE), Cross-Entropy (CE), Area Under Curve (AUC), and Average Precision (AP). Table 5 presents a summary of the most relevant solutions and the values of their metrics.

During the process of experimentation, the solutions have used different algorithms on their models to determine the performance of each one of them. Thus, it was observed that the most common algorithms used by the solutions are Logistic Regression, Support Vector Machine and some variants of Neural Networks such as DNN, RNN, ANN, and LSTM NN.

Table 4. Solutions for security threats

Security aspect	Threat	S1	S2	S3	S4	S5	S6	S7
A1	Interception of communications	✓		✓				✓
	Message suppression							
	Sending of fake messages	✓	✓	✓		✓		
	Sending of tampered messages	✓	✓	✓		✓		
	Replay attack							
A2	Transmission of fraudulent data				✓		✓	
	Sending of fake aggregate data				✓		✓	
	Wormhole attack				✓		✓	
A3	Tampering of device hardware							
	Tampering of device software							
	Replacing of device							
	Denial of service (DoS)		✓			✓		
	Car remote control							
	Tampering of device firmware							
	Unauthorized over-the-air update							
A4	Masquerade attack	✓		✓	✓	✓		
	Sybil attack	✓		✓	✓	✓		
	Repudiation attack				✓			

5 Challenges and Future Work

On one hand, much research about analysis and prediction of accidents has been achieved in recent years. The solutions are varied and, in some cases, very promising; however, most of them present shortcomings to be resolved. For instance, Yuan et al. [4] affirm that some solutions have not considered spatial heterogeneity within study areas and the time information. It implies that certain factors may affect traffic accidents in different ways, depending on the homogeneity of the study zone and the time of day. Meanwhile, Moosavi et al. [21] affirm that other solutions are entirely dependent on the availability of information, inapplicable for real-time models due to the lack of continuous data, and limited because of the use of small-scale datasets. Also, Park et al. [7] affirm that other solutions have had to resolve the issue related to information imbalance. Generally, the amount of data that is not related to traffic accidents is higher than useful data. The use of such information without preprocessing could cause high dimensionality in data. And also, the issue related to the use of significant computing capacities for processing and analysis of a big amount of traffic data.

On the other hand, Agarwal et al. [13] suggest as future work that the secret keys, shared during the key exchange process into the authentication protocol for VANETs, are stored as hash values instead of being transmitted without being previously encrypted. This strategy would accelerate the query and change of the anonymous id and improve the level of security. Meanwhile, Yadav et al. [14]

Table 5. Performance of proposals

Proposals	Evaluation Metrics											Compared with
	MAE	MRE	RMSE	MSE	CE	PAR %	FPR %	TPR %	F1	AUC	AP	
PA01	0.55	0.35	0.63									- LR, SVM, and DT
PA02	0.96	0.39	1.0									- Logistic Regression (LR), Support Vector Machine (SVM), and Decision Tree (DT)
PA03						76.35	12.76					- LR and SVM
PA04			0.116	0.013	0.010							- SVM, DT, Random Forest (RF), and Deep Neural Network (DNN)
PA05				0.019		81.58	0.34					- Convolutional Neural Network (CNN), Long Short-Term Memory (LSTM) Neural Network, ANN (Artificial Neural Network), and Gradient Boosting Regression Tree (GBRT)
PA06						61.11	38.16					- LR, ANN, and Bayesian Network (BN)
PA07						92.0		94				- SVM and ANN
PA08						94.0						- Multilayer Perceptron (MLP) Neural Network
PA09									0.681	0.786	0.712	- Support Vector Regression (SVR), LR, DNN, and Recurrent Neural Network (RNN)
PA10			0.40	0.16		88.89						- Auto-Regressive Integrated Moving Average (ARIMA) and Convolutional Long Short-Term Memory (ConvLSTM) Neural Network

proposes as future work the reduction of time for the authentication process into VANETs of their security framework.

6 Conclusions

Concerning the first state of the art, it was determined that most of the authors have conceived the traffic accident prediction as a classification problem and very few as a regression problem. About the algorithms, the most commons are Logistic Regression, Support Vector Machine, and some variants of Neural Networks however, the algorithms with better performance are those based on Neural Networks such as Convolutional Long Short-Term Memory (LSTM) Neural Network. This fact could suggest that the research direction of traffic accident prediction models points toward Deep Learning Models. About the data sources, almost all proposals have used weather and traffic conditions as data sources and very few have used data sources such as vehicle data, driver's data, external factors, and pedestrian mobility. This could be an indication that the research should be focused on exploring other data sources to correlate with the most common ones. This state of the art also demonstrated that the development of solutions in real-time is viable, but there are some issues to be resolved before; for instance, the use of large-scale datasets, integration of spatial heterogeinity, dependency on the availability of information, and solution for high dimensionality in data are some of them.

Respecting the second state of the art, it was observed that recent research has focused on proposing solutions for the interception of communications, the sending of fake/tampered messages, and masquerade and sybil attacks. By contrast, very few solutions point to resolve threats related to the security of devices. The growing trend about researching the security of communication links is the reason why at present it is increasingly common to find keywords such as IoV (Internet of Vehicles), VANET, V2V (Vehicle To Vehicle), or V2I (Vehicle to Infrastructure) in the literature. In fact, a significant number of solutions have proposed improvements such as architecture models, security frameworks, and authentication protocols for VANETs.

References

1. World Health Organization: WHO—Global status report on road safety2018. WHO (2018). http://www.who.int/violence_injury_prevention/road_safety_status/2018/en/
2. Zhang, X., Huang, F., Zheng, C.: Causes analysis of the serious road traffic accidents cases. In: 2016 5th International Conference on Energy and Environmental Protection (ICEEP 2016). Atlantis Press (2016)
3. Zhou, Z., Chen, L., Zhu, C., Wang, P.: Stack resnet for short-term accident risk prediction leveraging cross-domain data. In: Chinese Automation Congress (CAC), pp. 782–787. IEEE (2019)

4. Yuan, Z., Zhou, X., Yang, T., Tamerius, J., Mantilla, R.: Predicting traffic accidents through heterogeneous urban data: a case study. In: Proceedings of the 6th International Workshop on Urban Computing (UrbComp 2017), Halifax, NS, Canada, vol. 14 (2017)
5. Ren, H., Song, Y., Wang, J., Hu, Y., Lei, J.: A deep learning approach to the citywide traffic accident risk prediction. In: 2018 21st International Conference on Intelligent Transportation Systems (ITSC), pp. 3346–3351. IEEE (2018)
6. Chen, Q., Song, X., Yamada, H., Shibasaki, R.: Learning deep representation from big and heterogeneous data for traffic accident inference. In: Thirtieth AAAI Conference on Artificial Intelligence (2016)
7. Park, S.-H., Kim, S.-M., Ha, Y.-G.: Highway traffic accident prediction using VDS big data analysis. J. Supercomput. **72**(7), 2815–2831 (2016)
8. Bao, J., Liu, P., Ukkusuri, S.V.: A spatiotemporal deep learning approach for citywide short-term crash risk prediction with multi-source data. Accid. Anal. Prev. **122**, 239–254 (2019)
9. Lin, L., Wang, Q., Sadek, A.W.: A novel variable selection method based on frequent pattern tree for real-time traffic accident risk prediction. Trans. Res. Part C Emerg. Technol. **55**, 444–459 (2015)
10. Dogru, N., Subasi, A.: Traffic accident detection using random forest classifier. In: 15th Learning and Technology Conference (L&T), pp. 40–45. IEEE (2018)
11. Sharma, B., Katiyar, V.K., Kumar, K.: Traffic accident prediction model using support vector machines with gaussian kernel. In: Proceedings of Fifth International Conference on Soft Computing for Problem Solving, pp. 1–10. Springer (2016)
12. Huang, C., Zhang, C., Dai, P., Bo, L.: Deep dynamic fusion network for traffic accident forecasting. In: Proceedings of the 28th ACM International Conference on Information and Knowledge Management, pp. 2673–2681 (2019)
13. Agarwal, R., Pranay, S.S., Rachana, K., Sultana, H.P.: Identity-based security scheme in internet of vehicles. In: Smart Intelligent Computing and Applications, pp. 515–523. Springer (2019)
14. Yadav, A., Gupta, H., Khatri, S.K.: A conceptual framework for improving the software security of self-driven vehicles. In: 2019 Amity International Conference on Artificial Intelligence (AICAI), pp. 893–897. IEEE (2019)
15. Sharma, P., Möller, D.P.: Protecting ECUs and vehicles internal networks. In: 2018 IEEE International Conference on Electro/Information Technology (EIT), pp. 0465–0470. IEEE (2018)
16. Li, H., Lu, R., Misic, J., Mahmoud, M.: Security and privacy of connected vehicular cloud computing. IEEE Netw. **32**(3), 4–6 (2018)
17. Contreras-Castillo, J., Zeadally, S., Guerrero-Ibañez, J.A.: Internet of vehicles: architecture, protocols, and security. IEEE Int. Things J. **5**(5), 3701–3709 (2017)
18. Duc, T.L., Leiva, R.G., Casari, P., Östberg, P.-O.: Machine learning methods for reliable resource provisioning in edge-cloud computing: a survey. ACM Comput. Surv. (CSUR) **52**(5), 1–39 (2019)
19. Brownlee, J.: Master Machine Learning Algorithms: discover how they work and implement them from scratch. Machine Learning Mastery (2016)
20. Othmane, L.B., Weffers, H., Mohamad, M.M., Wolf, M.: A survey of security and privacy in connected vehicles. In: Wireless Sensor and Mobile Ad-hoc Networks, pp. 217–247. Springer (2015)
21. Moosavi, S., Samavatian, M.H., Parthasarathy, S., Teodorescu, R., Ramnath, R.: Accident risk prediction based on heterogeneous sparse data: new dataset and insights. In: Proceedings of the 27th ACM SIGSPATIAL International Conference on Advances in Geographic Information Systems, pp. 33–42 (2019)

Prediction of University Dropout Using Machine Learning

Aracelly Fernanda Núñez-Naranjo[1], Manuel Ayala-Chauvin[2(✉)], and Genís Riba-Sanmartí[3]

[1] Universidad Facultad de Humanidades y Artes, Universidad Nacional de Rosario, Rosario, Argentina
arafer@hotmail.es

[2] SISAu Research Group, Facultad de Ingeniería y Tecnologías de la Información y Comunicación, Universidad Tecnológica Indoamérica, Ambato, Ecuador
mayala@uti.edu.ec

[3] Centro de Diseño de Equipos Industriales, Universidad Politécnica de Cataluña, Barcelona, Spain
genis.riba@upc.edu

Abstract. University dropout is a complex issue that affects all higher education institutions worldwide. This phenomenon is shown by the high proportion of students that never finish their university training, with the associated economic and social costs. The challenge for higher education institutions is to design and improve policies to increase student retention, specially within the first years. This study uses data mining to find patterns and student clustering that help explaining university dropout. The data for the analysis was gathered from the students that signed up on two admission perioof the Universidad Tecnológica Indoamerica of Ambato, Ecuador. A k-means algorithm is used to classify and define the performance patterns, and predictions for new students are made using a support-vector machine (SVM) model. The results allow institutions and the faculty to focus in high risk groups during the first terms and amend their future learning behaviour. To sum up, this study presents a models to explain and predict university dropout, and to design actions to reduce it.

Keywords: University dropout · K-Means · Support vector machine

1 Introduction

University dropout is a phenomenon occurring in higher education institutions all over the world. Due to this ubiquity, it has been studied from multiple theoretical and methodological angles, as well as analysed from both social and economic perspectives [1–3]. University dropout constitutes a social issue [4–7] that affects not only the students themselves, but also their families and governments [8], since training a university student all the way to graduation entails a significant amount of resources (economic and otherwise), and dropping out in effect wastes most of these resources.

Á. Rocha et al. (Eds.): ICITS 2021, AISC 1330, pp. 396–406, 2021.
https://doi.org/10.1007/978-3-030-68285-9_38

This can lead to social inequality and increased poverty [9, 10]. Furthermore, it affects the learning process of university students [11].

University dropout is defined as the abandonment, prolongation or interruption of the university training [12] due to an intentional or unintentional personal circumstance [13] in which the student decides to abandon their studies [14].

Developed countries such as Germany or Switzerland had a dropout rate between 20% to 25% and 7% to 30% respectively in 2006 [15], although there is a high variability even among richer countries. For instance, Croatia had a 4.3% desertion rate in 2011, while Italy and the USA reached 55%.

Dropout rates directly affect graduation rates, which range from as low as 30% up to above 90% [16].

Latin America featured high desertion rates in 2017, around 47%, although with large differences from country to country. For instance, Dominican Republic had a desertion rate of 76%, followed by countries such as Bolivia (73.3%), Uruguay (72%), Brazil (59%), Chile (53.7%), Mexico (53%), Venezuela (52%), Honduras (40%), Argentina (49%), and Colombia (40%) [17]. At the other end of the spectrum, Cuba only had a 25% desertion rate thanks to the implementation of education policies such as tutoring, collaborative learning, learning communities and academic support, all of which have increasing retention as a main goal [18, 19].

The socioeconomic environment is one of the main drivers of dropout. Factors such as marital status, father's education level, career counselling, family wealth, or academic performance have proved to be of interest when studying this issue [20–23]. Furthermore, the widespread poverty and pervasive conflict situation prevalent in many Latin American countries adds up to the many social, academic, and economic variables that increase university dropout.

In Ecuador, university dropout rate was 26% in 2017. This situation can be considered a risk factor for families, society and the higher education network as a whole [24]. To address that, universities and government agencies are making an effort to improve retention and graduation rates through policies and regulations [25].

Assessing university dropout is a very complex problem, with many variables and factors that may contribute in different capacities in different contexts, and that are often difficult to quantify or evaluate, such as adaptation to the academy, motivation, age, gender, sex, parents' education, sociocultural background, or place of origin, among many others [26].

Most dropouts occur during the first two of university life, which may be caused by the onset of new goals or life projects and influenced by the personal, familiar and social factors of each student. On the other hand, motivated students with realistic expectations, clear goals, and acceptable academic performance do not abandon their training and have the highest chance to successfully achieve graduation [27].

Ramirez-Rueda et al. characterized the dropout risk as an integration of variables related to the socioeconomic, institutional, academic and personal status. Likewise, Moncada-Mora created a mathematical model of the dropout risk that integrates academic and non-academic variables, such as age, gender or housing. This model estimates the likelihood of a student to drop out, or otherwise stay in their university training. This study also determined that the decision to abandon the studies is often driven by factors related to the inadequate interaction between students and faculty [28].

In the light of the above, it is revealed that there is a lack of academic integration between student body and faculty due to the lack of vocation of the professors [29], and the lack of career counselling offered to the students, together with a lacklustre education beginning as early as middle school [30–32]. All of this affects the quality, pertinence and efficiency of university training [33].

One approach to analyse higher education dropout is the use of data mining techniques. These techniques enable multivariate studies, where data sets are grouped in personal data, academic data, and other data generated during the school life. The variables used in the different studies were entry age, sex, passed courses, quantity, and origin. The analysis of these data yielded behavioural patterns that can help predicting dropouts, and therefore use retention strategies to better understand this phenomenon and reduce its incidence [34–36].

Data mining coupled with artificial intelligence (AI) techniques is a powerful tool to store, process, classify and recognize patterns. Furthermore, it enables the implementation of algorithms such as k-means to classify and explain the characteristics of dropout students [37, 38]. Dropout patterns can thus be determined from the personal, academic and work-related variables [39].

Support Vector Machines (SVM) have been widely used to classify information and predict dropout [40]. However, non-integrated models could generate inaccurate or incorrect outcomes [41]. Therefore, this research proposes an algorithm that integrates k-means, restriction criteria and SVM to explore patterns and clusters in order to explain university dropout.

The experimental data for this study was obtained from two student groups of the Universidad Tecnológica Indoamérica in Ambato, Ecuador. A k-means algorithm was used to classify and define performance patterns. Then, an SVM model is used to predict the behaviour of students in the first terms. Finally, there is an explicative and predictive analysis of the personal, academic and economic variables that have an influence on the students' decision, may it be voluntary or coerced, to abandon university training.

2 Methods

The goal of this study is to define a set of statistical measures to determine the factors that have an effect on university dropout in Ecuador, and more precisely the province of Tungurahua. The k-means method is used to sort and classify the data, and the result of this analysis will be used to improve the student retention ratio during the first year.

The sample universe are the university level students in the Tungurahua province. The sample are the 1,078 students admitted on the first term on the two periods of 2014, on seven different degrees (industrial engineering, general psychology, urban architecture, business administration, digital design and multimedia, accounting and auditing, system engineering) in two modes (blended and on-site learning) on two sites of UTI, a private Ecuadorian university.

The approach used consists of 5 steps. First, data collection. For the study case, data was obtained from the Academic Management System (SGA) database. Second, data processing and statistical normalization to allow for further analysis. Third, generate vectors of the main characteristics, which combine the available data for grades, terms taken, efficiency index, etc. Fourth, group students into behavioural categories with a clustering techniques. Finally, use the resulting data to make predictions using Support Vector Machine techniques.

Each student is described with a vector of quantitative data, including grade average, efficiency index, and number of terms, which are used to assess the academic performance. Equation (1) represents a group of students x:

$$x = (Progen, Propdr, Indefi, Promapro, Semcur, Mention, Econom)^T \quad (1)$$

Where: Progen – general grade average, Propdr – weighted grade average, Indefi – efficiency index, Promapro – average grade of passed subjects, Semcur – number of terms, Mention – career, Econom – economic situation.

Table 1 shows the algorithm created for this research. It is divided in three parts. The first part creates clusters with their centroids, and calculates the statistical values (mean, standard deviation, maximum and minimum values) of the graduates. The second part sorts the graduates based on a quantitative restriction. The third part generates a prediction with the data of the first year students. It includes the prediction functions kernel "lineal" and kernel "poly" from the Python suite.

Table 1. Algorithm to cluster, classify, measure and predict

Algorithm	K-means, statistical measures and prediction
Input:	$\boldsymbol{x} = (Progen, Propdr, Indefi, Promapro, Semcur, Mention, Econom)^T$
Output:	Number of clusters, statistical measures, classification and prediction
1	Initialize - K Clusters with their centroids of $\mu_1, \ldots, \mu_k$ randomly
2	While not converge:
3	for i in range ($dataset$):
4	$C_k = argmin\lVert x_i - \mu_k \rVert^2$
5	for j in range ($\boldsymbol{k}$):
6	$\mu_j = \frac{1}{N}\sum_{i=1}^{N} x_i$
7	end while
8	Return Clusters and statistical measures
9	Initialize – classification with restriction criteria
10	if $Propdr \leq 10$ and $Indefi \leq 0.7$
11	class A
12	Else
13	$Propdr > 10$ and $Indefi > 0.7$
14	class B
15	end classification
16	Return Table of classification of students using class labels A and B
17	Initialize – prediction model SVM
18	Model (Kernel = 'linear') - $f(x) = sgn(w^T x + b)$
19	Model (Kernel = 'poly') - $k(x_i, y_j) = (x_i \cdot y_j)^d$
20	Input Data for prediction
21	$\boldsymbol{x^*} = (Progen, Propdr, Indefi, Promapro)^T$
22	end prediction
23	Return Table of prediction of students using class labels 1 and -1

It is important to point out that measuring academic performance is a very complex and multidimensional problem, which measures in a determined way the benefits of the learning process. It involves the characteristics of the student, of their social environment, and the education system. Therefore, there are many definitions of academic performance, but all of them agree that one indicator are the grades, which quantitatively measure the student's achievements. The efficiency index is another academic performance indicator that is considered in this research.

Finally, supervised classification methods based on linear al polynomic models are used to produce the forecast for the first year students. These methods are used in parallel during the third part of the algorithm in order to compare and validate the results of the prediction

3 Results

This sections analyses the results obtained from the algorithm when applied to the data provided by the UTI. In particular, from 376 students who graduated in 2015. For each student, the file contains the following attributes: general mean grade, weighted mean grade, efficiency index, passed subjects mean grade, and number of terms.

Figure 1 shows the efficiency index (Indefi) compared to the number of terms (Semcur) data in $\mathbb{R}^2$.

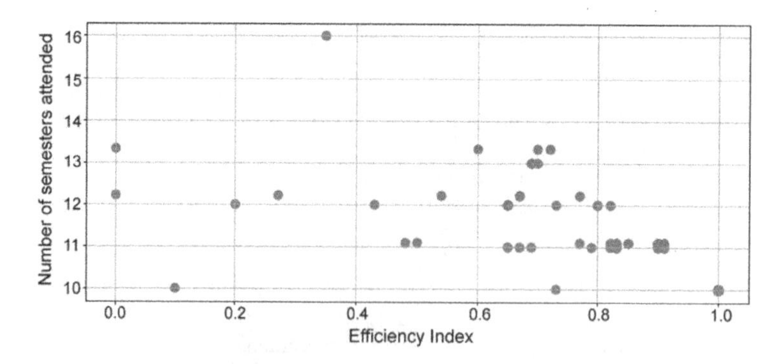

Fig. 1. Data display number of semester's vs efficiency index

3.1 K Clusters

The "elbow" method was used to determine the number of clusters for the analysis. This method allows to select the number of clusters adjusting the model with a range of k values, in order to find a value of k that causes a sudden change in the slope. In this case, k = 7 is a good fit. Then, the K-Means function was used to obtain the best configuration with 7 groups (1,000 iterations).

Figure 2 shows the results of applying the algorithm to the set of data, confronting the number of terms taken with the efficiency index.

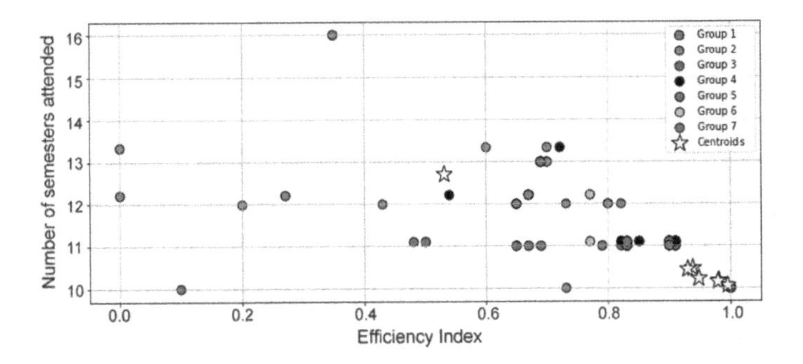

Fig. 2. Clusters with centroids.

Cluster 5 is the largest one, which includes 109 students with a very good grade average of 8.22. Clusters 6 and 7 have 18 and 23 students, with good average grades of 7.86 and 7.92, respectively.

3.2 Classification with Restriction Criteria

The statistical measures show that the weighted grade average and the efficiency index are both good parameters to classify the graduates. The clusters have distinct values of these parameters, which in turn allow for further classification:

Class A, labeled −1, includes all graduates that

$$\text{Propdr} \leq 10 \;\text{ and Indefi} \leq 0.7 \tag{2}$$

Class B, labeled 1, includes all graduates that

$$\text{Propdr} > 10 \;\text{ and Indefi} > 0.7.7 \tag{3}$$

The result of this restricted classification is 29 students in Class A, and 347 students in Class B. Table 2 shows the statistical measures for both Classes A and B.

Table 2. Statistical measures for Classes A and B

Statistical measures	Class A			Class B		
	Propdr	Indefi	Semcur	Propdr	Indefi	Semcur
ME	8.02	0.54	12.22	8.38	0.98	10.19
DE	0.43	0.21	1.09	0.50	0.06	0.46
Min	7.39	0.00	10.00	7.42	0.72	10.00
Max	9.10	0.70	16.00	9.83	1.00	13.33

3.3 SVM Prediction

Finally, once the graduates' data is sorted in two Classes, a Support-vector Machine (SVM) algorithm is used to obtain the prediction. SVM is a classifier formally defined by a hyperplane that separates the groups. In other words, given a set of labeled training data (supervised learning), the algorithm generates an optimal hyperplane that categorizes new students, with each class being on one side of the hyperplane.

The experiment was realized with a sample of 12 students attending the first year. A model was built using the SVM in order to predict which students will have a low performance. The results show that there is a single student in Class A. Therefore, teachers have to take corrective actions and design actions to reduce dropout. Furthermore, there are social, economic, institutional and cultural factors that can have an effect on the behaviour of the students.

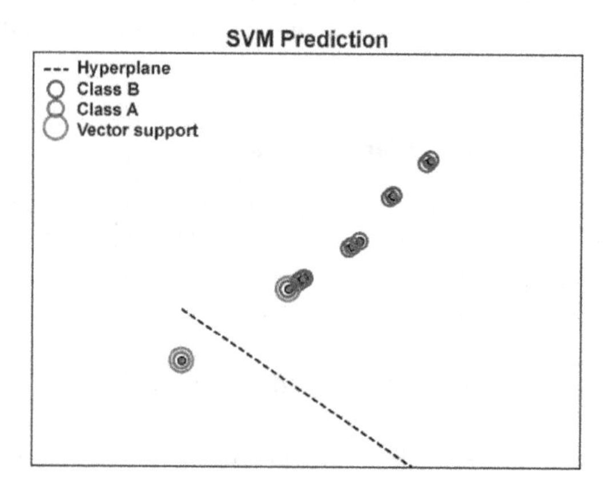

Fig. 3. Support-vector machine prediction.

Figure 3 shows the supervised classification of the sample. It displays the support vector (green), class B (blue) and A (red), and the hyperplane separator, which separates the student predicted to be in Class A.

A linear separation was applied using the Python library $kernel = '\ linear'$. In order to compare results, a second polynomial separation using the Python library $kernel = 'poly'$ was applied too. Both models, lineal and polynomic, produce the same results, that is that one student will have low performance due to economic and social issues during their college years.

4 Discussion

Results shown in previous sections display the power of this method applied to educational data mining. The application developed in Python is flexible and outputs the student performance results both graphically and numerically. With this information, teachers can identify low performance students, improve retention and obtain an overall better performance of the education system.

The models and configurations developed in Python are complex processes that create automatic learning solutions even without programming skills. Therefore, it is replicable by people working in education administration, in research or in learning support tasks, that may need to explore student performance in order to support education institutions.

The Universidad Tecnológica Indoamérica is a private higher education institution. Its dropout rate on 2014 was 44.9%, while its graduation rate in 2019 was 17.25%. This shows an academic lag in the graduation process due to failing and retaking courses, degree swaps, and re-entries. These rates result in a high academic failure rate when compared to similar institutions.

This research focuses on grouping, classifying and predicting using quantitative education data from the Academic Management System (SGA) of the UTI. The SGA was used to acquire data, using 5 data from each admission period in order to predict the students' performance. The total number of students was 376 between February 2014 and July 2015.

The assessment using the K-Means algorithm divides the sample in 7 clusters, which are further classified in classes A and B.

Compared to other algorithms [42, 43], this one requires less pre-processing, since its configuration accepts an input vector that contains all the characteristics of the sample. Therefore, it is very flexible, can be extrapolated, and has a fast processing speed. However, it also has limitations related to the difficulty to code the restrictions to determine student classes, since they are unique for each individual institution. On the other hand, outputs also depend on the purpose of the analysis. Therefore, the code needs to be adjusted for each analysis and educational institution.

Further development of the code should go into blending the prediction algorithm with the SGA of the university, so that student assessment can be obtained in real time, in order to improve retention, and teaching and learning processes.

5 Conclusions

The present research developed a technique to assess and classify the expected academic performance students admitted into the university system based on clustering techniques. The model is based on K-Means to determine the number of clusters, and a classification based on quantitative restrictions for the clustering. The predictions are obtained with supervised learning algorithms using the lineal model (kernel $=$ 'linear') and the polynomic model (kernel $=$' poly') of the Python suite.

The classification of the student groups based on restrictions improve the interpretability and explainability of the resulting models. Comparison between models produces as a result the value of low performance students in a quick and precise manner. Furthermore, the model is flexible but it also can be generalized. Therefore, we believe that these procedures and tools can be an asset in education data analysis.

The proposed method can be used to estimate the future performance of the students that are admitted to the university system, and thus prepare strategic actions in order to improve retention and graduation rates with learning and support activities according to each student's needs.

On the other hand, it is possible to expand the software so that it can be integrated within the Academic Management Systems to automatically run the analysis every term. It is paramount incorporate characteristics that can help to explain the results and understand why a certain student is more likely to drop out.

To sum up, using machine learning in education allows teachers to assess and analyse data from education environments and, in the end, improve overall academic results.

References

1. González, L.E.: Repitencia y deserción universitaria en América Latina. Inf. sobre la Educ. Super. en América Lat. y el Caribe 2000–2005. **2005**, 156–168 (2005)
2. Diaz, D.B., Garzón, L.P.: Elementos para la comprensión del fenómeno de la deserción universitaria en Colombia. Más allá de las mediciones. Cuad. Latinoam. Adm. **9**(16), 55 (2016). https://doi.org/10.18270/cuaderlam.v9i16.1248
3. Vargas, J.G.C., Rios, L.S.B., Laverde, R.M.: Propuesta Para Aumentar El Nivel Académico, Minimizar La Deserción, Rezago Y Repitencia Universitari. Sci. Tech. **11**(28), 145–150 (2005)
4. Zavala-guirado, M.A.: Factores internos, externos y bilaterales asociados con la deserción en estudiantes universitarios. **4**, 59–69 (2018). https://doi.org/10.24016/2018.v4n1.103
5. Ramírez-Rueda, M., Urrego-Velásquez, D., Páez-Zapata, E., Velásquez, C., Ramírez, E.H.: Perfiles de riesgo de deserción en estudiantes de las sedes de una universidad colombiana. **38** (1), 275–297 (2008) https://doi.org/10.18800/psico.202001.011
6. De Toluca, V.: Factores que inciden en el rendimiento académico de los estudiantes de la Universidad Politécnica del Valle de Toluca Factors affecting the academic performance of students (2017)
7. Tomás, J., Gutiérrez, M.: Aportaciones de la teoría de la autodeterminación a la predicción de la satisfacción académica en estudiantes universitarios Contributions of the self-determination theory in predicting university students' academic satisfaction. **37**, 471–485 (2019)
8. Carvajal-Muquillaza, C.M.: Modelación de la deserción universitaria mediante cadenas de Markov. **34**, 129–146 (2020)
9. Barrios, A.: Deserción universitaria en chile: incidencia del financiamiento y otros factores asociados. **14**(1), 59–72 (2011)
10. González-Ramírez, T., Pedraza-Navarro, I.: Variables sociofamiliares asociadas al abandono de los estudios universitarios. **35**, 365–388 (2017). http://dx.doi.org/10.6018/j/298651
11. Humphrey, K.B.: New sails for the recruitment, retention, and learning ship. About Campus. **13**, 2–3 (2008). https://doi.org/10.1002/abc.239
12. Cabrera, L., Bethencourt, J., Alvarez, P., González, M.: El problema del abandono de los estudios universitarios. **12**, 171–203 (2006)
13. Tinto, V.: Definir la desercion: Una cuestion de perspectiva. Rev. Educ. Super. **18**(71), 160 (1989). https://doi.org/10.1017/CBO9781107415324.004
14. Siegel, M.J.: Reimagining the retention problem : moving our thinking from end-product to by-product. **15**, 8–18 (2011). https://doi.org/10.1002/abc.20043

15. Villamizar-Acevedo, G., Bayona, L.P.: Identification of motivational and sociodemographic factors in deserters psychology estudents at university. **14**, 124–137 (2011)
16. Rico-Parada, A., Suárez-Correa, Y., González-Cárdenas, Y.: Factores relacionados con la permanencia estudiantil en programas de pregrado de una universidad pública 1. **19**, 155–170 (2017). https://doi.org/10.11144/javeriana.ie19-1.frpe
17. Martelo, R.J., Jimenez-Pitre, I., Villabona-Gómez, N.: Determinación de factores para deserción de estudiantes en pregrado a través de las técnicas lluvia de ideas y MICMAC. Espacios. **38**(20) 2017
18. Drake, J.K.: The Role of Academic Advising in Student Retention and Persistence. **16**, 8–12 (2011). https://doi.org/10.1002/abc.20062
19. Tinto, B.V.: Universities as learning organizations. **1**, 2–4 (1997)
20. Corengia, A., Pita, M., Mesurado, B., Centeno, A.: La predicción de rendimiento académico y deserción en estudiantes universitarios. **4827**, 101–112 (2012)
21. Flores-López, O., Gutiérrez-Emmanuelle, L., León-Corredor, O.: Centros de Apoyo y Desarrollo Educativo Profesional para la observación y disminución de la deserción universitaria. **18**(1), 48–62 (2016)
22. Barragán, D., Patiño, L.: Elementos para la comprensión del fenómeno de la deserción universitaria en Colombia. Más allá de las mediciones. **16**, 55–66 (2013)
23. Fonseca, G., García, F.: Permanencia y abandono de estudios en estudiantes universitarios: un análisis desde la teoría organizacional. Rev. la Educ. Super. **45**(179), 25–39 (2016). https://doi.org/10.1016/j.resu.2016.06.004
24. Vivas-Vivas, R.J., Cabanilla-Vasconez, E., Vivas, W.: Relación entre los estilos de aprendizaje y el rendimiento académico del estudiantado de la carrera de Ingeniería Agronómica de la Universidad Central del Ecuador. **43**, 468–482 (2019)
25. León, J.H.: El perfil del tutor de titulación de tercer nivel. caso: Facultad de Ciencias Administrativas de la Universidad de Guayaquil, Ecuador. **6**, 1–12 (2019)
26. Carvajal, C.M., González, J.A.: Variables Sociodemográficas y Académicas Explicativas de la Deserción de Estudiantes en la Facultad de Ciencias Naturales de la Universidad de Playa Ancha (Chile) Sociodemographic and Academic Variables Explaining Student's Dropout in the Faculty of N. **11**(2), 3–12 (2018)
27. Landry, C.C.: Self-efficacy, motivation, and outcome expectation correlates of college students' intention certainty (2003). LSU Doctoral Dissertations, 1254. https://digitalcommons.lsu.edu/gradschool_dissertations/1254
28. Moncada-Mora, L.: La integración académica de los estudiantes universitarios como factor determinante del abandono de corto plazo. Un análisis en el sistema de educación superior a distancia del Ecuador. **17**, 173–196 (2014)
29. Peña-Fernández, M.A.: Los factores pedagógicos influyen en la deserción universitaria Pedagogical factors influence university dropouts. INNOVA Res. J. . **4**(3), 108–115 (2019). https://doi.org/10.33890/innova.v4.n3.2019.996, ISSN:2477–9024
30. Almeida-Lara, L., Cueva, M.C., Romero-Palacios, A.: Análisis de la equidad de género en la Universidad Técnica de Cotopaxi. **VIII**, 259–268 (2017)
31. Rubio, M., Tocain, A., Mantilla, M.: La deserción universitaria en los primeros niveles y la inserción laboral de los graduados. Proyecto Alfa III, DevalSimWeb. Pomtificia Universidad Católica del Ecuador Sede Ibarra. **1**, 26–35 (2012)
32. Baquerizo, R.P., Tam, O.A., López, J.G.: La deserción y la repitencia en las instituciones de educación superior : algunas experiencias investigativas en el Ecuador. **6**, 102–107 (2014)
33. England-Bayrón, C.: Teoría Social Cognitiva y Teoría de Retención de Vincent Tinto: Marco Teórico para el estudio y medición de la auto-eficacia académica en estudiantes universitarios. **5**, 28–49 (2012)

34. Eckert, K.B., Suénaga, R.: Análisis de deserción-permanencia de estudiantes universitarios utilizando técnica de clasificación en minería de datos. Form. Univ. **8**(5), 3–12 (2015). https://doi.org/10.4067/S0718-50062015000500002
35. Palacios-Pacheco, X., Villegas-Ch, W., Luján-Mora, S.: Application of Data Mining for the Detection of Variables that Cause University Desertion. **1**(2017), 91–103 (2019). https://doi.org/10.1007/978-3-030-05532-5_38
36. Vila: Technology Trends Reading.pdf. **1**(2017), 91–103 (2019). https://doi.org/10.1007/978-3-030-05532-5
37. Vásquez, J., Miranda, J.: Data Science and Digital Business. Springer International Publishing (2019)
38. Treviño, M., Ibarra, S., Castán, J., Laria, J., Guzmán, J.: A framework to avoid scholar desertion using artificial intelligence. Lecturer Notes in Engineering Computer Science, vol. 3 LNECS, pp. 1493–1497 (2013)
39. Fernandez, J., Rojas, A., Daza, G., Gómez, D., Alvarez, A., Orozco, A.: Student Desertion Prediction Using Kernel Relevance Analysis. **1**, 201–209 (2018). https://doi.org/10.1007/978-3-030-01132-1_30
40. Mayra, A., Mauricio, D.: Factors to predict dropout at the universities: a case of study in Ecuador. In: IEEE Global Engineering Education Conference (EDUCON), vol. 2018, pp. 1238–1242, April 2018. https://doi.org/10.1109/educon.2018.8363371
41. Hegde, V., Prageeth, P.P.: Higher education student dropout prediction and analysis through educational data mining. In: 2018 2nd International Conference on Inventive Systems and Control (ICISC), pp. 694–699 (2018). https://doi.org/10.1109/icisc.2018.8398887
42. Manrique, R., Nunes, B.P., Marino, O., Casanova, M.A., Nurmikko-Fuller, T.: An Analysis of Student Representation, Representative Features and Classification Algorithms to Predict Degree Dropout. In: Proceedings of the 9th International Conference on Learning Analytics & Knowledge, pp. 401–410 (2019). https://doi.org/10.1145/3303772.3303800

Comparison of End-to-End Testing Tools for Microservices: A Case Study

Cristian Martínez Hernández(✉), Alexandra Martínez, Christian Quesada-López, and Marcelo Jenkins

Universidad de Costa Rica, San José, Costa Rica
{cristian.martinezhernandez,alexandra.martinez,cristian.quesadalopez, marcelo.jenkins}@ucr.ac.cr

Abstract. Microservices has emerged as a architectural style that provides several benefits but also poses some challenges. One such challenge is testability, since an application may have hundreds or thousands of services operating together, and each of them needs to be tested as they evolve. To overcome this challenge, test automation is key, and together with it, the use of effective and efficient testing tools. Hence, we aim to contribute to this area by evaluating two tools that support end-to-end (E2E) testing of microservices. E2E tests allow to verify if the system works well as a whole (particularly relevant for systems made up of microservices). In this work, we first surveyed E2E testing tools reported in academic literature and by industry practitioners. Then, we applied the IEEE 14102-2010 standard to evaluate those tools. The two top-rated tools, Jaeger and Zipkin, were selected for further evaluation of their effectiveness and efficiency. Results from our case study reveal that Jaeger is more efficient and effective than Zipkinin terms of execution and failure detection times, as well as information provided to detect faults, severity and coverage.

Keywords: End-to-end testing · Microservices · Tools · Automation · Case study

1 Introduction

Microservices is a popular architectural style for designing and implementing distributed systems, which has had a huge impact on the software industry [11,18]. When compared to monolithic architectures like SOA, microservices exhibits many benefits such as enabling a clearer application structure, simplifying the development, deployment and upgrade of large-scale applications, as well as increasing deployability and modifiability [5,14]. However, adopting microservices still has challenges associated to an increased number of services, evolving contracts among services, technology diversity, and testing [5].

In particular, testing microservices is difficult due to the potentially large number of services involved, their heterogeneity (developed with different programming languages and technologies), as well as their asynchronous and independent nature. The path to meet this challenge is automation [16], and for

Á. Rocha et al. (Eds.): ICITS 2021, AISC 1330, pp. 407–416, 2021.
https://doi.org/10.1007/978-3-030-68285-9_39

this we need tools that support microservices testing. Despite being a little investigated area [8], a recent study by Sotomayor et al. [22] compares several microservices testing tools, and concludes that the increasing use of microservices in software applications has evidenced the need for more suitable testing tools and techniques.

In this paper, we address end-to-end (E2E) testing of microservices because it helps diagnosing performance problems in distributed systems [13] as well as correctness of the system as a whole [14]. The increase in microservice complexity makes reliance on automated tests and tools critical for E2E testing [2]. Nevertheless, test automation is not trivial, and often generates monetary and time losses [9]. To help companies that want to automate their microservices testing, this study provides an evaluation of several E2E testing tools drawn from the literature and industry professionals. Using the IEEE 14102-2010 standard, two of these tools (Jaeger[1] and Zipkin[2]) are selected for further evaluation of their efficiency and effectiveness in the context of microservices. We believe that studies like this will aid software organizations to select the appropriate tool for their context, and thus adopt end-to-end testing automation.

The rest of the paper is structured as follows: Sect. 2 offers a background on microservices and E2E testing, Sect. 3 summarizes relevant previous works. Section 4 offers the first evaluation of tools based on IEEE 14102-2010 standard, Sect. 5 presents the second evaluation (case study) of Jaeger and Zipkin, and Sect. 6 outlines the conclusions.

2 Background

Microservices has emerged as an architecture alternative for the design and implementation of distributed systems, resulting in systems with low-coupled components that exhibit properties like flexibility, scalability, adaptability, and fault tolerance [11]. Martin Fowler, pioneer in the microservices architecture, defined it as an architectural style to develop a single application as a suite of small services, each running on its own process and communicating by lightweight mechanisms, often an HTTP resource API [7]. During the last decade, companies like Netflix, Facebook, Twitter, Amazon, Spotify, eBay, LinkedIn, Uber and SoundCloud have embraced microservices [3,12]. Yet, there are still challenges for its adoption, including testability [4].

End-to-end tracing has emerged in the past decade as a valuable technique to diagnose correctness and performance problems in distributed systems, increase coverage, model workloads, resource usage and timings [13,22]. Obtaining these metrics enables developers and operators to understand how their system works and why it fails. Meanwhile, the primary objective of the E2E tests is to ensure a consistent and reliable behavior, and to catch hard-to-test bugs before users do, when unit and integration tests are insufficient [14]. However, executing E2E tests for microservices by traditional (i.e., manual) methods is quite difficult [16],

[1] https://www.jaegertracing.io/.

[2] https://zipkin.io/.

hence automation seems to be the way to go. There is a need for more studies that expand the empirical evidence in this area [8].

3 Related Work

Here we present relevant related work, dealing especially with tools for microservice testing. In particular, Sotomayor et al. [22] compared the execution time of four testing tools for microservices (two E2E test tools and two test harness tools) on an open-source testbed application composed of multi-language microservices. They also characterized a set of 16 test tools for microservices, along several dimensions including test objective, test level, platform, interface, and supported languages.

Schreiber [20] introduced PREvant, a tool that provides a simple RESTful API for deploying and composing containerized microservices. This tool can automate E2E tests to verify the performance of microservices.

Harsh et al. [10] pointed out that complex and large software systems are proliferating due to the commodity of the cloud and the need for elastic applications, which push developers towards resilient software architectures like microservices. Consequently, practices and tools need to evolve to support testing in these new scenarios. They proposed a new testing service model called Testing as a Service, and a new tool named ElasTest, which enables E2E testing.

Meinke et al. [16] studied how learning-based testing (an emerging paradigm for fully automated black-box testing) can be used to evaluate the functional correctness and fault-robustness of a distributed system. In their experiment they used LBTest, a learning-based testing tool. They concluded that the performance of LBTest compared favorably to manual techniques for E2E testing.

Previous works mainly proposed new tools for E2E testing of microservices, or evaluated existing ones. Our study contributes to the field by first compiling a list of 40 test tools for microservices specifically geared towards end-to-end testing, which were extracted from the literature and industry experts. It also provides an evaluation of those tools, based on necessary features of E2E microservices test tools. Finally, it reports a case study where two E2E microservices test tools are evaluated for efficiency and effectiveness.

4 Tools that Support E2E Testing of Microservices

Here we explain how the E2E testing tools were identified and evaluated.

4.1 Tools Identification

In order to identify existing E2E testing tools, we first reviewed the academic literature, where 28 tools were found. Then we surveyed international experts in search of tools used in the industry. Specifically, during May and June of 2020 we sent a survey to 25 Docker's captains[3] and got response from 7 of them. This gave us 13 additional tools reported by industry practitioners. One of them, Selenium, had already been found in the literature survey. Hence, a total of 40 ($28 + 13 - 1$) tools were identified. The complete list can be found at https://tinyurl.com/y4sq58mj.

Some interesting findings from our survey follow. First, all participants deemed necessary to automate microservices tests. Second, most participants indicated that a major challenge in microservices testing lies in the interaction complexity among multiple services. Third, all but one of the participants typically perform E2E testing of microservices. Fourth, participants argued that there are differences between traditional software testing and microservices testing. On one hand, testing individual microservices is easier due to their small size, but new problems arise due to the complex interactions between microservices. On the other hand, it is easier to launch a monolithic system by calling one of its functions than to launch tens of interdependent microservices in order to test one of the functional flows they provide.

4.2 Tools Evaluation

In order to evaluate the previously identified tools, we followed the IEEE 14102-2010 standard and associated recommendations [15]. This is an international standard that provides guidelines for the evaluation and selection of CASE tools. Our evaluation criteria are based on the *tool specific* aspects typified by Powell et al. [17], which represent fitness for purpose of the tool. We selected a subset of these aspects that were considered relevant and even necessary in the context of E2E testing tools for microservices. Particularly, we used the metrics, documentation, support, integratibility, cost, and compatibility criteria. Table 1 shows the 5 tools with highest scores. The complete evaluation is available at https://tinyurl.com/y4sq58mj.

The *metrics* criterion evaluates the ability of the tool to generate useful information from E2E tests. A value of 30 is assigned if the tool generates tracing and timing information. A value of 15 is given to tools that generate only one useful metric; otherwise a 0 is assigned.

The *documentation* criterion gauges the amount of tool documentation available (videos, tutorials, forums, or formal documentation). A value of 20 is assigned when documentation is extensive, including that for E2E tests. A 10 is given when documentation is extensive but does not include E2E tests documentation. A 5 is given when documentation is scarce, and 0 when is non-existent.

The *support* criterion refers to the type of *technical support* available: provider (formal), community (informal), or none. We assigned a value of 15 to provider support, a value of 10 to community support, and 0 for no support.

The *integratibility* criterion assesses the tool's level of compatibility with existing languages or browsers (for Software as a Service). We assigned a value

[3] https://www.docker.com/community/captains.

Table 1. Top 5 testing tools to support testing end-to-end of microservices.

Tool name	Metrics	Documentation	Support	Integratibility	Cost	Platform compat	Total score
Jaeger	Tracing, Timing, Fails System, Resources use **30\30**	High **20\20**	Community support **10\15**	High (Go, Java, Python, Node, C++, C#) **10\10**	Free **20\20**	Both (Local and Cloud) **5\5**	**95\100**
Zipkin	Tracing, Timing, Fails System, Resources use **30\30**	High **20\20**	Community support **10\15**	High (C#, Go, Java, Javascript, Ruby, Scala, PHP) **10\10**	Free **20\20**	Both(Local and Cloud) **5\5**	**95\100**
Test Project	Fails System **15\30**	High **20\20**	Provider support **15\15**	High (Browser Support) **10\10**	Free **20\20**	Both(Local and Cloud) **5\5**	**85\100**
Spring Cloud	Timing, Resources Use **15\30**	High **20\20**	Provider support **15\15**	Low (Java) **2\10**	Free **20\20**	Both(Local and Cloud) **5\5**	**77\100**
Browser Stack	Fails System **15\30**	High **20\20**	Provider support **15\15**	High (Browser Support) **10\10**	Trial **10\20**	Both(Local and Cloud) **5\5**	**75\100**

of 10 if the tool supports 4 or more languages, a value of 5 if it supports 2 or 3 languages, and a value of 2 if only one language is supported.

The *cost* criterion considers only the licensing aspect of the tool. A value of 20 was assigned to free tools, a value of 10 was assigned if a trial version was available, and 0 if payment was required.

The *platform compatibility* criterion refers to platform dependence. If the tool is compatible with both local and cloud, a value of 5 is assigned. If it is only compatible with one platform, a 3 is assigned.

5 Case Study

We provide here details about the case study conducted, following the guidelines proposed by Runeson et al. [19]. The **objective** of our case study was to evaluate the efficiency and effectiveness of Jaeger and Zipkin as tools for E2E testing of microservices. These two tools were chosen for having the best scores from our previous evaluation.

5.1 Experimental Objects

Our experimental object was an open-source application called Socks Shop[4], built to aid research and testing of microservices technologies. This application

[4] https://github.com/microservices-demo/microservices-demo.

encompasses the user-facing part of an e-commerce website that sells socks [1], and uses microservices implemented in Java, .Net Core, Go, and NodeJs as front-end. Docker was configured to orchestrate these applications since it is a reference technology in containers virtualization [21].

5.2 Data Collection Procedure

We first prepared a virtual machine with 8 MB RAM, 2 cores, Ubuntu OS v18.04, Docker v19.03.5, Docker-Compose v1.23.2, Apache Maven v3.6.0, Jaeger-client v3.18.0 and Zipkin v0.22.0.

The architecture and ecosystem of the Socks Shop microservice application used in the case study are depicted in Fig. 1. Jaeger and Zipkin were configured on the following microservices: MS1-Order and MS5-Cart. The repositories were cloned and modified to include Zipkin and Jaeger, updating all their dependencies. Our MS1-Order and MS5-Cart containers were created, and the file docker-compose.yml was modified to use our containers.

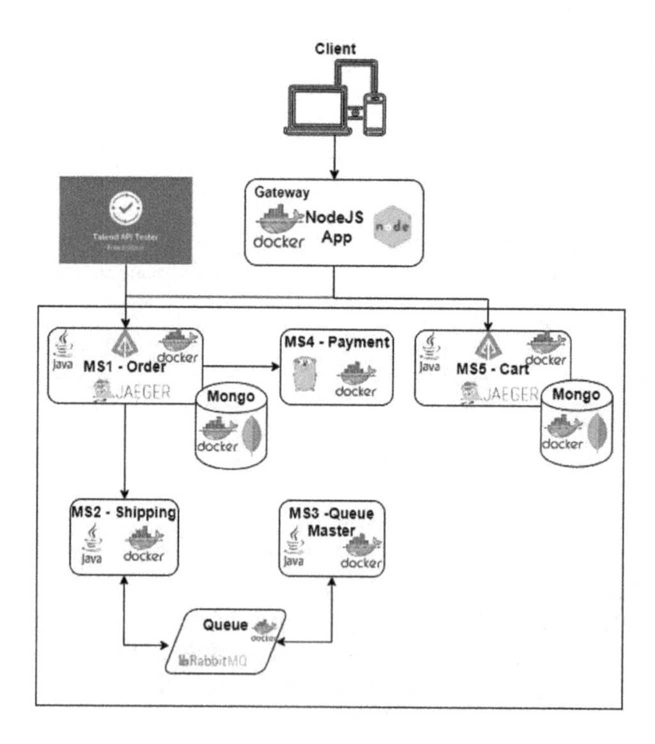

Fig. 1. Architecture of the evaluated microservices application.

Then, two E2E tests were designed and run (invoked by API Tester): the first one invokes the MS1-Order's microservice (POST), which in turns calls MS2-Shipping, MS4-Payment and MS3-Queue Master; the second one invokes only the MS5-Cart's microservice (GET). Both tests collected the following metrics: execution time, failure detection time, fault detection, error type identification, fault severity, and dataflow coverage. We forced the occurrence of failures by stopping the Mongo container used by MS5-Cart, as well as the MS2-Shipping container used by MS1-Order.

To assess and compare the efficiency and effectiveness of the tools, we partly relied on the recommendations by Eldh et al. [6]. Efficiency is measured by the execution and failure detection times, while effectiveness is measured by the ability to detect faults and the provision of information for inferring fault severity and dataflow coverage.

5.3 Results

Next we present the results of our case study. Table 2 shows the results for efficiency: it contains the average execution time and average failure detection time (in ms) for Jaeger and Zipkin, when executing the E2E tests MS1-Order and MS5-Cart. Each test was run 5 times using the API Tester. These results suggest that Jaeger is more efficient than Zipkin since both execution and failure detection times are better in all but one case: the second test case (MS5-Cart). In this case, the failure detection time for Jaeger is slightly higher than that of Zipkin, but this difference is not significant as it is triggered by timeout. From Table 2, we see that for the first test case (MS1-Order), Jaeger's failure detection time is about a third of Zipkin's, and its execution time is close to half of Zipkin's. For the second test case, Jaeger's execution time is 12 times less than Zipkin's.

Table 3 shows the results for effectiveness in terms of three features that help developers and operators to identify and characterize bugs in microservices: ability to detect faults, and provision of information that allows to infer both fault severity and data coverage. Our results found that both Jaeger and Zipkin exhibit these features, with some differences among them. In the case of *fault detection capability*, the tools have a tag with the http status code to present the error type or response from the microservice. Jaeger, however, has additional tags such as component, error indicator, http method, http url, hostname, IP address, among others. With respect to the information offered by the tools to infer the *severity of the fault*, both provide the error stack, from which severity can be inferred, but Jaeger offers more information: message, event, handler method, and handler class name.

Finally, the information provided by the tools to infer the *dataflow coverage* are traces. These are data/execution paths through the system, which form a graph of spans (logical units of work), allowing for test tracing. Jaeger gives further alternative views like the trace graph or trace JSON, and can generate a graph of service dependencies. In summary, although both tools satisfy the three effectiveness properties, we consider Jaeger to be more effective than Zipkin because it provides more information in all three features.

Table 2. Efficiency measurements for Jaeger and Zipkin.

	Test 1 (MS1-Order)		Test 2 (MS5-Cart)	
	Jaeger	Zipkin	Jaeger	Zipkin
Execution time	47.81 ms	80.26 ms	4.30 ms	53.149 ms
Failure detection time	46.25 ms	167.42 ms	10.64 s	10.03 s

Table 3. Effectiveness results for Jaeger and Zipkin.

Feature	Jaeger	Zipkin
Fault detection	Shows entire trace and identifies point of failure. Provides depth, duration, services, trace ID, total spans, trace start, http status, component, error indicator, http method, http url, hostname, IP address, among other data	Shows entire trace and identifies point of failure. Provides depth, duration, services, trace ID, total spans, and http status
Fault severity	Provides error message, event, handler method, handler class name, and stack	Provides the error stack
Dataflow coverage	Offers traces, which can represent dataflow coverage. Allows alternative views like trace graph, trace JSON, and service dependencies graph	Offers traces, which can represent dataflow coverage

6 Conclusions

In this paper, we have identified a set of 40 tools for end-to-end testing of microservices, after surveying the literature and industry experts. These tools were then evaluated following the IEEE 14102-2010 standard, based on criteria that assess their suitability as E2E microservices test tools. Furthermore, we performed an evaluation of the efficiency and effectiveness of Jaeger and Zipkin, the two top-rated tools. Execution time and failure detection time were used as measures of efficiency. For effectiveness, we used the ability to detect faults, and the provision of information to infer fault severity and dataflow coverage.

We found that a variety of tools are used to perform end-to-end testing of microservices, despite not being meant for it. Even more, none of the identified tools allowed both test execution and report generation (a highly desirable feature). We believe that performing end-to-end testing in a microservice environment thus requires specialized tools with appropriate features.

Results from our case study suggest that Jaeger is more efficient than Zipkin because of its shorter execution time and failure detection time in most test scenarios. With respect to effectiveness, both are able to detect faults and provide information that can help developers infer fault severity as well as dataflow coverage. However, Jaeger offers more information than Zipkin, generates tracing with alternate views, and creates useful dependency graphs.

As future work, we suggest to complement this work by studing available tools for unit and integration testing of microservices (E2E are system tests).

Acknowledgments. This work was partially supported by University of Costa Rica's projects No. 834-B8-A27 and 834-C0-726, financed by the Research Center on ICT (CITIC) and the Department of Computer Science (ECCI). We thank the Empirical Software Engineering Group (ESEG) for its valuable feedback and help.

References

1. Aderaldo, C.M., et al.: Benchmark requirements for microservices architecture research. In:2017 IEEE/ACM 1st International Workshop on Establishing the Community-Wide Infrastructure for Architecture-Based Software Engineering (ECASE), pp. 8–13 (2017)
2. Alvaro, P., et al.: Automating failure testing research at internet scale. Association for Computing Machinery, New York (2016). ISBN 9781450345255
3. Bogner, J., et al.: Microservices in industry: insights into technologies, characteristics, and software quality. In: 2019 IEEE International Conference on Software Architecture Companion (ICSA-C), pp. 187–195, March 2019. https://doi.org/10.1109/ICSA-C.2019.00041
4. de Camargo, A., et al.: An architecture to automate performance tests on microservices. In: Proceedings of the 18th International Conference on Information Integration and Web-Based Applications and Services, pp. 422–429. ACM, New York (2016). ISBN: 978-1-4503-4807-2
5. Chen, L.: Microservices: architecting for continuous delivery and DevOps. In: Proceedings - 2018 IEEE 15th International Conference on Software Architecture, ICSA 2018, pp. 39–46 (2018)
6. Eldh, S., et al.: A framework for comparing efficiency, effectiveness and applicability of software testing techniques. In: Testing: Academic Industrial Conference - Practice and Research Techniques, pp. 159–170 (2006)
7. Fowler, M., Lewis, J.: Microservices, pp. 1–15 (2018)
8. Ghani, I., et al.: Microservice testing approaches: a systematic literature review. Int. J. Integr. Eng. **11**(8), 65–80 (2019). https://doi.org/10.30880/ijie.2019.11.08.008
9. Gkikopoulos, P.: Data distribution and exploitation in a global microservice artefact observatory. In: 2019 IEEE World Congress on Services, vol. 2642-939X, pp. 319–322 (2019). https://doi.org/10.1109/SERVICES.2019.00089
10. Harsh, P., et al.: Cloud enablers for testing large-scale distributed applications. In: Proceedings of the 12th IEEE/ACM International Conference on Utility and Cloud Computing Companion, pp. 35–42. ACM, New York (2019). https://doi.org/10.1145/3368235.3368838
11. Heinrich, R., et al.: Performance engineering for microservices: research challenges and directions. In: Proceedings of the 8th ACM/SPEC on International Conference on Performance Engineering Companion, pp. 223–226. ACM, New York (2017). ISBN: 978-1-4503-4899-7
12. Heorhiadi V., et al.: Gremlin: systematic resilience testing of microservices. In: Proceedings - International Conference on Distributed Computing Systems, vol. 2016, pp. 57–66, August 2016

13. Las-Casas, P., et al.: Weighted sampling of execution traces: capturing more needles and less hay. In: Proceedings of the ACM Symposium on Cloud Computing, SoCC 2018, pp. 326–332. Association for Computing Machinery, New York (2018). https://doi.org/10.1145/3267809.3267841, ISBN: 9781450360111
14. Lei, Q., et al.: Performance and scalability testing strategy based on kubemark. In: 2019 IEEE 4th International Conference on Cloud Computing and Big Data Analysis (ICCCBDA), pp. 511–516, April 2019
15. Lundell, B., Lings, B.: Comments on ISO 14102: The standard for CASE-tool evaluation. In: Computer Standards and Interfaces 24.5. cited By 7, pp. 381–388 (2002). https://doi.org/10.1016/S0920-5489(02)00064-8
16. Meinke, K., Nycander, P.: Learning-based testing of distributed microservice architectures: correctness and fault injection. In: Software Engineering and Formal Methods, pp. 3–10. Springer, Heidelberg (2015). ISBN: 978-3-662-49224-6
17. Powell, A., et al.: A practical strategy for the evaluation of software tools, July 1996. https://doi.org/10.1007/978-0-387-35080-6_11. ISBN: 1475758243
18. Quenum, S., Aknine, J.G.: Towards executable specifications for microservices. In: Proceedings - 2018 IEEE International Conference on Services Computing, SCC 2018 - Part of the 2018 IEEE World Congress on Services (2018), pp. 41–48 (2018)
19. Runeson, P., Höst, M.: Guidelines for conducting and reporting case study research in software engineering. Empir. Softw. Eng. **14**(2), 131 (2009)
20. Schreiber, M.: Prevant (Preview servant): composing microservices into reviewable and testable applications. In: OpenAccess Series in Informatics, vol. 78 (2020). https://doi.org/10.4230/OASIcs.Microservices.2017-2019.5
21. Singh, C., et al.: Comparison of different CI/CD tools integrated with cloud platform. In: 2019 9th International Conference on Cloud Computing, Data Science Engineering (Conuence), pp. 7–12, January 2019
22. Sotomayor, J.P., et al.: Comparison of runtime testing tools for microservices. In: 2019 IEEE 43rd Annual Computer Software and Applications Conference (COMPSAC), vol. 2, pp. 356–361, July 2019

An Approach to Integrate IoT Systems with No-Web Interfaces

Darwin Alulema[1,2(✉)], Javier Criado[2], and Luis Iribarne[2]

[1] Universidad de las Fuerzas Armadas ESPE, Sangolquí, Ecuador
[2] Applied Computing Group, University of Almería, Almería, Spain
doalulema@sespe.edu.ec, {javi.criado,luis.iribarne}@ual.es

Abstract. The introduction of new technologies such as the Internet of Things (IoT) raises important problems that must be solved. One of these problems is the interoperability caused by the heterogeneity of the protocols and platforms. In this context, the web of Things is an architectural solution to this problem, because it takes advantage of Web technologies as a means to guarantee interoperability. Another significant challenge is developing and maintaining multiple variants of the applications to support different platforms. For this, Model Driven Engineering (MDE) can help in solving the current problem of developing and maintaining separate application variants. This is because MDE makes use of models to raise the level of abstraction in order to avoid addressing the details of the platforms and speed up the software development process by allowing the design and reuse of the code, and increasing the software quality. In this sense, this document describes a methodology based on models and services for the interoperability of emerging technologies that have a fundamental role in the emergence of the IoT. The result of this research is a concrete and abstract syntax along with a Model-to-Text transformation, to automatically generate software artifacts for IoT systems, thereby achieving the interoperability of mobile technologies, DTV, controllers, REST services and service coordination.

Keywords: Model-Driven Engineering (MDE) · Domain Specific Language (DSL) · Internet of Things (IoT) · Mobile · Digital TV (DTV)

1 Introduction

In recent years, the fourth industrial revolution is taking place and one of the main enabling technologies of this revolution is the Internet of Things (IoT) whose objective is to connect all kinds of devices to exchange information. IoT can be defined as a set of interconnected things (human beings, tags, sensors, etc.) through the Internet which have the ability to measure, communicate and act around the world. The main idea of the IoT is to obtain information about our environment to understand, control and interact with it [4]. However, the lack of a common standard and clear interoperability rules for the IoT can result in

Á. Rocha et al. (Eds.): ICITS 2021, AISC 1330, pp. 417–427, 2021.
https://doi.org/10.1007/978-3-030-68285-9_40

higher maintenance and development costs. The scalability issues of IoT could be expected to worsen if the current growth rate is maintained in the near future [3]. Thus, the Web of Things (WoT) is a recent approach in the IoT domain, focusing on solving the problem of IoT interoperability by using the Web as an integration layer. The Web, as one of the pillars of the Internet, offers a possibility of reaching a consensus to achieve the interoperability of IoT [8]. The W3C initiative for the specification of WoT aims to address the fragmentation of IoT through the use of existing web standards [7]. This opens up the possibility for industries to use the cyber workspace for collaboration from anywhere in distributed environments. For example, remote control of robots becomes important not only in rescue operations but also in domestic and industrial environments [10].

IoT is now one of the present most interesting research fields, and the most active applications are transportation, smart home, robotic surgery, aviation, defence, critical infrastructure, etc. [10]. In this regard, the homes contain more and more smart objects and will continue to grow. Proper orchestration between all connected objects in the home could save end users money. This technology will allow not only to know the average temperature of our home and the consumption of energy and water but also the quality of the air we breathe by being able to automate the ventilation of the home, the prediction of water or gas, leaks or any structural failure [4].

For example, when we get back home, we want a comfortable temperature both in summer and winter. However, the room temperature may change when the number of people increases or when the oven is turned on. In this context, the proposed Smart Home scenario (Fig. 1) is composed of multiple smart objects (sensors and actuators) connected locally and with the option of being controlled from the Internet, in which through an interface mobile, the home owner can monitor the status of its home or even activate some of the actuators from work. In addition, while other tasks are being carried out, the DTV (digital TV) interface is also available so that you can control all the smart objects in your home from multiple places from your sofa. This system also maintains communication with the Health services and can act before the recommended

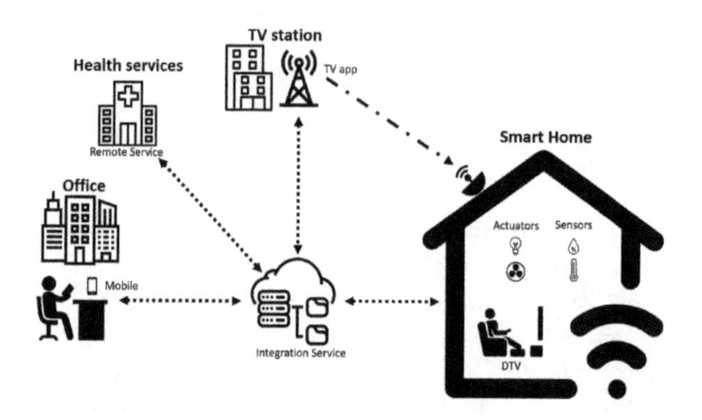

Fig. 1. Coexistence of heterogeneous systems in Smart Home.

protocols such as when there is a concentration of people in a closed environment and it can activate a fan or open the windows to maintain a healthy environment.

There are several difficulties in developing the vision of a Smart Home, and this work addresses two of them: (i) interoperability between systems; (ii) and development costs. With regard to interoperability, our approach proposes the use of: (a) an Event-Driven Architecture (EDA) for the publication and subscription of smart objects to a topic; (b) a Service Oriented Architecture (SOA) for the creation of RESTful services for each of the smart objects and thus can be accessed from the web; (c) a Bridge for interoperability between EDA and SOA; (d) an Orchestrator for the coordination of web services; and (e) a user interface on non-web platforms. In terms of development costs, our approach proposes the use of Model-Driven Engineering (MDE) together with Model-to-Text (M2T) transformations, to alleviate the problem of code generation for each of the platforms in addition to speeding up development processes and validation.

The rest of the article is organized as follows. Section 2 provides a review of relevant IoT interoperability solutions, focusing on those that employ MDE. Section 3 presents the methodology for the design of IoT applications by means of a metamodel and M2T transformations for code generation and it describes the scenario used to demonstrate the operation of the proposed methodology. Finally, Sect. 4 presents the conclusions and future work.

2 Related Work

There is a growing body of research on automated code generation for IoT systems as well as on the use of MDE for the IoT environment.

Cai *et al.* [3] propose a metamodel that integrates CIM (Computation Independent Model) with PIM (Platform-Independent Model) as a reference structure to encapsulate and manage commercial resources for the development of applications for Cloud of Things (CoT). The metamodel connects business requirements with executable components. Additionally, the authors provide three development patterns: role-driven, data-driven and process-driven for service configuration. These patterns allow the rapid development of mobile services based on service resources. The proposal does not contemplate the use of non-web interfaces nor does it consider the coordination of services.

The work of Sneps-Sneppe and Namiot [13] analyses the challenges for the detection and collection of data from various sources in the development of applications for smart cities. The authors propose a Domain-Specific Language (DSL) based on Java Server Pages. With this DSL, operations enable IoT communication between process instances and sensors. This approach does not automatically generate code or have a graphical editor. Also, it does not consider a mechanism for interaction with users.

The authors of [11] propose a specific language along with an Editor developed in JavaScript to design IoT devices. The proposal generates Java code and OpenHAB which allows the interconnection of software components and middlewares; hardware and software; and mapping the code of the devices to the software components. However, none service coordination mechanism is presented in this proposal.

In [14], the authors present an approach based on a UML profile for the integration of Cyber Physical Systems (CPS) and IoT. The approach automates the component generation process and a RESTful API is implemented. The technologies adopted by the authors are: Open Mobile Alliance (OMA) LWM2M application protocol and the Internet Protocol for Smart Objects (IPSO) Alliance. This work does not address the development of user interfaces or service coordination mechanisms.

The approach presented in [15] addresses the problem of mobile application maintenance because multiple native variants of applications must be developed and maintained to support different mobile operating systems, devices, and varying application functional requirements. This approach considers three types of variations in mobile applications: variation due to operating systems and their versions, the software and hardware capabilities of mobile devices, and the functionalities that the mobile application offers. However, it does not address the development of hardware nodes or object coordination mechanisms.

Although many solutions provide high-level interfaces to simplify IoT application development, they are often based on HTTP which is too costly for the scarce resources of embedded devices. In contrast, SOA-based solutions are more efficient but they suffer from inefficient access to physical resources: some of them require direct interaction with embedded devices; others expose resources but only through an intermediate database. Ultimately, no solution provides truly simplified tools that enable the development of IoT applications for both non-expert and expert users, spanning hardware, business logic and a non-web user interface, as given in our proposal.

3 Proposed Integration Approach

In this section we present our approach to the integration of heterogeneous platforms. Our proposed architecture is an extension of some of our previous work [2] and consists of three layers: (a) Physical layer, it corresponds to the sensors, actuators and controllers from which the hardware nodes are built; (b) Logical Layer, together with the Physical Layer, they are the Back-End of the system and allow the coordination of the information from the physical and application layers, for which it consists of: (i) MQTT Broker, it handles publication events or subscription from the physical layer, (ii) Bridge, it routes the messages from the Broker to the RESTful API or the Orchestrator, (iii) RESTful API, it deploys HTTP methods (GET, POST, PUT and DELETE) with communication to a database so that each physical node at which it is associated can be consumed by the orchestrator, the application layer or by remote services; (iv) Orchestrator, it coordinates the call sequences to the REST or Bridge services for the execution of the business logic; (c) Application layer, it corresponds to the Front-End of the system and allows user interaction in non-web interfaces (mobile and DTV). The Fig. 2 shows graphically the relationships that each of the layers of the proposed architecture have.

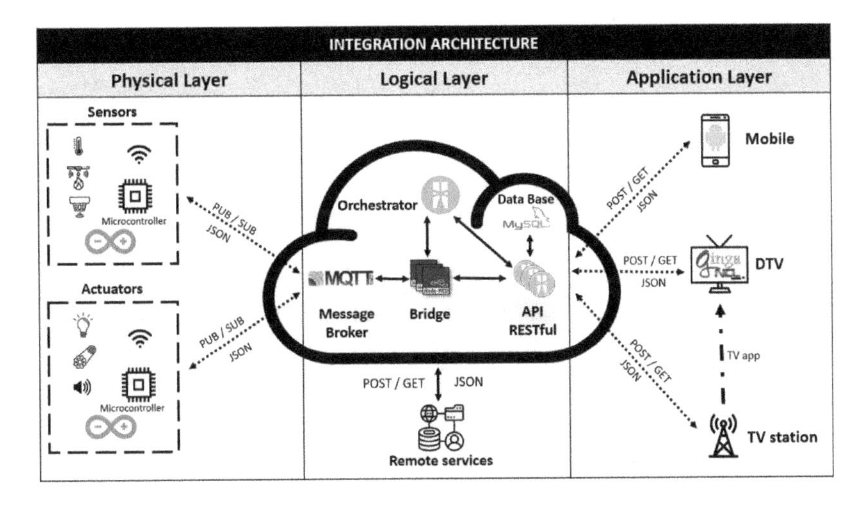

Fig. 2. Proposed integration architecture.

To allow the integration of the platforms, the Thing Description of the WoT [9] has been taken as a guide and a simplified version has been proposed that can be handled more easily by the hardware nodes, which have fewer computing resources; in addition to not saturate the Bandwidth if many nodes are deployed. The structure for the exchange of information consists of: (a) *Id*, identifier of the event; (b) *Date*, date of occurrence of the event; (c) *Time*, time of occurrence of the event; (d) *Location*, physical location of the node; (e) *Attribute*, logical value of the event; (f) *Artefact*, node name; and (g) *Property*, identifier of the data type of the attribute value.

For the development of the metamodel (Abstract Syntax) the Eclipse Ecore tool was used [5]. The metamodel of the Fig. 3 is a capture of the new section of our proposal in which the mobile device platform is incorporated in order to provide a new interface to interact with IoT systems. The meta-classes are grouped according to the layers of the proposed architecture as detailed below:

- Physical layer: This layer groups the meta-classes that allow the design of the hardware nodes and details the characteristics of the network infrastructure on which the IoT system is deployed. The main meta-classes are: (a) For hardware nodes, IoTNode, Sensor, Actuator, Controller and Communication; and (b) For the network infrastructure, WebServer, DataBaseServer, AccesPoint and MessageBroker.
- Logical Layer: This layer groups the meta-classes that allow integrating and coordinating all the nodes and services of the IoT system. The main meta-classes are: BridgeServer, REST, IntegrationPattern and Orchestrator.
- Application layer: This layer groups the meta-classes that allow designing the non-web user interface that allows interaction with the IoT system. The main meta-classes are: DTV, Interface, Mobile and Activity.

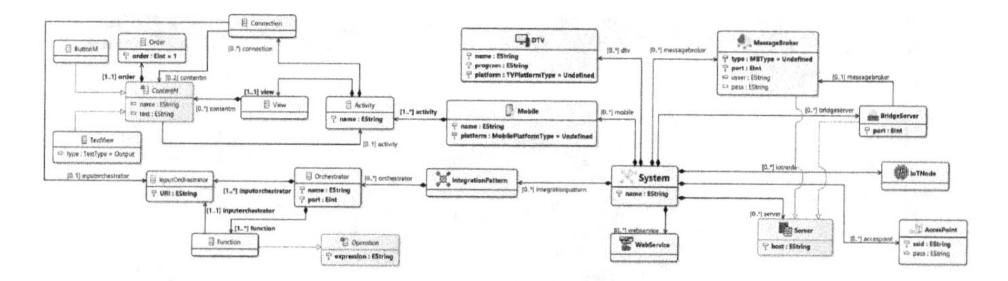

Fig. 3. Proposed metamodel for the integration of heterogeneous systems.

3.1 Graphic Editor for the Design of IoT Systems

For the development of IoT systems, a previous graphical editor (Concrete Syntax) has been created in Eclipse Sirius [12] based on the metamodel described in the subsection. For the development of the Graphic Editor, the VSM (Viewpoint Specification Model) was configured with the meta-classes and their relationships. In addition, the visual components were created to allow the developer to create applications graphically.

The graphical editor (Fig. 4) has 3 main areas: (a) Tool palette (right panel) from which developers can select the elements to be incorporated into their applications; (b) Canvas (central panel) in which developers can drag and drop the visual components and their relationships for an IoT application; and (c) Properties (bottom panel) to add or edit information related to the different components of the model. As can be seen in Fig. 4, the Tool Palette allows modelling the Network Infrastructure, the hardware of the IoT nodes, the integration patterns of the services and the control interface for the Mobile and DTV interfaces.

3.2 Model-to-Text Transformation for Code Generation

To automate the development process of the IoT applications, Model to Text transformation was used in the Eclipse Acceleo tool [1]. To exemplify the usefulness of MDE techniques, the following were used as test platforms: (a) on Android mobile phones (b) in the case of DTV, the ISDB-T standard [6] with NCL-Lua; (c) for the controller, Node MCU ESP 8266 compatible with Arduino; (d) for REST services, Ballerina; (e) for the Bridge, Node-Red and (f) for the integration of services, Ballerina. These platforms were selected for the following reasons: (i) they are platforms that allow at least one textual programming language; (ii) they are platforms that have Internet connectivity; and (iii) they are platforms with an active development community.

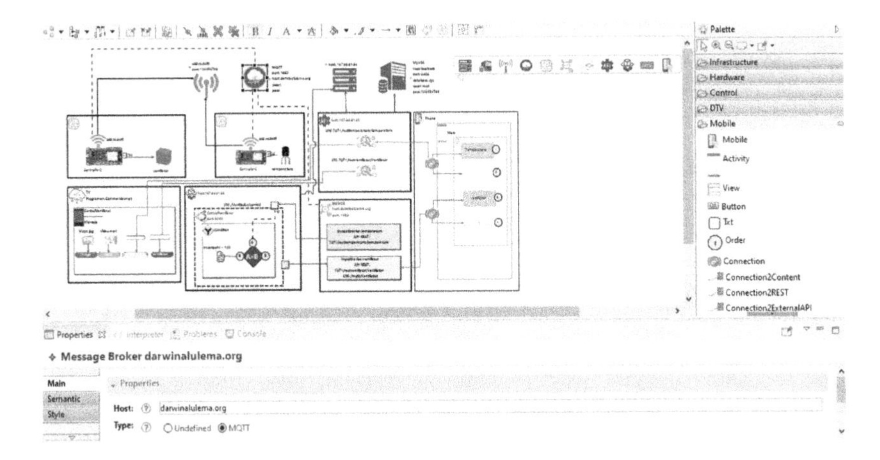

Fig. 4. Screenshot of the graphical editor (The lower section corresponds to the system properties. The section on the right corresponds to the tool palette. The middle section is the canvas for the system design).

The Acceleo code shown in Listing 1.1 contains the shows the general code blocks that generate the software artifacts that will be deployed on each of the system platforms: (a) Creation of all the RESTful APIs associated with each of the sensors and actuators (.bal); (b) Creation of all the Bridges associated with each of the sensors (OutputBridge) and actuators (InputBridge) (.json); (c) Creation of the controller configuration for sensors and actuators (.ino); (d) Creation of the integration pattern based on Sagas (.bal); (e) Creation of the user interface for DTV (.ncl and .lua); and (f) Creation of the user interface for Mobile (.java and .xml).

```
[comment encoding = UTF-8 /]
[**
 * The documentation of the module generate.
 */]
[module generate('http://www.example.org/wsiotvm')]
[**
 * The documentation of the template generateElement.
 * @param aSystem
 */]
[template public generateElement(aSystem : System)]
[comment @main/]
[comment API RESTFULL/]
[for (webservice : WebService | aSystem.webservice)]
 [for (rest : REST | aSystem.webservice.rest)]
  [file (rest.device.name.toString().concat('.bal'), false, 'UTF-8')]
  ...
  [/file]
 [/for]
[/for]
[comment BRIDGE  MQTT/]
[for (broker : BridgeServer | aSystem.bridgeserver)]
 [for (out : OutputBridge | broker.bridge)]
  [file (out.sensor.name.toString().concat('.json'), false, 'UTF-8')]
  ...
  [/file]
 [/for]
```

```
[for (input : InputBridge | broker.bridge)]
 [file (input.actuator.name.toString().concat('.json'), false, 'UTF-8')]
  ...
 [/file]
[/for]
[/for]
[comment HARDWARE/]
[for (nodo : IoTNode | aSystem.iotnode)]
 [for (controller : Controller | nodo.device)]
  [file (controller.name.toString().concat('.ino'), false, 'UTF-8')]
   ...
  [/file]
 [/for]
[/for]
[comment ORCHESTRATOR/]
[for (integration : IntegrationPattern | aSystem.integrationpattern)]
 [for (orchestrator : Orchestrator | integration.orchestrator)]
  [file (orchestrator.name.toString().concat('.bal'), false, 'UTF-8')]
   ...
  [/file]
 [/for]
[/for]
[comment TV APP/]
[for (tv : DTV | aSystem.dtv)]
 [file (tv.name.concat('.ncl'), false)]
  ...
 [/file]
[file ('tcp.lua', false)]
 ...
[/file]
[file ('ConnectorBase.ncl', false)]
 ...
[/file]
[comment MOBILE APP/]
[for (movil : Mobile | aSystem.mobile)]
 [for (actividad : Activity | movil.activity)]
  [file (actividad.name.concat('Activity.java'), false, 'UTF-8')]
   ...
  [/file]
   [file ('activity_'.concat(actividad.name.toLower().concat('.xml')), false, 'UTF-8'
        ↪ )]
   ...
  [/file]
 [/for]
 [file ('AndroidManifest.xml', false, 'UTF-8')]
  ...
 [/file]
[/for]
[/template]
```

Listing 1.1. M2T transformation for generating the software artifacts.

3.3 A Case Study Scenario

In order to demonstrate the functionality of our proposal, the Smart Home domain was chosen for its high degree of diffusion. In this case, a simplified implementation of the scenario proposed in Fig. 1 was made, which was modelled with the graphic editor developed as shown in Fig. 4. The scenario consists of (Fig. 5): (a) a temperature sensor (LM35); (b) an actuator (Relay) that controls the switching on of a fan; (c) an interface for Android mobiles with which the user can check the status of the sensor and activate the actuator and; (d) an interface for DTV that is transmitted via Broadcast by the TV provider and is controlled with the interactivity buttons. In addition, a control was established to automate the ambient temperature by means of an ON/OFF control algorithm implemented in the Orchestrator that coordinates the REST services and the sensor and actuator Bridges. The users of both interfaces can interact with the temperature sensor to observe its sensed value or with the actuator that controls the fan relay to switch it on or off.

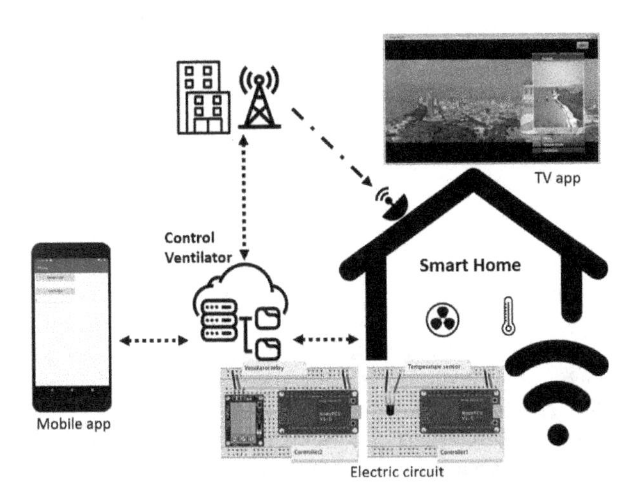

Fig. 5. Coexistence of heterogeneous systems in Smart Home.

4 Conclusions and Future Work

IoT is transforming the way modern systems will develop and operate. However, the adoption of IoT imposes a paradigm shift for systems development and complicates the design process and for this reason, effective approaches are required to handle the complexity introduced by this transition. In this article, we address the challenges that IoT application developers face in achieving interoperable software solutions against highly heterogeneous platforms, protocols, and data. To do this, we have proposed and implemented a solution based on models and services that allows the integration of software and hardware platforms. Thus, avoiding that developers have to delve into the programming languages of all the technologies that intervene in the system. The tools created are: (a) a graphical editor for the design of IoT applications and; (b) an M2T transformation for the generation of code for each platforms. To achieve this goal, we have expanded the functionalities of several of our tools developed in previous work [2] to facilitate the creation of mobile interfaces that are integrated into IoT systems. As a result, we obtained a semi-automatic process for the generation of Ballerina code for RESTful and orchestrator services, Arduino code for the deployment of IoT nodes, NCL-Lua code for the DTV interface and Android code for Smart Phones. The proposal includes a manual step before executing the application that involves the developer by having the multimedia components available for the DTV application (Images and videos) and copying the Android files (.java and .xml) in Android Studio for compilation. Finally, the electrical connections of the components (sensors and actuators) must be made in the controller.

With regard to our future work, we consider several relevant issues: (a) Developing an indexing and discovery service that will allow the use of multiple remote networks to implement more complex and intelligent applications; (b) Extending our proposal to go beyond REST and web services to include AMQ and Kafka; and (c) Incorporate all the W3C WoT Thing Description recommendations.

Acknowledgments. This work has been funded by the EU ERDF and the Spanish Ministry MINECO under the Project TIN2017-83964-R.

References

1. Acceleo - The Eclipse Foundation. https://www.eclipse.org/acceleo/. Accessed 26 August 2020
2. Alulema, D., Criado, J., Iribarne, L., Frenámdez-García, A., Ayala, R.: A model-driven engineering approach for the service integration of IoT systems. Cluster Comput. **23**, 1937–1954 (2020)
3. Cai, H., Yizhi, G., Athanasios, V., Boyi, X., Jun, Z.: Model-driven development patterns for mobile services in cloud of things. IEEE Trans. Cloud Comput. **6**(3), 771–84 (2018)
4. Díaz, M., Martín, C., Rubio, B.: State-of-the-art, challenges, and open issues in the integration of internet of things and cloud computing. J. Netw. Comput. Appl. **67**, 99–117 (2016)
5. Eclipse Modeling Project: The Eclipse Foundation. https://www.eclipse.org/modeling/emf/. Accessed 26 August 2020
6. El-Hajjar, M., Lajos, H.: A survey of digital television broadcast transmission techniques. IEEE Com. Surv. Tutor. **15**(4), 1924–1949 (2013)
7. Iglesias-Urkia, M., Gómez, A., Casado-Mansilla, D., Urbieta, A.: Automatic generation of web of things servients using thing descriptions. Pers. Ubiquit. Comput. 1–17 (2020)
8. Kovatsch, M., Matsukura, R., Lagally, M., Kawaguchi, T., Toumura, K., Kajimoto, K.: Web of Things (WoT) Architecture, W3C Recommendation (2020)
9. Kaebisch, S., Kamiya, T., McCool, M., Charpenay, V., Kovatsch, M.: Web of Things (WoT) Thing Description, W3C Recommendation (2020)
10. Lihui, W., Törngren, M., Onori, M.: Current status and advancement of cyber-physical systems in manufacturing. J. Manufact. Syst. **37**, 517–27 (2015)
11. Salihbegovic, A., Eterovic, T., Kaljic, E., Ribic, S.: Design of a domain specific language and IDE for internet of things applications. In: International Convention on Information and Communication Technology, Electronics and Microelectronics, pp. 996–1001 (2015)
12. Sirius - The easiest way to get your own Modeling Tool. https://www.eclipse.org/sirius/. Accessed 26 August 2020
13. Sneps-Sneppe, M., Dmitry, N.: On web-based domain-specific language for internet of things. In: International Congress on Ultra Modern Telecommunications and Control Systems and Workshops, pp. 287–292 (2015)

14. Thramboulidis, K., Vachtsevanou, D., Kontou, I.: CPuS-IoT: a cyber-physical microservice and IoT-based framework for manufacturing assembly systems. Ann. Rev. Control **47**, 237–248 (2019)
15. Usman, M., Zohaib, M., Uzair, M.: A product-line model-driven engineering approach for generating feature-based mobile applications. J. Syst. Softw. **123**, 1–32 (2017)

Computer Networks, Mobility and Pervasive Systems

Simplified Path Loss Lognormal Shadow Fading Model Versus a Support Vector Machine-Based Regressor Comparison for Determining Reception Powers in WLAN Networks

Mauricio González-Palacio[1(✉)], Lina Sepúlveda-Cano[1], and Ronal Montoya[2]

[1] University of Medellín, 050026 Medellín, Colombia
{magonzalez,lmsepulveda}@udem.edu.co
[2] University of Antioquia, 050011 Medellín, Colombia
ronal.montoya@udea.edu.co

Abstract. A common task when planning a wireless network is the analysis of propagation losses between two transceivers. Different models, from theoretical and empirical natures are proposed in the literature; however, some of them are difficult to be parametrized, and others are thought for very specific scenarios. In this work, we perform a comparison between the simplified path loss lognormal shadow fading model versus a Support Vector Machine (SVM) regressor in a WLAN network. Results show that SVMs are more accurate predicting shadow fading effects.

Keywords: WLAN · SVM · Path loss · Shadow fading · Lognormal distribution

1 Introduction and Problem Statement

Path loss in wireless channels is a mandatory step when designing radio links, to ensure good metrics, such as long distances, high signal-to-noise ratios, among others. Such models can be characterized by physical-based models and also from observation (empirical basis) [1]. In the first case, the best results are obtained in the ideal case; however, the environment must be characterized completely, for example, geometries, materials, electromagnetic properties (permittivity, permeability), among others, making it difficult to use in practical applications [2]. On the other hand, empirical models can be adjusted according to the particular conditions of the environment, such as urban macrocells, urban microcells, indoor environments, semi-rural and rural environments, etc. [3]. However, such models are based on distance, frequency, and characteristics of buildings or geographic regions, which makes their generalization not viable or questioned by the scientific community [1].

Therefore, baseline models are proposed for simplifying the design tasks, and they serve as a first approximation in the subsequent performance analysis [1]. The most and

Á. Rocha et al. (Eds.): ICITS 2021, AISC 1330, pp. 431–441, 2021.
https://doi.org/10.1007/978-3-030-68285-9_41

applied scheme is known as Simplified Path Loss Model with Lognormal Shading [1]; nonetheless, its application suggests that measurements are normally distributed, which cannot be always guaranteed. Besides, it is possible that, due to shading, heteroskedasticity at different distances is presented, causing non-acceptable predictions. In that way, new regressors based on Machine Learning (ML) can be exploited to overcome this issue. Particularly, Support Vector Machines (SVM) [4] are a widespread set of algorithms, that deals with an optimization problem to evaluate regression and/or classification problems. A powerful feature of SVM is related to its low computational complexity, which makes them suitable to little databases and easiness of training and evaluation [5], against other ML algorithms. Thus, this work shows a comparison between a classical Simplified Path Loss Model against a regressor based on SVM. Preliminary conclusions indicate that the SVM model outperforms the simplified loss model.

The rest of this work is organized as follows: first, Sect. 2 presents a basic theoretical framework; then, Sect. 3 shows the previous work; after that, Sect. 4 describes the experimental design and the methodology; in Sect. 5 the experiment is performed and the models are fitted; later, in Sect. 6, analysis over the results is carried out. Finally, conclusions, future work, acknowledgments, and references are presented.

2 Theoretical Framework

The main idea when implementing Path Loss models is to estimate the reception power in function of different independent variables, such as carrier frequency and distance. This section shows basic concepts about both Simplified Path Loss and SVM models.

2.1 Simplified Path Loss Lognormal Shadow Fading Model

The model to be adjusted [1, 6] is presented in Eq. (1):

$$\frac{P_r}{P_t}(dB) = 10\log_{10} K - 10\gamma \log_{10}\left(\frac{d}{d_0}\right) - \psi_{dB} \quad (1)$$

Where P_r is the received power, P_t is the transmit power, K is a dimensionless constant that depends on the characteristics of the antennas and the channel average attenuation, d is the distance, d_O is the far-field distance, γ is the exponent of path loss and ψ_{dB} is a random variable from a lognormal distribution, which characterizes shadow fading, whose Probability Distribution Function (PDF) is presented in Eq. (2):

$$p(\psi) = \frac{\xi}{\sqrt{2\pi}\sigma_{dB}\psi} exp\left[-\frac{\left(10\log_{10}\psi - \mu_{\psi_{dB}}\right)^2}{2\sigma^2_{\psi_{dB}}}\right], \psi > 0 \quad (2)$$

Where $\xi = 10/\ln(10)$, σ_{dB} is the standard deviation of ψ_{dB}, and $\mu_{\psi_{dB}}$ is the mean of ψ_{dB}. The methodology for determining the parameters in (1) and (2) will be presented in the Sect. 5. The constant K can be determined from Eq. (3) [1]:

$$K = 20\log_{10}\left(\frac{\lambda}{4\pi d_0}\right) \tag{3}$$

2.2 Support Vector Machines (SVM)

The goal of this supervised learning strategy is to find one (or more) hyperplanes that separate previously tagged classes [7]. By finding the support vectors, it is possible to predict, in case of regression, the value of a new continuous variable based on a set of inputs. The objective is to find the hyperplane that separates two classes with the maximum margin between them, therefore, finding the model is reduced to an optimization problem. However, in many cases, the classes are not linearly separable, so, it is recommended to use kernels to perform a transformation that increases the number of dimensions, allowing separating such classes. To choose the kernel, linear, sigmoidal, and RBF types are evaluated. The tuning of hyperparameters is performed through cross-validation, and finally, the RMSE that exhibits the best results with both metrics will be chosen.

3 Previous Work

Different proposals are found in the literature for modeling losses and shadowing effects in wireless networks. The most general theoretical model is based on Maxwell's equations [8], which consists in closed and deterministic expressions to characterize wave phenomena such as diffraction, reflection, scattering, among others.

General Ray Tracing is the most known method that belongs to deterministic methodologies [9, 10]. It is commonly used to characterize losses and delay spread of any wireless transmission, taking into account geometry, dielectric properties, and distances [1]. Simplified methods use the direct line-of-sight ray and other reflected rays to understand how received signals are affected by propagation phenomena, and the most used are Two-Ray Model [11] and Ten-Ray Model (Dielectric Canyon) [9]. Because the nature of these models demands very specific environment features, it is difficult to be implemented for general scenarios.

To overcome the limitation exhibited by deterministic methods, empirical approximations have been developed to fulfill generality in a wider number of scenarios by using a set of extensive measurements. Thus, Okumura's model [12] is commonly used for urban environments in UHF from 1 km up to 100 km, and it is based on the median attenuation relative to free space loss in non-regular environments, considering typical path loss, gain factors of the base station and mobile antennas, gain associated to the environment and the median attenuation. It does not provide a closed-form expression, but it considers piecewise functions. Due to the standard deviation of the Okumura's model can be up to 14 dB, its accuracy is not the best. Another common model was proposed by Hata [13], and can also be applied in UHF, but it is simpler since it does provide a closed-form, depending on the same parameters of Okumura's model; besides, it offers corrections for suburban and rural environments. Nonetheless, since Hata works well for distances >1 km, it is not applicable for most modern wireless designs. For frequencies from 1.5 to 2 GHz, the COST 231 model [14] can be used as an extension of the Hata model. However, the most used scheme is known as the Simplified Path Loss Model, which optionally can include lognormal shadowing, and its theoretical framework was shown in Sect. 2.1. Explorations over this approximation show that, although this model is quite simple, its predictions are sometimes biased from field measurements (for this analysis, refer to Sect. 5).

4 Methodology and Experiment Design

To fit both chosen models, it is needed to implement a measurement phase, where the P_r/P_t ratio is measured 100 times, using the distance d as the independent variable. After that, a model fitting is performed in both cases, to obtain a closed form of them; and finally, an evaluation is carried out to see how the models' performance was. Subsequent subsections will show how each phase was executed.

4.1 Measurement Phase

An Access point (AP) WLAN (Wireless Local Area Network), brand Dragino, reference LG308, was used as a base station. A smartphone Xiaomi Redmi Note 8 Pro was linked to the access point. Furthermore, the smartphone was used to measure the reception power, by using the Android app Network Cell Info Lite. Both devices were fixed at different heights. The smartphone was shifted over distances in a range of one meter to ten meters, in increments of one meter, taking ten measurements for each point, to ensure repeatability of measurements. The WLAN channel and the frequency were recorded, and the following parameters were calculated: the wavelength λ, the far-field distance d_O (to meet the criteria for applying the simplified path loss model $d_O >> 2\ \lambda$ [2].

4.2 Model Fitting Phase

To fit both models, R Studio V 1.2.5019 software was used. In principle, a preliminary inspection was made through a path loss versus distance plot, and each point was statistically characterized, to observe the distribution in quartiles, means, medians, and standard deviations (represented in boxplot graphs), to understand the nature of the data. Because the simplified path loss model incorporates a lognormal term, it must meet the normality assumptions, this analysis was performed for each subset of measurement points, and the Shapiro Wilk normality test was applied [15] to clean the database of points that affect this assumption. Next, the database was divided into a training set equivalent to 70% of the total measurements, and a test set with the remainder [16]. Such subsets were chosen randomly.

Once the data preprocessing was performed, the curve fitting for the simplified path loss model [6] was executed, using the training set, and applying the least-squares strategy [17] to determine the value of the path loss exponent γ (see Sect. 2.1). Having guaranteed the normality of the data, the lognormal distributions for each of the subsets were determined (using the ten measurements to determine the mean and the variance). By determining the path loss component and shadow fading (random component) [18], the model can be evaluated, as will be discussed later.

In turn, a simple machine learning strategy was chosen to make the comparison of the simplified loss model, guaranteeing that the order of computational complexity was low, both from training and of the subsequent execution of the model. The strategy based on Support Vector Machines (SVM) is selected [4], given that its computational complexity is of the order of $O(n^2)$ [5], compared to other regression strategies. Different dimensionality transformations were evaluated for hyperplane separation using clusters: linear, sigmoidal, and radial base functions (RBF) [19]. Once the parameters of the chosen model were obtained, the support vectors were determined to carry out the respective evaluation.

4.3 Results Analysis Phase

After the models were fitted, the regression was carried out with both models, by using the test set. An initial graphical inspection was performed, and two performance metrics are used: RMSE (Root Mean Square Error) and R^2. Both metrics and the graphical approximation served to determine the accuracy of the models regarding field measurements. The pertinent analysis was developed.

5 Experiment Execution

We worked according to the parameters in Table 1. The distance in the far-field was 0.24 m away according to the wavelength, therefore, the measurements were started from distances from one up to ten meters. Line of sight was guaranteed; however, it was not possible to ensure there were no objects (such as walls, chairs, among others), which could generate shadow fading. The distance between the AP and the smartphone was adjusted using the Pythagorean theorem. Measurements are shown in Table 2, where d_g is the distance in the ground (in meters), P_r/P_t is the power ratio between the received one and transmitted one at the smartphone, and d is the corrected distance (in meters).

Table 1. Parameters used in the experiment.

Parameter	Value	Parameter	Value
Freq (Hz)	2.462×10^8	WLAN Channel	11
C (m/s)	3×10^8	AP height (m)	0.88
λ (m)	0.1218	Instrument height (m)	0.27
d_O (m)	0.2437	Relative height (m)	0.61

5.1 Model Fitting Phase

A preliminary graphical inspection of the data is performed, as shown in Fig. 1. According to the theoretical prediction, it would be expected that the trend would always be negative; however, it is noted that particularly for distances greater than six meters there is a behavior that does not follow a descending pattern, possibly caused by reflections over surrounding walls. It is not possible to attribute such behavior to the measurement errors in the instrument (smartphone), because, although there is dispersion in each of the distances, visually they seem to have similar standard deviations, except in $d = 8$ m, where a greater deviation is higher. Such facts can be seen in the boxplot diagram in Fig. 2.

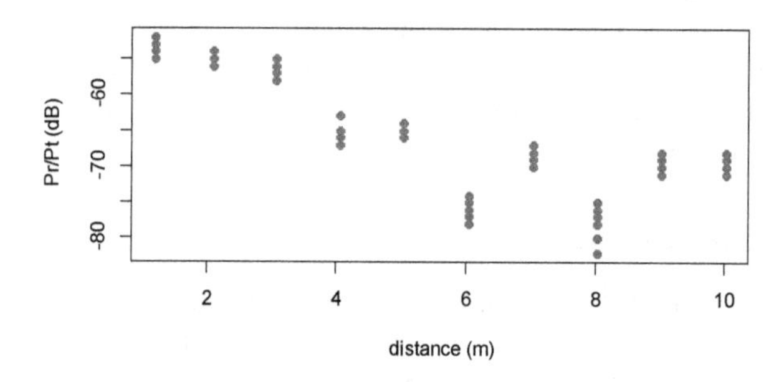

Fig. 1. P_r/P_t vs. Distance plot. Each distance d relates a set of measurements of P_r/P_t

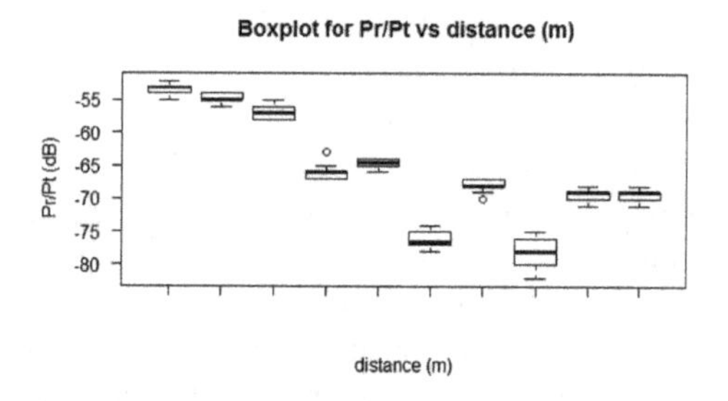

Fig. 2. Boxplot P_r/P_t vs. Distance plot. Each distance shows a boxplot to see how the variability of measurements is.

From Fig. 2. it can be noticed that in most cases problems of normality could appear, while skewness problems are observed qualitatively, since there is a concentration of the data, in some cases, in the first quartile ($d = 1, 4, 9, 10$), and the third quartile ($d = 2, 6, 7$). Outliers can also be observed at $d = 4, 7$. So, the diagram indicates that some normality tests must be performed before adjusting the lognormal shadow fading model, fulfilling this assumption. However, when observing the boxplot graph, it can be noted that the standard deviations are different, and thus, the PDFs will be different for each subset of data at each distance. This analysis will be carried out in the next section.

Simplified Path Loss Model Fitting

First, it is necessary to verify that, for each distance d, the set of measures of the power ratio is normal. The Shapiro Wilk test [15] is run, setting a significance value of 5%. In this test, the null hypothesis is normality and the alternative hypothesis is non-normality. In case the test statistic is less than 5%, the null hypothesis is rejected, indicating non-normality. Table 3 shows the results of the test. As predicted from the boxplot in Fig. 2 there are non-normal subsets, so we proceeded to remove them from the set.

Table 2. Measurement set. Each distance relates a power ratio P_r/P_t in function of the ground distance d_g and the corrected distance d.

d_g	P_r/P_t	d	d_g	P_r/P_t	D	d_g	P_r/P_t	d	d_g	P_r/P_t	d	d_g	P_r/P_t	d
1	−53	1.17	3	−57	3.06	5	−65	5.04	7	−68	7.03	9	−69	9.02
1	−52	1.17	3	−58	3.06	5	−64	5.04	7	−67	7.03	9	−70	9.02
1	−53	1.17	3	−56	3.06	5	−66	5.04	7	−68	7.03	9	−69	9.02
1	−53	1.17	3	−55	3.06	5	−65	5.04	7	−67	7.03	9	−70	9.02
1	−52	1.17	3	−56	3.06	5	−64	5.04	7	−70	7.03	9	−69	9.02
1	−54	1.17	3	−58	3.06	5	−64	5.04	7	−68	7.03	9	−69	9.02
1	−53	1.17	3	−57	3.06	5	−65	5.04	7	−67	7.03	9	−68	9.02
1	−54	1.17	3	−56	3.06	5	−64	5.04	7	−69	7.03	9	−68	9.02
1	−55	1.17	3	−58	3.06	5	−66	5.04	7	−68	7.03	9	−70	9.02
1	−53	1.17	3	−57	3.06	5	−64	5.04	7	−67	7.03	9	−71	9.02
2	−56	2.09	4	−66	4.05	6	−76	6.03	8	−75	8.02	10	−69	10.02
2	−55	2.09	4	−65	4.05	6	−77	6.03	8	−76	8.02	10	−70	10.02
2	−55	2.09	4	−67	4.05	6	−74	6.03	8	−75	8.02	10	−68	10.02
2	−55	2.09	4	−66	4.05	6	−77	6.03	8	−77	8.02	10	−69	10.02
2	−54	2.09	4	−67	4.05	6	−75	6.03	8	−78	8.02	10	−70	10.02
2	−54	2.09	4	−66	4.05	6	−75	6.03	8	−82	8.02	10	−69	10.02
2	−55	2.09	4	−67	4.05	6	−77	6.03	8	−80	8.02	10	−69	10.02
2	−54	2.09	4	−66	4.05	6	−76	6.03	8	−80	8.02	10	−70	10.02
2	−55	2.09	4	−67	4.05	6	−78	6.03	8	−78	8.02	10	−69	10.02
2	−55	2.09	4	−63	4.05	6	−77	6.03	8	−80	8.02	10	−71	10.02

The values to determine K (Eq. 3) were documented in Table 1, except for d_O, which is set at one meter, obtaining $K = -40.26$. Besides, it is necessary to determine the optimal value of the path loss exponent γ, which is obtained minimizing the expression in Eq. (1) with the least-squares method, according to Eqs. (4) and (5):

$$F(\gamma) = \sum\nolimits_{i=1 i\neq 2,4,5,7}^{10} \sum\nolimits_{j=1}^{10} \left[M_{measured_j}(d_i) - M_{model}(d_i)\right]^2 \quad (4)$$

$$F(\gamma) = \sum\nolimits_{i=1 i\neq 2,4,5,7}^{10} \sum\nolimits_{j=1}^{10} \left[M_{measured_j}(d_i) - K - 10\log_{10}\left(\frac{d_i}{d_0}\right)\gamma\right]^2 \quad (5)$$

Table 3. Shapiro Wilk normality test results. For each distance, a set of power measurements indicates if they are normally distributed.

d (m)	p-value	Conclusion	*d* (m)	p-value	Conclusion
1	0.14	Normality	6	0.39	Normality
2	0.012	Non-normality	7	0.032	Non-normality
3	0.19	Normality	8	0.465	Normality
4	0.0068	Non-normality	9	0.286	Normality
5	0.0084	Non-normality	10	0.17	Normality

For simplicity, R is used to perform the reduction of the quadratic polynomial as a function of γ, applying the *rSymPy* package. The polynomial obtained, using the data in Table 2 (filtered according to its normality), is shown in Eq. (6):

$$F(\gamma) = 34774.63 - 18305.18\gamma + 2562.36\gamma^2 \tag{6}$$

By differentiating Eq. (6) with respect to γ and equaling 0, the exponent exhibits a value $\gamma = 3.57$. This value agrees with the empirical data presented by various authors [1, 20, 21], according to the environment scenario where the measurements were taken.

To obtain the complete model, including the random component of the lognormal distribution, the mean and the standard deviation are analyzed at each distance, and thus determine the PDFs, as shown in Eq. (2). The values found per distance are shown in Table 4.

Table 4. Parameters of lognormal PDFs $p\ (\psi)$ according to ground distance d. Means μ and standard deviations σ are calculated to determine the random component of shadow fading in the Simplified Path Loss Model

d (m)	μ (dB)	σ (dB)	*d* (m)	μ (dB)	σ (dB)
1	−56	0.91	8	−78.1	2.37
3	−56.8	1.03	9	−69.3	0.94
6	−76.2	1.23	10	−69.4	0.84

So, the model in Eq. (1) is parameterized in an experiment as shown in Eq. (6):

$$\frac{P_r}{P_t}(dB) = -40.26 - 35.7 \log_{10} d - \psi_{dB} \tag{7}$$

Support Vector Machine-Based Regressor Fitting

Because the previous model requires normality in the data subsets and taking into account that four of the ten subsets had to be removed, a regression model is applied through SVM [4]. In this case, R library *e1021* was used for setting a generic SVM, and the method *tune* was deployed to generate a grid with different Cost *C* and kernel size values, thus determining the best values through cross validation.

For the specification of the model with the radial base kernel [22], the cost C, the kernel width γ, and the support vectors are required. In this case, the hyperparameters C and γ took values of 5 and 5, in such a way that the RMSE was the least, through the cross-validation process. Based on the advantages exhibited by the radial-based kernel, it is possible to map the data in the input space to a new space with infinite dimensions (if necessary).

6 Results and Discussion

A visual inspection of the performance can be observed in Fig. 3. To assess the quality of the fitted models, both RMSE and R^2 will be evaluated for the test set, as well as a plot that will show measurements and model predictions. For the path loss model, the RMSE found was 6.498, and the R^2 was 0.722, and for the SVM regressor, the RMSE was 1.71, and the R^2 was 0.977. In the case of the simplified path loss model, the expression used is shown in Eq. (7), using the lognormal parameters shown in Table 4. To find the value of γ, a random value is obtained (mapped from a uniform distribution), which will simulate a random probability that it will be an input argument to the lognormal inverse PDF. In this way, it will be possible to perform the regression and calculate the errors. In the case of the SVM regressor, it is possible to use all the database measurements to carry out the training, since the subsets of measurements for each distance must not meet the normality assumption. It can be seen, from both the graph and the metrics, that the fitting was better by using the SVM strategy than by using the simplified path loss model.

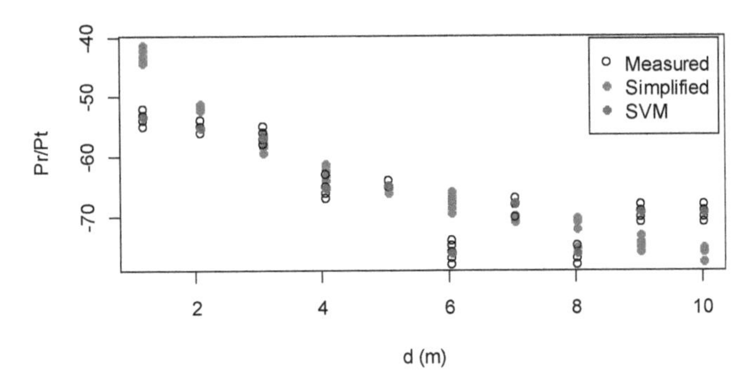

Fig. 3. Performance of simplified path loss model and SVM regressor vs. Measured data

Regarding the performance of the path loss model, it can be caused by: *i)* the inaccuracy of the measuring instrument (smartphone); nonetheless, it is noted that it is repeatable since there is a concentration of the data (low standard deviations, as shown in Table 4, besides, the pattern of accuracy errors did not follow an additive or multiplicative regimen (like other measurement instruments) [23], so it is possible that replacing the instrument does not eliminate errors; *ii)* constructive interference by multipath (floor, walls, and ceiling); *iii)* violation of the assumption of normality in four of the ten subsets of measurement set, causing the reduction of samples in the database, which could affect the quality of the model fit; and/or *iv)* heteroskedasticity in the standard deviations of the remaining subsets causes the term ψ of Eq. (1) to be mapped to different probability density distributions, which causes the regression values at certain points (for example in d = 8) are more deviated and induce greater errors.

On the other hand, it is important to note that although the SVM regressor provides better results, the model could not be extrapolated to other scenarios, so the training of the algorithm should include other measurements to fulfill those restrictions.

7 Conclusions

In the present work, a field comparison was developed to fit a simplified path loss model with lognormal shadow fading, as well as an alternative model using Support Vector Machines (SVM), to determine the reception power in a WLAN network. First, measurements were taken whose independent variable was the distance between an AP and a smartphone, and the dependent one the power ratio P_r/P_t. Subsequently, the respective models were adjusted, specifying the needed parameters in each case. Finally, the analysis of results was performed, concluding that the simplified model exhibited large errors, which could be caused by: constructive multipath interference, the reduction of the samples in the database (due to the violation of the normality assumption), or by heteroskedasticity in the subsets corresponding to each distance. On the other hand, the SVM model exhibited better results; however, the generalization of the model cannot be achieved with the hyperparameters found, if the application scenario is not similar. A research opportunity would be to fit the model with an additional categorical variable that indicates the measurement scenario, making enough measurements so that the database becomes more representative.

8 Future Work and Acknowledgments

A similar experiment will be carried out in a LoRaWAN Internet of Things Network. We would like to thank the University of Medellín for financing the Ph.D. studies of one of the authors, as well as the project: “Esquema algorítmico para la transmisión de información que minimice el consumo energético y mejore el uso del espectro radio-eléctrico en dispositivos de Internet de las Cosas”, code 1039.

References

1. Goldsmith, A.: Wireless Communications. Cambridge University Press, Cambridge (2005)
2. Erricolo, D., Uslenghi, P.L.: Propagation path loss-a comparison between ray-tracing approach and empirical models. IEEE Trans. Antennas Propag. **50**(5), 766–768 (2002)
3. Singh Y. Comparison of okumura, hata and cost-231 models on the basis of path loss and signal strength. International journal of computer applications. 2012;59(11).
4. Noble, W.S.: What is a support vector machine? Nat. Biotechnol. **24**(12), 1565–1567 (2006)

5. Abdiansah, A., Wardoyo, R.: Time complexity analysis of support vector machines (SVM) in LibSVM. Int. J. Comput. Appl. **128**(3), 28–34 (2015)
6. Robertson, I., Somjit, N., Chongcheawchamnan, M.: Microwave and Millimetre-Wave Design for Wireless Communications. John Wiley & Sons, Hoboken (2016)
7. Fine, S., Scheinberg, K.: Efficient SVM training using low-rank kernel representations. J. Mach. Learn. Res. **2**(Dec), 243–264 (2001)
8. Sharma, P.K., Singh, R.: Comparative analysis of propagation path loss models with field measured data. Int. J. Eng. Sci. Technol. **2**(6), 2008–2013 (2010)
9. Green, D., Yun, Z., Iskander, M.F.: Path loss characteristics in urban environments using ray-tracing methods. IEEE Antennas Wirel. Propag. Lett. **16**, 3063–3066 (2017)
10. Athanasiadou, G.E., Tsoulos, G.V.: Path loss characteristics for UAV-to-ground wireless channels. In: 2019 13th European Conference on Antennas and Propagation (EuCAP), pp. 1–14. IEEE (2019)
11. Cheffena, M., Mohamed, M.: Empirical path loss models for wireless sensor network deployment in snowy environments. IEEE Antennas Wirel. Propag. Lett. **16**, 2877–2880 (2017)
12. Okumura, Y.: Field strength and its variability in VHF and UHF land-mobile radio service. Rev. Electr. Commun. Lab. **16**, 825–873 (1968)
13. Hata, M.: Empirical formula for propagation loss in land mobile radio services. IEEE Trans. Veh. Technol. **29**(3), 317–325 (1980)
14. Damosso, E., Correia, L.: Urban transmission loss models for mobile radio in the 900 and 1800 mhz bands. The Hague (1991)
15. Royston, P.: Approximating the Shapiro-Wilk W-test for non-normality. Stat. Comput. **2**(3), 117–119 (1992)
16. Liu, H., Cocea, M.: Semi-random partitioning of data into training and test sets in granular computing context. Granular Comput. **2**(4), 357–386 (2017)
17. Burden, R.L., Faires, J.D.: Numerical Analysis. Brooks, Cole (1997)
18. Panic, S., Stefanovic, M., Anastasov, J., Spalevic, P.: Fading and Interference Mitigation in Wireless Communications. CRC Press, Boca Raton (2013)
19. Deng, N., Tian, Y., Zhang, C.: Support Vector Machines: Optimization Based Theory, Algorithms, and Extensions. CRC Press, Boca Raton (2012)
20. Perez-Vega, C., Garcia, J.L.G.: A simple approach to a statistical path loss model for indoor communications. In: 1997 27th European Microwave Conference, pp. 617–623. IEEE (1997)
21. Srinivasa, S., Haenggi, M.: Path loss exponent estimation in large wireless networks. In: 2009 Information Theory and Applications Workshop, pp. 124-129. IEEE (2009)
22. Kavzoglu, T., Colkesen, I.: A kernel functions analysis for support vector machines for land cover classification. Int. J. Appl. Earth Obs. Geoinf. **11**(5), 352–359 (2009)
23. Vasilevskyi, O.M., Kucheruk, V.Y., Bogachuk, V.V., Gromaszek, K., Wójcik, W., Smailova, S., et al.: The method of translation additive and multiplicative error in the instrumental component of the measurement uncertainty. In: Photonics Applications in Astronomy, Communications, Industry, and High-Energy Physics Experiments 2016, p. 1003127. International Society for Optics and Photonics (2016)

Opportunistic Networks with Messages Tracking

Jorge Herrera-Tapia[1](✉), Jefferson Rodríguez[1], Enrique Hernández-Orallo[2], Leonardo Chancay-García[3], Juan Sendón-Varela[1], and Pietro Manzoni[2]

[1] Universidad Laica Eloy Alfaro de Manabí, Manta, Ecuador
{jorge.herrera,e1313689265,juan.sendon}@uleam.edu.ec
[2] Universitat Politècnica de València, Valencia, Spain
{ehernandez,pmanzoni}@disca.upv.es
[3] Universidad Técnica de Manabí, Portoviejo, Ecuador
leonardo.chancay@utm.edu.ec

Abstract. Sharing information or simply keeping in touch has become necessary, usually through smartphone messaging applications. However, using these applications is difficult in places where there is no telecommunications infrastructure or transmission channels are saturated. In this context, the use of opportunistic networks is a feasible alternative to provide communication. This type of network takes advantage of the opportunity for contact between mobile devices, as long as users are willing to collaborate. In this article we present a disconnection tolerant messaging application, based on opportunistic networks, using WiFi-Direct, that send/receive text and multimedia messages, allowing the track of messages. The evaluation was carried out considering the transmission time for each type of message in line-of-sight (LOS) and non-LOS (NLOS) scenarios between the devices. The results show the feasibility of using opportunistic networks in smart devices.

Keywords: Opportunistic networks · Vehicular networks · Vanets · Epidemic protocol · Delay tolerant networking · Wireless ad hoc networks

1 Introduction

Instant messaging systems, due to their versatility, are the most used application for communication between people through the Internet. Unfortunately, there are scenarios where this communication is not possible or this data transmission infrastructure does not work properly, as in the case of places that have been impacted by a natural disaster, or in certain events where there is overcrowding of people and the telecommunications infrastructure is saturated. The messages that people send vary according to the circumstances, and can be short text messages, concerning a request, alert or suggestion, as well as photos and videos that show certain situations, for example, a fair or concert.

Á. Rocha et al. (Eds.): ICITS 2021, AISC 1330, pp. 442–451, 2021.
https://doi.org/10.1007/978-3-030-68285-9_42

In this paper, we present the design of a disconnection-tolerant messaging application for Android devices, capable of keeping messages in memory if the devices lose connection, and that tries to send the information using intermediary devices until the messages arrive at their destinations. We evaluate the efficiency of this approach using WiFi in an open place without obstacles and in a closed place with obstacles ar. To determine the feasibility of communication through this type of wireless networks, the dissemination of messages was analyzed using The ONE (The Opportunistic Network Environment) simulator [1].

Our proposed communication system is based on opportunistic networks. The authors of [2], in their work, explain in detail the operation of this type of networks, based on the opportunity to transmit information when devices establish contact with each other. The evaluation of the proposal has been carried out considering the data transmission-time for messages of different size and distance between the mobile devices (nodes) that act as sender and receiver.

The outline of this paper is as follows: an overview of related works about opportunistic networks and message diffusion is presented in Sect. 2, the proposal is explained in Sect. 3, the evaluation and findings of the experiments are presented in Sect. 4. Finally, Sect. 5 contains some conclusions and future work.

2 Related Work

The opportunistic networks for their diffusion model can be considered as a subclass of DTN (Delay Tolerant Networks) [3,4]. DTNs are based on the principle of *store, transport and forward* the messages that the nodes receive when they come into contact with each other.

The performance of opportunistic networks depends on the number, mobility, and level of collaboration of mobile devices users, which determines the duration of contact between the nodes. This contact duration, as well as the type of radio or interface wireless used in the nodes, affects the amount of data transmitted, The authors of [5–8], analyzed the diffusion of messages considering the contact duration according to the user's attitude of collaboration or friendship. To examine the performance of opportunistic networks using simulations, the authors of [9] explain in-depth how to run and analyze this type of networks using different tools. Regarding the application of opportunistic networks, the authors of [10,11], expose the scenarios where their deployment would be feasible. The experiments were conducted in real settings over several weeks, establishing the term *social distance*.

Many protocols have been proposed for the dissemination of information in opportunistic networks.

Among these protocols, the most widely used is the Epidemic, which resembles its dissemination of messages to a viral disease epidemic. Authors such as [12,13] evaluate the performance of this algorithm, which has become the basis for other routing algorithms.

Summing up, the aforementioned research looks to opportunistic networks as a viable alternative for the transmission of messages in places where users of mobile devices do not have access to the Internet. The proposal that we present makes use of the advantages of opportunistic networks combined with the properties of WiFi and GPS sensor that cellular devices have in order to provide instant messaging applications.

3 Proposal

In this section we detail the most relevant aspects of our application (app), named *ImmediateChat*.

3.1 Diffusion Model

Considering the operation of opportunistic networks, we have designed and implemented a mobile-based messaging system that supports their disconnection, without losing the messages that users wanted to send. These messages are re-transmitted when there is a direct opportunity to establish a connection again or using intermediate nodes, who are willing to collaborate with the re-transmission of the messages, producing the epidemic diffusion of the messages.

Based on the transmission properties of WiFi, especially the transmission range, our messaging app will use this wireless standard but in ad-hoc mode through WiFi-Direct, thus expanding the connection range and contact time between devices. Wi-Fi Direct is an standard for peer-to-peer wireless connections that allows two devices to establish a direct Wi-Fi connection without an intermediary wireless access point, router, or Internet connection.

Epidemic routing is one of the most efficient ways used for broadcasting messages. Authors of [14], state that the operation of this routing is based on the fact that a node sends the same data packet to all other nodes connected to the same network. This packet will be available to be sent to any new device that connects, so this device will disseminate the packet to other devices that establish a future connection (see Fig. 1). Only the destination node can access this information unless the sending user indicates that the message is for everyone. In the example in Fig. 1 we can see that $D1$ is the sender of the message, with destination $D3$, the same one that has used other devices as a means of propagation until reaching its destination. The app also saves the GPS position when forwarding the message, allowing messages tracking.

The following software tools and technical considerations were used to implement the functional prototype of the system: a) Operating system: Android from version 4.4 on-wards. b) IDE: Android studio 3.0. c) Database: SQLite. d) Connection method: WiFi-Direct. e) Type of dissemination: Epidemic.

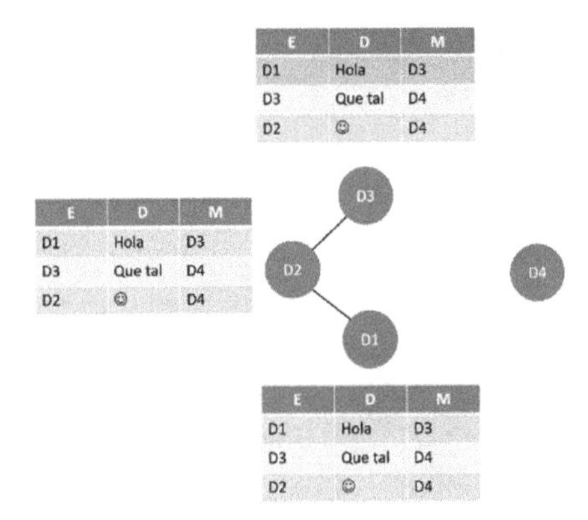

Fig. 1. Epidemic diffusion of a message to a specific user. Messages in the buffer of the app (E: Emisor (Source), D: Destination, M: Message).

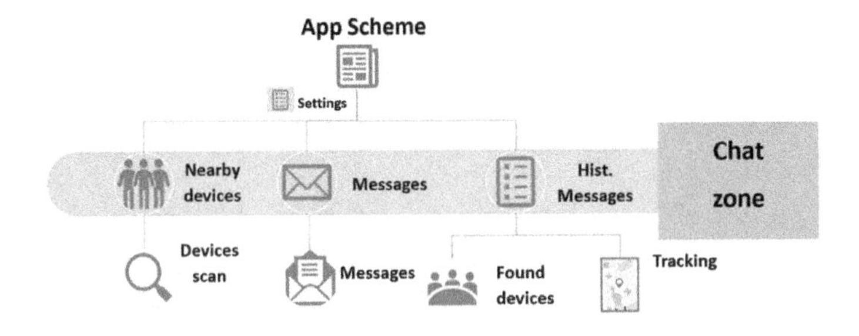

Fig. 2. Application operating scheme.

3.2 How Does the Application Work?

Some of the application's functionality is shown in Fig. 2, such as device settings, discovery devices, and message management. Once the application is installed, the *Nickname* of the user is written, and the search for other devices begins. A list of nearby devices appears, (see Fig. 3). A connection request is sent to a remote *device*, and if it is accepted, the connection is established.

The *remote* device receives from the *requestor* all the messages that it has stored in its memory, but the owner of the device will only be able to see those messages which are destined to the owner. The device is always in the mode of discovering other nearby devices to transmit the messages. The *remote* device sends all message packets whenever it connects to other *devices*. Finally, the devices delete all messages that have expired from their archive based on the TTL (Time to Life) configuration.

Figure 4a shows a *chat* between two users. The application can also be used to send files of different types such as audio, drawings, images, and videos. Figure 4b displays the date, hour, and GPS position where the message was forwarded, just touching over the position indicator.

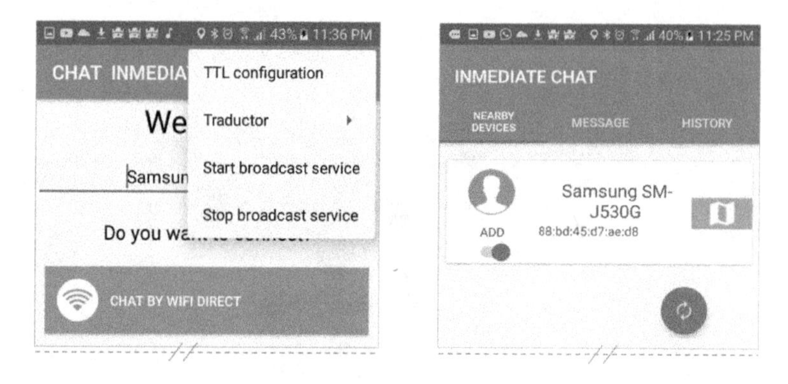

Fig. 3. Setting and discovering devices in the communication range.

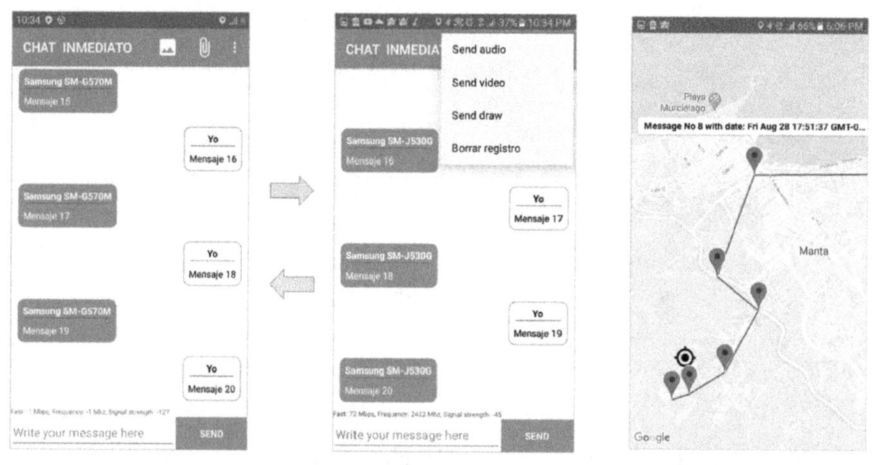

Users chatting. Message tracking.

Fig. 4. Sending and message tracking.

4 Evaluation

Firstly, we tested the functionality of the application, reaching the objective of managing messages when mobile devices do not Internet connection. Messages are stored in memory until the mobile comes into contact with other devices to follow the sending process. In addition, it was verified that the nodes help in the dissemination of the messages, even if they are not the final destination.

Apart from verifying the operation of the application, we proceeded to measure the transfer-time of different files considering its size and type. Average values of 16 measurements were obtained from each transmission, in an open environment (outdoor) with *Line of Sight* (LOS) between the mobiles, and within a building with obstacles, that is with *Non Line of Sight* (NLOS) between the devices (indoor). WiFi-Direct was used in the 2.4 GHz frequency, with Samsung smartphones.

The results obtained in open field (outdoor LOS) are showed in Table 1. It is evident that the larger the file size and distance between nodes, the longer the transfer time. As you can see, the range with WiFi-Direct with line of sight exceeded 300 meters (m). Transmitting a simple text (32 bytes) to 18 m takes 42 milliseconds (ms) (it is almost instantaneous). Likewise, we see that an 8KB file (low-resolution image) or audio takes a little more than 500 ms. Finally, transferring a short video file takes almost 10 s at a distance of 306 m. Based on the values obtained from these real measurements (Table 1), we empirically derived the following expression 1:

$$t = (398 + x) * \frac{d}{183 * (10^n)} \tag{1}$$

Where t is the transfer time, x is the size in bytes of the file, 398 is a metadata constant, 183 is smoothing constant, d is the distance between devices. Finally, the value of n varies according to the size of the file to send as follows: $n = 0 : 1$B – 1kB, $n = 1 : 1$kB – 100kB, $n = 2 : 100$kB – 1MB.

Figure 5, shows the values when the formula is applied for different types of messages. In the results we can notice another important aspect of wireless networks: due to their nature, it is difficult to control the interference of some invisible factors. For example, in the graph, we can see that at 144 m away transmission-time goes down when it is supposed to follow an ascending pattern.

Table 1. Real measurements in milliseconds (LOS).

Type	Text	File	Drawing	Audio	Image	Video
Dist (m)	32B	215B	8kB	10kB	312kB	570kB
18	42	70	100	108	324	645
36	98	100	140	212	749	1183
54	130	190	260	314	1023	1893
72	196	205	342	402	1229	2130
90	204	280	422	498	1536	2805
108	376	324	540	616	1905	3462
144	338	444	679	876	2644	4472
162	394	583	765	910	2564	3618
180	454	503	900	1001	3124	5794
198	490	679	987	1027	3678	6172
216	524	752	992	1345	3687	6453
234	567	777	1098	1452	3995	7722
252	600	893	1192	1486	4302	8165
270	612	908	1200	1578	4722	8176
288	645	985	1392	1669	5006	8899
306	714	1046	1720	1818	6593	9782

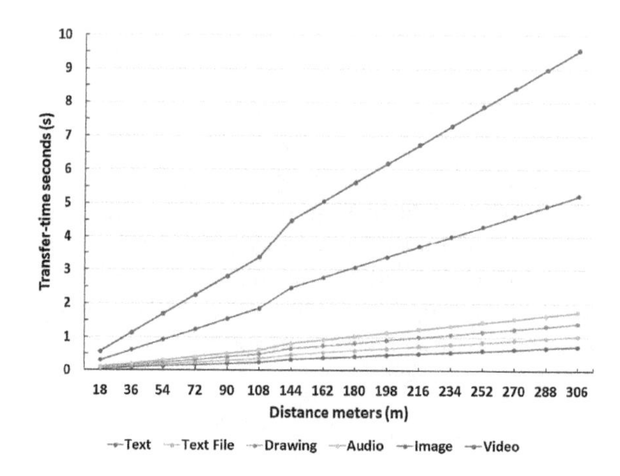

Fig. 5. File transfer-time, using Formula 1. LOS measurements.

And this irregularity is observed in the rest of the measurements with more or less incidence.

Regarding the measurements in an indoor environment (NLOS), the walls were considered as obstacles between mobile devices. The first obstacle was a cement block wall (22 cm thick), with a distance of 3 m between the devices, after 2 walls were considered at a distance of 5 m between the devices, and finally 9 m with 3 walls between the mobiles. In addition to the walls, there was an average of 14 people in these areas. Figure 6 shows the results. Obviously, the obstacles interfere with wireless communication, for example, transmitting a message of 20kB with a wall in the middle takes about 0.5 s, while with 3 *obstacles* they exceed 12 s.

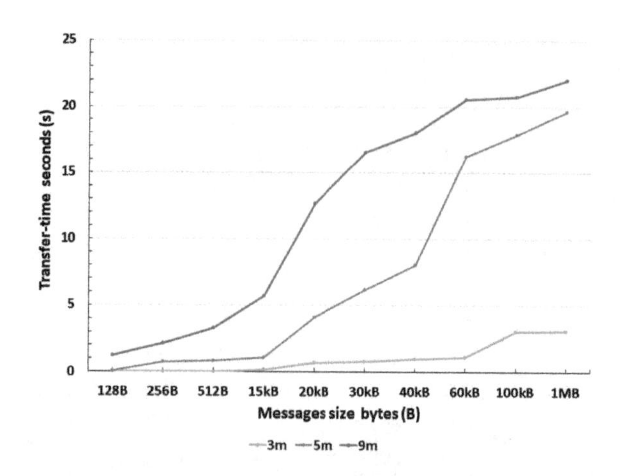

Fig. 6. Transfer-time of different types of files. NLOS measurements.

Table 2. Simulation parameters.

Parameter	Value
No. users	25, 50, 100
Range Tx	30 m
Bandwidth	2 Mbps
Protocol	Epidemic
Message sizes	32B, 215B, (8, 10, 312, 570)kB, (1, 2, 4, 6, 8, 10)MB

To evaluate the feasibility of our messaging application regarding the dissemination of messages, we proceeded to carry out simulations using The ONE (The Opportunistic Network Environment) simulator. The main simulation parameters are shown in Table 2. The scenario is a city with a simulation time of 6 h, with a user speed between 0.25 m/s–1.5 m/s. The messages are generated with a frequency between 1–3 min. Furthermore, due to the current memory capacity of phones, the buffer size is considered unlimited.

According to the specifications, the WiFi-Direct transmission range is 100 m and bandwidth of 54 Mbps. But, in this case, to obtain more realistic results, and according to the experiments described, we are going to consider lower values in the simulation: a range of 30 m and a bandwidth of 2 Mbps, to take into account possible interference effects and obstacles. In order to know the degree and time of dissemination of the messages during the simulation process, we will vary the number of users and the size of the messages that are disseminated on the network. We explain the results below.

The delivery probability depending on the size of the generated messages is shown in Fig. 7. For 100 users or nodes participating in the network, the delivery possibility is close to 85% for small messages (up to 10kB). As the size

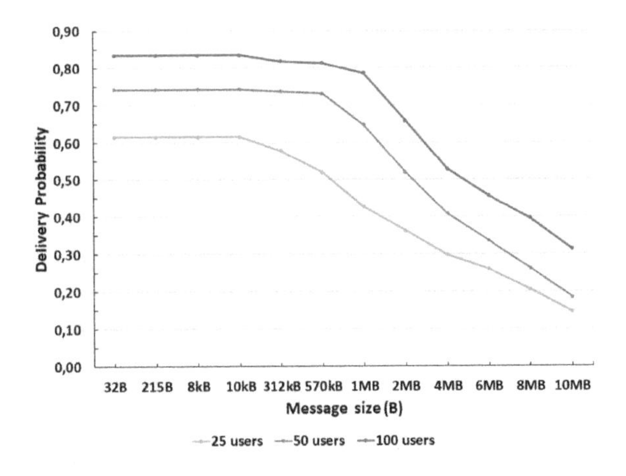

Fig. 7. Delivery probability of messages according to their size and number of users.

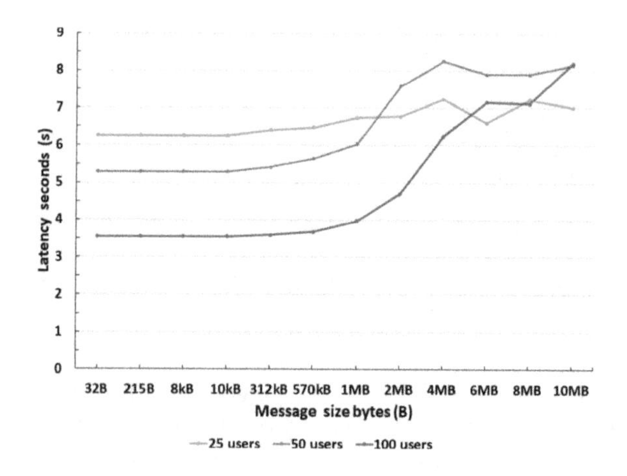

Fig. 8. Average delivery time (seconds) of messages (Latency), according to their size and number of users.

of the messages increases, the probability decreases, reaching almost 30%. This is because the contact time between nodes is not long enough to complete the transfer of large messages. This drop effect is similar in the cases of 50 and 25 nodes, with a difference of approximately 10% between them.

Regarding the delivery time of the messages, the results are shown in Fig. 8. The effect of the number of users and the size of the messages is opposite to the probability of delivery. It can be seen that with a smaller number of nodes the delivery time increases, and the effect is maintained when the size of the messages increases. For example, for a short text message, it takes around 6300 s to be disseminated on the network, that is, more than 1.75 h, while for 100 users, it is distributed at approximately 3600 s (1 h).

5 Conclusions and Future Work

In this work, we have introduced an instant messaging application, which supports temporary disconnection between devices, based on the opportunistic network model. According to the performed tests and evaluations, it is evident that the distance and the setting (obstacles) impact the transmission capacity of the information, and that there are fluctuations during the transmission process. The results of the simulations show that the number of users and the size of the messages affect the dissemination of information.

Summing up, the use of opportunistic networks is feasible, becoming a viable alternative to share information through messages in places where there is not telecommunications and Internet infrastructure. For future work, we are planning to implement a system that uses the different interfaces of a mobile device depending on where the user is. Additionally, we will analyze memory management mechanisms to optimize message delivery.

References

1. Keränen, A., Ott, J., Kärkkäinen, T.: The ONE simulator for DTN protocol evaluation. Proceedings of the Second International ICST Conference on Simulation Tools and Techniques (2009)
2. Pelusi, L., Passarella, A., Conti, M.: Opportunistic networking: data forwarding in disconnected mobile ad hoc networks, pp. 134–141 (2006)
3. Cha, S., Talipov, E., Cha, H.: Data delivery scheme for intermittently connected mobile sensor networks. Comput. Commun. **36**, 504–519 (2013)
4. Ito, M., Nishiyama, H., Kato, N.: A novel communication mode selection technique for DTN over MANET architecture. In: 2014 International Conference on Computing, Networking and Communications (ICNC), pp. 551–555 (2014)
5. Herrera-Tapia, J., Hernández-Orallo, E., Tomás, A., Manzoni, P., Tavares Calafate, C., Cano, J.C.: Friendly-sharing: improving the performance of city sensoring through contact-based messaging applications. Sensors **16**(9), 1523 (2016)
6. Thilakarathna, K., Viana, A.C., Seneviratne, A., Petander, H.: Mobile social networking through friend-to-friend opportunistic content dissemination. In: Proceedings of the Fourteenth ACM International Symposium on Mobile ad Hoc Networking and Computing - MobiHoc 2013, p. 263 (2013)
7. Chancay-Garcia, L., Herrera-Tapia, J., Manzoni, P., Hernandez-Orallo, E., Calafate, C.T., Cano, J.C.: Evaluation of routing protocols for opportunistic networks in scenarios with high degree of people renewal, pp. 228–235, May 2018
8. Herrera-Tapia, J., Hernandez-Orallo, E., Manzoni, P., Tomas, A., Calafate, C.T., Cano, J.C.: Evaluating the impact of data transfer time and mobility patterns in opportunistic networks. In: 2016 Intl IEEE Conferences on Ubiquitous Intelligence & Computing, Advanced and Trusted Computing, Scalable Computing and Communications, Cloud and Big Data Computing, Internet of People, and Smart World Congress (UIC/ATC/ScalCom/CBDCom/IoP/SmartWorld), pp. 25–32 (2016)
9. JDede, J., Förster, A., Hernández-Orallo, E., Herrera-Tapia, J., Kuladinithi, K., Kuppusamy, V., Manzoni, P., bin Muslim, A., Udugama, A., Vatandas, Z.: Simulating opportunistic networks: survey and future directions. IEEE Commun. Surv. Tutor. **20**(2), 1547–1573 (2018)
10. Cabero, J.M., Molina, V., Urteaga, I., Liberal, F., Martín, J.L.: Acquisition of human traces with Bluetooth technology: challenges and proposals. Ad Hoc Netw. **12**, 2–16 (2014)
11. Förster, A., Garg, K., Nguyen, H.A., Giordano, S.: On context awareness and social distance in human mobility traces categories and subject descriptors, pp. 5–12 (2012)
12. Herrera-Tapia, J., Manzoni, P., Calafate, C.T., Cano, J.C.: Selecting the optimal buffer management for opportunistic networks both in pedestrian and vehicular contexts. In: 14th IEEE Annual Consumer Communications Networking Conference (CCNC 2017), pp. 395–400 (2017)
13. Hernández-Orallo, E., Herrera-Tapia, J., Cano, J.C., Calafate, C.T., Manzoni, P.: Evaluating the impact of data transfer time in contact-based messaging applications. IEEE Commun. Lett. **19**, 1814–1817 (2015)
14. Boldrini, C., Conti, M., Passarella, A.: Modelling data dissemination in opportunistic networks. In: Proceedings of the Third ACM Workshop on Challenged Networks, pp. 89–96 (2008)

Metamaterial-Based Energy Harvesting for Wi-Fi Frequency Bands

Sandra Costanzo[1,2,3,4(✉)] and Francesca Venneri[1,4]

[1] University of Calabria, 87036 Rende, CS, Italy
{costanzo,venneri}@dimes.unical.it
[2] CNR – Institute for Electromagnetic Sensing of the Environment (IREA), 80124 Naples, Italy
[3] ICEmB, Inter-University National Research Center on Interactions Between Electromagnetic Fields and Biosystems, 16145 Genova, Italy
[4] CNIT, Consorzio Nazionale Interuniversitario per le Telecomunicazioni, 43124 Parma, Italy

Abstract. A dual-band metamaterial cell, based on the use of the Minkowski fractal geometry, is proposed for ambient power harvesting applications within the Wi-Fi bands. The simulated fractal-based cell offers very high absorption peaks and quite good efficiencies in correspondence of 2.45 GHz and 5 GHz frequencies. Thanks to its appealing features, the proposed metamaterial configuration could be very attractive for the implementation of high efficiencies and environmentally friendly harvesting systems for mobile devices/sensors and low power portable electronics.

Keywords: Energy harvesting · IoT · WSN · Metamaterials · Wi-fi

1 Introduction

Nowadays, the increasing demand for mobile devices and portable/wearable electronics has led to strong interests in finding solutions to meet their energy needs. At the same time, Internet-of-Things (IoT) applications and wireless sensor networks (WSNs) are becoming increasingly pervasive in our daily lives, requiring the deployment of a large number of sensor nodes and low-power wireless devices. The major challenges facing both IoT as well as WSNs are related to the limited and non-renewable energy supply of deployed sensors and wireless devices.

Up now, conventional batteries represented the major energy source for portable devices and wireless sensors. However, this approach suffers from several intrinsic drawbacks, namely limited lifespan and fixed energy rate of conventional batteries, maintenance costs and pollution issues associated with periodic battery replacement.

A promising solution to overcome the above issues could be the adoption and the implementation of energy harvesting systems [1].

Energy harvesting is essentially a conversion process from different free energy sources into electrical energy. In the last decade, the energy harvesting technology has attracted huge attention, due to its ability to produce electricity from various environmentally friendly and free energy sources, such as thermal, solar, wind, kinetic, and radio frequency (RF).

Á. Rocha et al. (Eds.): ICITS 2021, AISC 1330, pp. 452–457, 2021.
https://doi.org/10.1007/978-3-030-68285-9_43

In particular, RF energy harvesting [2] is very attractive for low power devices and consumer electronics. Nowadays, several RF broadcasting infrastructures could provide RF energy, such as AM/FM radio, analog/digital TV, GSM and Wi-Fi networks. The rapid increase in the number of RF emitters, due to the development of new wireless technologies, offers the opportunity to exploit the unused RF energy, available in the surrounding environment, for efficiently powering small electrical devices and sensors.

In this work, a dual-band metamaterial-based harvesting system, operating within the Wi-Fi frequency bands (i.e. 2.45 GHz and 5 GHz), is presented. The unit cell consists of two pairs of miniaturized Minkowski fractal elements printed on a thin-grounded dielectric substrate. The fractal shape, already adopted by the authors for reflectarray antennas design [3, 4] and the realization of metamaterial absorbers operating within the UHF-RFID band [5, 6], allows to achieve very small resonators with respect to standard geometries. Furthermore, as demonstrated in the following, it can be fruitfully exploited to obtain multiband operation skills, thus allowing to simultaneously harvest electrical energy from different RF-sources.

A preliminary analysis of the proposed metamaterial-based harvester is presented and discussed in the following sections.

Very high absorption percentages and good unit cell efficiency are demonstrated in correspondence of the Wi-Fi frequencies (i.e. 2.45 GHz and 5 GHz). Thanks to the above features, the proposed metamaterial configuration could be very attractive for the implementation of energy harvesting systems for mobile devices/sensors, within the Wi-Fi frequency bands.

2 Metamaterials for RF-Energy Harvesting

A conventional RF harvesting system (Fig. 1) is composed by a receiving antenna and a rectifying circuit (i.e. a rectenna), which is able to harvest high-frequency energy into free space and convert it to DC power.

The antenna is the key element of an RF-energy harvesting system, able to efficiently capture and convert the incident RF energy into AC power, by conveying it to the rectifier input for the final AC-to-DC power conversion stage (Fig. 1).

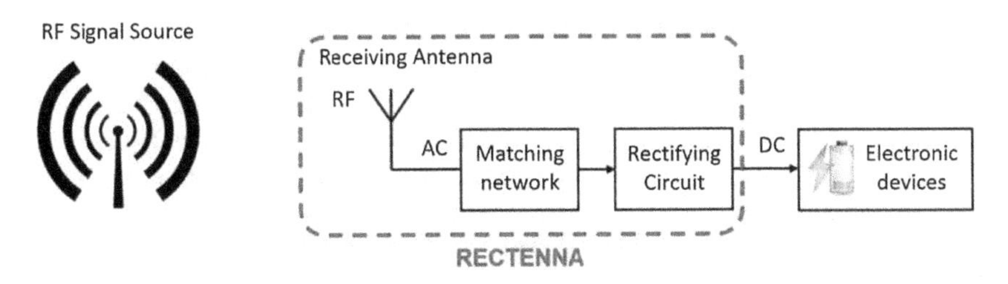

Fig. 1. Schematic diagram of a basic RF harvesting system.

In order to increase the rectifier's sensitivity (i.e. the ability to operate at low power density), some researchers have designed rectenna arrays [7].

Recently, metamaterial structures have been also investigated as a promising alternative to conventional rectennas, with the key advantage of higher efficiencies in the preliminary conversion stage from RF-to-AC power [8–10]. Metamaterial harvesters comprises an array of electrically small metallic resonators printed on a grounded dielectric substrate (Fig. 2). Each resonator effectively couples to the incident electromagnetic (EM) wave at the resonance, thus capturing the EM power from the ambient. Similarly to metamaterial absorbers, the metallic pattern must be designed to fulfill the perfect absorption condition (i.e. $\Gamma \cong 0$) around a given frequency and/or a frequency band. Afterwards, most of the energy captured by each metamaterial element is channeled through one or more vias, which deliver the collected AC power to a rectification circuit [11–13].

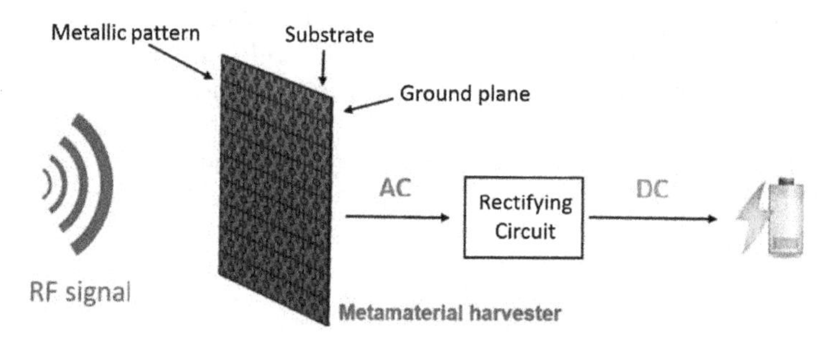

Fig. 2. Schematic diagram of a metamaterial harvester.

3 Metamaterial Unit Cell Layout and Design

The dual band metamaterial unit cell proposed in this paper for energy harvesting applications is depicted in Fig. 3. It consists of two pairs of Minkowski fractal metallic resonators printed on a thin grounded dielectric slab. A via hole is properly integrated to each resonator, in order to channel the captured RF signal to a resistor load, representing the downstream rectification circuit. The proposed cell is an extension of the single frequency configuration, already proposed by the authors in [10] for ambient power harvesting at 2.45 GHz.

The metamaterial cell is synthesized for perfect absorption, namely to perform the matching between unit cell and free space impedances at the two design frequencies $f_1 = 2.45$ GHz and $f_2 = 5$ GHz, namely $Z_{cell}(f_1) \cong Z_{cell}(f_2) \cong \zeta_0 = 376\Omega$ (i.e. $\Gamma(f_1) \cong \Gamma(f_2) \cong 0$). Both patch lengths ($L_1$ and L_2) and inset size (S_1L_1 and S_2L_2) are accurately chosen to satisfy the above condition. Furthermore, the vias positions and the resistive value of the two loads are properly fixed to maximize the percentage rate of the absorbed RF energy.

Following the synthesis rules outlined in [5], the cell is designed to achieve two absorption peaks in correspondence of about $f_1 = 2.45$ GHz and $f_2 = 5$ GHz. At this scope, a commercial full-wave code (Ansoft Designer), based on the periodic infinite array approach, is adopted.

The cell size is fixed to $0.23\lambda \times 0.23\lambda$ at f_1. In order to minimize dielectric losses, a Rogers TMM10i dielectric substrate is adopted, having a dielectric constant $\varepsilon_r = 9.8$, a loss tangent equal to 0.002 and a thickness h = 1.524 mm. The resistive loads are properly fixed to the following values: $R_1 = 80\ \Omega$ and $R_2 = 350\ \Omega$. The sizes of the Minkowski patches are equal to (L_1 - S_1) = (14.2 mm - 0.28) and (L_2 - S_2) = (8.6 mm - 0.1).

As it can be observed in Fig. 4(a), where the simulated absorption coefficient of the cell (i.e. $A(f) = 1-|\Gamma(f)|^2$) is reported, two absorption peaks, equal about to 99.5%, are achieved at the two design frequencies.

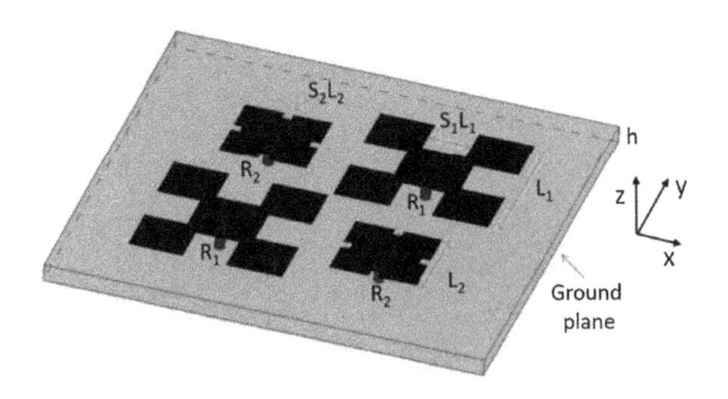

Fig. 3. Metamaterial unit cell for power harvesting.

Finally, a comprehensive analysis of unit cell loss balance is performed, by adopting the full-wave frequency domain solver CST Microwave Studio. The structure is excited by a Floquet port giving a plane wave normally incident along the z direction (Fig. 3), with the incident power set to a value equal to $P_{inc} = 0.5$ W.

The RF-to-AC efficiency of the cell ($\eta_{RF\text{-}to\text{-}AC}$) is evaluated as the ratio between the total power delivered to the loads (P_{del_loads}) and the power incident onto the metamaterial surface (P_{inc}) [10]. Fig. 4(b) illustrates the amount of incident power which is delivered to the loads, corresponding to an efficiency respectively equal to 87% @ 2.45 GHz and 84% @ 5 GHz. Only a smaller amount of power is dissipated within the copper and the dielectric.

In conclusion, the above results make the proposed structure very appealing for energy harvesting application within the Wi-Fi frequency bands.

As future development, the proposed fractal structure will be optimized to improve the RF-to-AC efficiency of the cell. Furthermore, the high versatility of the proposed fractal structure will be exploited to achieve a broadband behavior, by properly tailoring the separation distances between the MA absorption peaks.

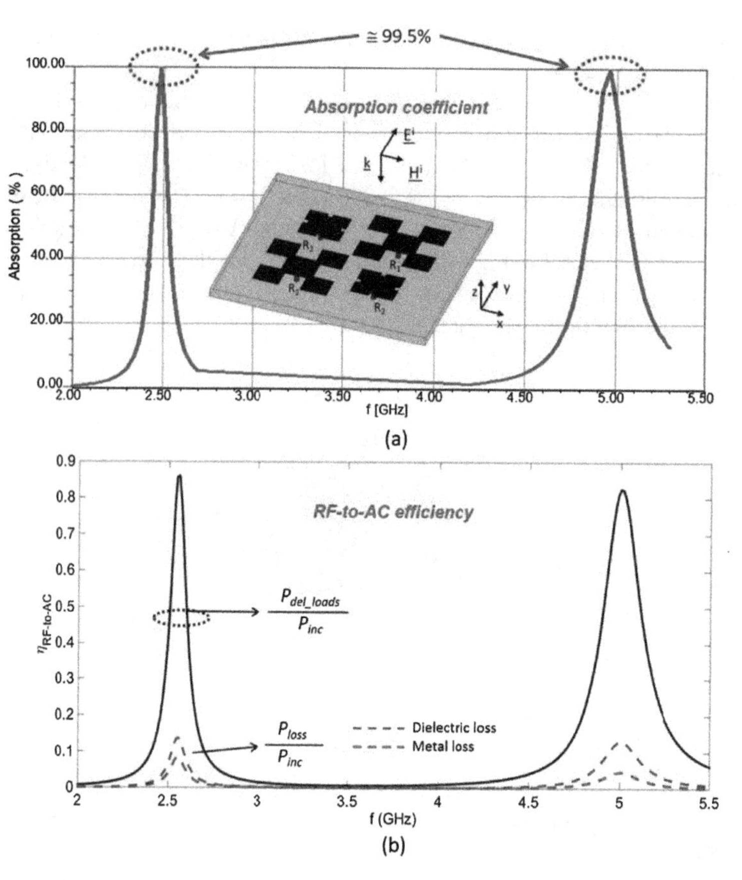

Fig. 4. Simulation of the designed metamaterial unit cell: (a) absorption coefficient; (b) RF-to-AC efficiency.

4 Conclusion

Power supply challenges for IoT systems, WSNs and wearable electronics applications have been discussed. The energy harvesting concept has been addressed as the most promising solution to overcome the power supply issues of sensors and wireless devices. In particular, RF-energy harvesting has been explored for low power devices and for consumer electronics.

Metamaterial-based harvesting systems have been proposed as a promising alternative to conventional rectennas, with the key advantage of higher efficiencies.

Finally, the potentialities of a fractal-based metamaterial harvester have been illustrated, for operations within the Wi-Fi band. A high versatility has been demonstrated in achieving high absorption percentages and good unit cell efficiencies, also showing the ability of the proposed metamaterial cell to offer a multi-band behavior.

References

1. Patel, A.C., Vaghela, M.P., Bajwa, H., Patra P.K.: Power harvesting for low power wireless sensor network. In: 2009 Loughborough Antennas & Propagation Conference, Loughborough, pp. 633–636 (2009)
2. Kim, S., Vyas, R., Bito, J., Niotaki, K., Collado, A., Georgiadis, A., Tentzeris, M.M.: Ambient RF energy-harvesting technologies for self-sustainable standalone wireless sensor platforms. Proc. IEEE **102**, 11 (2014)
3. Costanzo, S., Venneri, F.: Miniaturized fractal reflectarray element using fixed-size patch. IEEE Antennas Wirel. Propag. Lett. **13**, 1437–1440 (2014)
4. Costanzo, S., Venneri, F., Di Massa, G., Borgia, A., Costanzo, A., Raffo, A.: Fractal reflectarray antennas: state of art and new opportunities. Int. J. Antennas Propag. Article ID 7165143 (2016). https://doi.org/10.1155/2016/7165143.
5. Venneri, F., Costanzo, S., Di Massa, G.: Fractal-shaped metamaterial absorbers for multireflections mitigation in the UHF band. IEEE Antennas and Wirel. Propag. Lett. **17**(2), 255–258 (2018)
6. Venneri, F., Costanzo, S., Borgia, A.: A dual-band compact metamaterial absorber with fractal geometry. Electronics **8**, 879 (2019). https://doi.org/10.3390/electronics8080879
7. Assimonis, S.D., et al.: Sensitive and efficient RF harvesting supply for batteryless backscatter sensor networks. IEEE Trans. Microw. Theory Tech. **64**(4), 1327–1338 (2016)
8. Ramahi, O.M., Almoneef, T.S., AlShareef, M., Boybay, M.S.: Metamaterial particles for electromagnetic energy harvesting. Appl. Phys. Lett. **101**, 173903 (2012)
9. Alavikia, B., Almoneef, T.S., Ramahi, O.M.: Electromagnetic energy harvesting using complementary split-ring resonators. Appl. Phys. Lett. **104**, 163903 (2014)
10. Costanzo, S., Venneri, F.: Polarization-insensitive fractal metamaterial surface for energy harvesting in IoT applications. Electronics **9**, 959 (2020)
11. El Badawe, M., Almoneef, T.S., Ramahi, O.M.: A metasurface for conversion of electromagnetic radiation to DC. AIP Adv. **7**, 035112 (2017)
12. Ghaderi, B., Nayyeri, V., Soleimani, M., Ramahi, O.M.: Pixelated metasurface for dual-band and multi-polarization electromagnetic energy harvesting. Sci. Rep. **8**, Article Number 13227 (2018)
13. Almoneef, T.S., Erkmen, F., Ramahi, O.M.: Harvesting the energy of multi-polarized electromagnetic waves. Sci. Rep. **7**, 14656 (2017)

Ethics, Computers and Security

Risk Analysis and Android Application Penetration Testing Based on OWASP 2016

Thomás Borja(✉), Marco E. Benalcázar, Ángel Leonardo Valdivieso Caraguay, and Lorena Isabel Barona López

Departamento de Informática y Ciencias de La Computación (DICC), Escuela Politécnica Nacional, Ladrón de Guevara E11-253, Quito, Ecuador
{thomas.borja,marco.benalcazar,angel.valdivieso, lorena.barona}@epn.edu.ec

Abstract. Mobile Applications have become part of our daily lives so that almost every web or desktop application can be executed from a smartphone, i.e., social networking, Internet shopping, on-line banking, gaming applications, among others. Furthermore, most of the existing mobile applications in digital stores are Android-based applications. Security in these kinds of applications is an issue that must be addressed because they handle sensitive personal information exposed to be exploited or misused by malicious agents. In this context, we have performed a complete security penetration testing on several Android applications following the most common risks according to OWASP mobile 2016 and using different tools such as Drozer, Dex2jar, Android Debug Bridge, among others. We describe the vulnerability, type of attack, application analyzed, and external tools used for each scenario. Once the vulnerabilities are exposed, we show a summary of the performed attacks, a risk analysis, and provide security recommendations for each layout. This work's novelty is the provisioning of a risk matrix that resumes each attack's main points and the whole vulnerability analysis in mobile devices, as it does not exist on the official site of OWASP.

Keywords: Information security · Mobile application penetration testing · OWASP mobile 2016

1 Introduction

The mobile application market has grown significantly in the last decade, where the number of mobile users has surpassed by eight times the number of fixed Internet users, according to the Ericsson Mobility Report [1]. In this landscape, Android is the preferred operative system for developers to program their mobile applications or apps since Android is the most used platform around the world, followed by Apple [2]. As the smartphone capabilities are evolving in terms of processing, storage, displays, cameras, and more, new applications such as high definition streaming, on-line gaming, or virtual reality apps begin to gain popularity. As a result, the app market grows, and it will keep growing in the next decade. Information security is one of the main concerns

Á. Rocha et al. (Eds.): ICITS 2021, AISC 1330, pp. 461–478, 2021.
https://doi.org/10.1007/978-3-030-68285-9_44

among mobile users as their apps have to deal with personally sensitive information that can be disclosed by malicious agents and could harm end-user devices, identities, as well as companies' trust and reputation.

Security flaws in mobile applications could be platform or application-specific; other problems include server-side, client-side, and network-related attacks. There are code-level bugs and misconfigurations, application data that can be hijacked while in transit or at rest, and others. Throughout the study presented here, it can be found that many security problems are existing in many mobile applications, and it is necessary to have tests that let us evaluate the robustness of the apps against these problems. Although penetration tests are not the ultimate tool to prevent security problems, they help detect and discover security weaknesses existing in the applications.

In this context, the OWASP Foundation has gathered many of these approaches and the most common vulnerabilities in applications. As a result, OWASP produced the top ten mobile risks in its last version of 2016 [3] as a starting guide for developers and security researchers to test their mobile software. Although the list of the ten risks is explained in the subsequent sections, the vulnerabilities are categorized as Insecure transmission of data, Insecure data storage, Lack of binary protections, Client-side vulnerabilities, Hard-coded passwords, and Leakage of private information [2].

Some works analyze mobile security from different approaches such as OWASP vulnerabilities [4], provisioning of some security guidelines from the software development perspective [5], and partial risk analysis of the OWASP list [4, 6–8]. Although these works cover some of the mobile security concerns, they do not consider the whole list of OWASP risks, or they only analyze these risks with a previous version of the OWASP list. Furthermore, some researchers do not provide further analysis of the vulnerabilities, a risk matrix, or they do not detail the process of security penetration testing.

Therefore, it is necessary for software developers and security experts to know which are the most common, widespread, and dangerous vulnerabilities and their risks in mobile applications and a detailed process of penetration testing. For this reason, we present a study that covers all these aspects and the missing concerns in the literature reviewed. This paper presents a complete penetration testing of all risks of the last OWASP list. Also, we give a risk matrix that summarizes the problems. This risk analysis may help other researchers to focus on the vulnerabilities that should be fixed and avoided in the future.

The rest of this document is structured as follows: In Sect. 2, we present a literature review concerning mobile application security, a quick review of some Android components, and the OWASP list. Then, Sect. 3 shows each of the scenarios created and how penetration testing was performed. In Sect. 4, we summarize each layout's relevant elements and make a security analysis of the used applications. Finally, Sect. 5 displays the conclusions of the present research.

2 Background and Related Work

This section describes the elements required to understand the penetration tests carried out in this research and their related risks. We also provide a quick review of the last version of the OWASP mobile list (2016). We finally present a complete literature review related to mobile application pentesting and its relation to OWASP mobile risks.

2.1 The Android Operating System

Android is an open-source Linux-based operating system for mobile devices such as smartphones, tablet computers, smartwatches, and recently, cars and televisions. It was initially developed by the Open Handset Alliance and acquired by Google after. Android OS is Linux-based, and it can be programmed in C/C++, but most of the application development is done in Java or Kotlin [2]. Statista [9] reports that Android has occupied more than 80% of the global market share since 2014, and it will keep growing in the upcoming years. Statista presents the market share for mobile operative systems for recent years and a prediction to 2023, as shown in Fig. 1.

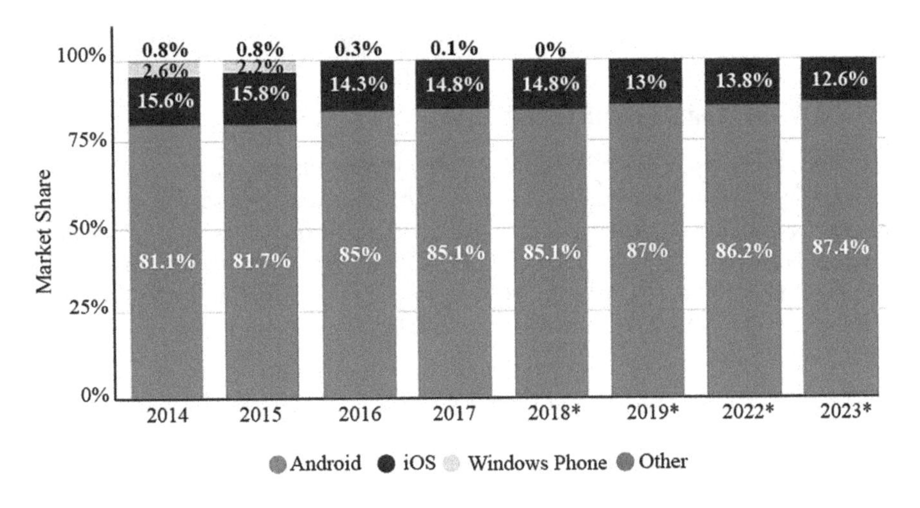

Fig. 1. Global smartphone shipments by operating system from 2014 to 2023 (Statista.com)

Android has an absolute market occupancy (81,1%) followed by iOS, with approximately 15%, and other operative systems (Fig. 1). Since applications are platform-dependent, vulnerabilities are categorized by the operating system. At the moment of writing this paper, 2563 vulnerabilities are counted for the Google Android platform, as reported by the CVE site [10]. Figure 2 shows the Android vulnerabilities classified by type and year.

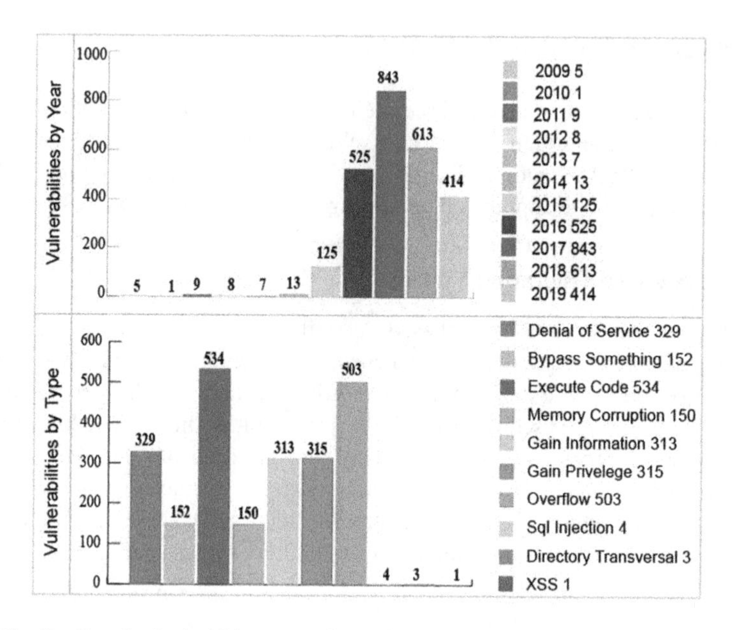

Fig. 2. Google Android known vulnerabilities count (Source: cvedetails.com)

2.2 OWASP Top 10 mobile Risks 2016

The Open Web Application Security Project (OWASP) is a non-profit organization founded in 2001 to help others design, operate, and maintain secure software. The project was initially released to help to secure web applications. With the emergence of new applications, they also released the mobile version containing the ten most common mobile risks. The present work covers all risks displayed in the OWASP list in its latest version of 2016, summarized in Table 1.

Table 1. OWASP top 10 mobile risks 2016

No	Risks
M1	Improper Platform Usage
M2	Insecure Data Storage
M3	Insecure Communication
M4	Insecure Authentication
M5	Insufficient Cryptography
M6	Insecure Authorization
M7	Client Code Quality
M8	Code Tampering
M9	Reverse Engineering
M10	Extraneous Functionality

2.3 Literature Review

In order to analyze the security of Android applications, a literature review has been done. The following topics were considered: Android Application Security with OWASP Mobile Top 10 2014 [4], A Conceptual Exploration for the Safe Development of Mobile Devices Software Based on OWASP [5], Mobile Application Security Penetration Testing Based on OWASP [8], and Security Analysis Protocol for Android-Based Mobile Applications [11].

In [4] a security penetration testing is performed in several Android Applications following the previous OWASP list (2014). It is worth mentioning that since 2014 some of the OWASP risks were updated, deleted, or moved within the list, and it is necessary to make new penetration tests according to the latest list of 2016. Moreover, it does not cover up all ten risks. In [5], OWASP risks from a development perspective are studied. This work analyzes software development practices that should be followed to create secure mobile software products. It provides some code obfuscation techniques, authentication, password-session management, and other recommendations. However, the applications are not tested until they are released.

The research presented in [11] shows a security protocol to perform penetration testing on Android-based devices. Besides, some techniques and tools for information gathering, static analysis, and dynamic analysis are also given. However, this work only provides a kind of protocol with the steps that should be followed to test applications, and it does not consider the test penetration. For its part, in [8], OWASP risks are studied in its latest version, but they are not subjected to a detailed review. Summarizing, Table 2 highlights some of the main aspects covered in each paper used for this literature review.

Table 2. Literature review

Item	King [4]	Gil [5]	Alanda [8]	Palacios [11]
Pentesting	✓	–	✓	–
Latest OWASP version	–	✓	✓	✓
Full coverage of risks	–	–	✓	–
Development perspective	–	✓	–	–
Risk analysis	✓	–	–	–

As shown in Table 2, none of these works provide a complete analysis considering the OWASP list. For this reason, our research proposes different scenarios to do penetration testing for each risk considering the most recent OWASP list. Furthermore, new applications, tools, and methodologies are used. This paper's main contribution is further analysis and a scored risk matrix for each list item.

3 Implementation

This section covers the elements used to perform the security analysis, the applications tested, and how the scenarios were created and exploited. In this section, we briefly resume the applications, tools, and penetration tests that have been carried out. This section also includes a summary of the scenarios.

3.1 Applications and External Tools

Four applications and six tools were mainly utilized in the elaboration of this security analysis. Table 3 shows a summary of the applications, the security used tools, and a description of each one.

Table 3. Applications and security testing used tools

	Name	Description
Applications	Android-InsecureBankv2	It is a banking training application with a back-end server component written in python. It allows users to create accounts, make transfers, and check status balances [12]
	FourGoats	It simulates a social network application. Users of this application are allowed to add friends, make check-ins at locations visited, and claim special rewards for social activity [13]
	HerdFinancial	It is another banking-like training application working together with a Java-based back-end server. It also allows account management and transactions [13]
	Damn Insecure and Vulnerable Application (DIVA)	This application was made specifically for security testing purposes. Some of the issues that can be found are insecure loggings, insecure data storage, hardcoding issues, among others [14]
Testing tools	Android Debug Bridge (ADB)	It is a command-line tool that allows communication between real or emulated mobiles and PC hosts. It helps perform actions such as installing apps, pushing/pulling files, managing activities, and others [15]
	Androwarn	It is a static code analyzer for Android applications. It provides plain text reports containing permissions, activities, application information, among others [16]
	Burp Suite	It is a desktop application containing a port scanner, a customizable proxy, and other features to help explore security flaws [17]
	Dex2Jar	It converts Dalvik Executables (.dex) to class files (.jar) [18]
	Drozer	It is a security testing framework for Android that searches for security vulnerabilities by interacting with Dalvik MV, IPCs, and the OS [19]
	Java Decompiler	JD-GUI is a graphical interface that displays Java source codes of.class files into readable structured code [20]

3.2 Security Testing Android Applications

M1 – Improper Platform Usage. This risk refers to the inadequate usage of mobile platforms such as Android or iOS due to a lack of controls or weak security policies related to these platforms. One example of how this risk can be exploited is by sniffing intents of the applications. Intents are used to change from one activity to another, where the activities represent a unique graphical interface of the application. In order to test this vulnerability, an audit tool called Androwarn is used. Androwarn shows the activities that are used by the app, so with a simple command from the Android Debug Bridge Console, we can jump into any desired activity without performing a login process. Figure 3 shows the activities found in the Android-InsecureBankv2 application (left) and the result of abusing Improper Platform Usage (right).

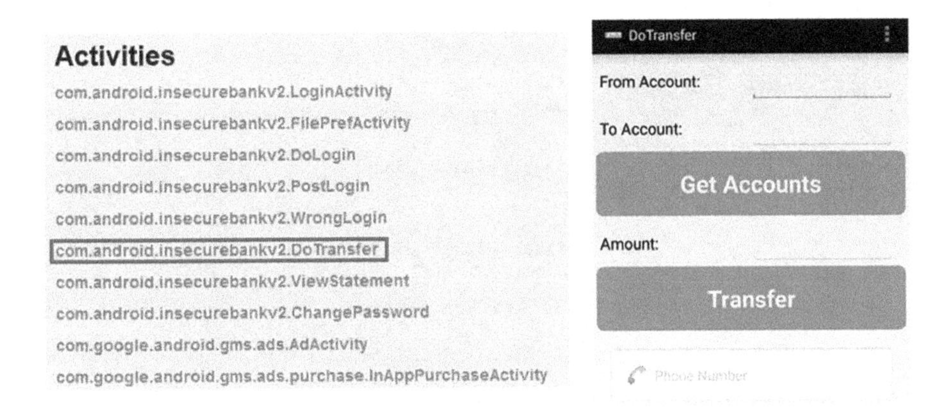

Fig. 3. M1 – Improper Platform Usage

M2 – Insecure Data Storage. Insecure Data Storage refers to a non-authorized access to sensitive user information that resides in the mobile file system. Information can be leaked through identity theft, tampered applications, man in the middle attacks, among others. After gaining access to the device, attackers may snoop around the file system and catch data that holds personal information linked to any application. Our test illustrates how an unprotected database of the FourGoats application can be retrieved using a couple of ADB commands. Figure 4 shows part of the check-in history of a given user of this application.

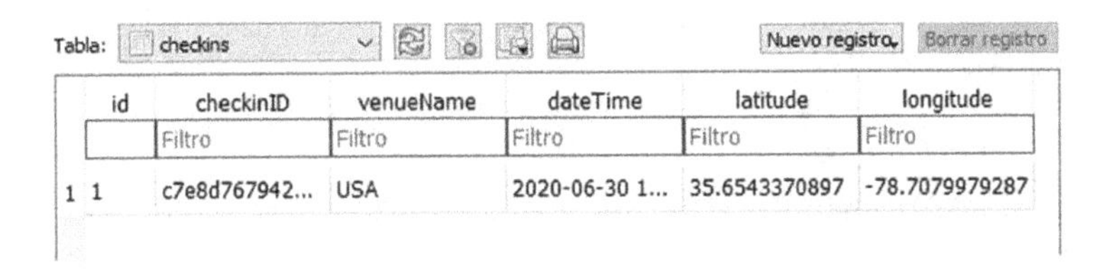

Fig. 4. M2 – Insecure Data Storage

M3 – Insecure Communication. Applications are always communicating and exchanging information to operate correctly, and many of them do it in a client-server fashion. When apps communicate, they use a network (Wi-Fi, LTE, USB) potentially exposed to intruders that may modify, steal, or disrupt normal data flow. In order to test M3, a proxy listener has been set up between the client of the FourGoats application and its corresponding server using Burp Suite. Figure 5 shows the result of having an Insecure Communication where the proxy catches personal information after a new user has been registered in the FourGoats app.

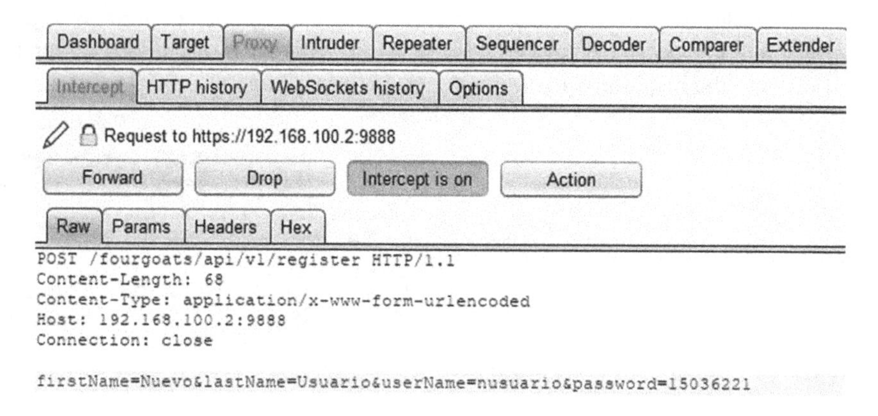

Fig. 5. M3 – Insecure Communication

M4 – Insecure Authentication. Many mobile applications perform an authentication process before users can execute any action within the app. Applications must identify users and securely keep sessions to prevent an anonymous user perform non-authorized actions that may harm legitimate users or the system itself. Insecure Authentication has been tested using the HerdFinancial app, where users may select the option "Authorize

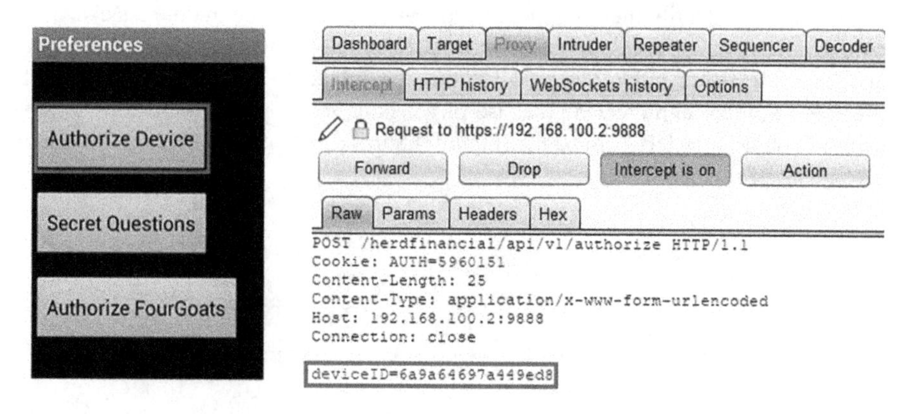

Fig. 6. M4 – Insecure Authentication

Device" so they do not have to authenticate every time they use the app. Figure 6 shows how this feature is exploited by sniffing the deviceID, a unique physical identification for every phone.

M5 – Insufficient Cryptography. Cryptography is a central issue in any security branch because it keeps information hidden or unintelligible to unauthorized persons. M5 refers to the necessity of implementing protection mechanisms of data through secure schemes of encryption. The mobile applications must be secured while traversing a communication network or residing in the phone's file system. For this example, we have decompiled the HerdFinancial application using Dex2Jar and Java Decompiler. A hard-coded string "hammer" was found and turned out to be an encryption key that can be used to unencrypt a protected database of this banking application. Figure 7 shows a decryption tool used to change to clear text the userinfo database of the HerdFinancial application.

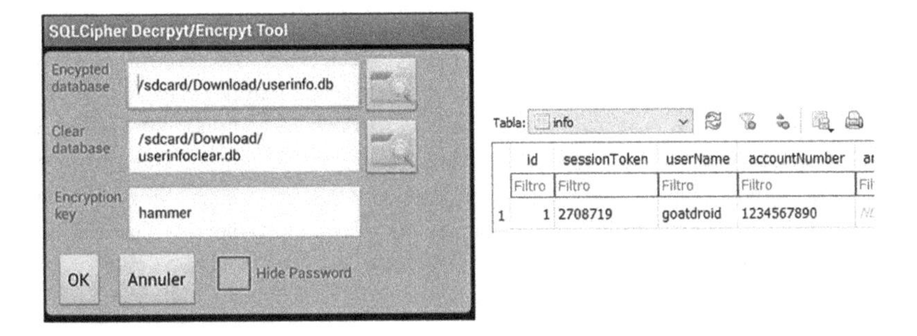

Fig. 7. M5 – Insufficient Cryptography

M6 – Insecure Authorization. While authentication refers to how a system validates the identification (a user is whom they claim to be), authorization allows users to perform an action after they have already been identified. A process of Insecure Authorization appears when a given user executes actions that are exclusively assigned to other roles (administrative tasks). For this experiment, a username of the GoatDroid application has been modified using the Burp Suite proxy into another username that belongs to an administrative role (androidguy93). With this change, an average user could perform non allowed tasks such as see other user check-ins in this application. Figure 8 shows the check-in history of a regular user "iLov3Cyan" (left) and the check-in history of an administrator "androidguy93" (right).

Fig. 8. M6 – Insecure Authorization

M7 – Client Code Quality. Historically, code injection has been a recurrent problem in software applications, so that it has remained in the top 1 of the web risks of OWASP since its first version. Mobile applications may be vulnerable to code injection as well as other code-level weaknesses. That is why a low code quality appears as the seventh most common risk of the OWASP mobile list. In order to test M7, the DIVA application has been used. After decompiling and analyzing the code of this app, an inappropriate SQL sentence may be found. If we use the correct statement, some confidential login information will be returned. Figure 9 shows one of the many strings that can be run to exploit Client Code Quality, and the message returned on the same screen.

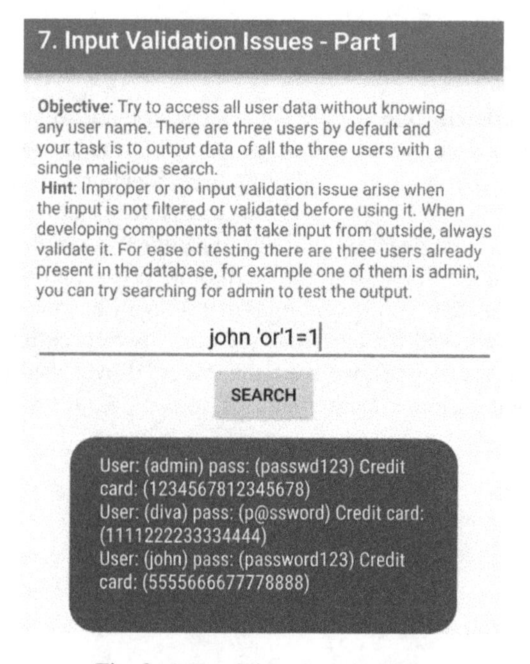

Fig. 9. M7 – Client Code Quality

M8 – Code Tampering. Application data can be statically analyzed and modified to execute some actions that were not intended for the original application. Attackers may inject malicious code into applications and make them behave as they wish. After applications are modified, they can be uploaded to digital stores offering some benefits such as a free version of a paid app to make it attractive. In order to test M8, the InsecureBankv2 application has been used. Using a tool called Metasploit, a malicious payload has been inserted into this app to create InsecureBankv3. InsecureBankv3 is the modification of the original app but containing a trojan that can steal personal information such as the user's contact list. The final application will look exactly like the original, but the personal data is stolen from a remote console behind when it is used. Figure 10 shows a dump of the contact list after the tampered application has been used.

```
/usr/local/bin/contacts_dump_20200707180043.txt - Mousepad
File  Edit  Search  View  Document  Help
Warning, you are using the root account, you may harm your system.

=====================
[+] Contacts list dump
=====================

Date: 2020-07-07 18:00:43.307009765 -0400
OS: Android 8.0.0 - Linux 4.4.34-genymotion (i686)
Remote IP: 192.168.100.2
Remote Port: 52542

#1
Name    : Thomás Borja
Number  : (099) 359-6889
Email   : xtratech1406@hotmail.com
```

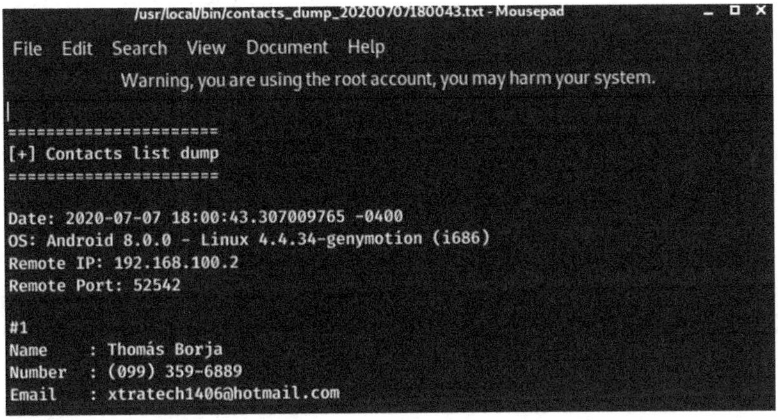

Fig. 10. M8 – Code Tampering

M9 – Reverse Engineering. Reverse Engineering refers to the process of using the binary code of an application to study its libraries, algorithms, and any other resource that may lead to the execution of actions that are generally not allowed to run. Using reverse engineering, an attacker may reveal general information about the app, server connection, cryptographic resources, execute non-authorized actions, or use the existing code to modify and rebuild the app. In this case, the FourGoats application has been used. The application is decompiled using Dex2Jar, Java Decompiler, and analyzed using the Drozer Tool. The Drozer tool reveals that this application has an unprotected broadcast receiver, which is in charge of sending global notifications to the users. After executing a command from the Drozer console, the SendSMSNowReceiver broadcast receiver is exploited to send a hidden SMS. Figure 11 shows the notification after the SMS has been sent (left) and the Inbox SMS screen of the phone (right).

Fig. 11. M9 – Reverse Engineering

M10 – Extraneous Functionality. The last risk of the OWASP list is called Extraneous Functionality. It refers to vulnerabilities that have been discovered (frequently after the product has been released). Usually, these functionalities should not exist in the app. According to OWASP, backdoors are commonly left for developing purposes and not removed later. This is how Extraneous Functionalities were tested in this work. The application InsecureBankv2 has been used, and after analyzing its code, a hard-coded credential was found. Using the username "devadmin" and, without introducing any password, the login screen can be skipped. Figure 12 shows a login process using "devadmin" as a username and no password (left), and the PostLogin activity indicating successful authentication (right).

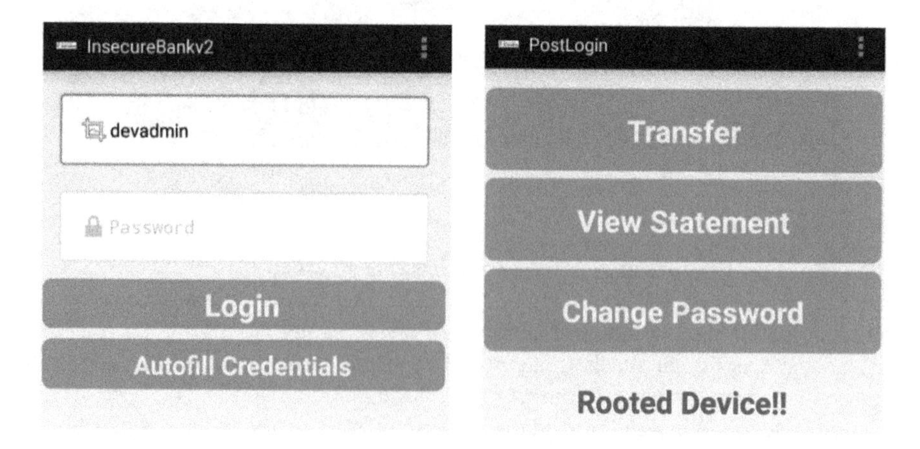

Fig. 12. M10 – Extraneous Functionality

3.3 Summary of OWASP Top 10

In Table 4, a summary of the attacks is presented. It contains all ten risks categorized by the affected area: Client-side, server-side, network-related, or a combination of them according to the previous section's scenarios.

Table 4. Risks categorized by the affected area

Risks	Client	Network	Server
M1 - Improper Platform Usage			
M2 - Insecure Data Storage			
M3 - Insecure Communication			
M4 - Insecure Authentication			
M5 - Insufficient Cryptography			
M6 - Insecure Authorization			
M7 - Client Code Quality			
M8 - Code Tampering			
M9 - Reverse Engineering			
M10 - Extraneous Functionality			

4 Results and Discussion

4.1 Risk Analysis

Following the OWASP Risk Rating Methodology [21], a risk matrix has been created to present on a scored scale all the risks regarding the latest version of OWASP mobile. As in the web version of OWASP, a graphic matrix is presented. The matrix contains prevalence, detectability, and impact with their respective score. Figure 13 shows the scores assigned to each aspect on a scale from 0 to 3. This scale will be used later in the calculation of the risk severity.

Threat Agents	Exploitability	Weakness Prevalence	Weakness Detectability	Technical Impacts	Business Impacts
Application Specific	Easy: 3	Widespread: 3	Easy: 3	Severe: 3	Business Specific
	Average: 2	Common: 2	Average: 2	Moderate: 2	
	Difficult:1	Uncommon: 1	Difficult:1	Minor: 1	

Fig. 13. Scoring scale for risk analysis (Source: owasp.org)

Figure 14 shows how a risk score can be calculated for a single item on the OWASP list.

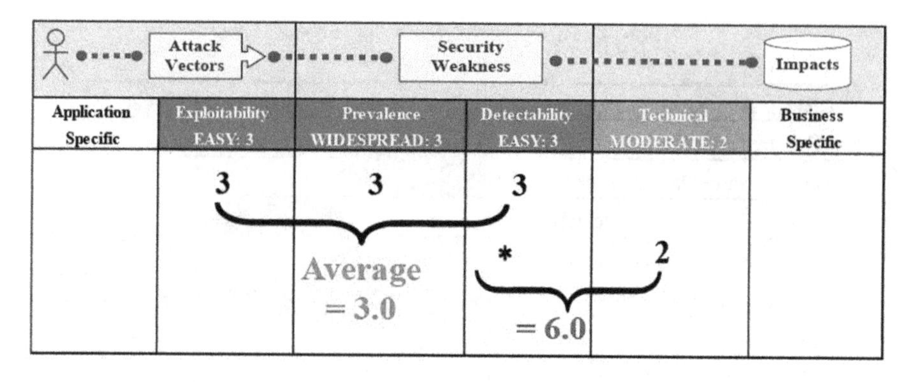

Fig. 14. Example of risk calculation (Source: owasp.org)

With this information and the results gathered from the previous section, the Top 10 Risk Factor Summary for OWASP mobile 2016 version is presented (Fig. 15). It is important to note that the last column's score is rated up to 9 points according to the calculation presented in Fig. 14.

RISK		Attack Vectors	Security Weakness		Impacts		
	Threat Agents	Exploitability	Prevalence	Detectability	Technical	Bussiness	Score
M1:2016-Improper Platform Usage	App Specific	EASY: 3	COMMON: 2	AVERAGE: 2	SEVERE: 3	App Specific	7.0
M2:2016-Insecure Data Storage	App Specific	EASY: 3	COMMON: 2	AVERAGE: 2	SEVERE: 3	App Specific	7.0
M3:2016-Insecure Communication	App Specific	EASY: 3	COMMON: 2	AVERAGE: 2	SEVERE: 3	App Specific	7.0
M4:2016-Insecure Authentication	App Specific	EASY: 3	COMMON: 2	AVERAGE: 2	SEVERE: 3	App Specific	7.0
M5:2016-Insufficient Cryptography	App Specific	EASY: 3	COMMON: 2	AVERAGE: 2	SEVERE: 3	App Specific	7.0
M6:2016-Insecure Authorization	App Specific	EASY: 3	COMMON: 2	AVERAGE: 2	SEVERE: 3	App Specific	7.0
M7:2016-Client Code Quality	App Specific	DIFFICULT: 1	COMMON: 2	DIFFICULT: 1	MODERATE: 2	App Specific	2.7
M8:2016-Code Tampering	App Specific	EASY: 3	COMMON: 2	AVERAGE: 2	SEVERE: 3	App Specific	7.0
M9:2016-Reverse Engineering	App Specific	EASY: 3	COMMON: 2	EASY: 3	MODERATE: 2	App Specific	5.3
M10:2016-Extraneous Functionality	App Specific	EASY: 3	COMMON: 2	AVERAGE: 2	SEVERE: 3	App Specific	7.0

Fig. 15. Risk factor summary for OWASP mobile 2016

Finally, it is possible to know how dangerous a risk can be. It is calculated following the same methodology suggested by OWASP in [21]. The scale used to measure the severity of the risk, from 0 to 9, is shown in Fig. 16.

Severity of the Risk	
0 to <3	Low
3 to <6	Medium
6 to 9	High

Fig. 16. Severity of the Risk (Source: owasp.org)

From the data taken from Figs. 15 and 16, it is concluded that eight of the risks presented have a high severity (6 to 9), and only two items present a medium or low severity (0 to <6).

4.2 Vulnerability Analysis and Security Recommendations

This section further mentions some other problems that could exploit the risks existing in the OWASP list. The main issues and some security recommendations for every risk in the OWASP list are presented in Table 5.

Table 5. Vulnerability analysis and security recommendations

Risk	Problems found	Security recommendations
M1 – Improper platform usage	• Lack of platform security controls • Intentional misuse due to the existence of bugs in the code	• Implement authentication mechanisms when moving from an Intent to another • Follow published development guidelines for every platform (Android, iOS, Windows Phone)
M2 – Insecure data storage	• Unprotected personal information in the file system • Devices lost or stolen grant direct access to unauthorized audiences	• Encrypt databases that contain sensitive information • Avoid the storage of unnecessary user or app information. Implement sanitization of data after usage

(*continued*)

Table 5. (*continued*)

Risk	Problems found	Security recommendations
M3 – Insecure communication	• Information sent in clear text through any communication network • Diverse channels are exposed to sniffing: Wi-Fi, 3G/4G, Bluetooth, NFC	• Implement secure transmission of data by using SSL/TLS protocols • Use certificates signed by a trusted CA provider
M4 – Insecure authentication	• Authentication based on physical identifications • Weak password policies are implemented	• Implement multifactor authentication methods • Enforce more robust password policies
M5 – Insufficient cryptography	• Use of obsolete or well-known encryption schemes • Inadequate management of encryption keys	• Use strong encryption algorithms with robust key lengths • Encrypt any sensitive data regarding the application or users' data
M6 – Insecure authorization	• Application does not persistently authenticate users • System components with low or no security (interfaces, activities, receivers)	• Verify users' identity when changing to relevant activities • Configure correctly roles and actions that are allowed to any user
M7 – Client code quality	• Low protection against buffer overflow attacks • Problems related to format-string vulnerabilities (i.e., injection)	• Validate lengths of incoming buffer data to prevent overflow • Test third-party libraries when necessary before adding them to your code
M8 – Code tampering	• Applications do not check integrity at runtime to know if code has been modified • Cellphones that have been rooted are more vulnerable to code tampering	• Implement integrity check techniques such as checksums or digital signatures • Use rooting/jailbreak detection before executing untrusted applications
M9 – Reverse engineering	• Source code of most applications may be easily decompiled for analysis • Important information is commonly stored within the source code of an app	• Increase code complexity to protect algorithms or any app information • Use code obfuscation techniques to deter reverse engineering
M10 – Extraneous functionality	• Hidden functionalities are deliberately included for testing or development purposes • Information about cryptography schemes or superuser passwords is commonly stored within the code	• Check for testing code that it must not be included in the final version of the app • Carefully manage logs. They do not leak any descriptive information on how the system works

5 Conclusions

The mobile application market is growing, and security problems appear along with it. Security is a concern among developers, testers, security experts, and smartphone users. Security is no longer an optional topic because mobile applications handle sensitive information that may be misused. Information stolen may lead to further problems such as identity theft, fraud, loss of trust, among others.

Concerning mobile security, the OWASP foundation has released the list of the ten most common risks of mobile platforms as a starting point for security testers to design, operate, and maintain secure software. Following this list, a complete penetration security testing has been carried out on several applications using different tools. After performing all security tests on ten different scenarios, it can be seen that security problems are present in all studied applications. In general, multiple vulnerabilities are present in each application analyzed, and these vulnerabilities are relatively easy to exploit, leading in all cases to several security problems. Based on the results gathered in the penetration testing stage, a scored risk matrix has been presented.

The risk matrix resumes the main points of each vulnerability existing in the OWASP list. To the best of our knowledge, this is the first proposal that presents a graphic matrix to evaluate mobile risks. Finally, a vulnerability analysis to resume some security flaws together with security recommendations to fix and prevent these problems is presented.

References

1. Telefonaktiebolaget LM Ericsson: Ericsson Mobility Report (2020)
2. Kumar, V.: Mobile Application Penetration Testing, pp. 3–10. Packt Publishing Ltd (2016)
3. Open Web Application Security Project: Top 10 Mobile Risks 2016 (2016). https://owasp.org/www-project-mobile-top-10/. Accessed 31 Aug 2020
4. King, J.: Android Application Security with OWASP Mobile Top 10 2014 (2014)
5. Gil, C., Baquero, L., Hernández, M.: A conceptual exploration for the safe development of mobile devices software based on OWASP. Int. J. Appl. Eng. Res. **13**(18), 13603–13609 (2018)
6. Pinos, S.: Diseño de un esquema de seguridad para la autenticación de teléfonos inteligentes Android en aplicaciones web corporativas, que utilizan un servicio de directorio (2018)
7. Colorado, P., Torres, I.: Análisis de seguridad de aplicaciones móviles nativas para el sistema operativo Android versión jelly bean 4.1.2 en dispositivos móviles Smartphone (2015)
8. Alanda, A., Satria, D., Mooduto, H., Kurniawau, B.: Mobile Application Security Penetration Testing Based on OWASP (2020). https://doi.org/10.1088/1757-899X/846/1/012036
9. Statista: Share of global smartphone shipments by operating systems from 2014 to 2023 (2020). https://www.statista.com/statistics/272307/market-share-forecast-for-smartphone-operating-systems/. Accessed 31 Aug 2020
10. CVE Details: Google Android Common Vulnerabilities and Exposures (2019). https://www.cvedetails.com/product/19997/Google-Android.html?vendor_id=1224. Accessed 31 Aug 2020

11. Palacios, J., López, G., Sánchez, F.: Security analysis protocol for android-based mobile applications. In: Revista Ibérica de Sistemas e Tecnologias de Informação; Lousada N.º E19, April 2019, pp. 366–378 (2019)
12. Shetty, D.: Android-InsecureBankv2 (2019). https://github.com/dineshshetty/Android-InsecureBankv2. Accessed 31 Aug 2020
13. Manino, J.: OWASP Goatdroid project (2012). https://github.com/nvisium-jack-mannino/OWASP-GoatDroid-Project. Accessed 31 Aug 2020
14. Jakhar, A.: Damn Insecure and Vulnerable Application (2016). https://github.com/payatu/diva-android. Accessed 31 Aug 2020
15. Google Inc.: Android Debug Bridge (2020). https://developer.android.com/studio/command-line/adb?hl=es-419. Accessed 31 Aug 2020
16. Debize, T.: Androwarn: Yet Another Static Code Analyzer for malicious Android applications (2012). https://github.com/maaaaz/androwarn/. Accessed 23 Aug 2020
17. PortSwigger: Burp Suite Community Edition (2020). https://portswigger.net/burp. Accessed 31 Aug 2020
18. Pan, X.: Dex2Jar (2016). https://sourceforge.net/projects/dex2jar/. Accessed 31 Aug 2020
19. F-Secure: Drozer (2019). https://github.com/FSecureLABS/drozer. Accessed 31 Aug 2020
20. Dupuy, E.: Java Decompiler (2019). https://java-decompiler.github.io/. Accessed 31 Aug 2020
21. Williams, J.: OWASP Risk Rating Methodology (2020). https://owasp.org/www-community/OWASP_Risk_Rating_Methodology. Accessed 31 Aug 2020

Interface Diversification as a Software Security Mechanism – Benefits and Challenges

Sampsa Rauti(✉)

University of Turku, Turku, Finland
sjprau@utu.fi

Abstract. Interface diversification is a proactive approach to combat malware. By uniquely diversifying critical interfaces on each computer, the malicious executable code can be rendered useless. This paper discusses the advantages and challenges of interface diversification as a software security mechanism in order the gauge its feasibility and also gives some ideas for practical implementations. An analysis of strengths and drawbacks related to this security scheme will hopefully facilitate its adoption in practical systems.

Keywords: Diversification · Obfuscation · Interfaces · Software security

1 Introduction

The malware authors rely on the known interfaces in operating systems, libraries and languages. By exploiting the fact that these interfaces are similar in millions of computers, they can launch massive large-scale attacks. The main idea of interface diversification is to make interfaces used on each machine unique. After diversification has been applied, the adversary does not know the "language" used in the system anymore. Any malicious program is considered foreign code and it will malfunction in a diversified system, because it does not know how to use the secret, diversified interfaces.

The approach is particularly useful against automated large scale attacks and self-propagating malware. Malware cannot make use of the computer's services and is rendered useless. Interface diversification is a viable countermeasure when malicious instructions are injected into an existing process (injection attacks) but it also works when a piece of malware tries operate from an independent process that well-known interfaces of the system.

The contributions of this paper are as follows. Based on our practical experiences and literature on the topic, we discuss the benefits and challenges pertaining to interface diversification. In this sense, this paper can be seen as a review in which the key properties, advantages and drawbacks of the new proactive software security approach are highlighted in order to gauge its feasibility.

Á. Rocha et al. (Eds.): ICITS 2021, AISC 1330, pp. 479–488, 2021.
https://doi.org/10.1007/978-3-030-68285-9_45

Moreover, this work gives guidance and ideas for practical implementations of interface diversification schemes in different modern application areas. Therefore, the paper aims to discuss the most important characteristics of software diversification on a general level without any fixed execution environment.

The rest of the paper is organized as follows. In Sect. 2, we first present the general theoretical scheme used in interface diversification. The fundamental basic properties of the approach are then discussed in Sect. 3. Sections 4, 5 and 6 then move on to describe performance factors, implementation issues and the most important considerations related to deployment and usability of interface diversification. Section 7 concludes the paper.

2 Interface Diversification

Internal interface diversification aims to reduce the amount of knowledge an adversary possesses about a particular execution environment [12]. In other words, diversification modifies applications' and operating systems' internal interfaces in order to make them difficult to predict for malware authors. Interface diversification often takes advantage of simple code obfuscation techniques like renaming, but more complex obfuscation techniques can also be used.

The term *interface* should be interpreted broadly here. In this context, the term does not only refer to traditional interfaces provided by software modules and libraries. For example, command sets of different languages [5] and memory addresses [14] should also been seen as diversifiable interfaces in order to prevent malicious programs from exploiting them. For our purposes, an interface is any collection of entry points that can be utilized by malware to abuse the critical resources of a computer.

A practical example of interface diversification is diversifying a shell language such as Bash in the Linux operating system. This diversification process follows the general idea illustrated in Fig. 1. We use a secret diversification key to rename all tokens in the shell command language. This change naturally has to be performed both for the interpreter (execution environment) and for the scripts in the system (executable code). By using the secret unique key, the modified interpreter can execute diversified scripts. Now the trusted scripts are compatible with the system but malicious scripts or script fragments are not. Script execution can be halted when erroneous commands are issued by malicious code. Note that a unique key can (and usually should) be used for each separate script or program.

In a similar fashion, we can diversify the system call numbers of an operating system [6]. System calls are a mechanism that provides programs with an access to computer's essential resources through operating system's kernel [22]. For instance, in Linux we can replace a few hundred original system call numbers with new ones [17]. Secret diversification is then propagated to the trusted binaries in the system: the system call numbers have to be accordingly modified in all trusted libraries and programs that make use of system calls. Other examples of interface diversification include diversifying instructions in a

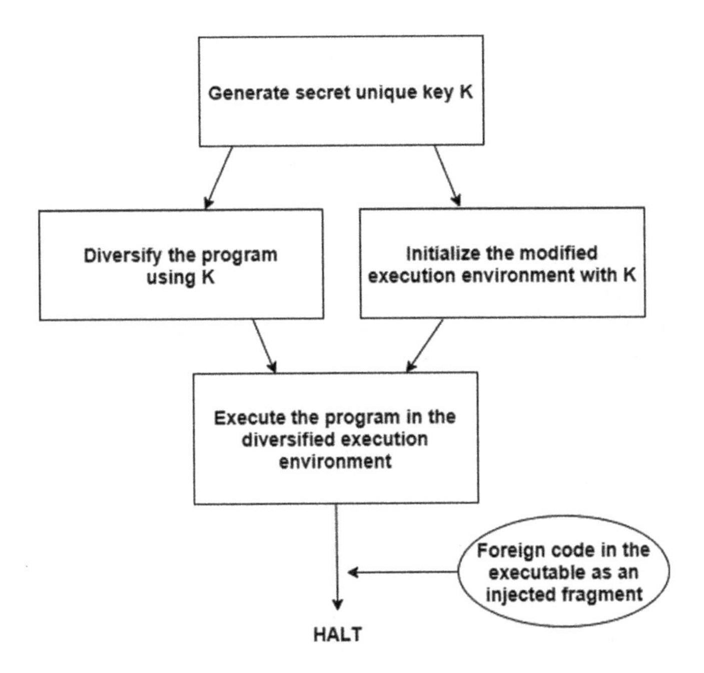

Fig. 1. The general idea of internal interface diversification.

machine language command set, altering keywords in the JavaScript language or renaming the SQL commands.

3 Basic Properties

This section presents some typical characteristics of interface diversification. Most of these properties can also be seen as advantages. When applying diversification, the malicious code does not have to be known beforehand. Also, diversification does not exclude other security measures but works in combination with them. Diversification can also be utilized in various application areas and fares well against a multitude of attacks.

Proactiveness. Without a doubt, one of the greatest properties of interface diversification is that it is a proactive countermeasure. The exact threat that is being defended against does not have to be known beforehand like e.g. in fingerprint-based malware detection approaches. This is because diversification does not aim to detect the malicious program. Instead, malware is allowed into the system but is not able to consume any resources or use services.

More and more malware is continuously being churned out. According to AVTest [4], hundreds of thousands of new pieces of malware are released into the wild every day. It is therefore hard for traditional malware prevention approaches to keep up with the current development. Novel proactive approaches are needed. This still does not mean the traditional defence mechanisms are useless. They can be used together with interface diversification and other proactive countermeasures. This brings us to the next desirable property of diversification: orthogonality.

Orthogonality. In many cases, interface diversification can be used in combination with other security measures. For example, if we want to first diversify the executable code and then encrypt it (for example, in order to safely store it when it is not being executed), this is perfectly possible. There is also nothing preventing antivirus programs from functioning in a diversified system, as long as the antivirus program itself is compatible with the diversified system. An antivirus program could still check downloaded programs or updates when they arrive into the system – and in some scenarios, it could even cooperate with the diversification engine.

Another example could be integrating an anomaly detection system with a diversification scheme. This is very possible by employing "fake original" interfaces [16]; for instance, when we diversify the mapping of system call numbers, we can still leave the old original system call interface there as a bait. With this dual interface approach, we can detect and log all the processes that invoke the original system calls – this behaviour is always suspicious in the schemes in which the whole system should be completely diversified and the trusted programs only use the diversified interfaces!

Wide applicability. As we already saw in Sect. 2, interface diversification can be applied at several software layers and in multiple different execution environments. It can therefore be seen as a global scheme that covers all the important software layers and key interfaces of a computer system (see also [15]). What makes interface diversification unique as a security mechanism is the fact that it is easily applicable in so many different application areas. Interface diversification may not always be as perfect and effective as some environment specific security measures, but the idea can readily be applied almost anywhere.

Another issue that has to do with wide applicability of interface diversification is its capability to prevent or mitigate a broad range of different attacks. In many contexts, interface diversification (or instruction set randomization) has been advocated as a security measure against injection attacks specifically [5]. However, we believe its application area is even wider in this sense as well: it is a good method not only against injection attacks but also against the cases in which a malicious binary executable has somehow been slipped into the system. Of course, this is not to say that interface diversification is a silver bullet that works against any exploit. For example, many software design flaws and logic bugs are beyond the help of diversification.

Additional Security. Interface diversification makes many insecure systems more secure. For example, in the pieces of software for Internet of Things devices, security and privacy have not yet received the attention they truly deserve [3]. These devices also often do not receive regular security updates. To relieve this problem, interface diversification can patch up the security of insecure embedded systems at least to some extent.

Local and Distributed Diversification. Depending on the execution environment which the diversification is being applied to, diversification can be either local or distributed. For instance, when we modify the system call numbers, this only changes things on one computer. On the other hand, when we diversify the language interface of JavaScript on some web page, the client machines have to be able to interpret this new language. One way is to share the diversification key to the client in the HTTP header [2]. However, the adversary might be able to learn the key in a simple scheme like this. Such distributed and public diversification schemes are therefore quite challenging. Note that diversifying the language interface of SQL, although often distributed, is not public in the same sense, because the client programs that require the knowledge about the diversified language are usually known beforehand. Therefore, not all distributed schemes are equally challenging.

4 Performance

Performance is an important factor in any system. Together with security and resilience, it is also one of the most important quality attributes pursued in diversification schemes. The performance losses incurred as a result of interface diversification are quite modest in most cases [11,18], but performance depends on many design decisions made when outlining a diversification scheme. We will delve into these issues in this section.

Runtime Performance. Usually, interface diversification does not have significant effects on runtime performance. For example, simply renaming library functions or changing the mapping of system call numbers has no effect on execution time whatsoever [17]. Then again, if we want a more flexible scheme where the system call numbers can be different for separate processes, the diversified system calls have to be decoded at runtime, which leads to somewhat degraded performance [13].

An important implication of modest performance loss is good energy efficiency. In the era of Internet of Things devices, tablet computers and smartphones, and given the importance of cloud computing today, this is an important consideration in any diversification scheme [9]. Running antivirus programs on energy efficient devices and taking up a huge proportion of available processor cycles and energy for this is often not an option. On energy efficient devices, interface diversification is therefore a viable choice for a security scheme.

Time Taken to Diversify. The time loss caused by diversification can also occur before runtime, if the diversification takes place before the program's loading and execution phases. Static diversification that does not change at runtime falls into this category. The only waiting time is when the binaries or the program code are being diversified, so there is no additional delay at runtime. Still, if the whole system is often re-diversified, time spent to diversify it becomes important, even if the diversification does not happen at runtime. For example, diversification performed on each system start-up would cause an additional delay.

Dynamic Diversification. One important idea related to diversification is that it does not need to be static but the diversification of a program can continuously change [7,19]. This improves security, but degrades performance and possibly even makes the whole system unavailable for some time.

While a complete system probably cannot be re-diversified continuously, it might be good to run the diversification process regularly. To beef up security and retain good performance, some parts of the system can have quicker diversification cycles than others; for example certain critical files and libraries can be re-diversified more often. The frequency of re-diversification also depends on the inconvenience caused to the users of the system. However, as the cloud-based environments become more popular, changing to a differently diversified copy of the same system should be easy by utilizing several virtual machines. Uniquely diversified copies of the same system could be readily available.

5 Other Implementation Issues

Along with performance, there are other important key considerations when designing a diversification framework. These include the question of key storage model and placement of the diversification engine.

Key Management and the Diversification Engine. A diversification tool and a secret key are essential parts of any diversification scheme, but where should they be stored? The simplest option is to have the diversification engine running as a local process (probably in the kernel space so that the user space processes cannot disrupt it) and also store the key locally (e.g. using the in-kernel key management utilities available in Linux). If the whole system is diversified, this should be relatively safe, since malware cannot get access to the key storage easily.

An app store on a dedicated server is likely to provide even more security, as the keys and the diversification engine would not be needed locally at all and could not be compromised that easily. Note that diversification keys cannot be deduced from diversified interfaces; the key is not a part of the diversified program code or scripts. That being said, regular re-diversification is still a good idea. In general, the app store based approach seems like an enticing idea, especially for closed ecosystems (such as Apple's ecosystem and app store).

Algorithms and Token Space. What kind of function or algorithm should be used for diversification? There are several choices in the literature. The XOR function has been utilized in many schemes because of its simplicity and effectiveness [5]. Again, this is one trade-off point between performance and security.

Other simple and popular diversification functions are appending something to the end of original tokens [15] and using simple hash functions to create diversified tokens. Of course, sometimes the keyword space is fixed (like in the case of 32-bit system call numbers in Linux). In many cases, the token length can be changed, but excessively long diversified tokens result to the loss of disk space which can be a real issue in embedded devices. Somewhat more complex diversification functions that depend on the location and order of the (in addition to the original token itself) are recommended, as we shall see next.

Granularity of Diversification. An issue strongly related with the diversification function is the granularity of diversification. In other words, do we use the same diversification key for the whole system (e.g. for all the applications and script files) or do we provide a key specific to each application? The latter increases security of the scheme, but leads to a larger performance loss and causes issues with shared libraries.

Inside one file (or application), we can still continue diversifying parts of the code with different keys. For example, the line number or previous tokens could affect the diversification of a specific token in the code. For example, the token SELECT that is a first token in an SQL query might be diversified differently as the exact same token occurring later in the same query [20]. Obviously, finer granularity leads to a more complex diversification process. However, it also makes it more challenging for the adversary to inject his or her own malicious code fragments into the diversified code.

The Extent of Diversification. The target of diversification does not always need to be a complete system. Instead, we can only diversify some of the programs. Proxy-based diversification schemes (such as [13]) allow some applications to continue functioning without being diversified. We have usually advocated diversification of the whole system because of better security provided by this scheme [17]. After all, the programs outside of diversification can probably somehow be compromised and an adversary may be able to circumvent a diversification proxy. However, an advantage of proxy-based scheme is that the programs that are allowed to function normally without diversification suffer no performance penalties.

A completely different question related to the extent of diversification is what parts of an interface or an instruction set should be diversified. For example, leaving some less important keywords of an interpreted script language (such as Bash) undiversified might provide better readability of scripts while still securing the language by partial diversification. Usually, though, diversifying the complete set of commands is the best bet when we require decent security.

6 Usability and Deployment Issues

While diversification is an effective and promising technique to counter malware, it still comes with several challenges related to usability and deployment. These issues include enforcing transparency to users and developers, handling shared libraries and managing updates received by the system.

Transparency to Users. A fundamental goal in any diversification scheme is that uniquely diversified copies of the same program should still behave in an identical way after diversification [8,10]. Only some modest performance loss is acceptable. Other than that, the user should not notice anything out of ordinary when using a diversified program.

However, this is not the reality in all practical schemes. From time to time, the diversified interfaces pose some challenges for users. If a user wants to write a shell script in a diversified system, we would probably have to provide the user with an interface that allows writing scripts in their undiversified original form so that he or she could easily accomplish this task. The problem is that this user interface could also become a vulnerable spot targeted by malware. Most users, however, do not need to fiddle with internal interfaces that are targets of diversification. Also, the configurations of many systems (such as some embedded systems with limited set of applications and functionality) are quite stable and rarely need this kind of changes. Therefore, the applicability of diversification also strongly depends on the needs of the user and how the system is used.

Transparency to Developers. Ideally, diversification does not affect the software development process. This is often the case; we have shown previously that the vast majority of binary executables [17] and interpretable scripts [15] are quite easy to diversify automatically without human intervention after the code has been written. The developer usually does not need to be concerned with diversification.

However, there are some cases that may require the developer to step in. For example, when a PHP script dynamically generates an SQL query at runtime, it is quite hard to statically diversify the query in the source code with an automatic tool. In these cases, the programmer can perform the diversification with the help of a tool that diversifies the required tokens for him or her. This makes the adoption of diversification easier and more pleasant. Dynamically created Bash scripts in Linux constitute a similar problem. In this case, though, we can argue that this kind of programming practice (executables creating shell scripts dynamically) should be avoided anyway and the programs could be rewritten not to use dynamically generated scripts.

Shared Libraries. When each (dynamically linked) executable or shell script has its own unique diversification key, shared libraries become a problem. One solution is to create new versions of these libraries for each executable (time and space is lost but better security is gained). This solution is probably better than e.g. using undiversified libraries. Libraries can be diversified for each program or script before or during execution. In any case, diversification of shared libraries only has to be performed once (for each re-diversification cycle).

Updates. Software updates are a challenge for diversified systems. Each arriving patch has to be accordingly diversified in order to be compatible with the respective software and if the whole program is updated, it has to be compatible with the interfaces it depends on. Therefore, the same update issued by a software vendor does not readily fit the diversified systems, but has to be diversified either on an app store server or on the user's machine, depending on where the diversification engine (and the secret key used for diversification) is located. Of course, an arrival of a large update could be a good opportunity to re-diversify the whole system.

7 Conclusions

We have covered several considerations that developers and users of any interface diversification scheme have to take into account. As a malware prevention approach, interface diversification poses some challenges, but still shows a lot of promise because of several desirable properties such as proactiveness and low performance penalties. Diversifying internal interfaces of the system is an especially fitting protection mechanism for securing growing number of Internet of Things devices, which often have poor security and require security mechanisms that are not resource intensive (e.g. in terms of computational power).

The future work in the area involves implementing and experimenting with practical diversification schemes. More work is also needed to solve challenges of interface diversification. For instance, improving transparency to developers and solutions handle the updates are important topics. Finally, diversification has lots of potential when used in combination with other security measures such as anomaly detection systems [1] and machine learning [21]. These possibilities should be further explored and tested with proof-of-concept implementations.

We believe diversification should be seen as a global scheme covering all important layers of the system. This way, its effectiveness as a countermeasure against malware can be maximized. At the same time, orthogonal use with other security measures is still needed. After all, the adversary's imagination can always surpass our expectations: they may exploit implementation details that nobody has thought to diversify.

References

1. Ahde, H., Rauti, S., Leppanen, V.: A survey on the use of data points in IDS research. In: International Conference on Soft Computing and Pattern Recognition, pp. 329–337. Springer (2018)
2. Athanasopoulos, E., Krithinakis, A., Markatos, E.P.: An architecture for enforcing javascript randomization in web2.0 applications. In: International Conference on Information Security, pp. 203–209. Springer (2010)
3. Atlam, H.F., Alenezi, A., Alassafi, M.O., Alshdadi, A.A., Wills, G.B.: Security, cybercrime and digital forensics for IoT. In: Principles of Internet of Things (IoT) Ecosystem: Insight Paradigm. Springer, pp. 551–577 (2020)
4. AVTest: Malware statistics. https://www.av-test.org/en/statistics/malware/ Accessed 13 Sept 2020
5. Boyd, S.W., Kc, G.S., Locasto, M.E., Keromytis, A.D., Prevelakis, V.: On the general applicability of instruction-set randomization. IEEE Trans. Depend. Secure Comput. **7**(3), 255–270 (2008)
6. Chew, M., Song, D.: Mitigating buffer overflows by operating system randomization. Technical Report CMU-CS-02-197, Carnegie Mellon University, Pittsburgh, USA (2002)
7. Collberg, C., Martin, S., Myers, J., Nagra, J.: Distributed application tamper detection via continuous software updates. In: Proceedings of the 28th Annual Computer Security Applications Conference, pp. 319–328 (2012)

8. Collberg, C., Thomborson, C., Low, D.: A taxonomy of obfuscating transformations (1997)
9. Hosseinzadeh, S., Rauti, S., Hyrynsalmi, S., Leppänen, V.: Security in the Internet of Things through obfuscation and diversification. In: 2015 International Conference on Computing, Communication and Security (ICCCS), pp. 1–5. IEEE (2015)
10. Hosseinzadeh, S., et al.: Diversification and obfuscation techniques for software security: a systematic literature review. Inf. Softw. Technol. **104**, 72–93 (2018)
11. Larsen, P., Brunthaler, S., Franz, M.: Security through diversity: are we there yet? IEEE Secur. Privacy **12**(2), 28–35 (2013)
12. Larsen, P., Homescu, A., Brunthaler, S., Franz, M.: SoK: automated software diversity. In: IEEE Symposium on Security and Privacy, vol. 2014, pp. 276–291. IEEE (2014)
13. Liang, Z., Liang, B., Li, L.: A system call randomization based method for countering code injection attacks. In: International Conference on Networks Security, Wireless Communications and Trusted Computing, NSWCTC, pp. 584–587 (2009)
14. Marco-Gisbert, H., Ripoll Ripoll, I.: Address space layout randomization next generation. Appl. Sci. **9**(14), 2928 (2019)
15. Portokalidis, G., Keromytis, A.D.: Global ISR: toward a comprehensive defense against unauthorized code execution. In: Moving Target Defense. Springer, pp. 49–76 (2011)
16. Rauti, S.: Towards cyber attribution by deception. In: International Conference on Hybrid Intelligent Systems. Springer, pp. 419–428 (2019)
17. Rauti, S., Laurén, S., Hosseinzadeh, S., Mäkelä, J.M., Hyrynsalmi, S., Leppänen, V.: Diversification of system calls in linux binaries. In: International Conference on Trusted Systems, Springer (2014) 15–35
18. Rauti, S., Laurén, S., Mäki, P., Uitto, J., Laato, S., Leppänen, V.: Internal interface diversification as a method against malware. J. Cyber Secur. Technol. 1–26 (2020)
19. Rauti, S., Leppänen, V.: Internal interface diversification with multiple fake interfaces. In: Proceedings of the 10th International Conference on Security of Information and Networks, pp. 245–250 (2017)
20. Rauti, S., Teuhola, J., Leppänen, V.: Diversifying SQL to prevent injection attacks. In: IEEE Trustcom/BigDataSE/ISPA, vol. 1, vol. 2015, pp. 344–351. IEEE (2015)
21. Shaukat, K., Luo, S., Varadharajan, V., Hameed, I.A., Chen, S., Liu, D., Li, J.: Performance comparison and current challenges of using machine learning techniques in cybersecurity. Energies **13**(10), 2509 (2020)
22. Tanenbaum, A.S., Bos, H.: Modern Operating Systems. Pearson (2015)

Method for Implementation of Preventive Technological Tools for Control and Monitoring of Fraud and Corruption

Tannia Cecilia Mayorga Jácome[1,3](✉), Ronald Fernando Coloma Andagoya[2,3](✉), Marianela Edith López Veloz[1,3](✉), and Juan Alberto Toro Álava[1,3](✉)

[1] Universidad Tecnológica Israel, Francisco Pizarro E4-142 y Marieta de Veintimilla, Quito, Ecuador
{tmayorga, jtoro}@uisrael.edu.ec, marianela_lopez@hotmail.com

[2] CELCO Cía. Ltda., Elia Liut N45-26 y Edmundo Chiriboga, Quito, Ecuador
ronaldfernandocoloma@gmail.com

[3] Interregional Academy of Personnel Management, Frometivska Street, Kyiv 03039, Ukraine

Abstract. The purpose of this work is to present a method for the implementation of preventive technological tools like "The Fraud Explorer" for control and monitoring of fraud and corruption; The nature of this system is based on Donald Cressey's fraud triangle and for this it uses a dictionary of words that are related to this type of wrongdoing; It is installed in each workstation silently in Windows or Linux and has a Linux server with an OpenVZ container that manages a dictionary of words and compares what users type in their working hours on the institution's computers, in case of finding a match issues an alert that must be investigated and thus prevent illegal acts.

It is essential to resolve the gaps in the implementation of this tool, such as the legal regulations that allow this process to be carried out successfully without incurring problems of a violation of privacy, the way in which the entity should organize for the implementation and monitoring to which is presented a proposal in four phases in such a way that it provides solidity to carry out this process successfully.

Keywords: Fraud · Corruption · Anti-fraud tools · Control and monitoring · Fraud prevention

1 Introduction

Understanding that we live with fraud at different scales is worrisome as a society, so it is essential to have a strategic approach when managing the risk of fraud in any country in the world.

Á. Rocha et al. (Eds.): ICITS 2021, AISC 1330, pp. 489–499, 2021.
https://doi.org/10.1007/978-3-030-68285-9_46

Throughout history it is an evil that seems impossible to eradicate as mentioned by different authors, but nevertheless in countries such as Singapore, Hong Kong [1, pp. 59 - paragraph 1], they are exceptions with certain peculiarities in their behavior.

In Latin America, the most corrupt countries according to the report [2] are Venezuela, Bolivia, Paraguay and Ecuador.

There are many definitions of fraud and corruption that affect the image and growing of each country in relation to the economy [3, 4], [5, p. 45], financial sphere [6, p. 315]; as well as many analyzes of the possible causes that focus on the cultural [7, p. 65], [8, p. 268], [9, p. 869], technological [10, pp. 137–138], [11, p. 27], political field and more [12].

It is also important to mention the existence of many instruments or tools have been used to control fraud and corruption, such as: conflict of interest, audits, forensic audits [4], standards antibribery ISO 37001, internal informant [13]. Donald Cressey's hypothesis mentions that "trusted people become trust violators when they are conceived as having a problem that is not shareable, they are aware that this problem can be solved in secret by violating the position of trust and they are capable of apply to their own behavior in this situation, verbalizations that allow them to adjust their conceptions of themselves as trustworthy people." [14].

Author Akerman mentions that "Democracy can help limit corruption if it provides people with alternative ways to complain and offers those concerned an incentive to be honest" [15, p. 173] or author like as Eigen is more specific in indicating to go to the ombudsman, or make calls [16, p. 127] is a way of protecting the informant.

These approaches were applied in 2018 the public sector in Ecuador with the implementation of communication channels and incentives as a means of control in order to encourage reports of corruption and try to stop this behavior [17, p. 1], however there are other factors such as fear, harassment as described by several authors like [18, p. 6], [19, p. 12], or the use of judicial system as a means of evasion [18, p. 16], many times preferring to be accomplices than to expose oneself, leading to preventing this structure from working properly.

As an alternative to contribute it has been implemented, the figure of Technological Informant is born since this role cannot be intimidated, harassed, nor can it be an accomplice of anyone, constituting a form of control adaptable to anti-fraud management systems such as ISO 37001 [20].

This technological informant is a virtual figure that would collect information from devices such as computers, laptops within the network and thanks to Big Data technologies, this data collection by applying certain methodologies is possible to provide it with a structure so that they become information [21], same as allowing the generation of technological alerts; As in this case, the free software tool Fraud Explorer does it, by collecting all the words and then sending them to a customizable dictionary, to then generate an automatic report that alerts an attempt to fraud or corruption.

However, the installation of these tools to support fraud control would be invading the privacy of the official; Therefore, it is necessary to have an internal regulation, considering that everything that is written is captured and then subjected to an analysis, considering that the right to privacy involves correspondence, which among other instruments are email and chats, which would be the capture of the tool.

Therefore, without the internal regulations that regulate its application and use, there would be a possible violation of privacy. In different Latin American countries, the personal data protection law is being promoted [22] and specialized reports on cybersecurity such as [23] indicate the importance of protecting information, an action that goes hand in hand with security management.

With this background, it is necessary to propose the development of a method for the implementation of the preventive tool The Fraud Explorer for the control and monitoring of fraud and corruption in Ecuador for private or public entities to prevent any attempt at corruption that affects them.

This method involves four phases that are considered necessary to comply with the current legal regulations in the country, technical requirements that will help the better performance of both private and public entities for the control of fraud and corruption, which allows the assurance of information institutional.

Problem Approach

One of the recommendations of organizations such as Transparency International is to stay away from the political sectors and private organizations since the corruption perceived in Ecuador is related to this sector as can be publicly evidenced, it is documented in the press, in international organizations such as Transparency International, which together with the World Bank present an impact on the economic sector as the percentage of GDP loss due to corruption.

The problem is How to strengthen controls between the public – private or between private companies to control corruption.

2 Methodology

The research methodology used is mixed, since the qualitative paradigm is applied to assemble the methodological proposal, as well as the quantitative one when obtaining results after the implementation of the Fraud Explorer tool, where experimental research is applied, observing the actions that these triggered the committing of crimes within the organization, modifying those circumstances that are considered convenient in the development of the activities for the entity.

Once the Fraud Explorer software is implemented, numerical indicators are obtained with scores assigned by the Donald Cressey algorithm and previously configured by the systems administrator, representing the level of risk of the user connected to the institutional network, and which will be monitored through the tool.

A survey of incidents committed was carried out, where techniques were used as observation, and being part of the place where it was implemented became a participatory research method.

3 Proposed Model

A holistic approach is required with transparency agreements and long-term collaboration. The implementation of specific tools for the management of anti-bribery policies like is showed in the Fig. 1:

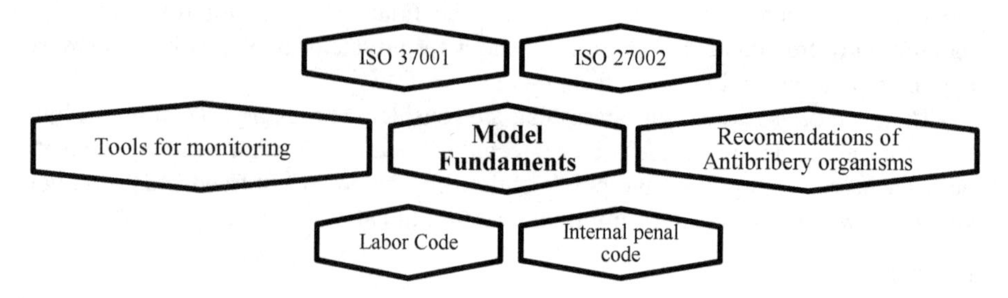

Fig. 1. Model Fundament **Source**: Elaborated by authors.

Tools for Monitoring

Table 1 describes a comparative chart of alternatives for monitoring user activities in order to detect and prevent internal threats within a company.

Table 1. Comparative table of alternatives for monitoring user activities.

Tool description	Ekran system	ObserveIT	Veriato 360	The fraud explorer
Business oriented	Large, medium and small	Big	Large, medium and small	Large, medium and small
Price	$6.500,00	$27.500,00	$5.073,00	$0,00
Word library	Yes	Yes	Yes	Yes
Server operating system	Windows, Linux, Mac, Citrix	Windows, Android, Iphone/iPad Mac	Linux	Linux
Database	MySql, Firebird, SQL Server	SQL Server	SQL Server	MariaDB
Support	Yes	Yes	Yes	Yes
Customer operating system	Windows, Linux y MAC	Windows, Linux, MAC	Windows Linux	Windows Linux
Real-time monitoring	Yes	Yes	Yes	Yes
Notifications and alerts	Yes	Yes	Yes	Yes

(*continued*)

Table 1. (*continued*)

Tool description	Ekran system	ObserveIT	Veriato 360	The fraud explorer
Reports	Advanced report generation system	Advanced report generation system	Advanced Report generation system	Standard Reporting System
Foundation year Origin	2013 EEUU	2006 Israel	1998 EEUU	2016 Colombia
Sectors that have implemented	Consultant auditorsPostal mail, Universities	Automotive Industry, Financial Sectors, NGOs	Technology sector, tire manuracturing, Provision of human talent and technology services, Aviation industry	Colombian Government

Source: Adapted of [24, pp. 32–33, 25–28]

To develop the model, The Fraud Explorer tool is chosen because it is a free software tool.

Standards ISO 37001, 27002

This tool hypothetically supports the following elements of ISO37001: in the sections 7.2.2.2 -b related with Employment procedures, 7.3 – e, f related Awareness and training, 7.4.1 referred with communications relevant the anti-bribery system, 7.5.3 Control of documented information, 8.1, Operational planning and control, 9.1 Monitoring measurement, analysis and evaluation; ISO 27002 in the chapters; Human Talent security, asset management, access control, relationship with suppliers, information security incident management, compliance monitoring.

Regulations that Support Implementation

The supreme norm that governs Ecuador is the Constitution of the Republic, which also has annexed and related regulations that govern throughout the national territory for its mandatory compliance, the same that cannot be violated because it carries unconstitutionality implications for both people, companies, entities in general.

It is also based on the Comprehensive Organic Penal and Labor Code, internal regulations that each entity must create for compliance, and on the organization's information security and confidentiality policies.

Constitution of the Republic of Ecuador

Duties and Responsibilities of Ecuadorians and Ecuadorians. (Constituent Assembly, 2008), in Title II Rights, Chapter Ninth Responsibilities, article 83, in number 12 it establishes "To exercise the profession or trade subject to ethics".

Fundamental Constitutional Principles. Likewise, it establishes fundamental principles that guide the entity, creating an environment that favors the effective, efficient, timely and integral action that, due to its observance, ensures an honest and ethical behavior of the workers and employees, a necessary condition to generate confidence in the products and services generated.

Principle of inclusion or non-discrimination, principle of equality, principle of distributive equity as a requirement to access good living, principle of respect for the rights of nature, principle of accountability to the public in a sufficiently broad, open, and systematic way, regarding the public management in their charge.

Integral Penal Code (COIP). The Integral penal code [29] establishes the different typologies of crimes for the state; from 6 months to one year Art 179 Secret revelation, from one to three years; Art 178 Violation of privacy, dissemination of information from restricted circulation Art 180, Abuse of confidence Art 187, fraudulent appropriation by electronic means Art 190, exchange, marketing or purchase of equipment information mobile terminals Art 192; and from five to seven years swindle Art 186 [30].

Labor Code. On September 26, 2012, the last modification to the Labor Code is carried out by the Legislation and Codification Commission of the National Assembly of Ecuador, with the purpose of keeping the labor legislation updated, in which the sanction is contemplated in the Art 172 causes by the employer may terminate the contract, 183 Approval rating.

Institutional Internal Standard

Internal Work Regulations. Each public or private entity is obliged to have an internal work regulation, in accordance with article 64 of the Labor Code, where the obligations, rights and sanctions of the collaborator and other people who are dependent or provide services to the organization, as well as the regulation of the interests of employers and workers, and full compliance with the relevant general and institutional legal regulations.

Code of Ethical Conduct. Ethics is the fundamental pillar to combat corrupt behavior [31] and the Code of Ethical Conduct for the organization has the purpose of establishing a reference framework of conduct in situations, for the actions of workers and employees in an ethical and proven manner, in observance of constitutional and ethical principles and the latter, science that deals with the duties and ethical principles that concern each profession or work environment, rules governing professional conduct and performance.

4 Results

As a result of the present investigation, the proposed and applied model is obtained for the case study of a company, whose identity is not mentioned for confidentiality purposes (Fig. 2).

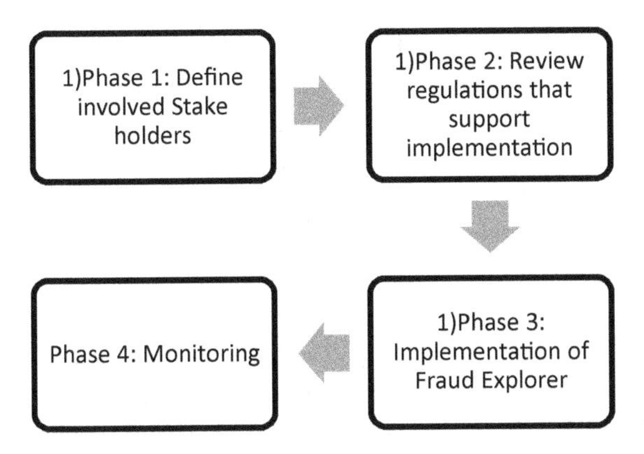

Fig. 2. Method to implement the technological informant

4.1 Phase 1: Define Involved Stake Holders

They are presented in Table 2 with the roles of the actors, activities, and description of the profile, then the revision of the regulations, to continue with the implementation where they take part: systems review, hardware revision, tests.

Table 2. Define stake holders.

Perfil/Rol	Activities	Perfil description
Managers legal representatives	Authorize implementation	It is the owner, or legal representative of the company who must authorize the installation This role must have autonomy to report to the highest authority, boards of directors, shareholders' meetings, collegiate bodies, and control bodies according to the type of organization
Systems auditor/fraud examiner	Maintain and improve the phrase library	Control of the library of phrases that must be adjusted to the needs and heading of the company You must fulfill the role of fraud examiner who understands Donald Cressey's triangle theory and helps feed, improve, or correct the main dictionary The Fraud Explorer will use to associate behaviors
	Set ranges of criticality of fraud	The Fraud Explorer solution has fraud categories: low, medium, high, and critical. Each alert score and average are defined by these categories, which vary according to the solution and needs defined by the fraud examiner

(*continued*)

Table 2. (*continued*)

Perfil/Rol	Activities	Perfil description
	Analyze alerts and communicate the results	The Fraud Examiner analyzes the alerts generated by the software, decides if they are relevant or if they are a false positive that must be marked and ignored, notifies the organization of any true alert so that the corresponding decisions are taken, such as conducting an investigation, a forensic analysis or request specialized legal assistance
Systems engineer/implementer	Install the solution, configure, and maintain it	Person in charge of install, configure and technically implement the solution, depending on the Linux infrastructure and PHP programming
	Deploy the endpoint	The person in charge implements and configures security, depending on the Windows and Active Directory infrastructure that will configure the GPO policy, for agent deployment on PCs

Source: Own elaboration based on the documentation presented by [32]

4.2 Phase 2. Review Regulations that Support the Implementation

These regulations may vary depending on the country where they are to be implemented and in the case of Ecuador they were described in the section on the foundation of the model.

4.3 Phase 3: Implementation of Fraud Explorer

This stage is the identification of the support and communication infrastructure, as well as systems, considering minimum requirements so that its execution and subsequent obtaining of results are satisfactory for the interested parties, the minimum characteristics at the hardware and software are; Active Directory, Virtualizer like Virtuozo, Virtualizor SolusVM, Proxmox Archipel or any other that support Open VZ; Linux Operating System like CentOS or Red Hat; Hard Disk 60GB and 16 GB of RAM, The File system used to upload the containers must be ext4 [24, p. 1, 33].

4.4 Phase 4: Monitoring

Once the tool is installed and configured, the deployment of technological controls is evaluated through communication monitoring tools with artificial intelligence techniques related to key issues for the organization by creating a database of terminology classified as vulnerable to bribery. The Fraud Examiner will carry out daily monitoring under the following the steps: monitoring, alert analysis, in case of there is evidence or suspicion of fraud must report findings, send report to the highest authority; in case of there isn't evidence or suspicion of fraud monitoring continue.

5 Discussion

Fraud management becomes relevant at different business levels, be they private or public, allowing transparent processes that are often linked to corruption issues.

The use of a method to implement fraud prevention tools is part of the integral management of organizations, be they even microenterprises; as well as part of the information security management considering the inclusion of controls in the information security policies and penalties, established in an internal legal tool, which go hand in hand for their application and implementation.

The design and implementation of a method that allows fraud prevention in the integral management of security in business processes, must include the integration of security policies, establishing control, monitoring and follow-up protocols.

Fraud exists in all organizational environments, so the determination of a control is required, which minimizes the possible results and risks of this kind of event.

Business processes do not have nor established norms that allow safeguarding your information, ignoring the need to establish and implement preventive technological tools to control and monitor fraud and corruption.

The evaluation of the proposed method was done in one company with excellent results.

The new term technological informant presented in this work helps organizations to control fraud and corruption before it occurs and is integrated into the ISO 37001, ISO 27002 standards.

Among the limitations we can find: lack of both physical and digital information to develop this type of project, especially when implementing the tool; the selection of the person who reports since the tool generates the alert, and the human being intervenes to inform the headquarters of the alleged found, thus this person has to be integrity, incorruptible; It is necessary to define a procedure to select the informant so that it is a systematic process and with measurable parameters and not a matter of subjective; and finally, the economic factor, since the implementation of a new system always requires financial resources and the training of personnel to develop a method of improvement.

For future research it is intended to report the findings found with the implementation of this method, as well as new proposals in favor of fighting corruption.

References

1. Cremer, G.: Corruption & Develoment Aid Confronting the Challenges, p. 169. LYNNE REINNER, Boulder (2008)
2. Transparency Internacional: Corruption Perceptions Index, 11 August 2019. https://www.transparency.org/research/cpi/overview. Accessed 11 Aug 2019
3. Campos, J.E., Sanjay, P.: The Many Faces of Corruption: Tracking Vulnerabilities at the Sector Level, p. 484. The World Bank (2007)
4. Arellano Gault, D.: ¿Podemos reducir la corrupción en México? Límites y posibilidades de los instrumentos a nuestro acance, Segunda ed., N. C. Larios, Ed., cide, Centro de Investigación y Docencia Económicas, A.C., p. 230 (2018)

5. Solimano, Tanzi, V., Del Solar, F.: LAS TERMITAS DEL ESTADO Ensayos sobre corrupción, transparencia y desarrollo, Primera ed., Salesianos Impresores S.A., p. 144 (2008)
6. Coello, A., Martín Rodrigo, M., Bertrand, A., García, D.R., Josep, J.M., Tafalla, A., Acosta, X., Huertz de Lemps, J.L.M., Dieste, Salem Seña, J.F.: LA JUSTICIA ROBADA Corrupción, codícia y bien público en el mundo hispánico (siglos XVII-XX), Primera ed., Icaria Antrazyt, p. 319 (2018)
7. Corporación Latinobarómetro: Latinobarómetro (2018). https://www.latinobarometro.org/lat.jsp. Accessed 6 Feb 2020
8. IDEA; NIMD; Instituto Clingendael, Briscoe, I., Perdomo, C., Burcher, C.U. (eds.): Redes Ilícitas y Política en América Latina, Trydells Tryckeri, p. 323 (2014)
9. Barr, A., Serra, D.: Corruption and culture: an experimental analysis. Public Econ. **94**, 862–869 (2010)
10. Segunda Cohorte del Doctorado en Seguridad Facultad de Ciencias Jurídica y Sociales de la Escuela de Estudios de Postgrado de la Universidad de San Carlos de Guatemala, Seguridad de la Información: Revista de la Segunda Cohorte del Doctorado en Seguridad Estratégica, p. 376 (2014)
11. Arroyo Jácome, R.P., Roman Vasquez, A.F.: Análisis de los delitos informáticos por ataque y acceso no autorizado a sistemas electrónicos, tipificados en los artículos 232 y 234 del Código Orgánico Integral Penal en el Ecuador, Quito (2016)
12. Aya Palencia, D.S., Fernández Gómez, N.A.: Technological surveillance for the identification of lines of research over corruption. Estrategia Organizacional **9**, 18 (2019)
13. Foro internacional: Instrumentar una política de informantes internos (whistleblowers): ¿mecanismo viable en México para atacar la corrupción? Foro Internacional **52**(1), 38–91 (2012)
14. NF Cybersecurity & Antifraud Firm (2017)
15. Ackerman, S.R.: LA CORRUPCIÓN Y LOS GOBIERNOS Causas, consecuencias y reforma, Primera, p. 366. SIGLO VEINTEUNO DE ESPAÑA EDITORES, España (2001)
16. Eigen, P.: LAS REDES DE LA CORRUPCIóN, p. 319. Planeta, Bogotá (2004)
17. Secretaría Anticorrupción del Ecuador: Secretaría anticorrupción del Ecuador - nueva heramienta tecnológica, 23 January 2019. https://www.anticorrupcion.gob.ec/secretaria-anticorrupcion-presenta-nueva-herramienta-tecnologica-para-la-lucha-contra-la-corrupcion/. Accessed 24 Feb 2020
18. Terán, A.J. (ed.): Comisión Nacional Anticorrupción, Corrupción en la Década Encubierta, Primera, Imprenta Don Bosco, p. 235 (2017)
19. Gordillo Montalvo, J.: DESHONESTIDAD VIA LAVADO DE SUCRES, Segunda ed., Quito, p. 14 (2001)
20. ISO: ISO (2016). https://www.iso.org/iso-37001-anti-bribery-management.html. Accessed 13 Sep 2020
21. Martinez Mosquera, D., Lujan Mora, S., Navarrete, R., Mayorga Jácome, T., Vivanco Herrera, H.R.: An approach to big data modeling for key-value NoSQL databases. RISTI, 519–530 (2019)
22. Mayorga Jácome, T.C., García Jiménez, M., Duret Guiérrez, J.F., Carrión Jumbo, J., Yarad Jeada, P.V.: Historia de la normativa reguladora de la Protección de Datos de carácter personal en distintos países Latinoamericanos. Dominios de la Ciencia **5**(1), 518–537 (2019)
23. Cisco Networking Academy: Introducción a cyberseguridad (2018)
24. Coloma Andagoya, R.F., Vaca Benalcazar, C.P.: IMPLEMENTACIÓN "THE FRAUD EXPLORER FTA" EN LA EMPRESA CELCO CÍA. LTDA. COMO MECANISMO DE PREVENCIÓN Y DETECCIÓN DE FRAUDE, p. 112. Universidad Tecnológica Israel, Quito, Pichincha (2019)

25. Ekransystem: Ekransystem. Ekransystem (2020). https://www.ekransystem.com. Accessed 12 Sep 2020
26. Veriato: Veriato, sn. https://www.veriato.com. Accessed 12 Sep 2020
27. ObserveIT: ObserveIT (2020). https://www.observeit.com/. Accessed 12 Sep 2020
28. NF S.A.S.: The Fraud Explorer, 12 September 2020. https://www.thefraudexplorer.com/index.html. Accessed 12 Sep 2020
29. Asamblea Nacional (2014)
30. Asamblea General de la República del Ecuador, Código Orgánico Integral Penal, Quito (2014)
31. Paredes Cabezas, M.d.R., Guachamboza Moposita, R.M., Velasteguí López, L.E.: Un análisis regional de la ética en la formación del Contador Público como pilar fundamental para combatir la corrupción. Ciencia Digital **4**, 45–69 (2020)
32. NF Cybersecurity and Antifraud Firm S.A.A.: The fraud explorer, 15 August 2019. https://www.thefraudexplorer.com/files/The_Fraud_Explorer_Userguide.pdf
33. GITHUB: The Fraud Explorer documentation (2019). https://github.com/nfsecurity/the-fraud-explorer/wiki/Requisitos. Accessed 2019
34. Locatelli, G., Mariani, G., Sainati, T., Greco, M.: Corruption in public projects and megaprojects: there is an elephant in the room! ScienceDirect **35**, 252–268 (2017)
35. Campos, J.E., Pradhan, S. (eds.) World Bank: The Many Faces of Corruption: Tracking Vulnerabilities at the Sector Level, p. 484. The World Bank, Washington (2007)

Cyberbullying and its Impact on Children and Adolescents in the City of Ibarra Ecuador

Daisy Imbaquingo[1(✉)], Erick Herrera[1(✉)], Bryan Aldás[1(✉)], Tatyana K. Saltos[1(✉)], Silvia Arciniega[1(✉)], and Gabriel Llumiquinga[2(✉)]

[1] Universidad Técnica del Norte, Av. 17 de Julio 5-21, EC100105 Ibarra, Ecuador
{deimbaquingo, epherrera, baaldasl, tksaltos, srarciniega}@utn.edu.ec

[2] Asociación Ecuatoriana de Ciberseguridad, EC170528 Quito, Ecuador
presidencia@aeci.org.ec

Abstract. Cyberbullying is a phenomenon that occurs worldwide, affects the youngest from a very early age and is increasing. Despite the fact that there is multiple research on the relationship between traditional bullying, psychological damage and the family and school context, there is very little scientific research on the influence of cyberbullying on the social environment of children and adolescents. Through a quantitative methodology, this research carried out in young students from 12 educational institutions in the city of Ibarra, Republic Ecuador, in the period 2019–2020, to determine which gender is the most vulnerable to suffer from cyberbullying, establish a range of ages with which it is possible to analyze in which stage of adolescence this crime occurs more frequently, to know in which educational institutions (prosecutors, private individuals, government agencies) there is a higher rate of presence of cyberbullying, to understand the dependence on the most used social networks of the affected people, on who has the great impact. The misuse of social networks has led to a prolongation of intervention against this type of harassment, and it is of utmost importance to include the family and educational institutions in cyberbullying prevention programas.

Keywords: Bullying · Social media · Prevention · Victim · Teens · Technology

1 Introduction

The close relationship that exists between the internet and technologies has not only helped humans to carry out different activities in a simpler way (Starcevic and Aboujaoude 2015). The internet is now easily accessible, widely available, and generally unregulated (Hinduja and Patchin 2019), allowing people to access an unknown number of topics, as well as breaking down communication barriers with people from all over the world (Mok et al. 2015).

While highlighting that information and communication technologies (mainly mobile phones and the Internet) have the main function of improving human

Á. Rocha et al. (Eds.): ICITS 2021, AISC 1330, pp. 500–514, 2021.
https://doi.org/10.1007/978-3-030-68285-9_47

relationships, it must be taken into account that they are not always used correctly (Domínguez et al. 2017), However, trying to limit its use is an impossible and futile task, especially in today's times, when its employment and dependency is constantly growing (Chan and Wong 2015).

Over the years and technological advancement, the use of electronic and smart devices has been on the rise, leading to new dangers (Durán and Martínez, 2015), which affect all people, mainly those who are in the adolescence stage (Amado et al. 2009), considered an adolescence period vulnerability (Kim et al. 2019).

Among all these new dangers due to excessive use and little control of the internet, cyberbullying or also known as cyberbullying has emerged (Garaigordobil et al. 2019). This issue has drawn the attention of leading researchers around the world, as a result of the increasing incidents of verbal and non-verbal aggressive acts on social media and the risk run by all those involved with these incidents (Alhujailli and Karwowski 2018).

This type of harassment has emerged recently and due to this, its conceptualization is still under construction, harbouring a plurality of meanings in its definitions (Della et al. 2015). Cyberbullying is a type of harassment that causes aggression and is generally carried out anonymously (Correa and Avedaño 2015), by a person or group of people, about a victim, who cannot easily defend himself (Ortega et al. 2016),-causing discomfort, and in many cases psychological damage, through the use of electronic and digital devices (Tokunaga 2010). This type of harassment is totally different from traditional bullying, due to the way and hours in which it can be presented, the potential exposure to which a victim may be subjected, as well as the proximity that exists between the bullies and the victims (John et al. 2018).

Research on cyberbullying indicates that this type of bullying has increased significantly in recent years (Fernández et al. 2015). Some authors describe that the highest number of cyberbullying victims are among female adolescents (Beckman et al. 2013), others among adolescent men (Durán and Martínez 2015), and on the other hand, there are authors who do not perceive differences in gender (Katzer et al. 2009).

With regard to age, the research carried out agrees that the main victims of cyberbullying are adolescents who are in the first cycle of compulsory secondary school, that is, between an average of 10 and 14 years (Young et al. 2016), there is gradually a decrease in cyber victimization in the second cycle of compulsory education, that is, between an average of 15 and 19 years (Kanyinga and Hamilton 2015).

The fact is that, despite the fact that there are no studies that address and deepen the issue of cyberbullying, the accumulation of evidence that has been collected demonstrates the effect and bad results that this type of harassment causes, both in the victim and in the aggressor, among these is low self-esteem, self-harm, anxiety, loneliness, and low academic performance (Foody et al. 2015).

Studies that have been carried out on cyberbullying victims have shown that they are more likely to suffer from agoraphobia, depression, anxiety, panic disorder and suicidal tendencies, around the age of 20 (Swartz and Bhattacharya 2017).

On the other hand, a study that belongs to the medical literature has found that people generally connect to the internet for more than three hours a day (Abreu and Kenny 2018), this directly affects young people, increasing their chances of suffering from cyberbullying four times (Arsène and Raynaud 2014).

Therefore, cyberbullying is a type of intimidation and violence (Berardelli et al. 2018), that little by little has become generalized over the years and the new technologies that have emerged (Viejo and Ortega 2015), This has led to this topic being included in the discursive field of health, associated between the practice of cyberbullying and the harmful results that it produces on the health of the aggressors and intimidated (Alim 2016).

Even considering the peaceful adolescent population, it is necessary to promote socialization and awareness programs, training and monitoring, which allow people to function in an environment of positive coexistence (Gairín et al. 2013).

Other authors suggest making a comprehensive prevention effort among students, school personnel, and parents, in order to obtain results, instead of failed attempts, which do not bring good individual results (Hébert et al. 2016).

Research suggests that students, parents, partners, and community leaders work together to achieve greater results and reduce cyber victimization among young people (Vaala and Bleakley 2015).

In this study, it has been chosen to examine cyberbullying according to the means by which it is carried out, the age range in which it occurs, gender most affected, and what psychological effect it causes on people in educational centers in the province of Imbabura.

2 Methodology

This study was developed within the framework of the Project for Liaison with the Collectivity of the Universidad Técnica del Norte, called "Promotion and dissemination of the safe use of the internet as a fundamental right of privacy and data protection in the city of Ibarra", whose objective was "Promote and disseminate the fundamental right of data protection to parents of the Educational Units of the canton Ibarra and also contribute to raise awareness about the value of privacy and responsible use of information."

To evaluate the levels of occurrence of cyberbullying in students in the city of Ibarra, a survey was carried out that was authorized by parents and students, who participated in the project's socialization workshops and authorized the application of the Reference survey, which consisted of 16 questions, which were divided into 8 categorical variables and 1 binary variable, where 1999 observations were obtained from secondary school students from 12 educational units of the eighth, ninth and tenth year of basic general education and first, the second and third year of unified general baccalaureate, the ages ranged from 10 to 19 years. This sample consisted of 1280 students from fiscal educational units, 371 students from fiscal-commissioned educational units, and 348 students from private institutions, of which a total of 942 female students and 1057 male students were obtained.La sección de la encuesta enfocada en el análisis de la ocurrencia de ciberbullying contenía 16 preguntas, que se divided into 8 categorical variables, 1 binary variable and 7 ordinal variables that were adjusted to the Likert scale.

The results of the surveys obtained were treated using the statistical programming language R and for the analysis of the occurrence of cyberbullying, as this is a binary variable, Pearson's χ^2 test with Yates continuity correction and the χ^2 test of Paired Nominal Independence as post-hoc.

Additionally, the reliability of the results of the ordinal variables was verified using Cronbach's Alpha. For this, it began by checking the additivity of the items of the unifactorial model, using the Pearson correlation matrix, where it was visualized that no pair of questions was perfectly correlated.

The parametric assumptions of the sample, linearity, normality, homogeneity and homoscedasticity were verified by means of a false regression analysis, based on the standardized, studentized and adjusted quantiles obtained from each observation of the database, the same ones that were compared. With the theoretical quantiles of the χ ^ 2 distribution. The results are presented in Fig. 1.

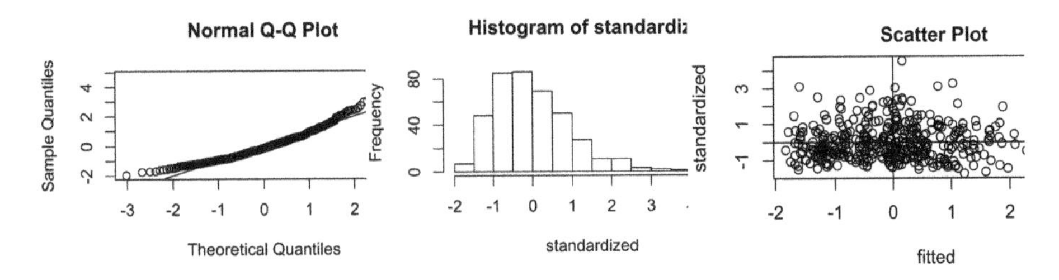

Fig. 1. Verification of parametric assumptions for databases

As can be seen in Fig. 1, the Q-Q plot shows a linear trend between the quantiles of the sample and the theoretical quantiles for the interval between –2 and 2, so the assumption of linearity was accepted.. Normality was visualized by means of a histogram of the standardized quantiles, which follow a normal distribution as can be seen in Fig. 1b, so the assumption of normality was accepted.. Homogeneity and homoscedasticity were verified using a scatter diagram of the adjusted and scaled quantiles, which, as shown in Fig. 1c, are homogeneously distributed in the four quadrants and do not present groupings or trends between them. Therefore, the assumptions of homogeneity and homoscedasticity were accepted.

Due to the fact that the database used for the analysis of the occurrence of cyberbullying only presented a binary variable, the same that was used for the analysis of different strata of the population, the reliability verification of the information extracted from the sample was carried out using Cronbach's Alpha reliability analysis. In this analysis, an $\alpha = 0.89$ value was obtained, so that according to the recommendation of (Nunnally 1975), this value is above 0.7, which is the recommended limit for tests of extension less than 20 items, for what was accepted reliability in the data obtained through the applied survey. In Fig. 2, the results obtained using Cronbach's Alpha are presented (Table 1).

Table 1. Results of the reliability analysis Cronbach's Alpha

raw_alpha	*std. alpha*	*G6 (smc)*	*average_r*	*ase*	*mean*	*sd*	*median_r*	*95% confidence boundaries*
0.89	0.89	0.81	0.81	0.033	0.45	0.16	0.81	$0.83 \leq \alpha \leq 0.96$

3 Results

Pearson's χ ^ 2 test allowed in the first instance to verify if there are significant differences in the frequencies of the population strata, defined by each categorical variable before the binary variable, defined by the occurrence of cyberbullying, where the zero value was used to define that the person had not experienced cyberbullying and one was used for those who had experienced it.

In this way, the first comparison between strata was made using the age of the students as a categorical variable, compared to the binary variable occurrence of cyberbullying. The frequency distribution and its graphic representation are presented in Fig. 2, and the results of the χ^2 and χ^2 Paired Nominal Independence tests are presented in Table 2.

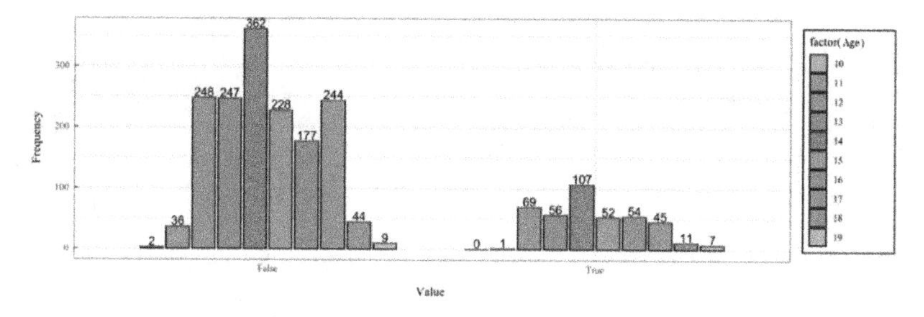

Fig. 2. Bar diagram of the occurrence of Cyberbullying at different ages

Table 2. Results of the χ^2 and χ^2 Paired Nominal Independence tests

Pearson's Chi-squared test					
χ - Squared	21.878	**df**	9	**p-value**	*0.009275 Significant*

Chi-squared Pairwise Nominal independence Test

Age	10	11	12	13	14	15	16	17	18
11	*0.0000 Significant*								
12	*0.0000 Significant*	*0.0111 Significant*							
13	*0.0000 Significant*	*0.0283 Significant*	*0.3581*						
14	*0.0000 Significant*	*0.0076 Significant*	*0.2040*	*0.0177 Significant*					
15	*0.0000 Significant*	*0.0280 Significant*	*0.3858*	*1.0000*	*0.02999 Significant*				
16	*0.0000 Significant*	*0.0075 Significant*	*0.7320*	*0.2013*	*0.0364 Significant*	*0.2210*			
17	*0.0000 Significant*	*0.0620 Significant*	*0.0650*	*0.4054*	*0.0230 Significant*	*0.4008*	*0.0323 Significant*		
18	*0.0000 Significant*	*0.0357 Significant*	*0.0929*	*0.0620*	*0.0205 Significant*	*0.0472 Significant*	*0.2797*	*0.5377*	
19	*0.0000 Significant*	*0.0006 Significant*	*0.0820*	*0.0313 Significant*	*0.0105 Significant*	*0.0331 Significant*	*0.1265*	*0.0099 Significant*	*0.1105*

As can be seen in Table 2, among the cases that did experience cyberbullying, the age of 14 stands out, in which a greater number of cases was found and through the χ ^ 2 test of matched nominal independence, it was identified that, this age as a stratum of the population, presents significant differences with respect to all the other ages of secondary school students, except for the 12-year-old age, which ranked second as the category with the highest frequency.

Next, the occurrence of cyberbullying was analyzed according to the type of educational institution, therefore, in this case, three types of institutions are analyzed: private, fiscal and fiscomisional. Again, since the occurrence of cyberbullying was tabulated as a binary variable, the tests run for this analysis were the χ ^ 2 and χ ^ 2 Paired Nominal Independence tests. The results for the frequencies of the different strata and the results of the statistical tests carried out are presented in Fig. 3 and Table 3.

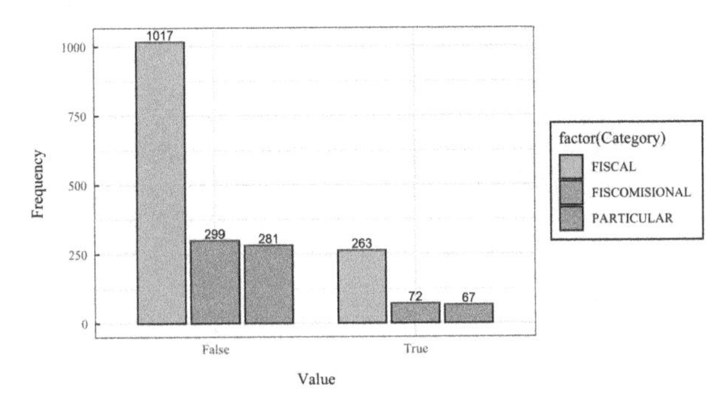

Fig. 3. Bar diagram of the occurrence of Cyberbullying for each type of educational institution

Table 3. Results of the χ^2 and χ^2 Paired Nominal Independence tests

Pearson's Chi-squared test					
χ - Squared	0.42532	**df**	2	**p-value**	*0.8084 Not significant*

Chi-squared Pairwise Nominal independence Test		
Educational institution type	Public	Government-dependent private
Government-dependent private	*0.6837 Not significant*	
Private	*0.6474 Not significant*	*0.9998 Not significant*

As can be seen in Fig. 3 and Table 3, in a descriptive way there is a greater number of students surveyed in the sample, belonging to fiscal institutions and it is for this reason that a high frequency value is visualized for the occurrence of cyberbullying in this type of institution. However, when applying the statistical tests, it was found that Pearson's χ ^ 2 test presented a retention of the null, suggesting that there are no significant differences in the groups within the population, which was ratified when executing the independence tests nominal, where all presented withholding of the null, so it cannot be said that there are differences between the strata. As shown in Fig. 3, this situation was expected, since, despite having a greater number of students from public institutions than students from fiscal and private institutions, the proportions between the non-occurrence and the occurrence of cyberbullying were similar with values of 3.86, 4.15 and 4.19 respectively, therefore, it is concluded that it cannot be stated that the occurrence of cyberbullying differs for each type of educational institution and has a generalized effect independent of this category.

Subsequently, the same analysis was carried out for the occurrence of cyberbullying based on the gender of the individuals that make up the database. The results are presented in Fig. 4 and Table 4.

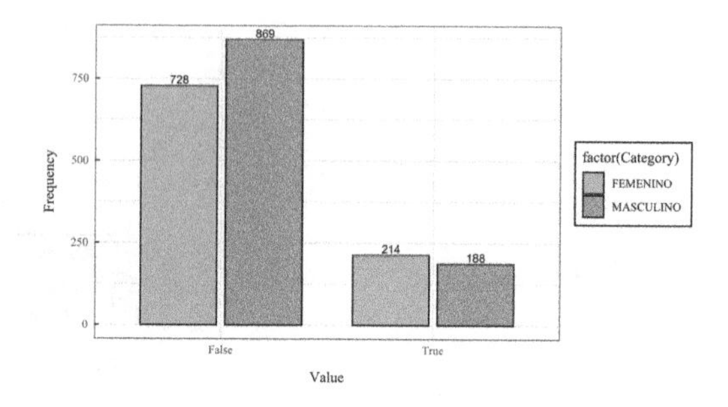

Fig. 4. Bar diagram of the occurrence of cyberbullying according to gender

Table 4. Results of the χ^2 and χ^2 Paired Nominal Independence tests

Pearson's Chi-squared test with Yates' continuity correction					
χ- Squared	7.2359	df	1	p-value	*0.007146 Significant*
Chi-squared Pairwise Nominal independence Test					
Gender			Female		
Male			*0.0071 Significant*		

As can be seen in Fig. 4 and Table 4, in the sample there are two categories that make up the gender of the students surveyed and the binary variable occurrence of cyberbullying, which determines the frequency of occurrence of this event. In this way, when executing the χ ^ 2 test on the contingency table, the proportion of positive cases of cyberbullying in the female stratum presents a significant difference over the male stratum and this is confirmed, when executing the paired test, obtaining significance in both cases with a p-value of 0.007146 which is well below the significance threshold. This, due to the fact that there was a greater occurrence in the female gender, even when this stratum is less numerous than the male gender, for which it is concluded that cyberbullying is a phenomenon that mainly affects the female gender.

Additionally, the occurrence of cyberbullying was analyzed for each of the grades surveyed in the educational units, which were made up of all secondary parallels that are: eighth, ninth and tenth year of basic general education and first, second and third year unified general baccalaureate. The results obtained from the experiment and the statistical tests carried out are presented in Fig. 5 and Table 5.

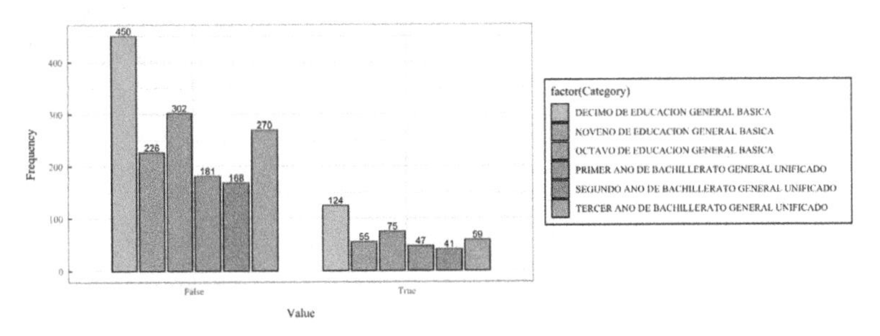

Fig. 5. Bar diagram of the occurrence of Cyberbullying for each grade

Table 5. Results of the χ^2 and χ^2 Paired Nominal Independence tests

		Pearson's Chi-squared test			
χ - Squared	11.0835	**df**	5	**p-value**	*0.04573 Significant*
		Chi-squared Pairwise Nominal independence Test			
Grade	Décimo EGB	Noveno EGB	Octavo EGB	Primero BGU	Segundo BGU
Noveno EGB	*0.4487 Not significant*				
Octavo EGB	*0.4193 Not significant*	*0.0027 Significant*			
Primero BGU	*0.1686 Not significant*	*0.1433 Not significant*	*0.0872 Not significant*		
Segundo BGU	*0.3855 Not significant*	*0.0012 Significant*	*0.0024 Significant*	*0.0177 Significant*	
Tercero BGU	*0.2171 Not significant*	*0.3209 Not significant*	*0.4290 Not significant*	*0.1115 Not significant*	*0.2929 Not significant*

In Fig. 5 and Table 5 it can be seen that there are some notable differences in the population strata. Although the number of students surveyed in the different grades is not the same, the ratio between their positive and negative cases is what determines the existence or not of significant differences. In this way, it can be seen that, in the ninth school year, despite having few students in the sample, it contains numerous cases of cyberbullying, which determine the significant differences that are visualized in Table 5, which are comparable in proportion to the tenth year cases, which is a more numerous stratum. In the same way, that, in the second year of unified general high school, despite being the least numerous stratum of the sample, it presents a number of cases of cyberbullying similar to the other strata. Therefore, it can be concluded that the aforementioned parallels are of special interest at the moment, since they present a high occurrence of cyberbullying cases, which was to be expected, since they coincide with the analysis carried out by age.

Continuing with the process of analyzing the categorical variables available in the survey, the study of the categorical variable that corresponds to the use that students give to electronic devices and media and the occurrence of the cyberbullying event was carried out, depending on the main activity for which these media are used. The alternatives considered as activities that students use on electronic devices were: education, recreation and social networks. Figure 6 and Table 6 present the results obtained.

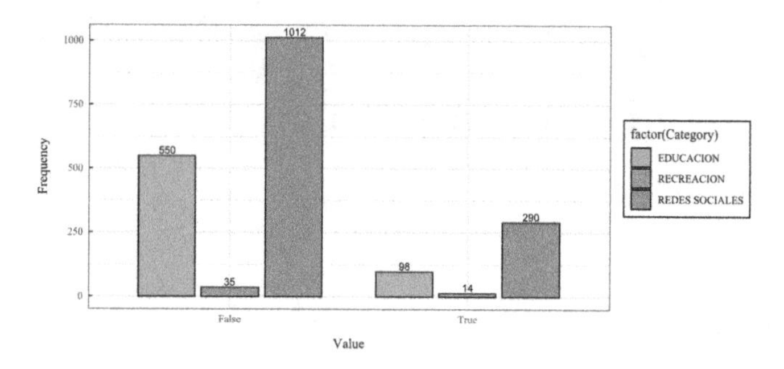

Fig. 6. Bar diagram of the occurrence of Cyberbullying before the main activities for which students use electronic devices

Table 6. Results of the χ^2 and χ^2 Paired Nominal Independence tests

		Pearson's Chi-squared test			
χ - Squared	16.006	**df**	2	**p-value**	*0.0003345 Significant*
		Chi-squared Pairwise Nominal independence Test			
Purpose		Education		Recreation	
Recreation		*0.0232 Significant*			
Social Networks		*0.0002 Significant*		*0.3886 Not significant*	

As can be seen in Fig. 5, high school students mostly use electronic devices to access social networks, secondly, they use them to develop educational activities and in fewer numbers for recreation. In addition, in Table 6 it can be verified that through the χ^2 test it was determined that there are significant differences of importance within the strata determined by the categorical variable, which is why the nominal independence test was executed, where it was found that there are significant differences in the occurrence of cyberbullying, in relation to the recreation - education and social networks - education strata. When analyzing the proportions between the positive and negative cases of the education, recreation and social networks strata, which are 5.61, 2.5 and 3.48 respectively, it can be concluded that students who use electronic devices mainly for educational activities are less likely to experience cyberbullying than students who use these means, mainly to access social networks or for recreation. Additionally, there is a null retention when comparing recreation activities with social networks, so it cannot be said that there is a difference between these strata and they may be prone to cyberbullying in a similar way.

Finally, the frequency of positive and negative cases of cyberbullying in the sample of 1999 students and their dependence on the social networks that the student mostly uses was evaluated.: Facebook, Instagram, Twitter, Whatsapp and Youtube. In this way, the different social networks were arranged as categories and the occurrence of cyberbullying as a binary variable. The results obtained through the statistical tests in the database are presented in Fig. 7 and Table 7.

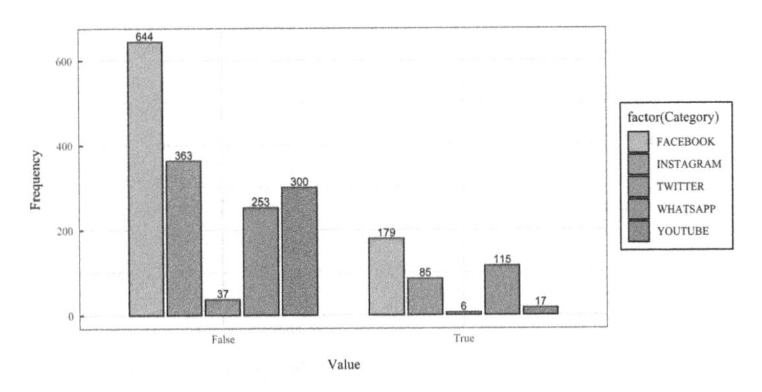

Fig. 7. Bar diagram of the occurrence of Cyberbullying for the social network most used by the student

Table 7. Results of the χ ^ 2 and χ ^ 2 Paired Nominal Independence tests

		Pearson's Chi-squared test			
χ - Squared	74.089	**df**	5	**p-value**	3.104×10^{-15} *Significant*
		Chi-squared Pairwise Nominal independence Test			
Social network	Facebook	Instagram	Twitter	Whatsapp	Second BGU
Instagram	*0.2742 Not significant*				
Twitter	*0.3053 Not significant*	*0.5460 Not significant*			
Whatsapp	*0.0005 Significant*	*7.0362e-05 Significant*	*0.0294 Significant*		
Youtube	*9.0273e-11 Significant*	*8.9277e-08 Significant*	*0.04735 Significant*	*2.4963e-17 Significant*	

As can be seen in Fig. 7, the social network most used by students is Facebook, but it should be considered that, in the survey, the student was asked to select the social network that he uses the most, so that the student can use more than a social network; However, that was not the objective of this research and it is left as future work to analyze the combinatorial alternatives that may exist. In this way, it was possible to establish a comparison of social networks and their frequency based on cases of cyberbullying and for the test χ^2 significance was obtained in the pooled sample, so all possible paired analyzes were carried out. Through the paired nominal independence test, the special case of the social network WhatsApp was detected, where, despite not being the most numerous social network selected by users, it was found that it presents a very high level of occurrence of cyberbullying, which is similar to the one presented by Facebook, which is three times more numerous. Based on the above, it is concluded that WhatsApp is the social network that provides the most favorable scenario for the occurrence of cyberbullying, and significance was obtained in all the paired tests with respect to the other social networks. YouTube stands out as the social network with the least occurrence of cyberbullying, reaching p-values of $9.0273e-11$, $8.9277e-08$, 0.04735 and $2.4963e-17$ with respect to other social networks. Finally, it should be mentioned that no significant differences were detected between the social networks Facebook, Twitter and Instagram, which suggests that they have a similar behavior.

4 Discussion

Cyberbullying is a problem that occurs in Educational Institutions, be they fiscal, fiscal, or private, whose occurrence is similar and is in proportion to the sample applied to these strata. Similarly, the results of this research show that cyberbullying is more frequent in 14-year-old students, and in second place are 12-year-old students, as pointed out by (Young et al. 2016), who states that the main victims of cyberbullying are adolescents between 10 and 14 years old. The age data are related to the school year; thus, the largest cases of cyberbullying are observed in the tenth year and in the second year of unified general high school.

According to the results of this research, adolescent women are the most vulnerable and cyberbullying is a phenomenon that affects them mostly, despite the fact that their number is lower compared to the male stratum of the sample obtained for this study. (Beckman et al. 2013), states that adolescent women are the most prone to cyberbullying, compared to men, however, both genders are victimized by this type of bullying. On the other hand, authors like (Katzer et al. 2009) they do not distinguish between gender.

Another of the results presented by this research is on the use of electronic devices, which are used by students mainly to access social networks, then to develop educational activities and finally for recreation. The results show that students who use electronic devices for educational activities are less likely to be victims of cyberbullying.

Finally, this study indicates that the most used social networks are Facebook, Instagram, Twitter, WhatsApp, and YouTube. The occurrence of a greater number of cyberbullying occurs mainly through the social network WhatsApp, however, the other social networks also present a favorable scenario for cyber-attack, and present a similar behavior among them, unlike YouTube with less occurrence of this type of attack. In accordance with (McFarlane 2014), cited in (Alim 2016), indicates that 75% of surveyed students use Facebook, and 54% of these report having experienced cyberbullying on this social network.

5 Conclusions

The use of the internet has spread in the educational field, and is a useful tool that offers a range of benefits, such as enabling psychosocial development and building social relationships. However, these interrelationships can lead to violent actions and so-called cyberbullying.

Cyberbullying occurs in fiscal, trust and private institutions, showing similar bullying figures among students belonging to these institutions. The most vulnerable age group is between the ages of 12 and 14, which coincides with the stage of physical, physiological and emotional changes, what makes teens the main focus for cyberbullying.

Analysis of the gender variable shows that adolescent women are the biggest victims of cyberbullying, however, teenage men do not escape being victims of cyberbullying, in fact there is research that does not differentiate gender around such attacks.

This type of intimidation is classified as an attack of extreme aggression, according to the scientific articles investigated, depending on the number of people online, this means that the more users are connected to the internet, the aggressor has a wider number of potential victims, forcing investigators to conduct extensive and multidisciplinary analysis of this issue.

Social media is used by teens who are unaware of the dangers they face, and the dangers of cyberbullying should be considered in linkage and research projects.

Future research could lead to a study on cyberbullying prevention mechanisms in educational institutions in the city of Ibarra, as well as the study of sexting and its impact on the lives of affected adolescents.

References

Abreu, R., Kenny, M.: Cyberbullying and LGBTQ youth: a systematic literature review and recommendations for prevention and intervention. J. Child Adolesc. Trauma **11**(1), 81–97 (2018). https://doi.org/10.1007/s40653-017-0175-7

Akkın, H.G., Demir, T., Gökalp, B., Kadak, M.T., Poyraz, B.Ç.: Use of social network sites among depressed adolescents. Behav. Inf. Technol. **36**(5), 517–523 (2016). https://doi.org/10.1080/0144929X.2016.1262898

Alhujailli, A., Karwowski, W.: Emotional and Stress Responses to Cyberbullying, vol. 776. Springer, Heidelberg (2018). https://doi.org/10.1007/978-3-319-94622-1

Alim, S.: Cyberbullying in the world of teenagers and social media: a literature review. Int. J. Cyber Behav. Psychol. Learn. **6**(2), 68–95 (2016). https://doi.org/10.4018/IJCBPL.2016040105

Amado, J., Matos, A., Teressa, P., Thomas, J.: Cyberbullying : Um desafio À Investigação E À Formação. **326**(13), 301–326 (2009)

Arsène, M., Raynaud, J.P.: Cyberbullying (ou cyber harcèlement) et psychopathologie de l'enfant et de l'adolescent: État actuel des connaissances. Neuropsychiatrie de l'Enfance et de l'Adolescence **62**(4), 249–256 (2014). https://doi.org/10.1016/j.neurenf.2014.01.012

Barry, C., Simon, F.: Cyberbullying and cyber law a Canadian perspective. Value Health **1**(4), 218–223 (2016)

Bastiampillai, T., Allison, S., Perry, S., Licinio, J.: Social network theory and rising suicide rates in the USA. The Lancet **393**(10183), 1801 (2019). https://doi.org/10.1016/S0140-6736(18)33048-4

Beckman, L., Hagquist, C., Hellström, L.: Discrepant gender patterns for cyberbullyiyng and traditional bullying - an analysis of Swedish adolescent data. Comput. Hum. Behav. **29**(5), 1896–1903 (2013). https://doi.org/10.1016/j.chb.2013.03.010

Berardelli, I., Corigliano, V., Hawkins, M., Comparelli, A., Erbuto, D., Pompili, M.: Lifestyle interventions and prevention of suicide. Front. Psychiatry **9**(NOV), 1–10 (2018). https://doi.org/10.3389/fpsyt.2018.00567

Bottino, S.M., Bottino, C., Regina, C., Lobo, A., Silva, W.: Cyberbullying and adolescent mental health: systematic review. Cadernos de Saúde Pública **31**(3), 463–475 (2015). https://doi.org/10.1590/0102-311x00036114

Castells, M.: El impacto de internet en la sociedad: una perspectiva global, pp. 1–6 (2013)

Chan, H.C., (Oliver), & Wong, D. : Traditional school bullying and cyberbullying in Chinese societies: prevalence and a review of the whole-school intervention approach. Aggress. Violent. Behav. **23**, 98–108 (2015). https://doi.org/10.1016/j.avb.2015.05.010

Correa, A., Avedaño, B.L.: Diseño y análisis psicométrico de un instrumento para detectar presencia de ciberbullying en un contexto escolar. Psychol. Soc. Educ. **7**(2), 213–226 (2015). https://doi.org/10.25115/psye.v7i2.534

Della, V., O'Neil, A., Craig, W.: Learning from traditional bullying interventions: a review of research on cyberbullying and best practice. Aggress. Violent. Behav. **23**, 61–68 (2015). https://doi.org/10.1016/j.avb.2015.05.009

Domínguez, J.A., Vázquez, E.V., Nuñez, S.L.: Cyberbullying escolar: Incidencia del teléfono móvil e internet en adolescentes. RELIEVE - Revista Electronica de Investigacion y Evaluacion Educativa **23**(2), 1–1 (2017). https://doi.org/10.7203/relieve.23.2.8485

Donegan, R.: Bullying and Cyberbullying: history, statistics, law, prevention and analysis. Elon J. Undergraduate Res. Commun. **3**(1), 33–42 (2012)

Durán, M., Martínez, R.: Ciberacoso mediante teléfono móvil e Internet en las relaciones de noviazgo entre jóvenes. Comunicar: Revista Científica Iberoamericana de Comunicación y Educación **44**, 159–167 (2015). https://doi.org/10.3916/C44-2015-17

El Asam, A., Samara, M.: Cyberbullying and the law: a review of psychological and legal challenges. Comput. Hum. Behav. **65**, 127–141 (2016). https://doi.org/10.1016/j.chb.2016.08.012

Fernández, J., Peñalva, A., Irazabal, I.: Hábitos de uso y conductas de riesgo en Internet en la preadolescencia. Comunicar **22**(44), 113–121 (2015). https://doi.org/10.3916/C44-2015-12

Foody, M., Samara, M., Carlbring, P.: A review of cyberbullying and suggestions for online psychological therapy. Internet Interv. **2**(3), 235–242 (2015). https://doi.org/10.1016/j.invent.2015.05.002

Fundación ANAR, Fundación Mutua Madrileña. II Estudio sobre acoso escolar y ciberbullying según los afectados. In *Fundación ANAR* (2017). https://www.anar.org/wp-content/uploads/2017/04/INFORME-II-ESTUDIO-CIBERBULLYING.pdf

Gairín, J., Armengol, C., Silva, B.: EL «BULLYING» ESCOLAR. CONSIDERACIONES ORGANIZATIVAS Y ESTRATEGIAS PARA LA INTERVENCIÓN. Educacion XX1 **16**(1), 161–190 (2013). https://doi.org/10.5944/educXX1.16.1.722

Garaigordobil, M., Mollo, J.P., Larrain, E.: Prevalencia de Bullying y Cyberbullying en Latinoamérica: una revisión. Revista Iberoamericana de Psicología **11**(3), 1–18 (2019). https://doi.org/10.33881/2027-1786.rip.11301

Hébert, M., Cénat, J.M., Blais, M., Lavoie, F., Guerrier, M.: Child sexual abuse, bullying, cyberbullying, and mental health problems among high schools students: a moderated mediated model. Depression Anxiety **33**(7), 623–629 (2016). https://doi.org/10.1002/da.22504

Hinduja, S., Patchin, J.: Connecting adolescent suicide to the severity of bullying and cyberbullying. J. School Violence **18**(3), 333–346 (2019). https://doi.org/10.1080/15388220.2018.1492417

John, A., Glendenning, A.C., Marchant, A., Montgomery, P., Stewart, A., Wood, S., Lloyd, K., Hawton, K.: Self-harm, suicidal behaviours, and cyberbullying in children and young people: systematic review. J. Medical Internet Res. **20**(4) (2018). https://doi.org/10.2196/jmir.9044

Kanyinga, S., Hamilton, H.: Social networking sites and mental health problems in adolescents: the mediating role of cyberbullying victimization. Eur. Psychiatry **30**(8), 1021–1027 (2015). https://doi.org/10.1016/j.eurpsy.2015.09.011

Katzer, C., Fetchenhauer, D., Belschak, F.: Cyberbullying: who are the victims? a comparison of victimization in internet chatrooms and victimization in school. J. Media Psychol. **21**(1), 25–36 (2009). https://doi.org/10.1027/1864-1105.21.1.25

Kim, S., Kimber, M., Boyle, M., Georgiades, K.: Sex differences in the association between cyberbullying victimization and mental health, substance use, and suicidal ideation in adolescents. Can. J. Psychiat. **64**(2), 126–135 (2019). https://doi.org/10.1177/0706743718777397

Mok, K., Jorm, A., Pirkis, J.: Suicide-related Internet use: a review. Aust. N. Z. J. Psychiatry **49**(8), 697–705 (2015). https://doi.org/10.1177/0004867415569797

Moreta, R., Poveda, S., Ramos, I.: Indicadores de violencia relacionados con el Ciberbullying en adolescentes del Ecuador. Pensando Psicologia **14**(24), 1–12 (2018). https://revistas.ucc.edu.co/index.php/pe/article/view/1895/2559

Ortega, J., Buelga, S., Cava, M.J.: Influencia del clima escolar y familiar en adolescentes, víctimas de ciberacoso. Comunicar: Revista Científica Iberoamericana de Comunicación y Educación **24**(46), 57–65 (2016)

Ortega, R., Elipe, P., Mora, J., Genta, L., Brighi, A., Guarini, A., Smith, P., Thompson, F., Tippett, N.: The emotional impact of bullying and cyberbullying on victims: a European cross-national study. Aggressive Behav. **38**(5), 342–356 (2012). https://doi.org/10.1002/ab.21440

Rodríguez, M., Arroyo, M.J.: Las TIC al servicio de la inclusión educativa. Dig. Educ. Rev. **25**(1), 108–126 (2014). https://doi.org/10.1344/der.2014.25.108-126

Sánchez, L., Crespo, G., Aguilar, R., Bueno, F., Benavent, R., Juan, V.: Los adolescentes y el ciberacoso. J. Chem. Inf. Model. **53**(9) (2013). https://doi.org/10.1017/CBO9781107415324.004

Starcevic, V., Aboujaoude, E.: Cyberchondria, cyberbullying, cybersuicide, cybersex: "new" psychopathologies for the 21st century? World Psychiatry **14**(1), 97–100 (2015). https://doi.org/10.1002/wps.20195

Swartz, M., Bhattacharya, S.: The Long-term impact of bullying victimization on mental health. World Psychiatry **16**(1), 26–27 (2017). https://doi.org/10.1002/wps.20393

Tokunaga, R.: Following you home from school: a critical review and synthesis of research on cyberbullying victimization. Comput. Hum. Behav. **26**(3), 277–287 (2010). https://doi.org/10.1016/j.chb.2009.11.014

UNESCO. A New Global Campaign to Address Cyberbullying (2017). https://www.unesco.org/new/en/media-services/single-view/news/a_new_global_campaign_to_address_cyberbullying/

UNICEF, Ministerio de Educación, & VisionWorld. UNA MIRADA EN PROFUNDIDAD AL ACOSO ESCOLAR EN EL ECUADOR. **13**(52), 661–670 (2013). https://doi.org/10.4321/S1139-76322011000600016

Vaala, S., Bleakley, A.: Monitoring, mediating, and modeling: parental influence on adolescent computer and internet use in the United States. J. Child. Media **9**(1), 40–57 (2015). https://doi.org/10.1080/17482798.2015.997103

Viejo, C., Ortega, R.: Cambios y riesgos asociados a la adolescencia. Psychol. Soc. Educ. **7**(2), 109–118 (2015). https://doi.org/10.25115/psye.v7i2.527

Young, R., Subramanian, R., Miles, S., Hinnant, A., Andsager, J.: social representation of cyberbullying and adolescent suicide: a mixed-method analysis of news stories. Health Commun. **32**(9), 1082–1092 (2016). https://doi.org/10.1080/10410236.2016.1214214

Nunnally, J.C.: Psychometric theory' 25 years ago and now. Educ. Researcher **4**(10), 7–21 (1975). https://doi.org/10.3102/0013189X004010007

Human-Computer Interaction

Factors Influencing the Adoption of Geolocation and Proximity Marketing Technologies

Elizabeth Ramírez Correa, Erika Pulido Arjona[1], Carlos Osorio[1](✉), and Stefania Pareti[2]

[1] Universidad de Manizales, Manizales, Colombia
cosoriot@umanizales.edu.co
[2] Universidad del Desarrollo, Santiago de Chile, Chile
s.pareti@udd.cl

Abstract. The use of geolocation technologies is an opportunity for businesses dedicated to retail the city of Manizales can approach their customers and offer different promotions, discounts, advertising and better and improve the customer experience, using its location. However, to ensure the proper design of a marketing strategy of proximity, it is necessary to know the factors that influence the acceptance that will have the use of this technology in users or end consumers, ensuring that stores in the city can offer appropriate services. Through the use of the UTAUT2 model an exploration of those factors that influence the adoption and use of this technology is made using a PLS-SEM model. The results confirm that the variables hedonic motivations, performance expectation, effort expectation and price are statistically significant for users. The use of this technology brings a better experience for and to the customer, providing added value to the final product.

This study provides a real time communication way for the marketing professionals to approach customers who use smartphones, having a better shopping experience.

Keywords: Proximity marketing · UTAUT2 · Consumer attitudes · PLS-SEM · Geomarketing

1 Introduction

Nowadays, smartphones have become very important, since they generate, store and process a great amount of data that is very useful for marketing, allowing you to know who the consumers are and how they behave, helping to predict their preferences; besides offering the possibility of connecting brands with their users in real time. Mobile Marketing Association (MMA) defines mobile marketing as "a set of practices that allows organizations to communicate and interact with their audience in an interactive and relevant way through any mobile device or network" [1]. This multichannel activity allows different actions, the most disruptive and innovative being geolocation, since from the coordinates that determine the location of a user, brands can

Á. Rocha et al. (Eds.): ICITS 2021, AISC 1330, pp. 517–525, 2021.
https://doi.org/10.1007/978-3-030-68285-9_48

deliver relevant messages of interest. All these channels allow for personalization by taking into account the habits and preferences of the users, time and location [2].

Geolocation is a technology that allows tracking the location of electronic devices with a good degree of accuracy and has evolved sustainably with mobile devices. With this data, marketing strategies reach levels of sophistication and efficiency desired by these industries, the great "Transmit the right message, to the right person, at the right time and in the right place" [3]. This technology is especially useful for the retail sector because it enables real-time delivery of promotions, discount coupons, catalogs, offers and other valuable communications, resulting in brand loyalty. In summary, the use of this technology allows to improve the communication between the commerce and the buyers, increasing the perception of the clients in the purchase experience. The use of this technology generates a better communication between the company and the customers who are using it.

When it comes to proximity marketing, or also called geomarketing, a new generation of low-cost devices allows marketing specialists to track the exact location of consumers through their mobile devices [4]. And it is precisely the increase in the availability of technologies and devices that makes small companies and establishments dedicated to commerce think about geomarketing as a feasible alternative. However, due to the recent availability of these technologies for small businesses, it is necessary to delve into the factors that influence the adoption of these technologies by them, this being the main purpose of this research.

The document is organized in the following way: in the following section, a literature review is presented that will provide the necessary elements to understand what is related to proximity marketing and the factors to be taken into account for the adoption of this technology based on the UTAUT2 model. Afterwards, the methodology to be worked on will be described, based on structural equations following the PLS-SEM model, followed by the analysis of the results. Finally, the findings and conclusions of the work are discussed.

2 Literature Review

According to Andrews [5] new technologies such as proximity marketing bring value by increasing revenues through strategies to attract new buyers, increase existing buyers, or generate greater consumer surplus. Additionally, Andrews and his group argue that through technology buyers "improve their perceptions of fairness, value, satisfaction, trust, commitment and attitudinal loyalty and evaluate the potential intrusion of technology into their personal privacy" [5].

Research in geomarketing and proximity marketing is still under development and more research is needed. Among the literature found is the research of Inman and Nikolova [6] which reviews the evolution of retail technologies, including beacons and other proximity marketing technologies to offer personalized offers and prices.

There is also the book edited by Cliquet [7] where the methods and strategies of geomarketing are worked out, and its use is proposed for retail location, consumer behavior, and marketing management. Being the first one where you can find a majority of literature. The consumer behavior is the focus of this research, taking as examples the research applied to the trade of Vyt [8] and Zaim and Bellafkih [9].

Another common approach in geomarketing research is related to communication protocols, where Cockrill [10] found at that time that one of the barriers for the adoption of proximity technologies based on gps and bluetooth, due to their high battery consumption. However, this obstacle has been re-evaluated by different researchers [9, 11, 12] proposing low energy bluetooth and other wireless technologies that improve this critique. Another issue is the perception of distrust by users regarding this technology due to privacy issues,

2.1 Factors Influencing Consumer Acceptance of Technology

Previous research in this area used the technology adoption model (TAM) [13] to evaluate the acceptance of proximity marketing strategies [14, 15]. However, this model has evolved to the Revised Unified Theory of Technology Acceptance and Use (UTAUT2) [16] which considers presents a revised version of behavioral models focused on the prediction of behavioral intent where the theory of planned behavior (TPB) and TAM stand out. Within the facts.

Table 1. Constructs and definitions from UTAUT2

Factor	Definition based on Venkatesh (2012) [16]	Hypotheses
Performance expectancy	Refers to the degree to which the user believes that using a specific technology will enhance performance when executing an task	H1. Performance expectation positively influences the intention to adopt Proximity Marketing technologies
Effort expectancy	Refers to the degree of ease or difficulty the individual feels associated with the use of	H2. Technology. Less effort when using a proximity marketing technology influences the intention of adoption and use
Social influence	Refers to the degree to which the individual believes that others consider it important that	H3. Technology be used. Social influence affects the intent of adoption and use of proximity marketing technologies
Hedonic motivations	Refers to the fun and/or enjoyment provided to the individual due to the use of the technology in question	H4. Hedonic motivations affect the intention of adoption and use of proximity marketing technologies
Price value	Refers to the cognitive exchange between consumers, the benefits of the applications and the monetary cost of using them	H5 Price Influences Intent to Adopt and Use Proximity Marketing Technologies
Habit	Refers to the automatism when you learn something, so there is a preference for the use of certain tool	H6. Habit influences the intention of adoption and use of proximity marketing technologies
Behavioral intention	Refers to the intention to consume a certain technological product or service	

Privacy
Considering the users' perception of privacy, it is considered appropriate to include this variable for the proposed model. Privacy refers to users' concerns about the collection and use of their location, Cordiglia [17], indicates that users have privacy concerns that their information is being misused and believe that location tracking is an invasion of their privacy.

H7. Privacy concerns negatively influence the intended adoption of proximity marketing technologies.

3 Research Methodology

Considering that the UTAUT2 has been selected as theoretical framework, re-search using this theory recommend using structural equations modelling as method to analyze the data. Likewise, Ramadani [18] used a similar approach for his research on geo-marketing. The sample used for this research was people over 16 years old, cell phone users and residents in the city of Manizales (Colombia). By 2018 the population of Manizales according to data provided by the Center for Information and Statistics was 400,136 people, of which according to the Colombian Regulatory Commission, it is estimated that 289,534 have smartphones; taking as parameters a 95% confidence level and 5% error, the minimum sample required for this population is 384 people. To collect the data, an invitation was sent by email and through Whatsapp with a link to the form to people who met the requirements. In total, 404 responses were received, of which 384 completed the form in its entirety. Of the people who filled out the form 60% were women and 40% men, in the following age ranges: 30% between 16 and 24 years, 18% between 25 and 34, 18% between 35 and 44 and 33% over 45 to 74.

To analyze the data, a Structural Equation Modeling (SEM) technique is proposed with the Partial Least Squares (PLS) method, which is a variance-based approach, using Smart PLS version 3 software [19, 20] for analysis. To evaluate a PLS model two steps are followed: a measurement model (or external model) that allows connecting the manifest variables (MVs) or elements to their own latent variables (LVs) and a structural model that links some endogenous LVs to the other LVs (or internal model) [10–21]. For the evaluation criteria of the model, the standards proposed by Hair [10–20] were followed, the classes are based on the evaluation of Cronbach's Alpha, composite reliability, Average Variance Extracted (AVE), for the validity and reliability of the constructs. For the discriminant validity it is evaluated by the Fornell-Larker criterion and the HTMT. For this research, several iterations of the model were run; starting with all the variables and items proposed in the theoretical model. The initial results showed problems with the user's behavior structure, which presented a high collinearity with use intention. Additionally, problems were found with loads of some of the items, which is evidenced in the validity and reliability coefficients. In subsequent iterations, the variable use behavior was eliminated. Likewise, item 13 of the performance expectation construct was eliminated. After these changes, the final model was run with 5000 iterations reaching adequate values for the evaluation criteria [10–20].

Figure 1 shows the loads of the items in relation to the constructs; as can be seen, these loads are above 0.7, the value recommended by Hair [10–20], except for Q34, which is close to 0.6; however, for theoretical consistency it is preferred to keep it, since it does not affect other indicators or quality of the model. In Table 1, we can see these values of validity and reliability for each of the LVs.

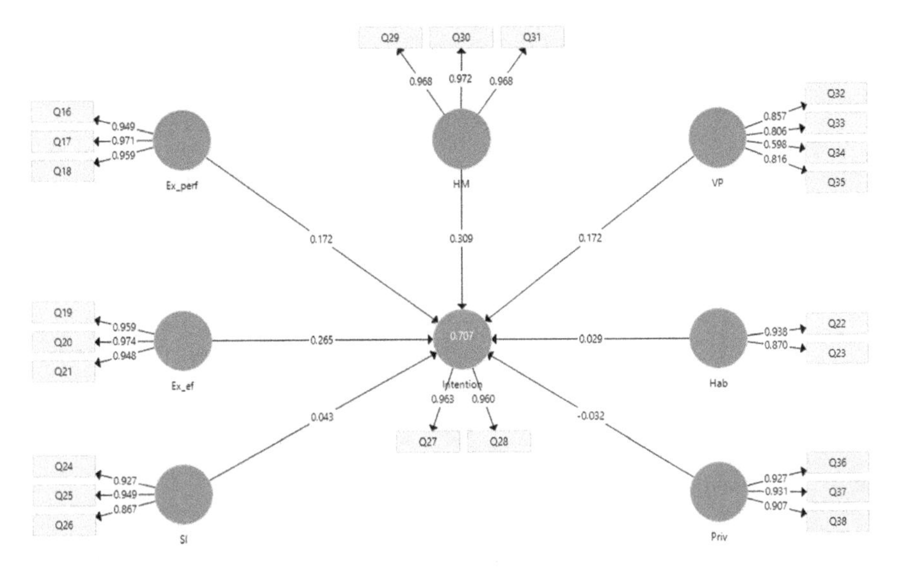

Fig. 1. Final version of the external model.

The reliability and validity construct, as its name indicates, measures the model's validity and reliability, all values exceeding. In the final model the questions Q40 of hedonic motivations were excluded and Q13 because it applied to the whole construct, generating confusion in the model, on the other hand, in spite of the fact that in the initial model the price was below 0.7, punctually with respect to question Q34, nevertheless for the criterion of quality it is well because in Cronbach's alpha-ha it validates it (Tables 2, 3, 4 and 5).

Table 2. Validity and reliability coefficients

	Cronbach's alpha (>0,7)	Composite reliability (>0,5)	Average variance extracted (AVE > 0,5)
Expected effort	0,958	0,958	0,973
Expected performance	0,957	0,957	0,972
Habit	0,785	0,854	0,900
Hedonic motivation	0,968	0,969	0,979
Intention	0,918	0,919	0,961
Privacy	0,912	0,924	0,944
Social influence	0,903	0,915	0,939
Value price	0,774	0,804	0,856

Table 3. Criterio de Fornell-Larcker

	Ex_ef	Ex_perf	Hab	HM	Intention	Priv	SI	VP
Expected effort	0,960							
Expected performance	0,790	0,960						
Habit	0,768	0,727	0,905					
Hedonic motivation	0,696	0,743	0,679	0,969				
Intention	0,756	0,751	0,687	0,756	0,962			
Privacy	0,571	0,518	0,493	0,444	0,464	0,922		
Social influence	0,496	0,545	0,565	0,560	0,531	0,308	0,915	
Value price	0,668	0,656	0,647	0,612	0,673	0,528	0,483	0,776

Table 4. HTMT

	Ex_ef	Ex_perf	Hab	HM	Intention	Priv	SI	VP
Expected effort								
Expected performance	0,825							
Habit	0,858	0,809						
Hedonic motivation	0,722	0,771	0,765					
Intention	0,806	0,801	0,788	0,801				
Privacy	0,606	0,547	0,552	0,466	0,500			
Social influence	0,530	0,582	0,671	0,599	0,580	0,338		
Value price	0,769	0,755	0,811	0,701	0,791	0,632	0,575	

Once the evaluation phase of the external model is concluded and secondly the internal model is evaluated to test the hypotheses raised by it, the internal model validates the hypotheses and the results of the external model by combining the weights of the routes. Hypotheses with a t-value less than 1.96 or a p-value greater than 0.05 are rejected.

Table 5. Internal model

	T Statistics (|O/STDEV|)	P Values	Decision
Expected effort -> Intention	3.29	0.001	Not rejected
Expected performance -> Intention	2.278	0.023	Not rejected
Habit -> Intention	0.45	0.653	Rejected
Hedonic motivation -> Intention	4.81	0	Not rejected
Privacy -> Intention	0.9	0.369	Rejected
Social influence -> Intention	1.013	0.311	Rejected
Value price -> Intention	3.346	0.001	Not rejected

4 Discussion and Conclusions

The UTAUT2 theory served as the basis for this study. Among the findings, it was found that the statistically significant factors for users in the city of Manizales are hedonic motivations, performance expectation, effort expectation and price, while variables such as habit, social influence and privacy were not relevant.

In terms of hedonic motivations, which were the most important variable, it is important for users to enjoy the use of technology, especially when it is useful, confirming what was planted by Cordiglia [17]. Along the same lines, Alalwan [22] found that if a user has a good experience when using an application, its continuous use will be guaranteed, since the user experience represents how a user feels when enjoying or not enjoying a technology. The second relevant issue is the effort expectation, which according to Cordiglia [17] is one of the most important factors in many technology adoption studies showing that less effort at the time of using a proximity marketing technology influences the intention of adoption and use of it. Performance expectation and value price are the other variables that follow in terms of the weight of the coefficient. The performance expectation is related to the degree of ease associated with the use of technology, validating the findings of Cordiglia [17] who stated that it is one of the most important factors in terms of technology adoption. Likewise, it is evident that users are interested in the use of this technology to generate some kind of benefit, which can be represented in saving time or money, this is supported by the fact that more than 70% of the sample wants to receive information that represents benefits. With respect to the price value, this variable follows the proposals of Venkatesh [17], users are concerned about the cost they must incur to use the technology, although it contradicts the findings of Cordiglia [17] for its study in South Africa,

The habit variable, it was found that for the inhabitants of Manizales is not rele-vant to have had the habit of using the technology of proximity previously; however, the theory suggests that it is more likely to use and acquire a technology of which you have a habit of previous use [17], where it was found that there is a direct relationship between the habit and the intention of use of the application. The social influence variable also did not represent importance for the inhabitants of Manizales, who did not find relevance between making use of a technology of proximity and a change in the way others see them, this result can be affected by the lack of knowledge of the technology and the benefits that this can represent at the time of visiting a commerce in the city where it can be used. Other research has shown the impact of social influence on the adoption of new technologies, for example, Gharaibeh [23].

4.1 Implications for Practice

The results of the study can be applied in stores in Manizales and other small cities, where the use of proximity marketing strategies is still incipient, constituting a new channel of communication in real time with consumers according to their location, and consequently, generating greater recognition, increased sales for the stores, and a better experience for the customers who visit them. Proximity technologies have great potential to provide added value when used in solutions focused on providing a better shopping experience to their customers; besides becoming a tool for marketing professionals who are looking for innovative ways to approach consumers, with the advantage that the global rate of users who do not authorize being located today is low. Future avenues for research are related towards the integration of geomarketing with other digital marketing strategies focusing on taking advantage of hedonic motivations, expected effort and value price factors to get better results. Likewise, from the technical side, more research is needed for making the geomarketing devices more accurate and compatible with other smart devices such as smartwatches.

References

1. Mobile Marketing Association: MMA Updates Definition of Mobile Marketing (2009). https://www.mmaglobal.com/news/mma-updates-definition-mobile-marketing
2. Yaniv, G.: Sold on mobile marketing: effective wireless carrier mobile advertising and how to make it even more so. Int. J. Mob. Mark. **3**(2), 86–91 (2008)
3. Mobile Marketing Association: Playbook MMA: Geolocalización aplicada al marketing (2016). https://www.mmaglobal.com/es/documents/geolocalizacion-aplicada-al-marketing
4. Newman, N.: Apple iBeacon technology briefing. J. Direct Data Digital Mark. Pract. **15**, 222–225 (2014)
5. Andrews, M., Luo, X., Fang, Z., Ghose, A.: Mobile ad effectiveness: hyper-contextual targeting with crowdedness. Mark. Sci. **35**, 218–233 (2016)
6. Inman, J.J., Nikolova, H.: Shopper-facing retail technology: a retailer adoption decision framework incorporating shopper attitudes and privacy concerns. J. Retail. **93**, 7–28 (2017). https://doi.org/10.1016/j.jretai.2016.12.006
7. Cliquet, G.: Geomarketing: Methods and Strategies in Spatial Marketing (2013)
8. Vyt, D.: Retail network performance evaluation: a DEA approach considering retailers' geomarketing. Int. Rev. Retail. Distrib. Consum. Res. **18**, 235–253 (2008). https://doi.org/10.1080/09593960701868522
9. Zaim, D., Bellafkih, M.: Bluetooth Low Energy (BLE) based geomarketing system. In: SITA 2016 - 11th International Conference on Intelligent Systems: Theories and Applications (2016)
10. Cockrill, A., Goode, M., White, A.: The bluetooth enigma: practicalities impair potential: awareness is high. Can usage be higher? J. Advertising Res. **51**(1), 298–312 (2011)
11. Papandrea, M., Giordano, S., Vanini, S., Cremonese, P.: Proximity marketing solution tailored to user needs. In: 2010 IEEE International Symposium on "A World of Wireless, Mobile and Multimedia Networks", WoWMoM 2010 - Digital Proceedings (2010)
12. Sadowski, S., Spachos, P.: RSSI-based indoor localization with the internet of things. IEEE Access **6**, 30149–30161 (2018). https://doi.org/10.1109/ACCESS.2018.2843325
13. Davis, F.D.: Perceived usefulness, perceived ease of use, and user acceptance of information technology. MIS Q. Manag. Inf. Syst. **13**, 319–339 (1989)
14. Kuan, F.-Y., Ho, Y.-P., Wang, R.-Y., Chen, C.-W.: Using RPC block adjustment models for the accuracy of environmental research, cartography and geomarketing: a new concept of cartography. Stoch. Environ. Res. Risk Assess. **27**, 1315–1331 (2013). https://doi.org/10.1007/s00477-012-0668-8
15. Palos-Sanchez, P.R., Hernandez-Mogollon, J.M., Campon-Cerro, A.M.: The behavioral response to location based services: an examination of the influence of social and environmental benefits, and privacy. Sustainability **9**, 1988 (2017). https://doi.org/10.3390/su9111988

16. Venkatesh, V., Thong, J.Y.L., Xu, X.: consumer acceptance and use of information technology: extending the unified theory of acceptance and use of technology. MIS Q. **36**(1), 157–178 (2012)
17. Cordiglia, M., Van Belle, J.P.: Consumer attitudes towards proximity sensors in the South African retail market. In: 2017 Conference on Information Communication Technology and Society (ICTAS), pp. 1–6. IEEE (2017)
18. Ramadani, V., Zendeli, D., Gerguri-Rashiti, S., Dana, L.-P.: Impact of geomarketing and location determinants on business development and decision making. Compet. Rev. **28**, 98–120 (2018). https://doi.org/10.1108/CR-12-2016-0081
19. Ringle, C.M., Wende, S., Becker, J.-M.: SmartPLS 3. Boenningstedt: SmartPLS GmbH (2015). https://www.smartpls.com
20. Hair, J.F., Black, W.C., Babin, B.J., Anderson, R.E.: Multivariate Data Analysis: International Version. Pearson, New Jersey (2010)
21. Ramírez-Correa, P., Rondán-Cataluña, F.J., Arenas-Gaitán, J., Martín-Velicia, F.: Analysing the acceptation of online games in mobile devices: an application of UTAUT2. J. Retail. Consum. Serv. **50**, 85–93 (2019)
22. Alalwan, A.A.: Mobile food ordering apps: an empirical study of the factors affecting customer e-satisfaction and continued intention to reuse. Int. J. Inf. Manag. **50**, 28–44 (2020)
23. Gharaibeh, M.K., Arshad, M.R., Gharaibeh, N.K.: Using the UTAUT2 model to determine factors affecting adoption of mobile banking services: a qualitative approach. Int. J. Interact. Mob. Technol, (iJIM) **12**(4), 123 (2018). https://doi.org/10.3991/ijim.v12i4.8525

Construction of a 3D Model to Computerized Training Centered in Patient: PerMed & HCI Approach

Eveling Castro-Gutierrez(✉), Christian Suca, and Elizabeth Vidal

Universidad Nacional de San Agustín de Arequipa, Arequipa, Perú
{ecastro, csucav, evidald}@unsa.edu.pe

Abstract. As a result of the pandemic in less than two months, humanity has managed to understand the importance of using information technologies in different areas and, even more so, the importance of developing software for care, monitoring, and training in the healthcare. Personalized Medicine or Precision Medicine is gaining more and more critical, and its implementation is necessary. Different gaps cannot be covered if the software continues to be developed in isolation without considering the user as the central axis of the conception, design, and implementation of the different solutions. For this reason, in this research work proposes the construction of a 3D model of the bone structure of the pelvis based on X-ray images of a patient and a volumetric template as a model that serves for training and simulation for subsequent use in a surgical. This model based on patient-centered personalized medicine and user-centered software development. As a result, we present the creation of a model that can be adapted to each patient's reality. Likewise, these results can be used in computerized training systems for future health professionals, either by printing the model in 3D or using the software to perform measurements on each patient.

Keywords: Human-computer interaction · Personalized medicine · Precision medicine · 3D model · Computer vision

1 Introduction

We live in times of constant changes that require us to meet specific challenges, and this offers us new opportunities in different areas. For example, in health areas, the traditional concept of medicine allows the specialist to diagnose and cure diseases; However, at present, the trend is that the medical specialist does not cure diseases but rather the patient, who is a subject with a lifestyle, genetic load, with particular characteristics that develop in an environment and is located in different cultures.

This approach is known today as personalized medicine (PerMed) or "precision medicine" it is a field that is in evolution where it is determined which medical treatments will work best for each patient [1]. At PerMed, it is moving from an emphasis on reaction to an emphasis on prevention.

There are many benefits for patients in the PerMed approach, such as a) increasing treatment efficacy, b) reducing patient side effects, c) using cell or gene therapies, d)

Á. Rocha et al. (Eds.): ICITS 2021, AISC 1330, pp. 526–533, 2021.
https://doi.org/10.1007/978-3-030-68285-9_49

increasing patient adherence to treatment. Treatment, e) reducing high-risk tests, f) improving the overall cost of healthcare, and f) supporting change and commitment to patient-centered care.

Medical care in terms of human-computer interaction (HCI - Human-Computer Interaction) presents the following paradoxes: first, we observed that there is a substantial investment in innovative health technologies, in terms of the analysis of "big data"; on the other hand, there are various interactive health technologies that are implemented on a large scale that are difficult to use, and there are no proposals for innovative solutions that have achieved significant market penetration.

While [2] Blandford and [3] determine that in HCI, studies have been reported on usability, user experience, and security in novel digital technologies in healthcare areas, he points out that systems design frequently does not consider human psychology. Blandford mentions that when a health application is designed, they consider that it is relevant: a) an observational approach to understanding the realities of the system; It has been detected that the result of these applications does not align with the preferences and objectives of the users, recommending a user-centered design and the respective evaluation, b) a focus on the interoperability of the systems and c) a focus on quality data, issues that deserve greater attention in HCI and which will generate an impact on the usability and usefulness of health technology.

Blandford [2] establishes that there is a term "Design X" to refer to user-centered design for complex socio-technical systems.

The proposal that we present in this research article focuses on the construction of a 3D model of the pelvis from x-ray images and a volumetric template, focusing on obtaining an appropriate model for each patient to face the challenge of medicine personalized and following the guidelines of the construction of the user-centered model, considering that in the process of elaboration of this model we have worked with experts in the area of trauma and specialists in computer vision in the processing of images in 2D and 3D.

The paper is organized as follows. Section 2 shows the main characteristic of PerMed and HCI approaches. Section 3 describes the new proposal of the project of construction of a 3D model of the pelvis from x-ray images and a volumetric template, focusing on obtaining an appropriate model for each patient. Section 3.2 describes the initial results and discussion. Finally, we show our conclusion.

2 Background

2.1 PerMed (Personalized Medicine)

The "Personalized Medicine" (PerMed) or "Precision Medicine" [4] began in the field of oncology, but it has spread to other fields such as cardiology, where information on pharmacogenetics and pharmacogenomics have raised various and important advances in the treatments of this sickness. PerMed encompasses diagnosis and therapy, as well as supporting the prediction and prevention of diseases.

The COVID-19 pandemic has highlighted the need for healthcare systems around the world, as well as increased use of telemedicine, with a permanent focus on the way patients are cared for. Telemedicine, in particular, maybe a critical part of personalized health care in the future. It also includes possible research topics related to artificial intelligence, data collection, data integration and interoperability, informed consent, and patient concerns regarding privacy and data access.

ERA PerMed [5] is a new ERA-Net Cofund, supported by 32 partners from 23 countries and co-funded by the European Commission. Its objectives are a) to align national research strategies, b) to strengthen the competitiveness of European actors, and to improve European collaboration with non-EU countries regarding Personalized Medicine.

The EULAC PerMed project [6] aims to integrate the Latin American and Caribbean countries (LAC) into the consortium and activities of ICPerMed (International Consortium of Personalized Medicine) and the ERANet ERAPerMed, as a means to expand the international scope of the policies of R&D related to Personalized Medicine (PM).

EULAC PerMed will also work to facilitate a PerMed that benefits patients, citizens, and society in general, with the ultimate goal of contributing to the United Nations Sustainable Development Goal No. 3 “Guarantee a healthy life and promote well-being for all in All ages.”

2.2 HCI (Human-Computer Interaction)

Human-computer interaction (HCI) is a discipline that deals with the design, evaluation, and implementation of interactive systems for human use [3].

Ekaterina in [7, 8] determines that poor human-computer interaction (HCI) or lack of experience with the system due to not having involved patients and/or specialists in the conception, design, and implementation of the systems could cause various errors. We assume that by improving HCI, we can, on the one hand, reduce the degree of data distortion, and on the other, improve interaction with the system and thus increase the satisfaction of specialists, patients, and other users of the systems.

Kimberly [9] presents a table of the categories found in studies about trends in healthcare systems, which is represented in Table 1.

Kimberly [9] identifies a challenge for health research; this will be HCI in training and simulation. She concludes that simulated training today focuses on the use of patient care mannequins, and there are few studies that consider smart models to simulate in this context, leaving a gap to fill. Adaptive tutoring methodologies can further streamline the learning process for clinicians and other healthcare professionals, as well as consider having more suitable models for patients’ different soft organs and bone structures.

Table 1. Categories about trends in healthcare systems [9].

Category	Description
Usability	• Usability analyses of a computerized product or system • Determining ease of use or efficiency of a device • Increasing ease of use for special population
Security, Privacy, & Trust	• Network information security • Doctor-patient information sharing & confidentiality
Automation	• Auto-response systems, such as alerts • Automated monitoring systems • Automated reminders
Training & Simulation	• Computerized training • Creating learning models for training
Information / Patient Records	• Displaying information via different displays / interfaces • Immediate or mobile accessibility • Tracking medical history or records
At-Home Healthcare	• Systems used by patients away from hospital / clinic • Computerized self-care / monitoring systems • Computerized portable life support systems
Human Factors / Machine Interaction	• Attention, perception, and cognitive workload • Communication, mental models, teamwork • Use of robotics to aid performance • Error reduction • Improve reaction / response time
Safety	• Directly examining patient and/or patient safety

3 Proposal

3.1 Description

Specialists require precise measurements to carry out preoperative planning in Total Hip Arthroplasty (THA), but they use complex medical images such as computer tomography (CT) or magnetic resonance imaging (MRI), which achieve a 3D model [10], exposing patients to certain disadvantages such as radiation to the patient, high costs in the acquisition of images, and there are not many hospitals that have the appropriate equipment for this acquisition.

Our proposal is to build a 3D model, comparing X-ray images in the lateral view and the anteroposterior view from every patient with the projections of a template volume. This comparison is applied to the registration method to align and deform the template volume according to the original X-ray images is shown in Fig. 1. The techniques for template volume (moving image) and the x-ray images (fixed images) by establishing a correct focal point describes in [11].

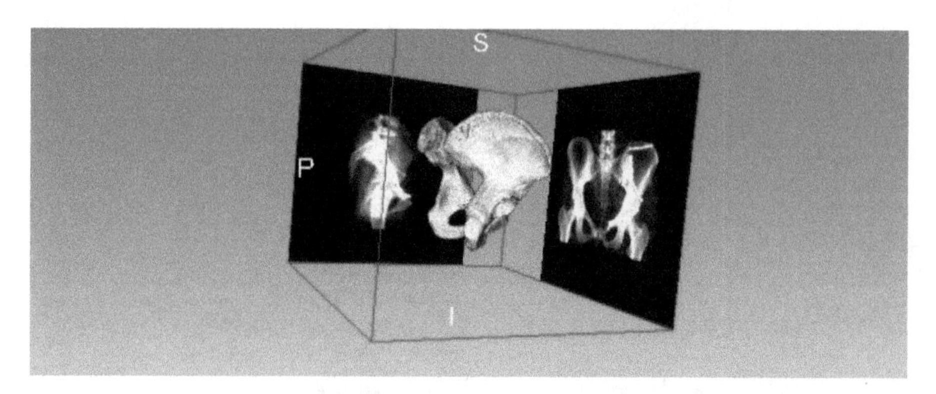

Fig. 1. View of 2 (two) x-ray images and the template volume, which will be aligned and deformed.

The pipeline of the project is represented in Fig. 2.

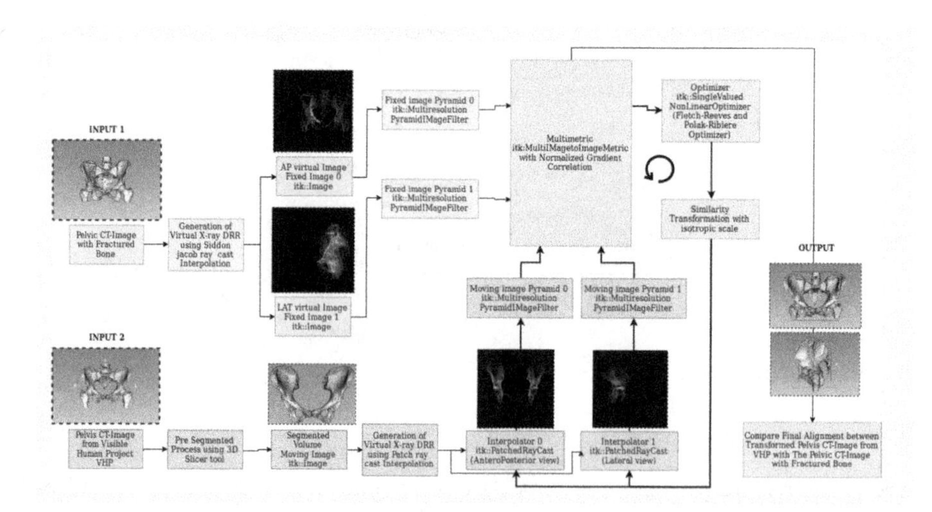

Fig. 2. The proposal's pipeline is divided into 2 phases, first phase focused in X-ray simulation by Ray-Casting method, and the second phase focused on alignment based on Volumetric template.

After applying the following process: a) Establish Region of Interest of Template Volume and b) Ray casting Simulation method (Lateral View, AnteroPosterior View), we obtained the Simulated image generated from the relocated volume with size: [333.245] mm resolution [1.1] mm. Focal Point [0, −1000.0] mm Distance from Volume to Image [−124] mm with orientation [90 0 0] degrees as we shown in Fig. 3.

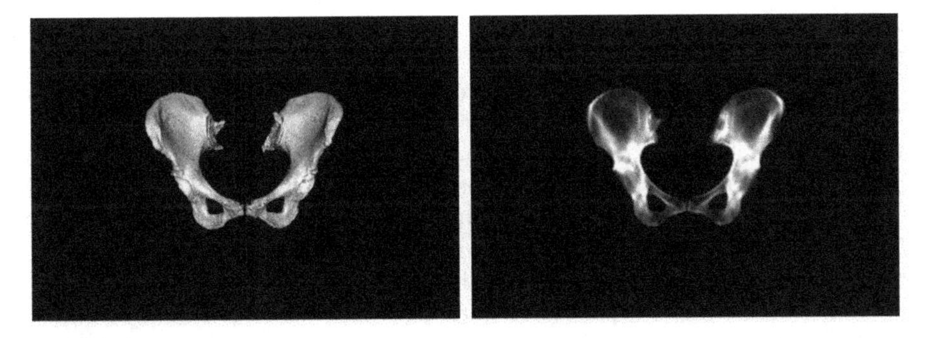

Fig. 3. X-ray simulation AP view. Right Figure is a Volume in Anteroposterior View, left figure is DRR in Anteroposterior View.

In the phase of Preparation of Randomly Positioned Standard Volume to relocate the model using a scale factor, it is necessary to transform the initial template to a different position than the original one and thus get to obtain the same position that was initially set (registration method), the same similarity transformation has been used to be able to alter the volume of the template see Fig. 4.

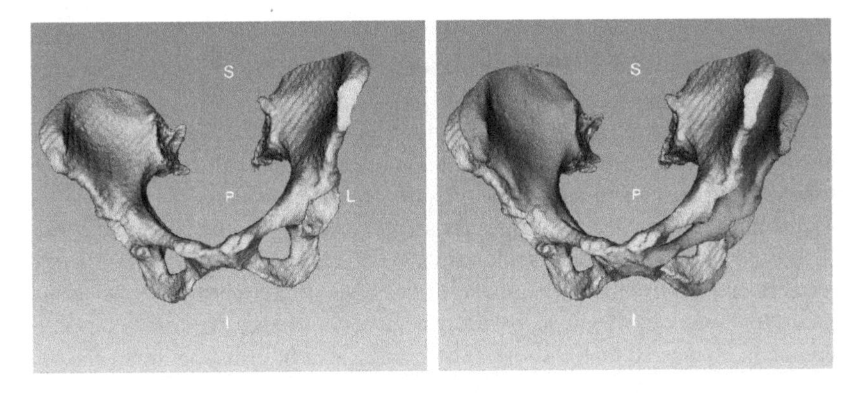

Fig. 4. Example of a Reference Volume in Positioning. Right figure is a volume repositioned template, left figure is volume repositioned.

3.2 Methodology, Results and Discussion

The methodology of the present work divided into four stages: a) data preparation, b) virtual image generation, c) Rigid Register, and d) Non-rigid Register.

In the case of the initial tests, there will be uses a computed tomography of a healthy person; in this case, a template from the "Visible Human Project" is used. This volume will be modified, applying a random transformation to become a target image, and the original volume will be used as the initial template. Then the template volume will be obtained from a simulated X-ray or DRR image [11]. These will be in AP anteroposterior views and an ML mediolateral view; with these images, the registration process will be carried out using the ITK framework. The types of registration required they will be rigid (alignment) and not rigid (deformation). With the final transformation parameters applied to the volume, the final transformation parameters will be compared. This will tell us if the registration was successful. There must be an initial volume that is clean, that is, it does not have abnormalities in the bone structure and does not have a surgical intervention instrument so that this volume would be handy for the experimentation process. With the volume obtained from both the abdomen and the hip, these two sets of images in Dicom format were joined to obtain a volume of both the femur's proximal part and the acetabulum of the hip. Nevertheless, this requires a previous semi-automatic segmentation when using methods of a tool like 3DSlicer.

A. Registration Tests. - Twenty synthetic data tests have been generated. With this data, we saw the errors between the values of the seven transformation parameters between the "Reference Volume" and the "Registration Volume". Furthermore, finally, the Hausdorff distance between the final volumes and the Reference volume is observed.

B. Validation Tests. - Applying the Hausdorff distance metric, the average separation between the Reference volume and the Register volume was 0.01855 mm, which shows that for different positions of the Reference Volume, the Template Volume's alignment is successfully achieved. The Root Mean Square Error was calculated, and a value of 0.05 mm was obtained.

4 Conclusions and Future Works

It is possible to generate a 3D model using X-rays images and an initial volume that allows adaptation to these two views. This model is oriented to cover the aspects of personalized medicine and user-centered software development.

Initial registration of X-ray images is useful; however, it would be more appropriate to establish a template model that allows instantiating the initial configuration of the X-ray images and, from this model, make the corresponding registration. Working with isolated structures allowed it to be detected that the metric we use is based on the difference in images so that with more extra or scarce information areas, the value of the metric will be lower compared to working with a template that has its entire structure.

The proposed methods were analyzed in the process of registration, from the configuration in the DRR image projection, and parameters used in the optimization process (step size, step tolerance, initial scales of transformations, and levels resolution), values that will subsequently determine the precision of the model. Likewise, in the evaluation of the model, the Hausdorff metric and its respective visualization graphs of the cost function were applied within a range of possible transformation parameters.

The present work used the focus centered-users and centered-patient, interacting with the specialist on traumatology and computer vision specialist.

It is necessary to probe this method with more clinician to improve the computational methods with a large dataset.

We could stablish that this template could be use like a custom-made implant when exist the enough clinical validations.

References

1. Personalized Medicine Coalition: Personalized Medicine Coalition - Precision Medicine Advocacy and Education. https://www.personalizedmedicinecoalition.org/. Accessed 13 Sept 2020
2. Blandford, A.: HCI for health and wellbeing: challenges and opportunities. Int. J. Hum. Comput. Stud. **131**, 41–51 (2019)
3. Gulliksen, J.: Institutionalizing human-computer interaction for global health. Glob. Health Action **10**(3), 1344003 (2017)
4. Medicine, A.P.: Moving beyond population averages, no. August (2020)
5. ERA PerMed: ERA PerMed. https://www.erapermed.eu/. Accessed 14 Sept 2020
6. EULAC: EULAC Permed. EULAC Permed. https://www.eulac-permed.eu/index.php/es/inicio/. Accessed 13 Sept 2020
7. Bologva, E.V., Prokusheva, D.I., Krikunov, A.V., Zvartau, N.E., Kovalchuk, S.V.: Human-computer interaction in electronic medical records: from the perspectives of physicians and data scientists. Procedia Comput. Sci. **100**, 915–920 (2016)
8. Liu, P., Fels, S., West, N., Görges, M.: Human Computer Interaction Design for Mobile Devices Based on a Smart Healthcare Architecture, p. 2 (2019)
9. Stowers, K., Mouloua, M.: Human computer interaction trends in healthcare: an update. In: Proceedings of the International Symposium on Human Factors and Ergonomics in Health Care, vol. 7, no. 1, pp. 88–91 (2018)
10. Kuhlmann, J., Halvorsen, T.: Precision medicine: integrating medical images, design tools and 3D printing to create personalized medical solutions. In: 2018 IEEE International Symposium on Medical Measurements and Applications, vol. 3528725544, pp. 1–5 (2018)
11. Velando, C.A.S., Gutierrez, E.G.C.: 2D/3D Registration with Rigid Alignment of the Pelvic Bone for Assisting in Total Hip Arthroplasty Preoperative Planning, vol. 11, no. 5 (2020)

Digital Twins vs Digital Trace in Megascience Projects

Artem A. Balyakin(✉), N. N. Nurakhov, and M. V. Nurbina

NRC Kurchatov Institute, 1, ac. Kurchatov sq., Moscow 123182, Russia
Balyakin_AA@nrcki.ru

Abstract. In presented study we discuss the operation of International Center for Neutron Research based on the PIK high-flux research reactor, and possible adaptation of successful foreign practices for this facility. We consider the phenomena of "digital twins" vs "digital trace". The first one is a model, possessing the main features of a real scientific facility, while "digital trace" deals with a lake of raw data and scientific outcomes are produced by multiple data processing with various methods. Both approaches lead to dynamic digital infrastructure development; it becomes a prerequisite for the functioning of any megascience facility. We argue this to be the next stage in the scientific facilities development: in nearest future digital twins and digital trace together would eliminate the uniqueness of the megascience, and would lead to the formation of a new supra-infrastructure. This object should be evidently based on taking into account the interaction of man, society, science and technology.

Keywords: Megascience · Digital twins · Digital trace · International scientific cooperation · Infrastructure

1 Introduction

Modern world presentation (and current ideas about the Universe) is largely based on the results of work obtained on unique scientific facilities. These installations are the only possible nowadays scientific research tool capable of solving urgent problems and responding to global challenges of our time, providing an opportunity for interdisciplinary research and projects in various areas [1, 2]. Projects of such a type are defined as a "supranational" organization with "independent representations" [3] or as organizational and managerial innovation [4].

Due to their complexity and the scale of the tasks to be solved, their successfully functioning is possible only in the format of international consortia. In practice, the point is that a characteristic feature of megascience is the fact that the cost of their creation and maintenance is beyond the capabilities of individual countries and/or entities [5]. Though to be constructed by one country solely (for example, ISSI-4 facility in Russia), their practical use upon completing the project is only possible by a number of scientific teams, and this approach has been already included at planning stage (access for international research teams, sale of time and/or data for external users, etc.).

Á. Rocha et al. (Eds.): ICITS 2021, AISC 1330, pp. 534–539, 2021.
https://doi.org/10.1007/978-3-030-68285-9_50

Thus, a megascience class project should be treated as a business-like project. It aims at obtaining new (unique) knowledge and technologies, the last to be later commercialized [4, 6] and adapted in various sectors of the real economy [7, 8]. For instance, super-bright X-ray beams are used for studying living complex systems (proteins and human brain cells). Megascience class facilities form differs from a set of infrastructural objects localized in one territory (single-sited) to a kind of distributed facilities (united under one "umbrella" project).

This gives rise to a whole range of features of their activities: from the philosophical questions and legal issues to maintenance of an appropriate infrastructure. One of the problems that gained additional relevance in 2020 due to the coronavirus pandemic was the organization of work on unique scientific facilities in remote access mode, and the development of appropriate changes to existing business processes. The possibility of remote scientific work on installations of this class has existed since the appearance of the technical possibility of realizing such access, and until recently it represented a specific highly specialized product (reducing costs and expanding access to equipment). In 2020, the social aspect was added: it could serve as health-protecting tool (so called scientific distancing).

The entire life cycle of megascience facilities can be divided into 2 related operation groups. First, the "engineering" task is the solution of applied problems arising during the creation, operation and modernization of the installation (the actual set of scientific equipment and its maintenance). The second, "scientific" task is the solution of scientific and practical challenges, setting up an experiment and analyzing the data obtained. While not regarding the "technical" side of organizing work on megascience-class facilities, we note the necessity to introduce a number of changes and/or corrections to some business processes in order to take into account the specifics of remote access in title documents.

In this work (carried out with the support of the RFBR grant No. 18-29-15015), we discuss some practices from different megascience facilities that can be implemented at International Center for Neutron Research on the basis of the high-flux research reactor PIK (ICNR PIK). ICNR PIK is now being constructed in the outskirts of St. Petersburg, Russia. It is a continuous flow type reactor and is intended for research in the field of condensed matter physics, nuclear physics and the physics of weak interactions, structural and radiation biology and biophysics, radiation physics and chemistry.

Its parameters are supposed to surpass the only analog in the world - the HFR reactor at the European Center for Neutron Research – International Institute Laue-Langevin. The facility will have 50 experimental stations fed by 10 horizontal, 6 inclined and 6 vertical experimental channels. It will possess 3 cold and 1 hot neutron sources. The volumetric energy release in the core will count up to 6,6 MW/l.

In our study we considered mostly possible cooperation with Republic of Belarus, BRICS countries, and European Union. This list, however, is not inclusive, and research infrastructure development projects are open for any participants. The most important thing here is the intention of the society and authorities to develop innovative infrastructure, and to participate in currently ongoing projects and/or initiate ones. Among Latin American countries, Mexico has the most consistent position on this issue. This country identifies several areas of cooperation: North America (USA, Canada), South America (Argentina, Brazil, Chile and Colombia), Europe (Germany,

Spain, France and the UK), and Asia (China, South Korea, India, Israel and Japan). Mexico has concluded about 100 bilateral and multilateral international agreements on scientific and technical cooperation with more than 25 countries, the EU and two multilateral organizations. In Brazil, interaction with foreign partners takes place within the framework of the Science without Borders (SwB) program, according to which Brazilian scientists have the opportunity to conduct research in the field of natural sciences and technology in leading foreign universities. Several directions can be mentioned: nanotechnology (e.g., CNPEM - Energy, LNNANO - Nanotechnology), fundamental physics (SIRIUS - The Brazilian Synchrotron Light Laboratory, LNLS). All other Latin American countries do not have specially formulated attitude towards cooperation in megascience field, but we expect this issue to draw much more attention in future.

2 Remote Access Mode for Megascience

It can be seen that remote access is a necessary element of many mega-science facilities, which has firmly entered into scientific practice. Our previous study revealed that remote access mode for megascience requires following adaptation of business process [9, 10].

First, it is necessary to reduce to a minimum the need for the physical presence of the customer at the facility where the unique scientific installation is located. For example, in the international project Borexino, there is an additional independent data collection system based on fast waveform digitizers. The complex allows to collect data remotely without the need for scientific teams to travel to the location of the detector.

Secondly, it is necessary to develop the appropriate digital infrastructure: it is required both the creation of new elements of the digital infrastructure (data centers, processing algorithms, etc.), and the integration of megascience installations into the existing e-Infrastructure facilities.

Thirdly, it is necessary to provide legal and methodological support for the operation of unique scientific facilities in remote access mode.

Fourth, it is necessary to provide engineering personnel to solve the related tasks of maintaining and ensuring the remote access mode.

The most interesting and challenging area of work that combines the fulfillment of these requirements is the development of digital infrastructure. The development of digital infrastructure gave rise to the new phenomenon – e-Infrastructure. This idea was first proposed and evolved in the EU, when digitalization of the scientific research was suggested as a method of organizing modern scientific research. The first step should be the unification of science, data collection systems and access to them. e-Infrastructure is thought to be the tool for implementing EU policies in science, when the achievements of the Internet, grid systems, cloud computing and databases are assembled in a new infrastructure. Europe started with the establishing of the open scientific portal EOSC (European Open Science Cloud), launched in 2016. Based on the EOSC work experience, the Go FAIR initiative is being prepared to put ideas and proposals related to digital science into practice [11].

To solve the applied problems of processing scientific information in the EU, a number of specialized data processing projects are planned: GEANT (management of scientific and educational network projects), EGI - Advanced computing for research (providing calculation options for CERN, EMBL projects), PRACE (providing computing power, 465 projects at the moment). In addition to the above, in order to codify and standardize digital infrastructure, within the framework of the Horizon 2020 program, the European Union launched the e-Standards project [12].

It is expected that the first consumers of e-Infrastructure will be representatives of the natural sciences, however, the greatest impact and the most significant results will be manifested in the field of humanitarian knowledge, which forces participants in the process to develop appropriate assessment methods and approaches today, simultaneously solving legal issues.

In this paper we present the dichotomy: phenomena of "digital twins" and "digital trace" (by analogy with digital footprint) from megascience installations. The first is a "digital reflection" (model), which carries the main features of a real scientific phenomenon. They can be used to work out the main experiments conducted on considered facility, and their interface allows to work on them in a remote mode and for unskilled users. This democratization of technology would provide easy access (including for non-specialists) to knowledge in technology and business without lengthy or expensive training. Called "citizen access", this policy is already used in application development, in data and analytics systems, and in solving design and knowledge problems [13]. The implementation of such a policy leads to a gamification of science. For example, it is discussed that the world of "megascience" can be transferred to the Minecraft game shell [10].

For the case of a digital trace, there is a maximum collection of raw information and its subsequent multiple processing by new methods (i.e., this, in fact, it is a digital data lake). This raw data are subject to repeated processing at different points in time by different methods and algorithms (i.e., a scientific discovery is possible on old data that have been processed with new methods). This is similar to extracting information about people from their digital footprint. Here we face an intriguing challenge: if the digital footprint is now right or duty? For the case of human beings, recent trends show the obligation of people to produce digital footprint that can be later analyzed and commercialized. With scientific data we notice the similar trend (the more data - the better), and it is accompanied with new data handling algorithms development.

Both approaches are widely used in CERN. For example, in 1998 the MONARC project (Models of Networked Analysis at Regional Centers for LHC Experiments) was launched. It yielded in the concept of a hierarchy of data processing, modeling and analysis centers. There are currently 4 tiers of processing centers. At the bottom (zero) level there is the Tier-0 CERN Computing Center that deals with the primary reconstruction of events, calibration, permanent storage and archiving of the complete set of "raw" and simulated data. Then there are Tier-1 (13 centers), Tier-2 (about 170 centers), and, finally, Tier-3 (about 50 centers), represented by university clusters, or centers that provide resources on a voluntary basis, where physical data analysis is carried out [14, 15]. A similar policy for working with data was implemented in the case of the global neutrino network (GNN) [16]. This network can be interpreted as a distributed scientific infrastructure, its elements being separate facilities and collaborations (for example, the Ice Cube collaboration with 47 organizations from 12 countries of Europe, America and Asia or the Dubna deep-sea neutrino telescope of multi-megaton scale).

The results of the work of the digital twin (in accordance with the FAIR principle actively implemented in the EU) have the same “rights” as the real installation. For example, ATLAS (CERN) uses data obtained by computer generated events (by Monte-Carlo method [13]) as an auxiliary tool during the “real” experiment.

3 Current Tendency and Conclusion

The described process fits into the general tendency in the development of scientific research systems, and the coronavirus pandemic acted only as a catalyst for the already existing trends. So, since the 2010s, the e-Science system is being formed and e-Infrastructure is being implemented [17]. At the same time, the digital infrastructure, which is an integral part of the e-Infrastructure, does not duplicate or replace the “physical” one. The idea of “digital twins” described in this paper also does not imply a complete replacement of “real” attitudes, but acts as a tool to support scientific research.

This approach leads to the active development of digital infrastructure, and its presence becomes a prerequisite for the functioning of any megascience facility. According to a number of experts, this process would lead to a universal withdrawal into the digital space. We argue, however, that this should be treated as the next stage in the development of unique scientific installations. The juxtaposition of the digital footprint and the digital twin is seeming: both of them are an integral feature of modern megascience facilities.

The next step would be to standardize the digital world, and the products it generates and/or possesses. From the point of view of scientific research, mega-science installations, translated into the format of remote access and acting in the form of digital twins or reproduced on their digital footprint, will lose their uniqueness and become a “common place” (a kind of a niche product). At the same time, a high degree of interconnectedness and inclusion would lead to the formation of a new supra-infrastructure.

The convergence of science and technology will bring together scientists from different countries and will facilitate the coordination of research and development aimed at overcoming global challenges, including the development of methods to prevent the spread of pandemics based on genetic research.

The format and principles of the new scientific digital infrastructure are not yet clear, but there is a consensus that scientific and technological progress is impossible without taking into account the interaction of man, society, science and technology. Therefore we could expect the adoption of a number of regulatory acts similar to the European Union General Data Protection Regulation [18].

Acknowledgments. This work was supported by RFBR grant № 18–29-15015.

References

1. Nurakhov, N.: The basic processes of creating a "megascience" project. In: International Conference on Integrated Science, ICIS 2019: Integrated Science in Digital Age, Batumi, Georgia, pp. 329–339 (2019)
2. Hallonsten, O.: History and politics. In: Big Science Transformed, pp. 43–98, Palgrave Macmillan, Cham (2016)
3. Crease, R.P., Martin, J.D., Pesic, P.: Megascience. Phys. Perspect. **18**, 355–356 (2016)
4. Karlik, A.E., Platonov, V.V.: Conceptual foundations of the study of megascience as an organizational and management innovation. Innovation **10**(228), 11–16 (2017)
5. Fotakis, C.: FP7 Interim Evaluation, Analyses of FP7 supported Research Infrastructures initiatives in the context of the European Research Area, Final Report 12 November 2010 (2010)
6. Lami, S.: Challenges and new requirements for international collaborations. Sci. Diplomacy **6**(2) (2017). https://www.sciencediplomacy.org/article/2017/mega-science-collaborations
7. Balyakin, A.A., Mun, D.V.: Formation of an open science system in the European Union Information and Innovations. In: Proceedings of the Conference "Sciencemetry and Bibliometry", pp. 33–37 (2017)
8. Gartner's Top 10 Strategic Technology Trends for 2020. https://www.pcmag.com/news/gartners-top-10-strategic-technology-trends-for-2020. Accessed 21 June 2020
9. Nurakhov, N.N.: Integrity of Innovation Management and INSO Inventory, p. 156. MST, Moscow (2010)
10. Nurbina, M.V., Nurakhov, N.N., Balyakin, A.A., Tsvetus, N.Yu.: Mega science projects for business. In: Ahram, T., et al. (eds.) Human Interaction, Emerging Technologies and Future Applications III Proceedings of the 3rd International Conference on Human Interaction and Emerging Technologies: Future Applications (IHIET 2020), Paris, France, pp. 488–492 (2020)
11. Dutch Techcentre for Life sciences. https://www.dtls.nl/fair-data/go-fair/. Accessed 10 July 2020
12. Report on the EU funded eStandards project. https://ec.europa.eu/digital-single-market/en/news/report-eu-funded-estandards-project. Accessed 14 July 2020
13. Ay, C., et al.: Monte carlo generators in ATLAS software. In: Journal of Physics: Conference Series, CHEP 2009, Prague, Czech Republic, vol. 219 (2010). https://iopscience.iop.org/article/10.1088/1742-6596/219/3/032001
14. Grigorieva, M., Golosova, M., Ryabinkin, E., Klimentov, A.: Exabyte repository of scientific data. In: Open Data Systems. DBMS, Moscow (2015). https://www.osp.ru/os/2015/04/13047963
15. Klimentov, A., Kiryanov, A., Zarochentsev, A.: Russian lake of scientific. In: Open Data Systems. DBMS, Moscow (2018). https://www.osp.ru/os/2018/04/13054563
16. Global Neutrino Network (GNN). https://www.globalneutrinonetwork.org/. Accessed 30 June 2020
17. Zhulego, V.G., Balyakin, A.A., Nurbina, M.V., Taranenko, S.B.: Digitalization of society: new challenges in the social sphere. Bull. Altai Acad. Econ. Law (9–2), 36–43 (2019)
18. The EU General Data Protection Regulation (GDPR). https://gdpr-info.eu. Accessed 18 Nov 2019. Accessed 17 Mar 2020

Organizational Models and Information Systems

Scoping Review of the Work Measurement for Improving Processes and Simulation of Standards

Gustavo Caiza[1], Paul V. Ronquillo-Freire[2], Carlos A. Garcia[2], and Marcelo V. Garcia[2,3(✉)]

[1] Universidad Politecnica Salesiana, UPS, 170146 Quito, Ecuador
gcaiza@ups.edu.ec
[2] Universidad Técnica de Ambato, UTA, 180103 Ambato, Ecuador
{pronquillo4211,ca.garcia,mv.garcia}@uta.edu.ec
[3] University of Basque Country, UPV/EHU, 48013 Bilbao, Spain
mgarcia294@ehu.eus

Abstract. Any company, regardless of size, must consider in order to overcome the economic situation caused by the pandemic the process innovation and resource-saving parameters. Methodologies such as work measurement should be used to generate efficient and reliable information through which timely decisions are made. The current research addressed a scoping review of the literature on the most used techniques for determining standard times. Therefore, thirty-three scientific articles were selected to answer the three research questions posed on this study. In addition to the results, it was determined that the use of predetermined time systems are instruments that allow the quantification of a job before executing its production. Indeed, it allows the combination of techniques to develop predetermined systems that are linked to a specific activity. Furthermore, with the use of these techniques, the expense used to develop a new product can be reduced, and the production and administrative processes can be optimized.

Keywords: Work measurement · Time studies · Predetermined time system · MOST · MODAPTS · MTM

1 Introduction

Competitiveness forces every company to develop systems and processes that meet the needs in the environment of each business and allow establishing strategies that control human and material resources. This is created to maintain or improve product costs, profitability and permanence in the market in the long term [42].

Currently, the competitive environment is not the only obstacle that companies must face, it must be borne in mind that the crisis caused by COVID 19 has caused a worldwide change in the economic, labor and productive fields [32].

Á. Rocha et al. (Eds.): ICITS 2021, AISC 1330, pp. 543–560, 2021.
https://doi.org/10.1007/978-3-030-68285-9_51

The measure taken for the mitigation of the virus cause an interruption of the productive activities therefore, the decrease in the demand for products and services. Almost no company was prepared to face a total disruption, affecting the gross fixed capital formation, loss of economies of scale, and even manufacturing smaller production batches [30].

Therefore, the crisis due to the pandemic accelerated the changes that companies should have adopted a decade ago. Today those changes were observed with virtual negotiations, the rise of e-commerce, and home office [8]. The last of the changes mentioned is a concrete reality that companies are already using today. However, what happens with the areas that need to record information directly from production? How can you issue data on products that are not developed or released? These are the questions that make companies research for ways through which they can issue efficient information to make timely decisions. That's when professionals must innovate in their processes and methodologies, leading to saving money and time.

The transformations that must be adopted require working together to adapt and improve existing methodologies. The study of work is a methodology used to determine standard times that have evolved and adapted to the environments of each company. Indeed. It has allowed the issuance of information without having timed the process [1]. Taking into account that some of the traditional methodologies may not be able to face the new reality in organizations, the knowledge of new techniques should be expanded, focusing on progressive digitization to achieve productive results [16].

The aim of this research is to present a scoping literature review to show updated information on work measurement techniques. These techniques, despite being developed long ago, could be adapted to current reality, allowing the establishment of standards, either by a system of predetermined times or by direct observation of activities.

The remainder of this article consists of 5 sections: Sect. 2 describes the methodology based on which the most relevant articles were chosen. Section 3 describes the literature review. Section 4 discusses and analyzes the results. And finally, in Sect. 5 the conclusions and future research implications are evidenced.

2 Article Selection Methodology

The review and selection of articles were carried out through the application of three phases: as an initial step, three research questions were asked, later, based on keywords a search for information was done, and finally, in the last step, by using a selection criteria the most relevant literature was determined.

2.1 Research Questions

Three research questions were posed (see Table 1). They became a guide in the literature search process. The questions focused on the measurement of work to emphasize its importance, as well as discovering the techniques which are currently giving results.

Table 1. Research Questions

	Questions	Goals
1	What is the authors' perspective on the measurement of work?	Demonstrate the importance of work measurement
2	What are the most widely used work measurement techniques today?	Identify the techniques with the greatest application
3	Can a job be measured and analyzed using various techniques?	Demonstrate whether joint work between techniques can be used for better analysis

2.2 Data Collection

The following keywords were used to search the literature corresponding to the measurement of work: work measurement, time studies, predetermined time system, methods-time measurement, Maynard operation sequence technique, the modular arrangement of predetermined time standards, preset timing standards, time formulas, production, and manufacturing. In the search engines, a variation or combination of the keywords was used in order to obtain articles referring to the use of work measurement techniques.

The literature search was performed in eleven databases (see Fig. 1), resulting in a total of sixty-four documents in English and Spanish corresponding to articles, articles published in conferences, books, and reports; The documents found correspond to articles published from 2000 to 2019.

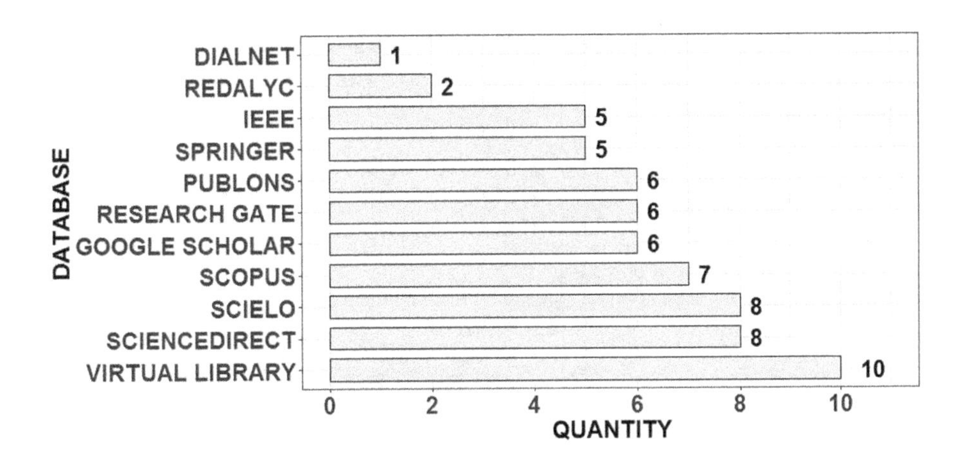

Fig. 1. Number of documents per database

2.3 Selection of Information

The selection criteria applied in the classification of documents are detailed below. First, articles that were not published in the last five years were initially discarded. Regarding books, this criteria was not used on them, since the methodology was discovered years ago. Therefore, the books used in this review correspond to the ones published from 2000 to 2009. The second classification criterion was based on giving relevance to the articles or books according to the answers of the three research questions.

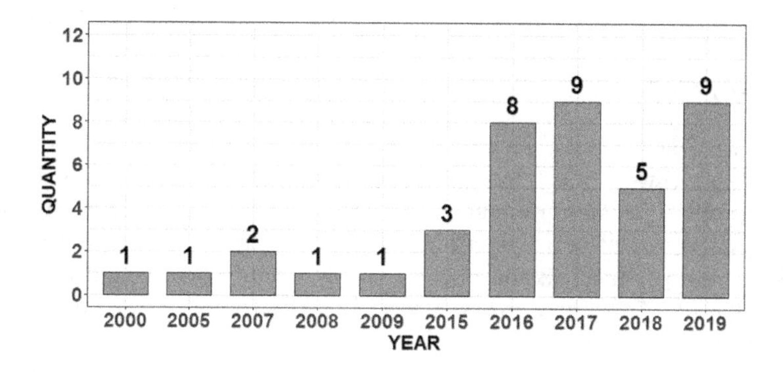

Fig. 2. Select articles by year

Based on the two criteria mentioned above, 33 articles published between 2015 and 2019 were selected (see Fig. 2), 24 from which are in English and 9 in Spanish. Besides, from a total of 10 books, 7 books were selected. Figure 3 shows that most of the articles were found in the databases: Scielo, ScienceDirect, and Scopus.

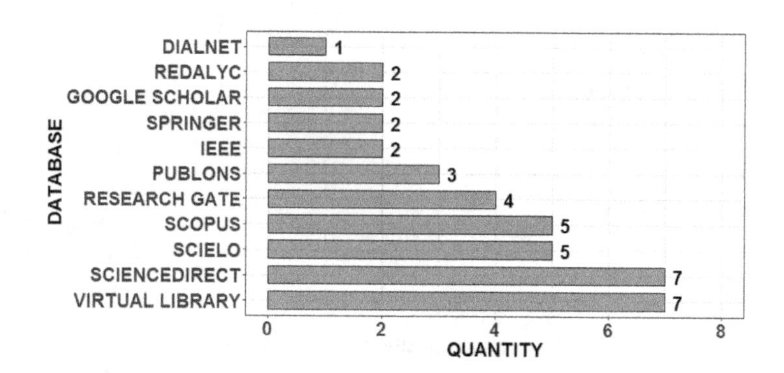

Fig. 3. Articles selected by database

3 Analysis by Topic of Literature

3.1 Contextualization of Work Measurement

All companies that offer a product or service need to measure, control, and improve the use of resources to have a proper management of them. Human resources play an important role in the industrial cost, thus the following questions arise: is the time of the operators being used efficiently? Is each operation carried out at the correct time? Does the administration have solid information for production plans and incentives? [13]. These questions have arisen throughout industrial evolution, which has led to the investigation and verification of a useful methodology for the administration and operation of human resources.

According to R. García, with the application of work measurement, the efficiency of workers can be controlled and increased [13]. A. Lago considers measurement as one of the necessary supports for increasing productivity [9]. A. Jananía mentions that a work measurement is a tool that has been continually improving and recognizes it as a necessary instrument to obtain better performance from machines and equipment [18]. Finally, F. Meyers believes that it is an aid for managers to make important decisions intelligently even before production starts [28].

It should be borne in mind that the measurement of work is a methodology that is complemented by the study of methods. Measuring work tries to reduce or eliminate activities that cause downtime while studying methods that aim to reduce the content of work by eliminating unnecessary movement [7]. In conclusion, the working method must be improved or standardized to be able to measure it.

The measurement of work can be defined in different ways and according to the perspective of each author, but in the definition of several authors, it can be seen that the following words are repeated: execution time, trained worker and established procedures [7,13,18,31]. For the development of this article, the definition issued by the International Labour Organization (ILO) will be considered: The measurement of work is the application of techniques focused on determining the time that a qualified worker executes a defined task, carried out according to a preset execution [19].

3.2 Benefits of Work Measurement

The information generated by the measurement of work and the establishment of time standards is not only helpful for the productive area, but the benefit is also evident for other systems of the company such as cost, planning, and even budgeting. The most representative benefits are discussed below:

The fundamental principle in the industry is that the operator has an adequate payment and that the company receives a reasonable workday. Therefore, setting a standard time with precision allows us to increase the efficiency of the machinery and the operating workers, hence the benefit of both participants is achieved [31]. In the same way, the implementation of standard times has a benefit in the goals about the consumption of materials, resource distribution, and the expectation of execution time [7]. That is, time and dependent variables are participants and executors in determining the standard cost.

Based on standard times, controls over labor can be established, thus maintaining standard costs [19]. The relationship between time and standard production sets cost as an indicator of productive efficiency in the company. Furthermore, using the standard labor time, the cost of a prototype that is planned to be produced, and whose operations will be similar to the current ones can be determined [13].

Standards can be used as indicators of performance levels, whether individual or collective, manual or intellectual, that operators can achieve considering human capabilities and personal restrictions [7]. Besides, they provide information with which the production program can be elaborated, including data on the machines and manpower necessary to fulfill the work plans and take advantage of the installed capacity [19].

In short, the measurement of work presents benefits to the entire company, and it can even reduce operating costs. According to Meyers, in organizations that operate without time standards, a performance of 60% is characteristic; if time standards are set, performance improves to an average of 85%, representing an increase of 42% [28].

3.3 Measurement of Work by Direct Observation

Time study is a direct measurement technique used to determine as accurately as possible, based on several observations, the time required to carry out a task with pre-established standards [13]. Here are some articles that have used the technique of direct observation by timing activities to obtain benefits:

E. Polanco et al. [34] reviews how companies look for the most appropriate way to integrate or improve direct observation tools. The research details the necessary steps to apply a time study and the possible implications when implementing it. The study concludes that when standardizing time there is a basis for work scheduling, standard cost calculation, and implementation of incentives. The information presented in the article serves as a guide for executing time and motion studies.

One of the objectives of the study of work is to increase productivity, for this reason in a footwear production company, when using management tools in conjunction with the study of methods and times, it was evident that the areas were not balanced. In this company, the work was balanced using standard times. As a result, an increase in productivity of 5.49% was obtained [2].

Another case study is presented in the company CIAUTO, in which time studies are carried out to determine production standards and balance the assembly line of the M4 model car. Based on the study of times, it was determined that the stations have a shorter time than the cycle time. Thus, the analyzes leaded to organize the workers so that the activities are carried out evenly on both sides of the vehicle [5].

Direct observation is not a specifically applied technique at a production level, that is why in Fratello Vegan Restaurant a study of times and movements is proposed. Its focus is to optimize resources and normalize operations in a service company. The purpose was to minimize the dissatisfaction of the clients concerning the delay on the service due to a long time in the preparation of the dishes. The research proposes improvements in the efficiency of the chefs, the location of the materials used in the kitchen. It mainly concludes that the cost analysis must be related to the time taking so that in this way the demand and the resource are necessary analyzed by determining whether or not it is appropriate to continue offering certain products [27].

Another case of time measurement application can be seen in a distribution system of a marketer. The research uses direct observation and determination of times with a stopwatch for 15 days. The data was collected and analyzed through descriptive statistics which allowed the development of a system that supports the planning and ordering of routes, thereby achieving delivery times to each of the clients [14].

The articles presented are a methodological guide on how to carry out a study of work with the direct observation technique by timing. Besides, they provide guidelines for calculating the adequate number of observations to ensure the reliability of standard or standard times.

3.4 Work Measurement Using Methods Time Measurement (MTM - Synthetic Method)

This section will state articles related to one of the first techniques of predetermined times. Methods time measurement (MTM) bases its methodology on decomposing the task into elements and measuring them using units of time measurements (UTM), then each UTM is added to obtain the total time of a process.

P. Kuhlang [23] and M. Koptak [22], in their investigations, use the MTM technique to determine times and improvements in logistics systems. In both articles, they selected this technique because of the notorious benefits of its application and the increase in productivity, especially in manual operations. P. Kuhlang concludes the work with a 43% increase in productivity and a reduction in the basic time from 18.35 to 13.69 min in the preparation and packaging of the spare parts operations, while M. Koptak obtained a reduction of 0.52 min in the cycle time.

On the other hand, M. Manns et al. [26], H. Tinoco et al. [41] and A. Bellarbi [4] focus their investigations on the application of the MTM methodology with digital analysis tools. The first researcher used time measurement methods in conjunction with data mining to estimate the time and sequence of operations on an assembly line, thereby successfully predicting assembly operations 237 times.

H. Tinoco et al. [41] In their research, they propose the automation of the MTM methodology using a motion capture system embedded in a virtual environment. In the case study, an assembly process is generated which does not consider the principles of the economy of movement. Each movement was recorded using an thermographic camera, passive markers, and video analysis software to subsequently analyze them using an algorithm. The automation of the methodology allows classifying some quantifiable aspects that are not easily identified in the traditional analysis of times and movements.

Similarly, A. Bellarbi [4] also focused their research on the automation of the MTM methodology. The article presented a system capable of automatically generating the MTM code using only the head and both hands in a virtual environment. Then, they divided the gestures made by the person and classified them into elementary movements using a decision tree algorithm. The purpose of the created system was to use it to train technicians or to help experts in the identification of the MTM code in industrial productions.

The articles presented above detail the application of the MTM methodology in real industrial systems as in virtual environments, but there is also research in which MTM is used to validate new time study methodologies. This is how M. Faber [10] in his article presents a revision on the accuracy of the methods time measurement human work design (MTM-HWD) with MTM-1. The case study has data from 62 German companies in which the two methods were used, and a strong linear relationship was obtained as a result. The method could be used in the design of human-oriented jobs, in fields of research on human-robot collaboration, exoskeletons, learning forecasts, and job rotation.

Similarly, F. Morlock et al. [29] In their article, they promote the use of the methodology in the design of workplaces. The research presents the teaching process used in the learning factories at Stellenbosch University in South Africa, which considered the basic and necessary knowledge to apply MTM with the objective already described while emphasizing the usefulness of the methodology to manage production times.

3.5 Measurement of Work Using Modular Arrangement of Predetermined Time Standards (MODAPTS - Synthetic Methods)

The following articles mention the results revealed with the application of the MODAPTS technique, it was considered appropriate to expand the theoretical framework on this technique because its use is beneficial to establish standards before starting production, even if there are no machines or work established cells.

The predetermined time systems are techniques used to establish standards. The systems describe the work according to the basic methods, the time of each movement is searched, and the individual times are added to obtain the necessary time for the entire work [28].

The predetermined time systems have evolved over the years. As a result, by using motion sensors, depth sensors, color cameras and a set of microphones, which provide captures of movements in three dimensions, the development of a software that analyzes movements and assigns a time using the MODAPTS technique has been developed. To evaluate the results, tests were carried out on the assembly of an electrical harness, and the results were compared with those obtained by an analyst. The hypothesis test determines that the data can be assumed to be similar [24].

B. Alkan et al. [1] conducted a research project, it proposes a method to evaluate the complexity of manual assembly using a description of operations based on the modular arrangement of predetermined time standards (MODAPTS), which together with virtual the manufacturing tool (vueOne) generated a model that helps to identify and compare the complexity of work to determine an optimal approach. The analysis using predetermined times helps to identify problem activities before start-up, making it possible to obtaine a reduction in costs and an increase in operating efficiency.

Similarly, S. Wu et al. [43] discusses a connection between motion analysis technology (PCA) and MODAPTS to optimize the predetermined time system. When using the PCA technology a precision of 80.08% was obtained in the determination of the times. In addition to a substantial saving in the execution time since the traditional method used 1 h, and with the implementation of technology, it is reduced to three minutes.

There are other application approaches, in which C. Erliana et al. [15] employs the MODAPTS technique to reduce working time in cement packaging. The research aims to compare the established processes with the standard time that can be determined by applying the system. The results gathered range from the decrease in cycle time to the fulfillment of the requested production.

In the same way, I. Siregar [39] carries out an analysis of the working method of an operator in a foundry company. The investigation is carried out since the company started to present losses due to the production process. The MODAPTS technique is used to improve the working methods and calculate the processing time. As a conclusion it was claimed that there is a difference of 126.65 s between the actual and the proposed method, the difference is produced by the operator making unnecessary movements in the work and consequently, there are losses and high production costs.

The articles reviewed have presented research on how the MODAPTS technique allowed to determine times and to analyze established methods to improve them. These methods are subsequently executed by trained and qualified workers therefore, the operational knowledge that workers acquire from years of experience is invaluable to a company.

Based on the high value of technical knowledge L. Liu et al. [25] proposes a concept of empirical operational knowledge directed to engineering (OEK) to describe and design a framework to acquire the operations of qualified technicians. The MODAPTS technique is used to segment the operating process into basic movement elements and the variable precision set algorithm (VPRS) to extract the technician's content. The objective achieved was to provide a system that focuses on the transfer of tacit knowledge.

The use of the MODAPTS technique in the different application approaches has given excellent results in previous investigations, but A. Chan et al. [6] conducted a research on the comparison in the estimation of times of two industrial assembly tasks. It was concluded that the times performed by estimation are lower than the real ones, but are higher than those determined using the MODAPTS technique. The researchers concluded that participants had difficulty determining times on tasks that required more precise handling.

3.6 Work Measurement Using Maynard Operation Sequence Technique (MOST - Synthetic Methods)

Another technique to determine times by using synthetic methods is the Maynard operations sequence technique (MOST), S. Rahman et al. [37] uses the technique in the sewing section of a garment company to identify non-value-added activities and minimize bottlenecks to improve productivity and reduce cycle time. The investigation reduced the total time from 139 s to 109 s, increasing the production from 600 pieces to 1600 pieces. The general sewing base (GSD) technique can also be applied in garment companies since this technique was developed with the MTM base data and has formulas and operations for sewing garments. Thus, it was possible to achieve a standard time that takes into account variations such as the number of stitches, seam length, revolutions per minute of the machinery [40].

D. Patel [33], A. Karim [21], Y. Fang [11] researched production lines, the objective they sought was to reduce operating times and control the cost of labor. Through the application of the MOST technique and the analysis of cycle times and bottlenecks, they achieved the stated objective. D. Patel concludes that most of the labor was saved time (65% reduction in time) and increased productivity, also it is emphasized that by applying the techniques, production time can be obtained before manufacturing really begins. A. Karim achieved a reduction of more than 2 min in cycle time and an increase in performance of 29.63%. Finally, Y. Fang affirms that the technique is very useful, but it is limited in the aspects of mitigating operator fatigue and improving processes by controlling working hours.

P. Karad et al. [20] analyzes the use of MOST in an automotive company for the assembly of the rear floor. This technique allowed to redesign the processes and analyze the proper flow, thus reducing the workforce from 17 to 11 people, because the required quantity can be met in a single production turn, and a reduction of 1.91 min was also achieved in cycle time. Similarly, M. Jadhav et al. [17] used the technique in an automotive company to analyze the real data in comparison with the estimates, achieving the reduction of a production shift because the demand is fulfilled in one turn.

3.7 Combination of Direct Observation Methods with Synthetic Methods

The separate application of synthetic methods and direct observation methods become an excellent tool to obtain improvements in productivity and economic savings. In this section the benefits that would be obtained by combining the techniques of the study of work will be presented. In that way, it will be possible to analyze production time and compare with a pre-established method.

A. Pusvanasvaran et al. [36] in his research highlights the use of the implementation of the Maynard operation sequence technique (MOST) associated with the direct measurement technique by timing to reveal hidden waste in the dry packaging operation in electronic industry. The study of times determines the real cycle time and with the help of the MOST technique, the equation of the worker's performance is calculated to be able to track or monitor the daily performance of the workers.

V. Polotski et al. [35] propose an approach that combines the MODAPTS technique with statistical techniques that use the real data obtained with time measurements with a stopwatch. The approach allows to establishe a system that can be used for estimating times in projects that contain operations of a different nature. The proposed method breaks manufacturing operations into a series of tasks with experimentally identified times, achieving not only point estimates but also performing linear regression models.

J. Ruíz et al. [38] employ the study of times by stopwatch for the determination of standards in the filling process, later the study of process optimization is performed through the application of MOST and MTM techniques. It is concluded that time analysis is an adequate instrument to set standards and analyze improvements that reduce downtime and therefore the cost of production.

The previous articles give a clearer view of what has been exposed in the work measurement techniques segment, due to the verification that can be done by combining each one (direct timing and predetermined systems) and the use of regression model systems that can be established to estimate the times of any type of activity and it is also possible to issue relevant information to productive capacities without the need to have a work cell implemented.

In addition to the benefits obtained by estimating times, it should be noted that these assessments are of vital importance for the costing of the product. A. Ganorkar et al. [12] propose an investigation in which the MOST technique and time formulas are used to determine the practical capacity of the activities. The information generated is used by the activity-based costing system to determine the cost of the product and analyze improvements. The results obtained range from the timely analysis of the information to the cost reduction due to the decrease in productive time.

Most of the articles that use the combination of methodologies have obtained good results, but it must be considered that the predetermined systems focus on estimating the time of manual operations, and the implemented system may not fully adapt to the analyzed process. This is how F. Assef [3] in his research combines the MODAPTS technique, direct observation, and simulation of processes in an automotive assembly industry. The work concludes that predetermined time systems are an excellent tool when the product is in the phase of design. However, other types of variables (operator conditions, mobility of tools, a field of vision) must be taken into account since they may limit the execution of the work and the increase in real-time.

4 Discussion and Results Analysis

The literature review presents a favorable panorama in the application of predetermined time systems to simulate standards without starting manufacturing. Hence these techniques can be considered as useful instruments to control the expense of companies when developing new products, to improve the operations already established, to serve as information for the cost analysis. What is more the systems discussed in this article could become a basis for developing proposals that meet the needs of each organization, such as the GSD system developed for companies of clothing.

Based on the second research question, it was possible to verify that the most widely used techniques at present are mostly made up of the classification of synthetic methods (see Fig. 4): MOST (8), MODAPTS (9), MTM (8), GSD (1). Last but not least, the traditional method of direct observation: the study of times (5). This means that organizations are focusing on optimizing their processes using pre-established methods.

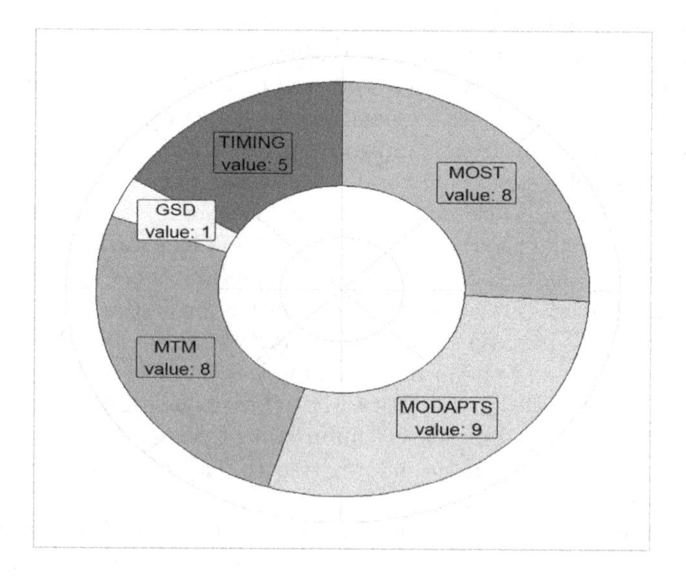

Fig. 4. Number of techniques per classification

In contrast it should be noticed in two articles [6,21], some recommendations were reflected. They were about the determination of standards for pre-established times since the calculated time was much less than the real-time. That why the operation analysis that require more meticulous treatment should be carefully done. Likewise, it was mentioned that predetermined time systems must be improved so that operator fatigue can be considered as a factor.

In addition to what has been already mentioned, it was possible to determine that the techniques corresponding to the synthetic method were investigated throughout the period with which the articles were classified, having an increase in 2019 with 4 research articles (see Fig. 5). The study of times as an individual application was only reviewed in articles published in the years 2017 and 2019. However, it is observed in Fig. 5 that the combination of methodologies has a wide field of research, that is to say that, having as a basis the technique of direct observation in conjunction with any of the predetermined time techniques can generate new systems.

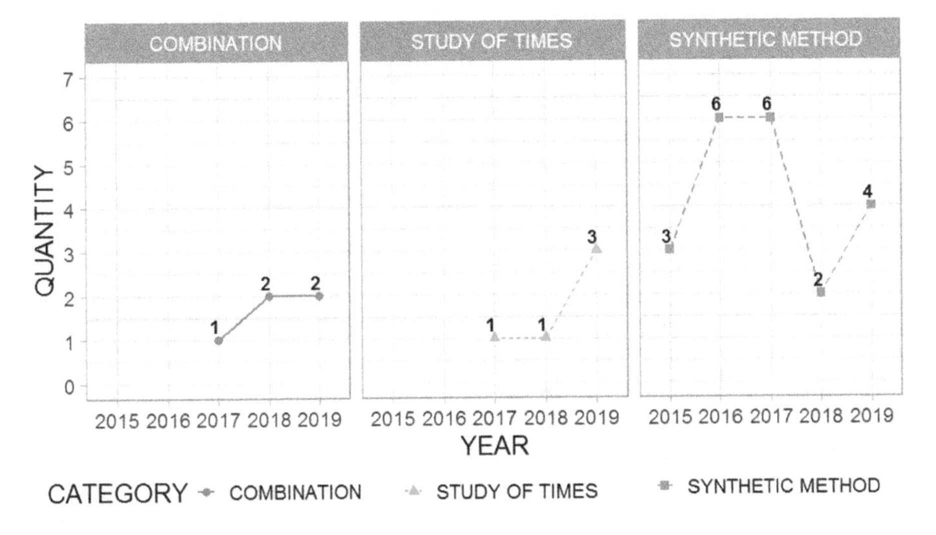

Fig. 5. Work measurement methodologies

Most articles were analyzed or applied worldwide, very few were investigated in South America, so it can be inferred that professionals in this sector are not familiar with the analysis and application of predetermined time systems. It would be very convenient to expand research in this field since the benefits are tangible in all areas of organizations, particularly in the control, optimization, and management of labor costs.

4.1 Challenges

All the research work mentioned in this article, applied methodologies that can become a fundamental part of any type of organization. Therefore, in this section, the articles which supported their research on real cases and which were verified in controlled environments or other than productive activities will be exposed. In the same way, the reasons for jointing the methodology with other tools will be exposed.

For the analysis method of work by direct observation, five articles were reviewed from which one of them [34] was used to methodically expose the study of times by chronometer. Therefore, this research does not have an application at the industrial level. The following articles [2,5,27], used the sequence of steps necessary for the correct execution of the methodology in real cases. To recap, they considered the formulas for calculating the optimal number of samples, as well as the selection of a qualified operator, the standardization of the process and the operations studied. Another aspect of an application in real cases was evidenced in the article [14], in which the methodology was correctly used, focusing on improving the activities carried out in a logistics process.

The article [29] did not apply MTM in a real case but it studied the theme and the necessary factors that should be considered when teaching the methodology. On the other hand, articles [10,26,38], correctly used this methodology in real cases of companies and even compared the results of productive activities with the use of new methodologies of predetermined times. By the same token, the use is evidenced in articles [22,23], since in these investigations each of the phases of MTM is executed correctly and systematically. The only difference is that the case studies were carried out in logistics operations.

On the contrary, the investigations [4,41], validate their objectives with the MTM methodology with simulated cases or in controlled environments, in which through digital applications they represented manual activities. The two articles correctly use the methodology, but in the last one, the activities that do not add value to the process (waiting) are not purified, thus causing the estimated time not to approximate reality.

Articles [1,6,24] are aligned to the same perspective. In these investigations the execution of the MODAPTS methodology in controlled environments is validated, the articles mentioned already focused their studies on the development of technological improvements that will optimize the correct application of predetermined time systems. In other words, this type of research allows the generation of tools that facilitate the application of any system in a real environment. Thus, reducing the execution error that a less-qualified technician may commit when using any system with predetermined times.

Seen in this way, technologies will provide substantial support for the study of work. However, at the moment articles [15,25,35,39] mention that the phases are still correctly used to estimate time using MODAPTS. In these investigations, the methodology is applied in real case studies, seeking the objective of continuous improvement. Another example of analysis is evidenced in the article [3], this study analyzes the estimated times in a real case by using MODAPTS and direct observation. In the investigation the correct application of each of the methodologies is observed, it is vital to take into account that for the execution any type of system must have prior knowledge of the process to be studied in order not to estimate or consider operations that do not add value.

Besides, it should be noted that in the review of articles corresponding to the MOST methodology, all the investigations have validations in real case studies, thus obtaining benefits with time and cost of production. In summary, from the total of thirty one articles, nineteen of them are investigations applied to real cases of production companies, three articles study real cases but of logistical processes, six focus their investigations in controlled environments and three do not apply the methodology in productive activities (See Fig. 6).

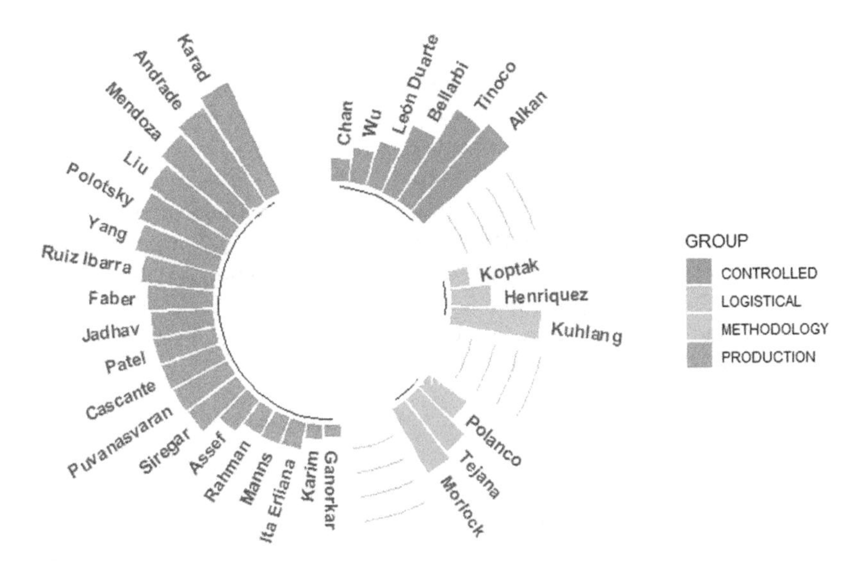

Fig. 6. Classification of cases

Turning now to analyzing the feasibility to combine methodologies with other types of tools. The literature review suggests that from the total number of articles analyzed in sixteen of them, the methodology is not combined with another type of tools, while in twelve articles there is a mixture of work-study methodologies with statistical tools, virtual analysis, meta-mechanisms, movement analysis, among others.

The need to integrate instruments to the case studies analyzed in this article depends on the type of objective being sought. On the one hand, it is evident that in investigations in which it is necessary to estimate a production time or compare existing times with a pre-established standard any of the described methodologies can be executed. On the other hand, when estimated times are the focus, improving processes or operations, innovating in existing methodologies, adapting the methodology to a specific process, a combination of techniques with tools must be carried out. As a result, it will improve research and generate new knowledge and applications in the industrial field.

5 Conclusion and Future Research Implications

The articles presented in this literature review emphasize the importance of the productive and managerial level of the implementation of work measurement with any of the existing techniques. Besides, to the benefits and applications described, it can be concluded that the techniques of predetermined times would be part of the industry and today would be an important instrument to face the new production horizon after the pandemic. The application cases demonstrated that increasing productivity is the consequence of continuous improvement and an innovative approach.

A very important issue that has been mentioned throughout the review is the cost of the product, all companies need to generate new offers with more competitive costs, and a main component of the cost is the time spent by the labor force. Therefore, companies need to determine the productive time. Therefore, the combination of techniques fosters the time estimation to be as reliable as possible. However, it is important to keep in mind that predetermined systems are designed to analyze manual operations by expanding research topics to develop tools that encompass machine conditions. In the case of the execution of MOST in the assembly lines, it could be studied how the system behaves in the mixed assembly models.

References

1. Alkan, B., Vera, D., Ahmad, M., Ahmad, B., Harrison, R.: A model for complexity assessment in manual assembly operations through predetermined motion time systems. Procedia CIRP **44**, 429–434 (2016). https://doi.org/10.1016/j.procir.2016.02.111
2. Andrade, A.M., Del Río, C.A., Alvear, D.L.: Estudio de Tiempos y Movimientos para Incrementar la Eficiencia en una Empresa de Producción de Calzado. Información tecnológica **30**(3), 83–94 (2019). https://doi.org/10.4067/S0718-07642019000300083
3. Assef, F., Scarpin, C.T., Steiner, M.T.: Confrontation between techniques of time measurement. J. Manuf. Technol. Manag. **29**(5), 789–810 (2018). https://doi.org/10.1108/JMTM-12-2017-0253. https://www.emerald.com/insight/content/doi/10.1108/JMTM-12-2017-0253/full/html
4. Bellarbi, A., Jessel, J.P., Da Dalto, L.: Towards method time measurement identification using virtual reality and gesture recognition. In: 2019 IEEE International Conference on Artificial Intelligence and Virtual Reality (AIVR), pp. 191–1913. IEEE, December 2019. https://doi.org/10.1109/AIVR46125.2019.00040. https://ieeexplore.ieee.org/document/8942288/
5. Cascante, G.M., Alulema, J.M., Mariño, C.S.: Tiempos estándar para balanceo de línea en área soldadura del automóvil modelo cuatro (2019)
6. Chan, A.H., Hoffmann, E.R., Chung, C.M.: Subjective estimates of times for assembly work. Int. J. Ind. Ergon. **61**, 149–155 (2017). https://doi.org/10.1016/j.ergon.2017.05.017
7. Durán, F.: Ingeniería de métodos. Globalizacion: Técnicas para el manejo eficiente de recursos en organizaciones fabriles, de servicios y hospitalarias (2007)
8. Elena, J.H.G.: Edición especial Impacto económico. Techical report (2020)
9. Escalante Lago, A.: Métodos y tiempos con manufactura ágil. Alfaomega Grupo Editor (2016)
10. Faber, M., Przybysz, P., Latos, B.A., Mertens, A., Brandl, C., Finsterbusch, T., Härtel, J., Kuhlang, P., Nitsch, V.: Empirical validation of the time accuracy of the novel process language Human Work Design (MTM-HWD®). Prod. Manuf. Res. **7**(1), 350–363 (2019). https://doi.org/10.1080/21693277.2019.1621785

11. Fang, Y.: A method of meta-mechanism combination and replacement based on motion study. Prod. Manuf. Res. **3**(1), 310–323 (2015). https://doi.org/10.1080/21693277.2015.1093437
12. Ganorkar, A.B., Lakhe, R.R., Agrawal, K.N.: Cost and productivity analysis of the manufacturing industry using TDABC & MOST. S. Afr. J. Ind. Eng. **30**(1), 196–208 (2019). https://doi.org/10.7166/30-1-1939
13. Garcia Criollo, R.: Estudio del trabajo. McGraw Hill (2005)
14. Henríquez-Fuentes, G.R., Cardona, D.A., Rada-Llanos, J.A., Robles, N.R.: Medición de Tiempos en un Sistema de Distribución bajo un Estudio de Métodos y Tiempos. Información tecnológica **29**(6), 277–286 (2018). https://doi.org/10.4067/S0718-07642018000600277
15. Ita Erliana, C., Abdullah, D.: Application of the MODAPTS method with innovative solutions in the cement packing process. Int. J. Eng. Technol. **7**(2.14), 470 (2018). https://doi.org/10.14419/ijet.v7i2.11249
16. Izquierdo, R., Novillo, L., Mocha, J.: El impacto de la cuarta revolución industrial en las relaciones sociales y productivas. Universidad y Sociedad **9**(2), 313–318 (2017)
17. Jadhav, M., Mungase, S., Karad, P.A.A.: Productivity Improvement through Maynard Operation Sequence Technique **3**(1), 565–568 (2017)
18. Jananía, A.C.: Manual de tiempos y movimientos. Limusa, México (2008)
19. Kanawaty, G.O.: Introducción al estudio del trabajo. Oficina Internacional del Trabajo, Textos académicos, cuarta edn. (2007)
20. Karad, P.A.A., Waychale, N.K., Tidke, N.G.: Productivity improvement by maynard operation sequence technique. Int. J. Eng. Gen. Sci. **4**(2), 657–662 (2016)
21. Karim, A.M., Tuan, S.T., Emrul Kays, H.: Assembly line productivity improvement as re-engineered by MOST. Int. J. Prod. Perform. Manag. **65**(7), 977–994 (2016). https://doi.org/10.1108/IJPPM-11-2015-0169
22. Koptak, M., Džubáková, M., Vasilienė-Vasiliauskienė, V., Vasiliauskas, A.V.: Work standards in selected third party logistics operations: MTM-LOGISTICS case study. Procedia Eng. **187**, 160–166 (2017). https://doi.org/10.1016/j.proeng.2017.04.428
23. Kuhlang, P., Sunk, A.: Productivity improvement in logistical work systems of the genuine parts supply Chain. In: 2015 IEEE International Conference on Industrial Engineering and Engineering Management (IEEM), pp. 280–284. IEEE, December 2015). https://doi.org/10.1109/IEEM.2015.7385652
24. León-Duarte, J., Aguilar-Yocupicio, L., Romero-Dessens, L.: A software tool for the calculation of time standards by means of predetermined motion time systems and motion sensing technology. In: Ahram, T., Karwowski, W., Taiar, R. (eds.) Human Systems Engineering and Design, pp. 1088–1093. Springer International Publishing, Cham (2019)
25. Liu, L., Jiang, Z., Song, B., Zhu, H., Li, X.: A novel method for acquiring engineering-oriented operational empirical knowledge. Math. Prob. Eng. **2016**, 1–19 (2016). https://doi.org/10.1155/2016/9754298
26. Manns, M., Wallis, R., Deuse, J.: Automatic proposal of assembly work plans with a controlled natural language. Procedia CIRP **33**, 345–350 (2015). https://doi.org/10.1016/j.procir.2015.06.079
27. Mendoza Novillo, P.A., Erazo Álvarez, J.C., Narváez Zurita, C.I.: Estudio de tiempos y movimientos de producción para Fratello Vegan Restaurant. CIENCIAMATRIA **5**(1), 271–297 (2019). https://doi.org/10.35381/cm.v5i1.267
28. Meyers, F.E.: Estudios de tiempos y movimientos para la manufactura ágil. México, segunda edn. (2000)

29. Morlock, F., Kreggenfeld, N., Louw, L., Kreimeier, D., Kuhlenkötter, B.: Teaching methods-time measurement (MTM) for workplace design in learning factories. Procedia Manuf. **9**, 369–375 (2017). https://doi.org/10.1016/j.promfg.2017.04.033. https://linkinghub.elsevier.com/retrieve/pii/S2351978917301518
30. Unidas, Naciones: América Latina y el Caribe ante la pandemia del COVID-19Efectos económicos y sociales. Technical report **1**, (2020)
31. Niebel, B.W., Freivalds, A.: Ingenieria industrial: Métodos, estándares y diseño del trabajo. McGraw Hill, duodécima edn. (2009)
32. Organización Internacional del Trabajo (OIT): El COVID-19 y el mundo del trabajo: Repercusiones y respuestas. Technical report (2020)
33. Patel, D., Tomar, P.: A review on optimization in total operation time through Maynard Operation Sequence Technique. Int. J. Sci. Technol. Eng. **3**(09), 13–16 (2017)
34. Polanco Vides, E.X., Lauren Andrea, D.J., Jorge Junior, G.R.: Análisis metodológico para la realización de estudios de métodos y tiempos. Methodol. Anal. Perform. Stud. Methods Times **1**, 3–10 (2017)
35. Polotski, V., Beauregard, Y.: Work-time identification and effort assessment: application to fenestration industry and case study. IFAC-PapersOnLine **51**(15), 569–574 (2018). https://doi.org/10.1016/j.ifacol.2018.09.217
36. Puvanasvaran, A.P., Yap, Y.Y., Yoong, S.S.: Implementation of Maynard operation sequence technique in dry pack operation-a case study **14**, 21 (2019)
37. Rahman, M.S., Karim, R., Mollah, J., Miah, S.: Implementation of Maynard operation sequence technique (most) to improve productivity and workflow - a case study. Int. J. Emerg. Technol. Innov. Res. **5**(6), 270–278 (2018)
38. Ruíz-Ibarra, J.I., Ramírez-Leyva, A., Luna-Soto, K., Estrada-Beltran, J.A., Soto-Rivera, O.J.: Optimización de tiempos de proceso en desestibadora y en llenadora. Ra Ximhai **13**(3), 291–298 (2017)
39. Siregar, I.: Quality engineering with taguchi loss function method and improvement of work method in anode changing. MATEC Web Conf. **296**(201 9), 02008 (2019). https://doi.org/10.1051/matecconf/201929602008
40. Tejana, N., Gisbert, V., Pérez, A.: Metodología De Estudio De Tiempo; Introdución al GSD. 3C Empresa, pp. 39–49 (2017)
41. Tinoco, H.A., Ovalle, A.M., Vargas, C.A., Cardona, M.J.: An automated time and hand motion analysis based on planar motion capture extended to a virtual environment. J. Ind. Eng. Int. **11**(3), 391–402 (2015). https://doi.org/10.1007/s40092-015-0107-9
42. Valderrama, Y., de Carmona, L.C., Colmenares, K., Jaimes, R.: Costo de la gestión laboral en el proceso productivo de una empresa manufacturera trujillana. Caso: Industrias Kel, C.A. Actualidad Contable FACES **2**(33), 96–111 (2016)
43. Wu, S., Wang, Y., BolaBola, J.Z., Qin, H., Ding, W., Wen, W., Niu, J.: Incorporating motion analysis technology into modular arrangement of predetermined time standard (MODAPTS). Int. J. Ind. Ergon. **53**, 291–298 (2016). https://doi.org/10.1016/j.ergon.2016.03.001

Chatbot to Simplify Customer Interaction in e-Commerce Channels of Retail Companies

Jean Martin Solis-Quispe(✉), Kathia Milagros Quico-Cauti, and Willy Ugarte

Universidad Peruana de Ciencias Aplicadas (UPC), Lima, Peru
{u201410947,u201414287}@upc.edu.pe, willy.ugarte@upc.pe

Abstract. Today, e-commerce has become the main channel of the retail sector, however, the abandonment of online shopping has also increased. This is because it is difficult for customers to navigate the website and they feel the unavailability of a sales consultant to guide them and resolve their doubts. The objective of this study is to propose a chatbot that allows to finish the purchases, improving the user experience in the process of buying online. In addition, it will be validated through acceptance tests to verify in what percentage it simplifies customers' interaction during the whole online purchase process and customer service.

Keywords: Chatbot · DialogFlow · E-commerce · Retail

1 Introduction

The e-commerce use has seen a 55% growth in 2020 due to global crisis originated by the covid-19 pandemic, since most of the people remain in quarantine, they prefer online shopping that allows them to have the products delivered to home and feel safer[1]. In the case of Peru, the impact of Internet sales has experienced significant growth in recent years, since in 2018 Peru has reached approximate sales of 4 billion dollars[2]. Likewise, in 2017 according to the study conducted by Ipsos, which reports that 10% of Peruvians have made an order, reservation or purchase of any product or service on the web[3].

Although e-commerce has increased and become the main channel in the retail sector, this does not allow the development of a personal relationship between the customer and the seller [1]. According to the study conducted by Klie, 77% of users leave the shopping cart before completing their purchase [2]. Therefore, the main issue is the abandonment in the process of purchase in e-commerce, this is reflected in the most recent study by Statista that according

[1] Guillem Sanz - https://cutt.ly/auyjDlS.
[2] Linio - https://www.linio.com.pe/sp/indice-ecommerce.
[3] Ipsos - https://www.ipsos.com/es-pe/comercio-electronico-2017.

Á. Rocha et al. (Eds.): ICITS 2021, AISC 1330, pp. 561–570, 2021.
https://doi.org/10.1007/978-3-030-68285-9_52

to statistics in 2019 the rate of abandonment of online shopping carts is 69.57% of customers who failed to complete the purchase[4].

This problem is due to factors in the purchasing process and the poor experience when interacting with the web e-commerce channel, among them we highlight that websites are difficult to navigate and the unavailability of an assistant during the purchasing process [3].

According to Concha [4] that surveyed 180 people that bought food through an application and website, 39% made it online in 1 to 2 h and 37% in 30 min to 1 h. In addition, structurally, most of websites display information by sections with links, difficulting visitors to open these links one by one, making the process less interactive and taking time for the client in finding the required information [5]. Because of that, the first stage of the online shopping experience, which is browsing product information and placing the order, becomes the most important of the online purchase.

Our main goal is to develop a chatbot that simplifies the interaction with the user when making an online purchases in the retail sector and thus reduce the time of purchase and the amount of interactions. A chatbot can help complement customer services by answering repetitive queries without any limitations and is available 24 h a day [6]. Virtual agents save time, facilitate the purchase procedure and influence the customer's decision making [7]. Many businesses are also using them to improve communication with their target audience so that they can recommend products and make purchases online [8].

For Gartner[5], in 2020, 85% of customer interactions with the business will be done through virtual assistants thus requiring the information in a more personalized way (text or voice) during the whole purchase process [9]. The focus of our work is on proposing a new e-commerce channel to the convenience stores in the retail sector in Peru. This sector was divided because, in recent years, convenience stores have had a great impact on the Retail sector, in 2019, 17.3% of households in the country visited this new channel and in 2018 sales were 60 million dollars and it is projected that by 2022, 220 million dollars in sales will be obtained according to the consulting firm Euromonitor International [10].

Our contributions are as follows:

- A framework for improving the user experience and streamlines the purchasing process by chatbot management.
- An improvement of user interactions number that leads to a faster service.
- An improvement of service time by retails.

For this work, in Sect. 2, a review of the concepts contained in the solution is presented then an explanation our main contribution. After, Sect. 3 validates the solution with proposed scenarios and list some related works and briefly discuss them. Finally, Sect. 4 describes the conclusions and future work.

[4] Statista - https://cutt.ly/ouyljXe.
[5] https://gtnr.it/2NE5vDZ.

2 Material and Method

A literature review was conducted to analyze and discuss fundamental concepts for a better understanding of the study.

2.1 Chatbot

A chatbot is a software implemented through artificial intelligence in which users can communicate by having conversations, the bots can receive as input text or uservoice, understanding the natural language [5]. Likewise, chatbots use a natural language processing system (NLP) that contains defined questions and answers. Also, chatbots are defined as systems that interpret and respond using case-based logic [9]. They can have animated images, interactive avatars or human faces that imitate a salesperson [11].

Definition 1 (Chatbot [12]**).** *Chatbots are computer programs that interact through NLP interfaces [12] that allow actions to be carried out (e.g., product information), being electronic agents that are available all the time, enabling to make better use of the customer time.*

Although their functioning is based on the interpretation of natural language, it also relies on defined conversational flows such as artificial intelligence or decision trees [12]. In the same way, chatbots respond through artificial intelligence to defined rules [13] adding that chat services are provided through mobile applications and messaging [14], and they could perform on e-commerce sites as service agents [15]. Chatbots are classified into three different types:

i) *Conversational chatbots*: supported by a knowledge base where information is analyzed to answer given queries.
ii) *Task-oriented chatbots*: also called service-based chatbots, intended to help customers to complete a defined activity such as buying a product or booking a service.
iii) *Social chatbots*: seek to communicate with other users to make recommendations [16].

2.2 Dialogflow

Dialogflow is a chatbot platform developed by Google Cloud, chosen after a comparative analysis[6] as the most suitable for the development of this project.

One of the main characteristics of Dialogflow is that it allows the user's query to coincide with the most appropriate attempt through the information contained in the user input. Another important feature is that it facilitates integration with various social network and messaging applications. In [17] the main chatbot platforms are compared, and it is concluded that Dialogflow is the most complete framework, highlighting the characteristics of the attempt and the

[6] Globalme - https://www.globalme.net/blog/lex-vs-dialogflow-vs-watson-vs-rasa/.

recognition of the entity. Dialogflow was also chosen as the best platform due to its compatibility with messaging applications such as Facebook [9].

The chatbot architecture is mainly composed of natural language processing (NLP) techniques relying on Machine Learning to understand a user's input and decide what intent they want to perform [10].

According to Icapps[7], for a computer to analyze, understand and extract meaning from human language it requires NLP. After the NLP technique is finished, another process called Natural Language Understanding (NLU) is performed to identify the user's intention [12]. Dialogflow's architecture is primarily composed of Entities, Intents, and Context.

i) Entities can be fields, data, or text describing just about anything—a time, place, person, item, number, etc.
ii) An Intent is the result of matching Entities through a recognition engine that is based on training sentences, in order to identify the user's intent. An intent contains information in the text that can be extracted according to defined rules [10], more precisely a greeting, display a menu, consult promotions, place orders, etc. [13].
iii) A Context has grouped parameters that define the assignment of Entities to training sentences that have information to extract and store [10]. A Context also helps to extract semantic parts of texts to find unknown information and provide a more appropriate answer [8].

2.3 Customer Experience

Chatbots help to improve customer satisfaction during the purchase and service process [15]. Nevertheless, customer satisfaction requires a further study of customer experience.

Definition 2 (Customer Experience [18]**).** *The subjective response of an actor to the elements of the service, which arise during the process of purchase and use, through imagination or memory.*

Likewise, Richardson [19] states that the customer experience is the quality of the experience he has when interacting with a website. Nevertheless, in [20], Jain et al. mention that the experience is the formation of attitudes, perceptions and feelings that are formed in the decision-making process and during the entire flow that leads to emotional and behavioral responses.

Chatbots act as a service interface that helps clients make decisions more quickly by impacting on their experience [11]. The perception of chatbot usage is key to determine the customer experience, being relevant the answers to the information requirements. It is also mentioned that they can always improve this experience with the feeling of being served by a virtual agent.

In [21], the authors claim that providing chatbot information and helping to save customer time is the main motivation for using chatbot and has a great

[7] Icapps - https://icapps.com/blog/linguistics-behind-chatbots.

impact on customer perception. The availability at any time is very important, as it is numerically shown that 91% of dissatisfied customers do not require services from the business again [16].

In [13], it is mentioned that chatbots can improve the customer experience in different sectors through process automation such as in e-commerce and education. This is the case for e-commerce, if they manage to fulfill their expectations in an immediate way, such as resolving complaints, returns and product exchange with the customer. Similarly, when customers find your services or products through the chatbot and exceed their expectations, customers perceive high satisfaction in their online shopping experience [7].

2.4 Implementation

For the implementation, the following phases were carried out:

- First, product and store location information were collected for convenience and the database was modeled.
- Secondly, the web application was created under the Laravel framework[8] and developed in PHP. The Web backend is connected to the MySQL database and develop all the functional requirements for an e-commerce.
- Thirdly, the integration and connection of the web page (chatbot interface) with Dialogflow's conversational agent through Google's API allowing authentication to make HTTPS requests was carried out. Likewise, entities, intentions and contexts were created and carried out in the training of the chatbot. Finally, it was deployed on the website and chatbot in a hosting hosted on Google Cloud Platform.

2.5 Solution Overview

Figure 1 shows the integrated architecture of the solution which is a chatbot integrated with a web system. When the user enters a website and interacts with the chatbot interface, when the user performs a query with the chatbot interface, this query is taken by the Backend Bot Request logic component which performs the query to the Dialogflow agent, authenticating itself through the Dialogflow API. This query enters the Data processing, where through the NLP that transforms from text to machine language, this component understands the captured query and transforms it into information.

Then, it enters the decision engine where the true intention of the user is identified according to the Entities, Intents and the Context where the user-expression is located [16]. Each intention has a defined Fulfillment, and when it has a match, Dialogflow will make an http webhook request[9] sending output data with a JSON object that contains information about the matching intention [17], this request is authenticated by the Dialogflow API to the website backend [9].

[8] https://laravel.com/.
[9] Webhook service -https://bit.ly/2YI4Qr9.

After the webhook component receives the request, it performs the assigned functionality in the backend (i.e., consult product, consult opening hours, show shopping cart, recommend nearest store, list orders, etc.).

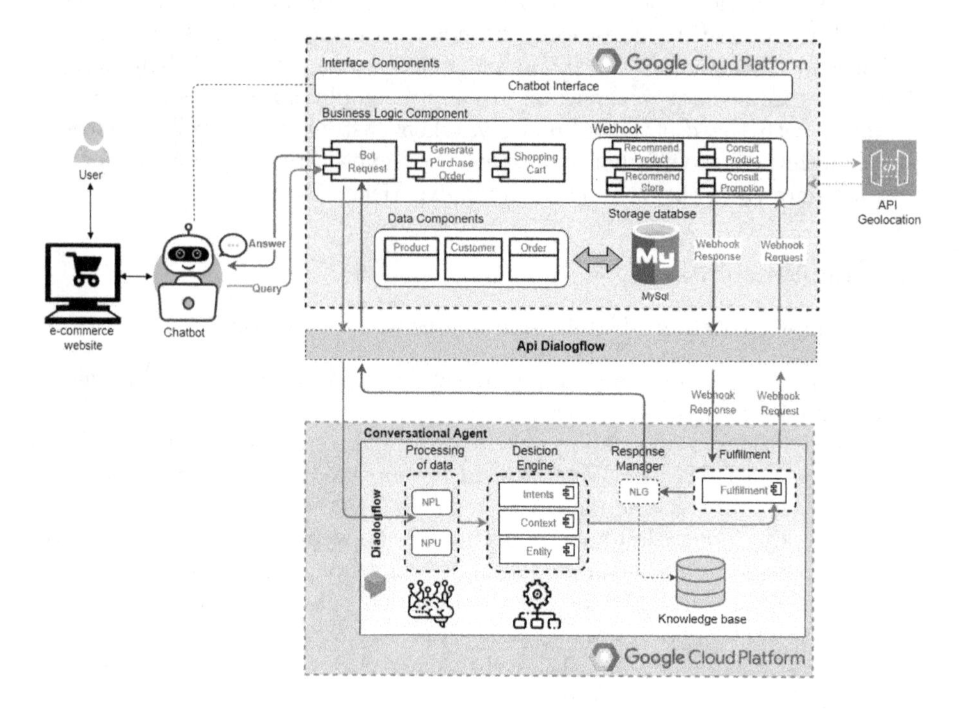

Fig. 1. Solution Overview

For each request, a query is made to the database supported by MySQL. The backend service then responds to the request with instructions on what it should do and show Dialogflow. Dialogflow can also answer simple text questions from the knowledge base without having to make a request.

Finally, the response manager gets the answer by transforming it from machine language to text and displaying the final answer to the user through the chatbot interface window.

3 Results and Discussions

Validating our goal of optimizing the purchase process, requires user acceptance tests, where the target audience tests the solution in a real environment. The objective is not to verify that it meets the functional requirements but to ensure that the software meets the customer's needs.

The indicators for this validation are:

- ID_1: Average user-time to perform a scenario.
- ID_2: Average amount of user-interactions for a scenario.

Table 1. Test scenarios

ID	Scenario	Description
E_1	Scenario 1	20 users made a product delivery price inquiry on the Plaza Vea site
E_2	Scenario 2	20 users made a purchase of 3 products on the Plaza Vea site
E_3	Scenario 3	20 users made a product delivery price inquiry through the chatbot
E_4	Scenario 4	20 users made a purchase of 3 products through the chatbot

3.1 Experimental Protocol

Our validation consists of performing user acceptance tests on 20 people between the ages of 25 and 34 who make online shopping at a supermarket website and usually attend the store for convenience.

This age range might seem arbitrary, but it has been proven by the *Lima Chamber of Commerce (CCL)*[10, 11] that 43.7% of online shopping in Peru is made by people in this range. These users will test 4 scenarios, where the goal is to compare traditional e-commerce purchase process with (or without) the chatbot. Table 1 shows 4 proposed acceptance test scenarios, where users used a traditional e-commerce website and the chatbot.

3.2 Comparison of Indicators by Scenario

The results of the indicators in each scenario are compared and in Fig. 2 you can see the difference between Plaza Vea e-commerce indicators and the average chatbot indicators. A customer takes around 20 s to make a query through the chatbot against 175 s of traditional e-commerce. Likewise, the average number of customer interactions for consulting the delivery price in a chatbot is 3 interactions, simplifying 13 interactions compared to the website.

Furthermore, in scenario 4 an online purchase of 3 products through the chatbot take 150 s (resp. 265 for scenario 2) and 19 interactions (resp. 28 for scenario 2). In Table 2, the reduction of the average purchase time by 43% and the amount of average purchase interactions by 32% are confirmed. Also, for the metrics in the process of customer service is reduced the average time by 88% and the number of average interactions by 80%, showing optimization of both processes through the chatbot.

Nevertheless, reducing interactions does not necessarily implies a better user acceptance. It is convenient to perform User Experience tests. To carry out this experiment we asked 20 students from the University of Applied Sciences (UPC). Most of the participants were still undergraduate students with an age range of 20 to 26 years. Most of these participants mentioned not having a

[10] Andina - https://andina.pe/agencia/noticia-internet-compras-online-llegarian-a-2800-millones-este-ano-757732.aspx.

[11] Gestión - https://gestion.pe/economia/compras-internet-llegarian-us-2-800-millones-ano-ccl-e-commerce-compras-online-nndc-272332-noticia/.

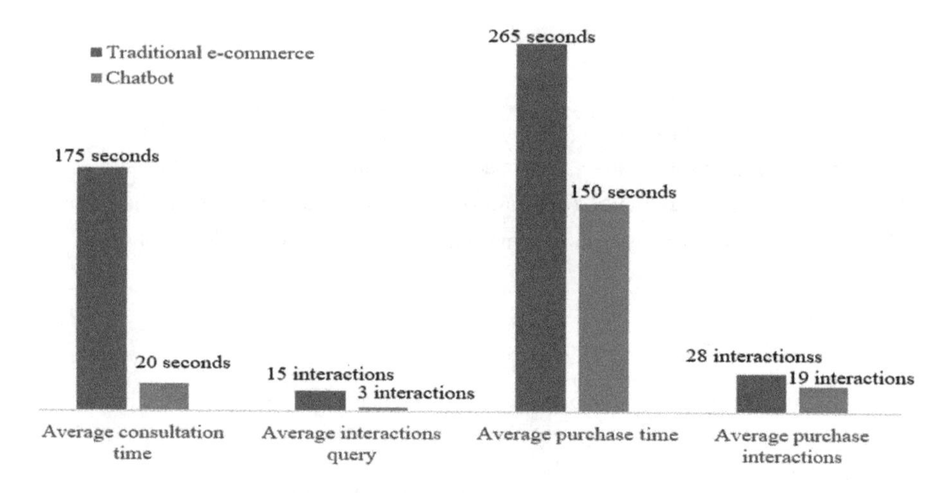

Fig. 2. Comparison of indicators

Table 2. Results for success metrics

Predefined metric	Expected	Result
Reduction of average customer service time	50%	88%
Reduction of average number of interactions in customer service	50%	80%
Reduction of average time in online shopping	30%	43%
Reduction of average number of interactions in online shopping	30%	32%

previous experience with chatbots in online shopping. Based on the previously conducted interaction, participants were asked to conduct a 10-question survey. These questions were taken to measure certain features of the interaction developed. We asked the participants what they thought was the interaction regarding the satisfaction of performed purchase tasks, did they find it satisfying or not (see Fig. 3).

3.3 Discussion

SuperAgent is a service chatbot that obtains public information about products, services, comments, questions and answers from various e-commerce sites in order to answer users' frequently asked questions (FAQ), freeing up the workload of the support staff [6]. Sambot is a chatbot of Samsung Iot Showcase website, it contains a botmaster that enhances its knowledge base capability to answer FAQ about Samsumg products, also recommends and self-completes questions [5]. Both, SuperAgent and Sambot are mainly used for answering FAQ. Unlike these, our proposal, additionally to FAQ, also guides the customer until finishing the purchase order, by recommending improvements to customer experience (e.g., the closest store to you according to your location). Although this chatbot is

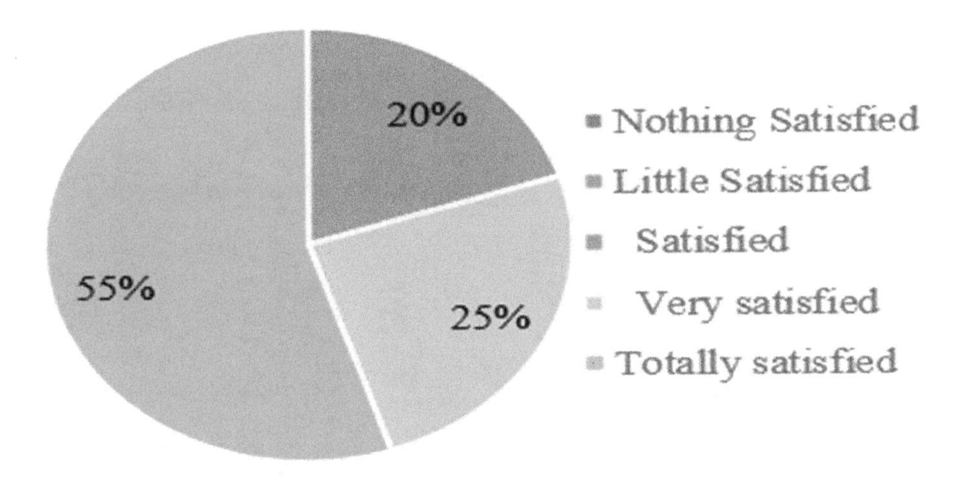

Fig. 3. Customer satisfaction for making purchases through the chatbot

hosted on a website, [22] recommends that chatbot be integrated into messaging and social networking applications to improve the quality and user experience.

4 Conclusion

The current pandemic will lead people to make more online shopping, thus sales in e-commerce will increase, this is a great opportunity for chatbots to help this by accompanying customers throughout the purchase process and solving queries instantly. As demonstrated, the chatbot improves the user experience and streamlines the purchasing process. Due to the conjuncture of the pandemic, people are doing more online purchases in e-commerce will increase, this is a great opportunity for chatbots to help make purchases online by accompanying throughout the purchase process and solving queries instantly. As shown in Fig. 2, the chatbot optimized the purchase process reducing the purchase time by 32% and the number of interactions by 43%, exceeding the expected value of the previously proposed metric.

Our proposal can also be integrated with messaging applications, such as Facebook, so purchases can be made directly from the application, without a web page. In addition, new applications and payment functionalities may be added for a future upgrade. Therefore, a future work aiming at expanding tests with the proposed chatbot should be performed, expanding the sample size and applying this proposal to other kinds of retails, or even using blockchain for retail databases [23].

References

1. Lee, Y.J., Dubinsky, A.: Consumers' desire to interact with a salesperson during e-shopping: development of a scale. Int. J. Ret. Dist. Man. **40**, 20–39 (2017)

2. Song, J.-D.: A study on online shopping cart abandonment: a product category perspective. J. Internet Commer. **18**(4), 337–368 (2019)
3. Devaney, E.: The state of chatbots report: How chatbots are reshaping online experiences. Drift **23** (2018)
4. Concha, A.: Comercio electrónico de alimentación en barcelona ¿de qué manera podrían los supermercados online incrementar las ventas en este sector? Master's thesis, Universitat de Barcelona (2019)
5. Pradana, A., Sing, G.O., Kumar, Y.J.: SamBot-intelligent conversational bot for interactive marketing with consumer-centric approach. Int. J. Comput. Inf. Syst. Ind. Manage. Appl. **6**, 265–275 (2018)
6. Cui, L., Wei, F., Huang, S., Tan, C., Duan, C., Zhou, M.: Superagent: a customer service chatbot for e-commerce websites. In: ACL (2017)
7. Chung, M., Ko, E., Joung, H., Kim, J.: Chatbot e-service and customer satisfaction regarding luxury brands. J. Bus. Res. **117**, 587–595 (2018)
8. Tran, O.T., Luong, T.C.: Understanding what the users say in chatbots: a case study for the Vietnamese language. Eng. Appl. Artif. Intel. **87**, 103322 (2020)
9. Perez, P., De-La-Cruz, F., Guerron, X., Conrado, G., Quiroz-Palma, P., Molina, W.: ChatPy: conversational agent for SMES. In: (CISTI). IEEE (2019)
10. Daniel, G., Cabot, J., Deruelle, L., Derras, M.: Multi-platform chatbot modeling and deployment with the jarvis framework. In: CAiSE (2019)
11. Pantano, E., Pizzi, G.: Forecasting artificial intelligence on online customer assistance: evidence from chatbot patents analysis. J. Retail. Consum. Serv. **55**, 102096 (2020)
12. Villegas-Ch, W., Arias-Navarrete, A., Palacios-Pacheco, X.: Proposal of an architecture for the integration of a chatbot with artificial intelligence in a smart campus for the improvement of learning. Sustainability 12, 1500 (2020)
13. Trivedi, J.: Examining the customer experience of using banking chatbots and its impact on brand love: the moderating role of perceived risk. J. Int. Commer. 18, 91–111 (2019)
14. Roca, S., Sancho, J., García, J., Alesanco, A.: Microservice chatbot architecture for chronic patient support. J. Biomed. Inf. **102**, 103305 (2020)
15. Go, E., Sundar, S.S.: Humanizing chatbots: the effects of visual, identity and conversational cues on humanness perceptions. Comput. Hum. Behav. **97**, 304–316 (2019)
16. Nuruzzaman, M., Hussain, O.K.: IntelliBot: a dialogue-based chatbot for the insurance industry. Knowl. Based Syst. **196**, 105810 (2020)
17. Shah, S.: A comparison of various chatbot frameworks. J. Multi-Criteria Decis. Anal. **6**, 375–383 (2019)
18. Jaakkola, E., Helkkula, A., Aarikka-Stenroos, L.: Service experience co-creation: conceptualization, implications, and future research directions. J. Serv. Manage. **26**, 182–205 (2015)
19. Richardson, A.: Using customer journey maps to improve customer experience. Harvard Bus. Rev. 15, 2–5 (2010)
20. Jain, R., Aagja, J., Bagdare, S.: Customer experience - a review and research agenda. J. Serv. Theory Pract. **27**, 642–662 (2017)
21. Brandtzæg, P.B., Følstad, A.: Why people use chatbots. In: INSCI (2017)
22. Van den Broeck, E., Zarouali, B., Poels, K.: Chatbot advertising effectiveness: when does the message get through? Comput. Hum. Behav. **98**, 150–157 (2019)
23. Cueva-Sánchez, J.J., Coyco-Ordemar, A.J., Ugarte, W.: A blockchain-based technological solution to ensure data transparency of the wood supply chain. In: IEEE ANDESCON (2020)

Towards the Integration of Internet of Things Devices to Monitor Older Adults Activities in a Platform of Services

Miguel Sousa Gomes[1], João Rainho[2], and Nelson Pacheco Rocha[3](✉)

[1] Microio - Serviços de Eletrónica, Lda, Aveiro, Portugal
miguelsousagomes@gmail.com
[2] Instituto Pedro Nunes, Coimbra, Portugal
rainho@ipn.pt
[3] Department of Medical Sciences, Institute of Electronics and Informatics Engineering of Aveiro, University of Aveiro, Aveiro, Portugal
npr@ua.pt

Abstract. The research study reported by the present paper aimed to integrate in a platform of services several Internet of Things (IoT) devices to monitor older adults' activities together with analytical procedures to infer relevant events. To guarantee the interoperability of the processed information, as well as its privacy, integrity, and confidentiality, the study applied several services provided by the platform, namely interoperability and security (i.e., authentication, authorization, logging and auditing) services. The results show the viability of the approach and it is foreseen the development and evaluation of additional algorithms for the detection of significant events that might be related to health conditions of older adults.

Keywords: Older adults · Internet of Things · Activities monitoring · Interoperability · Security mechanisms

1 Introduction

The advances in sensing and communication technologies such as Internet of Things (IoT) made possible the development of mobile and wearable devices able to continuously monitor physiological parameters in out-hospital conditions [1, 2]. Furthermore, monitoring physiological parameters together with activities and behaviors to assess, in a naturalistic and continuous way, health conditions, might help to automate assistance, to prevent the exacerbating of disease or accidents, and to properly react to emergency situations [1, 3].

The authors were involved in the development of a platform of services, the Social Cooperation for Integrated Assisted Living (SOCIAL) platform [4], that aims to deliver information services to support the care and assistance provided to community-dwelling older adults. Since this platform allows the integration of various data sources, the objective of the study reported by this paper was to apply the available services to

Á. Rocha et al. (Eds.): ICITS 2021, AISC 1330, pp. 571–580, 2021.
https://doi.org/10.1007/978-3-030-68285-9_53

integrate several IoT devices able to monitor older adults' activities together with analytical procedures to infer events that might characterize health conditions (e.g., identifying consistency and completeness of daily activities).

2 Background

The literature reports a diversity of studies comparing usual care with the use of various devices to support home monitoring of patients with chronic conditions [5]. For instance, both type 1 and type 2 diabetes conditions are being object of home monitoring [5, 6]. In turn, several devices are being used to remotely assess symptoms and vital signs of patients with congestive heart failure, as well as the transmission of automatic alarms [7]. Moreover, a broad range of devices is also being used to measure and transmit different types of information (e.g., weight, temperature, blood pressure, oxygen saturation, spirometry parameters, symptoms, medication usage or steps in 6-min walking distance) of patients with chronic obstructive pulmonary disease [8].

Activity recognition (i.e., the ability to recognize or detect current activity based on information received from different monitoring devices) is an active research topic in different areas, including healthcare [9, 10]. Several studies report the use of monitoring information to perform activity recognition in order to detect dangerous spatial or temporal configurations, to prevent anomalous situations, to infer human behaviors and emotions, or to confirm medication consumption [11–13].

Activity recognition can be based on several techniques. Some of these techniques use surveillance cameras together with computer vision algorithms to recognize activities being performed [14, 15], while others use dense sensing capacities resulting from the deployment in the environment of different devices such as motion, pressure, temperature and proximity sensors or radio-frequency identification tags [16, 17]. Other techniques use a hybrid approach by combining both wearable and object-tagged devices [10]. Moreover, the various sensing modalities might be integrated using IoT [18, 19].

However, the existence of a broad range of technological possibilities is not enough for their applicability. Mobilizing care in the community is a challenging issue and integrated solutions are required. This is even more relevant because the emphasis of research efforts related to the use of technological solutions to support health conditions of the individuals in their natural environment has been more oriented towards new ways of acquiring information (e.g., monitoring devices), and less on the development of new models and tools to facilitate communication and information access in order to improve functional care provision [20]. In this respect, the SOCIAL platform [4, 21] aims to contribute for care integration by delivering information services to support the care and assistance provided to community-dwelling older adults. For that, among other services, the SOCIAL platform is able to integrate IoT devices designed to monitor older adults' activities, as well as analytical procedures to infer events that might characterize health conditions, which is the objective of the study reported by this paper.

3 Platform of Services

The SOCIAL platform was designed to support the care of community-dwelling older adults. It aims to deliver information services to surpass the professional, organizational, and jurisdictional boundaries of healthcare, social care, and assistance provision. Therefore, the platform includes structural components required to support a range of coherent applications to ensure an integrated, consistent, and cross-cutting view of the information of the care receivers, and an efficient communication with and among caregivers. Some of its characteristics are: integration, when required, of health and social care information; empowerment of the care receivers and their informal caregivers by providing them the information they need; access by formal care providers to relevant care receivers' information; information sharing and exchange between multidisciplinary teams; improved communication between care organizations and local authorities; and comprehensive repository of health and social care information to allow analytical processing to obtain evidence on the effectiveness of actions and policies being developed and the respective impact in terms of quality of life of the older adults.

Important concerns are the interoperability and the privacy, integrity, and confidentiality of the care receivers' information.

The care providers need to have a comprehensive understanding of the care receivers' information. Moreover, this comprehensive understanding should not be limited by the boundaries of public and private institutions with different professional, organizational and jurisdictional natures. Therefore, reliable interoperability mechanisms are required to connect fragmented care receivers' information. This means that both the used technologies and the understanding of the collected information, processes, activities, and policies must comply internationally available standards. In this respect, the Fast Healthcare Interoperability Resources (FHIR) has become increasingly important to provide longitudinal records of care receivers' health conditions and healthcare [22].

The care networks of community-dwelling older adults are composed by a significant number of dispersed resources that can be accessed by a broad range of individuals that are distributed, both in geographical terms and in the administrative terms. Moreover, many of the available digital technologies use broadcast wireless communications, global connectivity and shared hosting of data and computations based on cloud computing [23]. This means that privacy, integrity, and confidentiality of the information need to be assured in a transparent way, be easy to maintain, and supported on a strong granularity in terms of the definition of different authorization levels. Therefore, following the applicable regulations (e.g., EU Directive 2016/679 on the protection of individuals that entered into force in May 2018) regarding the processing of personal data and the movement of such data, several principles were considered to guarantee the privacy of the individuals, namely openness, transparency, inviolability of the information being managed, and individual participation.

To provide common and shared software architectural standards, data formats, storage systems, modules, and information resources, as well as to optimize the development and integration of the required applications, a services-oriented architecture (SOA) was considered for the development of the platform of services. Moreover, since a 3-tier (i.e., application, business and data layer) SOA architecture was implemented, the platform presents a separation of business rules from the applications and technology that interpret them, which maximizes flexibility over the longer and minimizes the effort of accommodating changes in business rules.

The application layer is composed by several web and mobile applications to fulfil the requirements of the different entities involved in the care of community-dwelling older adults, while the data layer ensures the persistence of the required information, namely information related to the care receivers and the human and material resources available to carry out the different tasks, or the information to support the operations of the platform (e.g., terminologies).

The business layer is responsible for the logic related to the interactions between the different applications and the information being persisted. Concerning the study reported by this paper, the following business services were considered: FHIR Gateway, Authentication, Authorization, and Logging and Auditing.

The FHIR was the selected interoperability standard because it is regarded as the next generation of health information interoperability frameworks (i.e., by combining previous standards' features) and also due to its flexibility since it supports several web standards, such as the Representational State Transfer (REST) architectural style, the JavaScript Object Notation (JSON) serialization format or the Extensible Markup Language (XML).

The basic building blocks of a FHIR document are the FHIR resources, which are JSON or XML objects with healthcare related fields and values that, either by themselves or combined, satisfy most common requirements in terms of clinical information. Not fully compatible resources can be adapted to the desired context using FHIR extensions, which grants the ability to readjust the existent structure to one that fits specific requirements [22].

In terms of information model, a fundamental entity is the Person FHIR resource [22]. The individuals interacting with the SOCIAL platform (e.g., care receivers, informal care providers or formal care providers) have identities and, therefore, are considered Persons that might have several roles. Regardless of the role they play on the platform, their demographic information should be registered. The Person FHIR resource might be associated to several instances of the same individual, allowing the different roles the individual can assume within the platform. The care receivers can be referenced as Patient resources, where their demographic and other administrative information is structured as attributes (e.g., name or address). Moreover, other FHIR resources (e.g., Observation or Care Plan) are available to structure relevant clinical information.

In the case of care providers, they can also be characterized by FHIR resources, such as Practitioner (i.e., a person who is directly or indirectly involved in the provisioning of care), Practitioner Role resource (i.e., a specific set of roles, locations, specialties or services that a care provider may perform at an institution during a period of time), and Organization resource (i.e., a formally or informally recognized grouping of people or institutions organized for the purpose of achieving some form of collective action).

The FHIR Gateway provides Create, Read, Update and Delete (CRUD) operations to access and retrieve care receivers' information. Moreover, an advanced search is also allowed, with specific search parameters associated to each resource, to obtain specific information. All these operations also require the interaction of the FHIR Gateway with the various security services (i.e., Authentication, Authorization, and Logging and Auditing services).

The Authentication service allows the registration of specific users in the SOCIAL platform, the assignment of the respective roles and the verification of their identities. If an authentication is successfully validated, the Authentication service returns a valid JSON Web Token (JWT), which enables the secure exchange of information between parties and has the required attributes for the verification of identities so that specific information operations such as the CRUD operations of the FHIR Gateway can be authorized.

The main purpose of the Authorization services is to provide general mechanisms to control the access to the SOCIAL platform, which is an environment composed by a significant number of dispersed resources that can be accessed by a broad range of users that are distributed, both in geographical terms and in the administrative terms. Therefore, authorization is challenged due a diverse set of policies, complexity of workflows and high risk of denying access to key information.

The Authorization service implements role-based access control (RBAC) mechanisms. However, using a single identifying attribute (e.g., the users' role) to decide if someone should access a resource can create information misuse possibilities. Since the information access should not be too generic the access control analyses several different attributes, which means that attribute-based access control (ABAC) mechanisms were also implemented.

Finally, another important aspect is the need to guarantee the integration of the authorization mechanism with FHIR resources. This integration is possible since the FHIR resources are defined by attributes and these attributes can be used to manage the policies and rules. The FHIR resources have security labels for three different purposes: to indicate the permissions related to information operations, such as read or modify, to indicate what resources can be returned, and to indicate how specific information should be dealt with. Requests can also add their own security labels (e.g., break-the-glass protocol to allow a physician dealing with an emergency to access information of care receivers even when not having access permissions). The security labels can have different categorizations, such as confidentiality, sensitivity, compartment, integrity, or handling. Therefore, using the FHIR attributes together with the platform-specific attributes (e.g., subjects' id, role and number or list of care receivers) allows the handling of different types of access scenarios, which guarantees privacy, integrity, and confidentiality of the information.

The Logging and Auditing service aims the creation of logs from multiple entities, over a distributed execution environment, with distributed storage and validation, while providing integrity and auditing controls. For that a custom implementation of a blockchain framework was developed. This blockchain has attributes that are meant to be used with logs and events and prove their integrity, mainly when performing audit processes and verify the consistency of the data before handed. In turn, log audit is done by direct observation of the event stream stored in the database. Existing tools such as Kibana can be used and the integrity of the data observed can be assured. The integrity of the chain can be verified anytime to check for any inconsistencies.

To facilitate the interaction between the services of the platform, the architectures of the FHIR Gateway and the different security services are based on REST. RESTful web services allow the communication via Hypertext Transfer Protocol (HTTP), or optimally via HTTP Secure (HTTPS), if an adequate public key infrastructure is provided upon the provisioning process.

4 Integration of the IoT Devices

According the purpose of this study, the IoT devices should be able to monitor older adults' activities to detect otherwise imperceptible occurrences that might be related to health conditions deterioration. For that four types of IoT devices were developed: motion devices (i.e., passive infrared sensors, commonly used for individuals' motion detection in indoor environments, such as rooms and hallways); pressure devices (i.e., strain gauges whose electrical resistances vary with the applied pressures, which enables the measurement of weight variation); contact devices (i.e., reed switches that are operated by an applied magnetic field created by a magnet, which means that the devices and their auxiliary magnets are suitable for applications in doors and windows); and power devices (i.e., devices to be applied on power sockets, feature a 230VAC connector for appliances, as well as current and voltage sensors to monitor the power consumption and the activity states of the connected appliances). An additional device, the Communications Gateway, was also developed to bridge the connection between the monitoring devices and the SOCIAL platform.

In terms of specifications, the following characteristics were considered [24]: the IoT devices should be able to execute simple operational algorithms, be battery powered, and feature wireless communications, and the field communications technology must have low power requirements. According to the defined characteristics, the motion, pressure and contact devices are battery powered with rechargeable lithium polymer (LiPo) batteries, while the power devices are powered by 230VAC as it is meant to be applied in power sockets. Moreover, all devices feature a microcontroller and a Long Range (LoRa) transceiver for wireless communications with the Communication Gateway. LoRa was selected due to its adequacy for predicted use cases, with battery powered devices and a range requirements that can vary from a few meters in smaller homes to multiple dozens of meters including obstacles in bigger institutions, which impeded the usage of technologies such as Bluetooth Low Energy and Zigbee.

All the intelligence and decision procedures are centralized at a specific component, the Complex Event Processing, thus reducing the complexity of the devices and enabling the standardization of both the communications protocol and the analysis algorithm. Furthermore, this approach allows the optimization of autonomy of the devices' batteries with a non-continuous monitoring of the environment, interrupted by periods in which the microcontroller is put in sleep mode (i.e., a low-power consumption mode). The sample frequency and sleep mode were carefully defined considering the monitored parameters in real-life applications and validated in laboratory tests. Additionally, with the implementation of the 'I'm alive messages' it is possible to cope with any event loss due to the non-reception of device messages, by continuously comparing the device data on the database with the newly received data.

Data generated by the IoT devices are gathered by the Communication Gateway and forwarded to the Complex Event Processing, which continuously analyzes the events received from the IoT devices according to predefined rules [24]. Moreover, it is prepared to accept details of significant events rules, based on the type of events or their occurrence (or not) within a specific time frame or limit.

The Complex Event Processing is based in a pipe and filter architecture [24]. It is composed by filters that clean and transform the data, working all in parallel, and connectors to pass data to other components, known as pipes. The advantages of this architecture, in addition to the concurrent execution, are the simplicity and the ability to reuse the filters.

The structure of the Complex Event Processing is divided into three blocks: Data Source, Filters and Data Sink blocks connected by a Data Stream communication channel. The Data Source block feeds the Filters block, which contains diverse filters. The operation of this block depends on the previous configuration of an XML file that describes each of the filters used and the type of data that they treat. When events are determined, the processed data are passed to the Data Sink block. This can alarm a significant event if it is the case.

When a significant event matches the predefined rules, a notification is forwarded to the FHIR Gateway and the respective CRUD operation are used. The process starts with a request to the Authentication service. When a valid JWT is returned, its attributes, the permissions of the attributes of the FHIR resources and rules previously defined are used by the Authorization service to get a decision if the requested operation can or cannot be achieved. When receiving the request and the respective JWT, the FHIR Gateway follows the decision of the Authorization service (i.e., permit or deny) to execute or not the specific CRUD operation. Moreover, during all the different actions, notifications are stored by the Logging and Auditing service.

5 Validation

The validation was divided into five phases.

The first phase involved the experimentation on preliminary prototypes of the IoT devices to overall validated the decisions concerning the core technologies and modules, namely the experimentation on the capacity of data gathering and processing of the devices, the wireless communications over LoRa, and the Communication Gateway operation and connection to the Complex Event Processing.

With the satisfactory results of the first phase, it was possible to complete the specification of the IoT devices and the core functionalities of the Complex Event Processing and proceed to their implementation. This second phase was characterized by successive iterations and optimizations, namely involving the interaction with the motion devices, the definition of adequate sample times and sleep times for autonomy improvement, the interaction with the strain gauge and measurement of decision weight thresholds, which lead to the implementation of the reset procedure, the application requirements of the reed switch, particularly from the application standpoint which could jeopardize its accuracy, and the identification of the most adequate operation for the power devices, to avoid false-positives of devices with on-off controllers and standby power modes.

In the third phases the IoT devices were evaluated in conditions closer to the expected real-life applications. Therefore, two set of devices were installed in the home of two older adults. This experimentation brought up multiple improvements from application and operational standpoints, in terms of hardships experienced in devices deployment and initialization, remote firmware updates, communications range, battery autonomy, connectivity and optimizations of communication protocols.

Concerning the fourth phase, the Complex Event Processing was validated using a simulator to produce events, in order to solve the lack of data. The simulator was capable of injecting data for different types of devices similar to those that will be installed in real houses. For that it was configured with multiple house setups in terms of types of devices and their locations, multiple care receivers expected activities and data simulations for a specific number of days. Additionally, it was possible to have settings for different days of the week and to insert some randomness based on temporal windows as well as abnormal events on specific days.

The goal was to verify if the unit correctly identifies a set of care receivers' activities (e.g., in bed, not in bed, resting or not resting). The simulations were executed for a period of ten days and the results show that the expected events were detected.

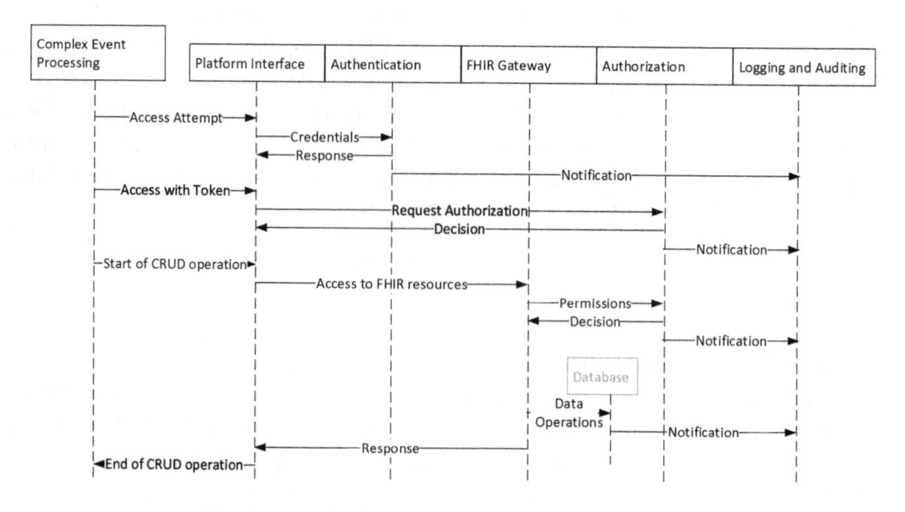

Fig. 1. Required flow to access the FHIR Gateway.

Moreover, the Complex Event Processing was object of stress tests, which validated its performance in critical operation circumstances (e.g., low communications signal, operation close to the defined thresholds or high number of successive events).

Finally, in the fifth phase, an experiment was conducted to validate the integration the services of the SOCIAL platform. A test scenario was created consisting of a random number of instances accessing the FHIR Gateway (between 100 and 200 instances) that for 60 min randomly executed CRUD operations. The goal was to evaluate the maximum delay to store a relevant event from the Complex Event Processing (i.e., the maximum delay between the and access attempt to store a relevant event and the end of the respective CRUD operation as represented in Fig. 1). As a result of the experimentation, a maximum delay of 121 ms was obtained.

6 Conclusion

The authors developed several IoT devices to monitor older adults' activities and performed their integration in a platform of services, the SOCIAL platform, which provides a set of common business services, including interoperability, authentication, authorization, and logging and auditing mechanisms to allow interoperable and secure information operations. The successful integration of IoT devices in the SOCIAL platform might contribute to optimize the development of innovative and integrated applications with impact on care provision of older adults, considering a diversity of settings.

The experimental results show the viability of the approach and, as future work, the authors envisage the development and evaluation of additional algorithms for the detection of significant events that might be related with health conditions of older adults.

Acknowledgments. This work was supported by Sistema de Incentivos à Investigação e Desenvolvimento Tecnológico (SI I&DT) of the Programa Portugal 2020, through Programa Operacional Competitividade e Internacionalização and/or Programa Operacional do Centro do FEDER - Fundo Europeu de Desenvolvimento Regional, under Social Cooperation for Integrated Assisted Living (SOCIAL), project number 017861.

The publication was financially supported by National Funds through FCT – Fundação para a Ciência e a Tecnologia, I.P., under the project UI IEETA: UID/CEC/00127/2019.

References

1. Rashidi, P., Mihailidis, A.: A survey on ambient-assisted living tools for older adults. IEEE J. Biomed. Health Inf. **17**(3), 579–590 (2012)
2. Arambepola, C., et al.: The impact of automated brief messages promoting lifestyle changes delivered via mobile devices to people with type 2 diabetes: a systematic literature review and meta-analysis of controlled trials. J. Med. Internet Res. **18**(4), e86 (2016)
3. Awolusi, I., et al.: Wearable technology for personalized construction safety monitoring and trending: Review of applicable devices. Automation Constr. **85**, 96–106 (2018)
4. Sousa, M., et al.: SOCIAL platform. In: Rocha, Á., Adeli, H., et al. (eds.) Trends and Advances in Information Systems and Technologies. WorldCIST'18 2018. Advances in Intelligent Systems and Computing, vol. 746, pp. 1162–1168. Springer, Cham (2018)
5. Queirós, A., et al.: Technologies for ageing in place to support home monitoring of patients with chronic diseases. In: HEALTHINF, pp. 66–76. INSTICC, Setúbal (2017)

6. Garabedian, L., et al.: Mobile phone and smartphone technologies for diabetes care and self-management. Curr. Diab. Rep. **15**(12), 109 (2015)
7. Nakamura, N., et al.: A meta-analysis of remote patient monitoring for chronic heart failure patients. J. Telemed. Telecare **20**(1), 11–17 (2014)
8. Lundell, S.: Telehealthcare in COPD: a systematic review and meta-analysis on physical outcomes and dyspnea. Respir. Med. **109**(1), 11–26 (2015)
9. Yang, J., et al.: Activity recognition based on RFID object usage for smart mobile devices. J. Comput. Sci. Technol. **26**(2), 239–246 (2011)
10. Wang, Y., et al.: A data fusion-based hybrid sensory system for older people's daily activity and daily routine recognition. IEEE Sens. J. **18**(16), 6874–6888 (2018)
11. Cacciagrano, D., et al.: Resourcehome: a rfid-based architecture and a flexible model for ambient intelligence. In: Fifth International Conference on Systems, pp. 6–11. IEEE (2010)
12. McConville, R., et al.: Vesta: a digital health analytics platform for a smart home in a box. Future Generation Computer Systems (2020)
13. Thomas, N.W., et al.: EVALUATE-AD and Tele-STAR: novel methodologies for assessment of caregiver burden in a telehealth caregiver intervention–a case study. Dement. Geriatr. Cogn. Disord. **47**(3), 176–184 (2019)
14. Xu, X., et al.: Exploring techniques for vision based human activity recognition: methods, systems, and evaluation. Sensors **13**(2), 1635–1650 (2013)
15. Yan, Y., et al.: Egocentric daily activity recognition via multitask clustering. IEEE Trans. Image Process. **24**(10), 2984–2995 (2015)
16. Liu, K.C.: Wearable sensor-based activity recognition for housekeeping task. In: 14th International Conference on Wearable and Implantable Body Sensor Networks, pp. 67–70. IEEE (2017)
17. Wang, L., et al.: Toward a wearable RFID system for real-time activity recognition using radio patterns. IEEE Trans. Mob. Comput. **16**(1), 228–242 (2016)
18. Hu, Y., et al.: Smart home in a box: usability study for a large scale self-installation of smart home technologies. J. Reliable Intell. Environ. **2**(2), 93–106 (2016)
19. Fafoutis, X. et al.: SPHERE deployment manager: A tool for deploying IoT sensor networks at large scale. In International Conference on Ad-Hoc Networks and Wireless, pp. 307–318. Springer, Cham (2018)
20. Queirós, A., et al.: Usability, accessibility and ambient-assisted living: a systematic literature review. Univ. Access Inf. Soc. **14**(1), 57–66 (2015)
21. Rocha, N.P., et al.: The social platform: Profiling FHIR to support community-dwelling older adults. J. Med. Syst. **43**(4), 86 (2019)
22. HL7 Organization, FHIR Release 3, https://www.hl7.org/fhir/resourcelist.html. Accessed 25 Aug 2020
23. Gunter, C.A.: Building a smarter health and wellness future: privacy and security challenges. In: ICTs and the Health Sector: Towards Smarter Health and Wellness Models. OECD Publishing, Paris (2013)
24. Gomes, M.S., et al.: The SOCIAL Platform and the integration of internet of things devices to monitor activities and behaviors of older adults. In: IoT and ICT for Healthcare Applications, pp. 255–274. Springer, Cham (2020)

Resilience by Digital – How Sociotechnical Helped Maintaining Operational and Recovery

Jussi Okkonen(✉)

Tampere University, 33014 Tampere, Finland
jussi.okkonen@tuni.fi

Abstract. Global health related disasters affect humans in several areas of life. COVID-19 pandemic in 2020 had great impact on people's everyday life, livelihood and health. It had huge impact on global economy as well as business dynamics. However, contemporary post-industrial society is more resilient due to digitalisation, turning operations spatially dispersed and even asynchronic. Thereby, most operations were maintained at least on satisfactory level. Taking the resilience to wider extension the burden on health care system was eased by quarantines and people still maintained most of their social contact by sociotechnical means. The aim of this paper is to address the topic of resilience by current global crisis, digital coping strategies and evident aftermath.

Keywords: Sociotechnical · Digital environments · Resilience

1 Introduction

Very soon after the COVID-19 outbreak in China and South East Asia local authorities took heavy measures in order to limit the extent and the speed of spreading the virus. The effect to societies was remarkable as for example mainland China was closed in February. Neighbouring nations had similar actions and by mid-March the situation was somewhat under control. As COVID-19 travelled along, other countries started to make similar actions to maintain control as well as ease the burden on health care system. In the beginning of March to mid-March there was an almost global shutdown, i.e. local curfews, traveling restrictions, social distancing executed, public venues and schools closed to mention the beginning. By the that time public discussion on the situation was about the dawn of global depression and disruption of civilisation, or at least that of the global economy. However, soon there was significant evidence of adaptation to new order as well as evidence of resilience. The aim of this paper is to discuss how resilience is built on sociotechnical modus operandi as well as on digital means on communication, interaction and working.

The history of mankind is history of disasters, epidemic, diseases, and conflicts. If natural disasters and conflicts are ruled out as local emergencies epidemic, or pandemic, has extensive effect on everyday life of people as well as economic structure and society. As long as there has been oral or written history there has been stories of how pandemic causes disruption in society. To reflect the current situation to previous outbreaks plague and in mid-17th century and Spanish influenza in early 20th century. In these two examples there was slow but wide outbreak due to lack of sufficient

Á. Rocha et al. (Eds.): ICITS 2021, AISC 1330, pp. 581–589, 2021.
https://doi.org/10.1007/978-3-030-68285-9_54

communications structure, limited mobility of people yet there was no way to bail out of physical contact. There are lot of literature on people during the plague, yet fiction but mostly historically accurate. The case of Spanish flu is different as science had taken long leaps and it was also researched widely among several disciplines. The question still remains what makes 21st century society more resilient that societies in 17th or 20th century. It was evident that 17th century as described e.g. in Decameron or Plague Year. According to the author of Plague Year Daniel Defoe [1] in mid-17th century there was no mass media, but news spread via hearsay or rumours and even administrative bodies made their decisions according to them. Also, scientific knowledge could not explain the sickness or epidemic but there were only assumptions on the mechanisms or the dynamics. As the knowledge was limited there was no true understanding how to stop the epidemic. Moreover, due to the nature of economic system people could not avoid contact if they needed to fulfil their needs for living. However, even this pandemic was slowed down by closing cities or setting people in quarantine, but it had massive effect on societies and economy.

Even in the 20th century as the Spanish influenza harvested population lack of possibilities to avoid physical contact caused large number of deaths. There are examples how it spread globally because people returning home from the World War I and other reasons for mobility. One reason for effective outbreak was established communications as virus reached even the remote parts of the globe by people working in postal and other services. Also, the extent of the outbreak hit by surprise as there was no means for quick communication as news travelled still slowly. Spanish influenza was fought same manner as the plague even recognizing the possible effect on development. As the several waves of the pandemic were over the aftermath was remarkable. Europe and other parts of the world were still recovering the great war and people had to deal with pandemic slowing it.

In January 2020 the closing societies started in China and quickly it spread through the South-East Asia and having its current peak in mid-March when most of the countries had restriction on their citizens as well as travelling restrictions. The aim was to slow the outbreak by quarantines, having the cases needing medical aid on sustainable level as well as gaining time for preparing better. In mid-March the digital communication and sociotechnical work environments started to have effect of the pandemic. Global information systems especially mass-media, social media and similar sources effectively delivered people the news on the status, information on desired behaviour, how to avoid the COVID-19, and most important it was possible to minimize social contacts in real life. This, at least to some extent, bought time and slowed the outbreak. There are several examples on how societies kept on going based on sociotechnical environments.

In this paper digital resilience is approached by perspective of the concept of resilience as a social and individual attribute. Moreover, it is pinned to sociotechnical environments, i.e. how resilience is facilitated and enhanced by digital. As proposed above the resilience is not caused by sociotechnical, yet significantly boosted by it. Resilience is human attribute, yet it is materialised in novel ways of utilising sociotechnical environments as well as human ability to adapt to new situations and creatively use the infrastructure and tools available. As working proposals this paper addresses resilience by sociotechnical as operational resilience, i.e. ability to adapt quickly to new outlining requirements, and organisational resilience, i.e. organisational capability to use resources and adapt and even excel in unexpected change.

2 Resilience by Digital

Different operation environments, especially ecological and economic environments of business activity are becoming more turbulent, operation modes or procedures to enable sustainability and support recovery gain importance. Especially in times of crisis organizations and individual employees need more than ever the capacities of acting on both unexpected threats and emerging opportunities on time as stated in Tsiapa et al. [2]. In these circumstances, the building of resilience into the socio-economic systems of different size is a core capacity. Organisational resilience is how those adapt to and drive emerging technologies that disrupt the status quo in areas of operating as well as they adapt.

Resilience is originally concept of ecology where it refers capability of individual, group or even the system, but its application has been expanded into social and behavioural sciences [3]. Recently, as logical extension the concept is applied in economics and organisation studies as stated in Sabahi et al. [4], Haase et al. [5] or Herbane [6]. In general, resilience denotes the capacity for continuous reconstruction [7]. Personal resilience refers to the capacity for maintaining or regaining psychological wellbeing in the face of challenge [8]. Organizational resilience is concerned with how organizations structure their activities in order to anticipate and circumvent threats and opportunities to their continued existence. As stated by Hillman [9] resilience is widely adapted, yet sometimes shallowly interpreted. Major characteristics of resilient organizations include (i) sensitivity to changes in the organization's operating environment; (ii) a flexible, adaptive decision-making process; (iii) a willingness to openly confront difficult issues such as power and control; and (iv) an organizational culture that is supportive of change [10, 11]. The organisational resilience is the capability to react, adapt and act according to internal or external signals or pressure [12].

The development of resilient organizations is not possible without reorganizing of work practices, reengineering operation modes and utilising adaptiveness or people within organisation. This might even require executing unconventional acts. This demand manifests itself in redesigning the contractual and accountability structures between individuals, organizations and diverse clusters of business actors. As stated in Burnard et al. [13] changing organisational mode enhances capability to adapt. Outsourcing, cloudsourcing, decentralization, individual contracting, job crafting, company-internal markets, and hyperspecialization are concrete examples of the ways in which companies have recently attempted to meet such requirements [14–17]. Since work task performance has increasingly become dependent on the seeking, using and sharing of information with means of information and communication technologies, a major part of employees of our time can be defined as knowledge workers and they mostly can work in sociotechnical environments. Therefore, the demands for reorganizing work practices mainly focus on the (re)organizing of knowledge related practices. Such practices, i.e. knowledge practices can be defined as a set of activities dealing with information seeking, acquisition, use and sharing, as well as environmental scanning and personal information management [18, 19]. A major characteristic of such knowledge practices should be that they are proactive, i.e., anticipatory, self-initiated, and future-oriented, pacing emphasis on agile action that introduces constructive changes. The above demands are also reflected in the ways in which resilient organizations organize knowledge management serving to the ends internal and external intelligence. For example, weak signal analysis has become an integral part of knowledge management [20]. Such change in operation mode implicitly promotes resilience as agility increases.

As discussed above resilience is a multilevel construct that has been investigated since the 1970s in diverse fields such as ecology and psychology [21]. Resilience at work is a positive developmental trajectory characterized by demonstrated competence in the face of, and professional growth after, experiences of adversity in the workplace [22]. Researchers have addressed a variety of issues related to organizational resilience [11] personal resilience and social resilience [23]. In this context, there is discussion about multitasking, availability management, awareness deficit, and temporal fragmentation of work task performance as stated e.g. [24]. In addition, the studies of job design have addressed issues of process reengineering [25]. The requirements for developing resilient organizations have also been reflected in the concurrent renaissance in studies of strategic planning, within the special context of strategic flexibility, agility emphasis and environmental scanning as an integral part of knowledge management [20, 26]. Especially in production engineering and industrial engineering are the domain on resilience, but resilience by digital is only marginal issue. For example, the studies of personal information management have largely neglected the issues of organizing knowledge practices for resilience [18, 27].

Digitalization of work enables better use of knowledge due to enhanced access, management and dissemination [28, 29]. It is expected to result in enhanced productivity [20, 30–34]. Sociotechnical work environments provide sufficient infrastructure for continuing operations and also provide suitable means for internal and external collaboration [35]. Digitalisation affects as well on efficiency. The great expectations are not easily fulfilled regarding digitalization on the employee's viewpoint. In fact, the effects of digitalization seem to be twofold: By bringing about ever more information systems, applications, user interfaces and operating systems to enhance productivity and efficiency of work, digitalization has led to increasing information load, hectic pace of work, multitasking, and interruptions [36]. Studies confirm that users can experience ICT as demanding and stressful [37–39]. Another rather negative result of digitalization is potential weakening of social ties and reducing social inclusion: by increased use of ICT people tend to have less face-to-face contacts [39]. In work context this may lead to weakening sense of community, and consequently issues with trust and motivation and during the crisis this might be non-optimal. There might also occur consequences of inadequate information systems, such as decreased job satisfaction and engagement with the organization [40]. Technical issues are evident hindrance of working, yet those can also lead to negative attitudes o working and therefore build communities of non-practice as negation of well-functioning hybrid organisation. Those can negatively affect work quality and productivity as stated in Franssila et al. [36]. However, the perverse use of ICT, I self-inflicted interruptions, excess communication and always on are side-effects of digitalisation and resilient individual can cope and by that even leverage organisational resilience especially when global crisis requires quick and extensive action.

3 Impact of COVID-19 and Digital Resilience

Globally the educational system in spring 2020 was in crisis as schools were closed and teachers had to change their modus operandi almost for one night. For example, in Finland grades from the 4th were totally closed and most preschool and on the grades 1st to 3rd stayed also home if possible. In Finland the situation was somewhat easy as there is high device penetration, people are digitally literate, and schools have suitable digital infrastructure and in general people were willing to adapt to the new situation. Tuition was provided on mixed platforms and it was managed mostly by the teachers. Agile action was proof of concept for resilience in semi-closed society. This kind of operational resilience occurred globally since need to keep children in school called for novel applications of technology and ways to work.

Operational ability of working organisation also relied extensively on utilising sociotechnical environments. As soon as pandemic was spreading all organisations that could closed their offices and staff took their laptops and smart phones and stayed home. The initial idea is to maintain operations by having people healthy, but as soon as officials pointed out the need social distancing the measures by organisations had dual impact. The need to maintain operations that are not production related yielded to novel ways of organising activities. In most countries forced or voluntary social distancing caused novel ways of working and utilisation of spatial dispersion. The effect will be permanent, and digitalisation of work takes the giant leap forward. There will be collateral effect too as COVID-19 will change work related traveling habits and time-space management. This is an example of organisational resilience, but not attribute of organisation per se yet sum of adaptiveness of personnel and sufficient infrastructure.

The mental confidence frame also explains why certain people in leadership positions are not acting as leaders. Creating strategies for an uncertain future requires a playful mind able to entertain multiple scenarios and hypotheses, and then ground those hypotheses through facts and organizational culture into strategies and action plans [41, 42]. This, however, carries the risk of individual hubris or groupthink: if the works are not grounded in the facts, the manager will enter the realm of magical thinking where his decisions "will work, because they are his decisions", and the organization is forced to carry out the ideas, either due to trust in the manager's skill, fear of being fired, or a combination thereof [43].

This points to the fact that the digital resilience of an organization, and on a wider scale, that of a society, is reliant on three factors: 1) a technological readiness, which has at least been experimented with, if not taken into everyday practice before a crisis takes place, 2) employers' trust of their work force and middle managers, and 3) top management's understanding and acceptance of the actual facts of the situation, and their willingness to let that acceptance trickle down in the organization as support for resilient, digital practices, and as a clear statement of trust towards the employees' competence and diligence. Moreover, resilience requires taking the critical steps without hesitation. Future textbooks will be full of examples from spring 2020 about how societies, organisations, and even individuals dismantled their operating models in order to maintain their functions, business or livelihood. Of course, the past months are example of the opposite, negations of resilience and how non-resilient have suffered greatly due to lack capability to adapt.

4 Concluding Remarks

The aim of this paper is to address the topic of resilience by current global crisis, digital coping strategies and evident aftermath. Interesting topic in the current context. The paper presents as introduction to the problem identified in summer 2020. In this paper digital resilience is a concept of resilient behaviour as a social and individual attribute. Social refers group action to maintain operational during the crisis. It could be seen as an implicit contract of joint endeavour to stay alive. As individual attribute it mostly refers to flexibility and grit. Flexibility is about adjusting to new ways of operating, adapting and accepting new tools, but it is also curious mind of discovering new. Grit is state of mind not to be afraid of changes and accepting unexpectable. Digital resilience is s pinned to sociotechnical environments, i.e. how resilience is facilitated and enhanced by platforms, tools, media, social media, digital convention, and digital practices. As proposed above the resilience is not caused by sociotechnical, yet it is significantly boosted by it. Resilience is human attribute, yet it is materialised in novel ways of utilising sociotechnical environments as well as human ability to adapt to new situations and creatively use the infrastructure and tools available. Put into action the resilience by sociotechnical is operational resilience as human ability to adapt quickly to new outlining requirements, and organisational resilience as organisational capability to use resources and adapt and even excel in unexpected change.

The last six months has been dramatic for all knowledge work, as all operations that could have been spatially dispersed have been spatially dispersed. Requirement for social distancing has challenged activities. Evidently routine work and most urgent issues have been saved, but peripheral activities and risky endeavour have been suffering. Social contacts with peers and clientele are minimised and only on need to basis. Social distancing blocks serendipity and kills creativity. People play it safe and new ideas are not nurtures during the quest for survival, but prolonged pandemic also calls for new vistas. Building capacity for renewal, creativity and innovation would be token of even higher order of resilience.

Ability to shift working paradigm will be the cornerstone of resilience in work organisations. It is dependent on cultural and organisational determinants. Resilience for example in Nordic countries was supported by low organisations and lesser hierarchy. Trust on personnel and minimal amount of micromanagement forecast success on paradigm shift. Cultural aspects are hard to scrutinize, but some initial evidence on work ethos a key factor is already discussed yet it needs more work.

This paper has discussed the resilience as concept in context of maintaining operational ability and staying in business. Resilience has more connotations, yet those need to be elaborated further as time goes by. This paper presents the notions of resilience during the first months of pandemic and further work is needed to see the different perspectives of resilience as organisational or individual attribute as well as capability promoted by sociotechnical. This is a study with a introduction and a identification of the problem. The state of the art regarding the concept of resilience and its definitions and particularities is evolving and by end of year 2020 or 2021 the whole concept of resilience is redefined because the people are resilient.

Acknowledgements. This paper has received funding from the European Union's Horizon 2020 programme H2020-WIDESPREAD-2018–2020 CSA Scaling Up Educational Innovations in Schools — SEIS under grant agreement No 856954. Dr. Heljä Franssila and Professor Reijo Savolainen are acknowledged for their elaboration on the concept of resilience.

References

1. Defoe, D.: Ruttovuosi. Otava, Finland (1997)
2. Tsiapa, M., Batsiolas, I.: Firm resilience in regions of Eastern Europe during the period 2007–2011. Post-Commun. Econ. **31**, 1 (2019)
3. Folke, C., Carpenter, S.R., Walker, B., Scheffer, M., Chapin, T., Rockström, J.: Resilience thinking: integrating resilience, adaptability and transformability. Ecol. Soc. **15**(4), 43–51 (2010)
4. Sabahi, S., Parast, M.: Firm innovation and supply chain resilience: a dynamic capability perspective. Int. J. Logist. Res. Appl. **23**(3), 254–269 (2020)
5. Haase, A., Eberl, P.: The challenges of routinizing for building resilient startups. J. Small Bus. Manag. **57**(sup2), 579–597 (2019)
6. Herbane, B.: Rethinking organizational resilience and strategic renewal in SMEs. Entrepreneurship Reg. Dev. **31**(5–6), 476–495 (2019)
7. Mannen, D., Hinton, S., Kuijper, T., Porter, T.: Sustainable organizing: a multiparadigm perspective of organizational development and permaculture gardening. J. Leaders. Organ. Stud. **19**(3), 355–368 (2012)
8. Ryff, C.D., Friedman, E.M., Morozink, J.A., Tsenkova, V.: Psychological resilience in adulthood and later life: implications for health. Ann. Rev. Gerontol. Geriatr. **32**(1), 73–92 (2012)
9. Hillmann, J.: Disciplines of organizational resilience: contributions, critiques, and future research avenues. Rev. Manag. Sci. **3** (2020)
10. Freeman, D.G.H., Carson, M.: Developing workplace resilience: the role of the peer referral agent diffuser. J. Workplace Behav. Health **22**(2), 113–121 (2006)
11. Välikangas, L.: The Resilient Organization: How Adaptive Cultures Thrive Even When Strategy Fails. McGraw-Hill, New York (2010)
12. Borekci, Y., Rofcanin, Y., Gürbüz, H.: Organisational resilience and relational dynamics in triadic networks: a multiple case analysis. Int. J. Prod. Res. **53**, 22 (2015)
13. Burnard, K., Bhamra, R.: Organisational resilience: development of a conceptual framework for organisational responses. Int. J. Prod. Res. **49**, 5581–5599 (2011)
14. Bernstein, A., Klein, M., Malone, T.W.: Programming the global brain. Commun. ACM **55** (5), 41–43 (2012)
15. Malone, T.W., Laubacher, R., Dellarocas, C.: The collective intelligence genome. MIT Sloan Manag. Rev. **51**(3), 20–31 (2010)
16. Oldham, G.R., Hackman, J.R.: Not what it was and not what it will be: the future of job design research. J. Organ. Behav. **31**(2–3), 463–479 (2010)
17. Wageman, R., Gardner, H., Mortensen, M.: The changing ecology of teams: new directions for tteams research. J. Organ. Behav. **33**(3), 301–315 (2012)

18. Jones, W., Teevan, J. (eds.): Personal Information Management. University of Washington Press, Seattle (2007)
19. Savolainen, R.: Everyday Information Practices. A Social Phenomenological Perspective. Scarecrow Press, Lanham (2008)
20. Kaivo-oja, J.: Weak signals analysis, knowledge management theory and systemic socio-cultural transitions. Futures **44**(3), 206–217 (2012)
21. Cote, M., Nightingale, A.A.: Resilience thinking meets social theory: situating social change in socio-ecological systems (SES) research. Prog. Hum. Geogr. **36**(4), 475–489 (2012)
22. Brooks, R., Goldstein, S.: The Power of Resilience: Achieving Balance, Confidence, Personal and Strength in Your Life. McGraw-Hill, New York (2004)
23. Caza, B., Milton, L.: Resilience at Work: Building Capability in the Face of Adversity. The Oxford Handbook of Positive Organizational Scholarship (2012)
24. Cacioppo, J.T., Reis, H.T., Zautra, A.J.: Social resilience: the value of social fitness with an application to the military. Am. Psychol. **66**(1), 43–51 (2011)
25. Carr, N.G.: The Shallows: What the Internet is Doing to our Brains. W.W. Norton, New York (2010)
26. Grant, A.M., Parker, S.K.: Redesigning work design theories: the rise of relational and proactive perspectives. Acad. Manag. Ann. **3**(1), 317–375 (2009)
27. Saritas, O., Smith, J.E.: The big picture - trends, drivers, wild cards, discontinuities and weak signals. Futures **43**(3), 292–312 (2011)
28. Pauleen, D.J., Gorman, G.E. (eds.): Personal Knowledge Management, Individual, Organizational and Social Perspectives. Gover, Farnham (2011)
29. Parida, V., Sjödin, D.R., Lenka, S., Wincent, J.: Developing global service innovation capabilities: how global manufacturers address the challenges of market heterogeneity. Res-Technol. Manag. **58**(5), 35 (2015)
30. Shujahat, M., Sousa, M.J., Hussain, S., Nawaz, F., Wang, M., Umer, M.: Translating the impact of knowledge management processes into knowledge-based innovation: the neglected and mediating role of knowledge-worker productivity. J. Bus. Res. (2017)
31. Michaelis, B., Wagner, J.D., Schweizer, L.: Knowledge as a key in the relationship between high-performance work systems and workforce productivity. J. Bus. Res. **68**(5), 1035–1044 (2015)
32. Chou, Y.-C., Chuang, H.H.C., Shao, B.B.M.: The impacts of information technology on total factor productivity: a look at externalities and innovations. Int. J. Prod. Econ. **158** (2014), 290–299 (2014)
33. Ferreira, A., Du Plessis, T.: Effect of online social networking on employee productivity. South Afr. J. Inf. Manag. **11**(1), 1–16 (2009)
34. Tuomi, I.: Economic productivity in the knowledge society: a critical review of productivity theory and the impacts of ICT. First Monday **9**(7) (2004)
35. Okkonen, J., Vuori, V., Helander, N.: Enablers and Restraints of Knowledge Work. Cogent Bus. Manag. (2018). https://doi.org/10.1080/23311975.2018.1504408
36. Franssila, H., Okkonen, J., Savolainen, R.: Developing measures for information ergonomics in knowledge work. Ergonomics **59**(3), 435–448 (2015)
37. Bordi, L., Okkonen, J., Mäkiniemi, J.P., Heikkilä-Tammi, K.: Employee-developed ways to enhance information ergonomics. In: Proceedings of the 21st International Academic Mindtrek Conference, pp. 90–96. ACM, Salanova (2017)
38. Tarafdar, M., Tu, Q., Ragu-Nathan, T.S., Ragu-Nathan, B.S.: Crossing to the dark side: examining creators, outcomes, and inhibitors of technostress. Commun. ACM **54**(9), 113–120 (2011)
39. Chen, W.: Internet use, online communication, and ties in Americans' networks. Soc. Sci. Comput. Rev. **31**(4), 404–423 (2013)

40. Ragu-Nathan, T., Tarafdar, M.. Ragu-Nathan, B., Tu, Q.: The consequences of technostress for end users in organizations: conceptual development and empirical validation. Inf. Syst. Res. **19**(4) (2008). https://doi.org/10.1287/isre.1070.0165
41. Normann, R.: Reframing Business: When THE Map Changes the Landscape. Wiley, Chichester (2001)
42. Ramírez, R., Mannervik, U.: Strategy for a networked world. Imperial College Press, London (2016)
43. Vesa, M., den Hond, F., Harviainen, J.T.: On the possibility of a paratelic initiation of organizational wrongdoing. J. Bus. Ethics **160**(1), 1–15 (2020)

Factors Influencing the Adoption of Enterprise Architecture in the South African Public Sector

Shaffique Patel[1], Jean-Paul Van Belle[1], and Marita Turpin[2](✉)

[1] University of Cape Town, Cape Town, South Africa
jean-paul.vanbelle@uct.ac.za
[2] University of Pretoria, Pretoria, South Africa
marita.turpin@up.ac.za

Abstract. Enterprise Architecture (EA) is touted to bring many possible benefits to private and public organizations. However, EA adoption in the public space has been lagging. This research paper looks at the factors influencing adoption of enterprise architecture in the public sector within South Africa as well as the perceived benefits. This is achieved by means of a qualitative research approach involving in-depth interviews with key decision makers and influencers of a South African national government department. The themes that emerged as factors were classified using the Technology-Organization-Environment (TOE) framework. Perceived potential EA benefits more specifically accruing in the public sector were also investigated. Given that EA adoption in the public sectors of other emerging economies is also lagging, some of the findings in this study are likely to resonate and may be applicable in government organizations of a similar Information and Communications Technology (ICT) maturity.

Keywords: Enterprise Architecture (EA) · Public sector · South Africa · EA adoption · EA benefits

1 Introduction

Enterprise architecture (EA) was introduced by John Zachman in 1987 as a logical structure for organizations [1]. The core of EA is alignment between business and IT which includes its interrelating components. The literature is replete with benefits which EA may bring to an organization in terms of their architecture plan, whether it be for profit or a public sector department. After decades of EA research and practice, EA implementation still remains an issue. Most national governments have official EAs but they are often not completely adhered to and enforced. This also appears to be the case in South Africa.

To investigate this situation, this research explored the following two research questions. (1) What are the factors which influence the adoption of EA in the public sector? (2) How do the identified factors affect the outcome of the adoption of EA? This research aimed to explain how the factors highlighted in literature, influences the adoption of an EA within the public sector in South Africa. The findings can provide an insight into the challenges or success of adoption organizations have faced and will

Á. Rocha et al. (Eds.): ICITS 2021, AISC 1330, pp. 590–602, 2021.
https://doi.org/10.1007/978-3-030-68285-9_55

hopefully resonate with those of other government departments in South Africa and, possibly, countries with a comparative government structure and maturity level.

The rest of this paper explores the literature on benefits and critical success factors in order to achieve these benefits. Then the qualitative research methodology adopted for the research is detailed. The analysis follows the TOE framework using a thematic approach and findings are summarized and critically evaluated in the conclusion.

2 Literature Review

2.1 EA Definition and Benefits

EA is a practice helping organizations attain alignment between business and IT goals. For this paper the following definition is used: "A coherent whole of principles, methods, and models that are used in the design and realization of an enterprise's organizational structure, business processes, information systems, and infrastructure." [1].

Table 1. High-level view of benefits.

Author/s	Standardization	Performance	Efficiency	Reduce Risk	Business Continuity	Optimization	Project Success	Organization Goals	Road Map	Organization Integration	Decision Making	Business IT Alignment	Increase ROI	Infrastructure Integration	Cost Reduction	Standardization	Reliability	Culture and Vision	Holistic View	Communication
	Operational			Managerial					Strategic					ICT				Organizational		
Jusuf & Kurnia [2]	X	X	X	X	X	X	X	X	X	X	X	X	X	X	X	X	X	X	X	X
Schmidt et al. [3]	X	X	X	X		X		X	X	X	X	X		X	X	X		X	X	X
Wan et al. [4]	X	X	X	X	X			X	X	X	X	X	X	X	X		X			X
Tamm et al. [5]	X	X	X	X	X			X	X	X	X	X	X	X	X		X			X

EA has many perceived or potential benefits: Jusuf & Kurnia [2] highlights 40 benefits which EA adds to organizations. These 40 benefits are then grouped into five categories namely operational benefits, managerial benefits, strategic benefits, IT infrastructure benefits and organizational benefits as illustrated in Table 1. EA has positive perceived benefits for changing business requirements; IT business alignment; frameworks; the complexity of IT and knowledge of EA amongst IT employees [3]. EA improves IT-business alignment, acts as an integration agent for the enterprise, improves decision making, reduces IT complexity and risk, improves the business structure and interoperability, optimizes resources and all of these benefits in turn leads to financial and economic benefits [4]. However, in order for these perceived benefits to result in tangible benefits, there are underlying critical success factors which need to be met.

2.2 Critical Success Factors (CSF)

EA needs to be able to integrate the strategy, technology and business formally and in a continuous manner [5]. In order to derive the benefits from EA, there are underlying prerequisites or critical success factors [5] as seen in Table 2. The CSF have been derived from literature and have been divided into 3 categories using the technological, organizational and environmental framework (TOE) [6] as a guideline.

The governance process should be able to integrate itself with all other management processes and should allow for exceptions [5, 7]. There should be a formal methodology which should be able to support the implementation and the maintenance of the EA plan [5, 7]. The framework needs to be able to establish the scope of the architecture holistically and link the sub-architectures and its components [5, 8]. There should be a wide range of integrated document artifacts [5]. The tools should be able to support modeling, configuration and should be in an online repository [5, 7]. A best practice should be used in guiding the EA plan and its use [5]. Alignment between business and IT is a high priority [2–4, 7, 9–12]. A clear understanding is required of the EA vision [7, 8]. The commitment and continuous use of the EA plan within the organization is required [7, 8, 13]. Communication is the enabling factor that binds all the other factors together [7]. Stakeholder buy-in plays a factor, especially senior management support [8, 13]. Regulation plays a factor in terms of the environment the organization positions itself in [5, 7, 8]. The structure and the size of the organization plays a role in the adoption. The more complex the organization, the more challenging the adoption becomes [8, 13]. Skilled employees play a factor in the overall success of adoption, whereas an EA program has the unintended consequence of changing job roles and benefits [8, 13]. With all the factors as previously mentioned, it means nothing if there is no way of monitoring and coordinating the progress of the EA plan [7].

Table 2. Summary of critical success factors grouped using TOE.

Technological Factors	Organizational Factors	Environmental Factors
Integration	Governance	Governance
Alignment to Business	Methodology	Government Policy
Size	Holistic View	Competition
Skills	Modeling tools	
Infrastructure	Best Practice	
	Alignment	
	Commitment	
	Clear Vision	
	Communication	
	Stakeholder Buy-In	
	Senior Management Support	
	Organizational Politics	
	Organizational Culture	
	Size and Structure	
	Monitoring & Coordinating	

2.3 Technology Organization Environment Framework (TOE)

The TOE framework has been used frequently to understand the adoption of information systems in organizations [14]. The TOE framework is an organizational theory which is used to explain the context of an organization in adopting decisions [15]. It is comprised of three elements, the technological, organizational and environmental.

The **technology** factor as shown in the framework is described as the collection of technologies available to the organization, whether used or available but not yet accepted [16]. The **organization** factor underlines the influence of the organizations' characteristics, resources and internal social network in relation to its behavior to adoption [17]. The organization element incorporates end-users as the adopters of innovation, be it a technology or a process [16]. This individual acceptance is referred to as intra-organizational acceptance. Within the element of the organization is the support from top management, the willingness of the organization and the information intensity of the product and its characteristics [14] (Fig. 1).

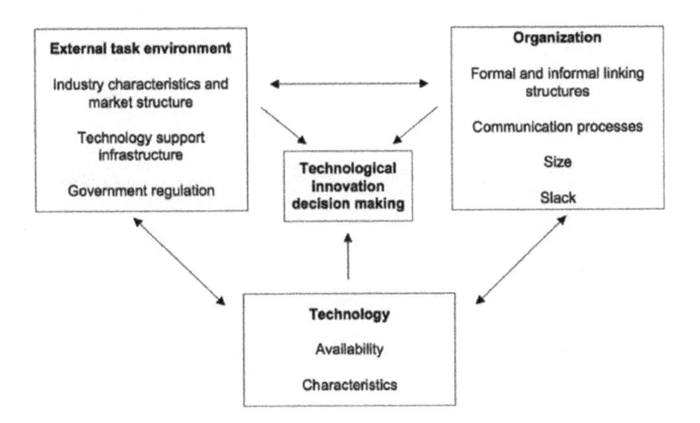

Fig. 1. Technology organization environment framework [6].

The **environmental** perspective describes that adoption is influenced by external factors which are out of the control of the organization [17]. These external factors concerns issues of government policies, trading partners and competitors. Correspondingly, [14] states that the environmental element concerns the industry pressure which relates to its competition in industry, the government policies and consumer readiness. Industry pressure has been shown to have a positive effect on adoption [15, 16]. Government legislation can either have a positive or a negative effect on the adoption of the innovation [15].

3 Research Methods and Design

This research adopted the positivist paradigm. The positivist paradigm uses causal relationships [18] as identified by the factors of the TOE framework and the literature review that has been conducted.

The data collection approach was qualitative, using semi-structured interviews to gather data. These in turn will be used to explore the factors highlighted in the framework and literature [19]. The target population identified were participants who had a direct or indirect role for the adoption of EA in a\South African national government department. 'Direct' participants are those who play a significant role in the development of the EA plan or the decision makers. The indirect participants are those who give input into the plan but does not play a significant role in the overall decision making. These targeted individuals would be able to give their insight into the questions derived from the factors previously mentioned. The sample population for this research was executive management, senior management, junior management and ICT staff within the public sector. This research gathered information from 14 participants from a national department based in Cape Town and Pretoria (Table 3).

Table 3. List of interview participants.

Participant	Role	Expertise
SM1	Chief Director: Strategy	Strategic Planning
SM4	Chief Director: Governance	Governance
EM1	Deputy Director General	Business
EM4	Deputy Director General	Business
EM3	Chief Financial Officer	Business
SM5	Chief Director: Strategy Programs	Strategy / Business
EM2	Chief Operating Officer	Strategy / Business
SM3	Chief Director: IT	Business
MM2	Director: IT	Service Delivery
MM3	Director: IT	Networking
MM1	Deputy Director: IT	Networking
SM2	Chief Director: Projects	Business
MM4	Deputy Director: IT	Networking
MM5	Director: Finance	Business

Thematic analysis was used to analyze the interviews [20]. The themes or codes were taken from literature and structured using the TOE framework. Ethics approval from the University and the National Department were obtained. Interviewees consented voluntarily to the study. The department as well as participant names remain anonymous.

4 Findings, Analysis and Discussion

The data collected was guided by the literature review, the research questions which resulted in the researcher deriving propositions. Main themes were grouped using the TOE framework, and subthemes were grouped using the *benefits of EA*, the *critical success factors* needed for EA to be successfully adopted and *challenges* organizations face in adopting EA plan. Fig. 2 shows a word cloud of the most recurring words. Systems, need, business and architecture are the most prominent. Although 14 interviews were conducted, data saturation [18] occurred by the 4th interview. The researcher continued with the remaining 10 in order to rule out additional emerging themes.

4.1 Environmental Factors

Table 4 shows which participants touched on each of the various subthemes. Interesting to note is that each subtheme either had a positive or negative effect depending on how the participant perceived the theme. Three subthemes were identified.

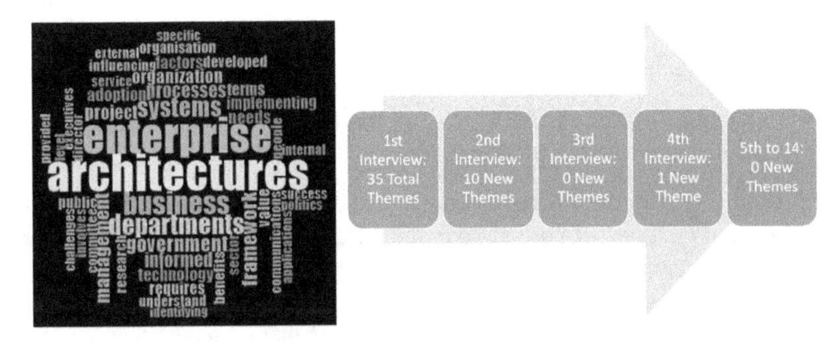

Fig. 2. EA word cloud (a) and interview saturation (b)

Table 4. Environmental factors.

	SM1	SM4	EM1	EM4	EM3	SM5	EM2	SM3	MM2	MM3	MM1	SM2	MM4	MM5
Lack of Competition	X		X			X	X				X	X	X	X
Governance	X	X	X	X	X	X	X	X	X	X	X			X
Government Policy	X	X	X	X	X	X	X	X	X	X	X	X	X	X

Lack of Competition. Within the TOE framework, the environmental context, competition plays a pivotal role in terms of adoption. Competition stimulates the adoption of innovation [15] and plays a positive role in terms of innovation and adoption. But due to the fact that the context of this study takes place within the public sector, there is no competition, therefore it has a negative impact. The lack of competition is caused by the State Information Technology Agency (SITA) act which states that procurement of IT-related goods and services and standards are done in line with the SITA act [21]. SITA's mandate is to provide ICT services to the South African public sector [21]. "...we are governed by SITA, we have a business agreement with SITA, and the DPSA requires us to be compliant ..." (EM2). Participants pointed to how government policies hinder adoption due to the lack of competition within the public sector. "...SITA is a problem in assisting the delivery of the things because SITA is not efficient..." (SM1).

Governance and Government Policy. The governance process should be able to integrate itself with all other management processes and should allow for exceptions [5, 7]. Governance within the public sector is regulated by government policy. The Department of Public Service Administration provides the frameworks in which government departments need to operate. "...your governance framework is prescribed the way of procurement as prescribed by the SITA Act, you are very limited in how you do things in government because it is prescribed and regulated..." (SM4). It can, therefore, be said by the information provided that the government policy and governance play a significant role in terms of the adoption of the EA within the department.

4.2 Organizational Factors

The organizational aspect of the TOE framework concerns the characteristics and resources of the department [15] (Table 5).

Clear Vision. As with the environmental context, the underlying success factor either had an enabling effect or was a barrier in the adoption process within the department. Vision refers to a clear understanding of the purpose of the EA [7, 8]. Even though the participants explained that EA has an enabling effect on the factor such as clear vision, "...what are the business needs, then what is the vision that we have because you need to have the vision to get there..." (EM2). This, however, contradicts the data as there is duplication in systems within the department, pointing to the fact that silos exist making the development of systems also being created in silos. "...we have a lot of Silo systems that we have which doesn't always necessarily respond to business needs..." (EM2). It can, therefore, be concluded that a clear vision plays a role in the adoption of EA.

Table 5. Organizational factors.

	SM1	SM4	EM1	EM4	EM3	SM5	EM2	SM3	MM2	MM3	MM1	SM2	MM4	MM5
Clear Vision					X		X		X		X			
Commitment							X				X			
Monitoring and Evaluation	X	X	X	X	X		X	X	X		X	X	X	X
Organizational Culture; Vision		X	X		X		X	X	X	X	X	X		X
Senior Management Support		X	X	X	X		X	X	X	X	X	X		X
Size and Structure												X		
Stake Holder Buy-In		X								X	X			X
Communication	X	X	X	X	X	X	X	X	X	X	X	X	X	X

Commitment. The commitment factor within the department is a barrier in terms of the adoption of the EA. Management commitment remains a challenge in many government institutions [5]. "I think there was a need identified for it (EA), the process map person was employed to do that, but they moved her to another unit, so they never resolved it. [...] It moved to IT but there's no one carrying it out." (MM1). So the initial EA planning was done, but it never moved further. The lack of continued commitment played a negative role in the adoption of EA.

Monitoring and Evaluation. Monitoring and evaluation within the department in certain sections are executed very well, but is lacking elsewhere. Without a mechanism for control and compliance, it is easier to go with a decision which brings less resistance and to bypass the architecture at certain points [9]. "Yes, there are monitoring and evaluation processes" (EM3). "A few years back there was a system developed for invoice tracking, till this day I know it has been developed but where is it? So where is the follow-up? Why are we not monitoring? ...We stop tracking once we developed the system." (EM4). In summary, monitoring and evaluation is done unevenly, well within certain business units and not in others where, once the project is completed, there is no retrospective analysis done.

Organizational Culture and Vision. After independence in 1994, the government started with the transformation of service delivery by introducing the public service act, this combined with the Batho Pele principles shaped the culture of the public sector [22]. The majority of participants touched on issues such as the department being reactive instead of proactive, processes existing in the minds of people instead of it being explicitly mapped, silos, lack of responsibility, a vague understanding of EA which relates to the organizational culture. By introducing customer intimacy or intra-communication, this barrier can be converted into a positive aspect as knowing your customer better and having a shared vision increases organizational alignment [11].

Senior Management Support. Senior management support plays a major critical success factor as mentioned by the most participants. Without senior management support, the project will not gain traction [8, 13]. EA in the department was initially driven by an audit finding, in turn, it was then driven by executive management and senior management. Consultation has been initiated with IT and business but a challenge exists where the planning is not done in an integrated manner. Even though the support exists from a senior level, staff from lower levels on the organogram expressed that more support is needed from an executive level. "... It is an executive management priority but it is not done in an integrated way..." (EM3). Without senior management support, the plan will not be driven from the top down and the EA plan will ultimately fail.

Size and Structure. The structure and the size of the organization plays a role in the adoption. The more complex the organization, the more challenging the adoption becomes [8, 13]. Within the department, the subtheme size and structure emerged only once, indicating that it does not play a critical factor in the adoption of EA within the context of this empirical report. "... the bigger the organization the more need for EA..." (SM2). Even though this theme only emerged once, it supports the literature.

Stakeholder Buy-In. In order for an enterprise architectural plan can be implemented within an organization, a hurdle exists in order to convince stakeholders of the perceived benefits [8, 13]. The department has a steering committee where stakeholders consult on the various projects within the department. The committee gives their advice, but there is a grey area on whether the committee is allowed do give approval or decline projects. Because of this grey area, the project owner may still proceed without the approval of the committee. This leads to silo systems and duplication within the department. "… you don't know if the IMS (Steering Committee) has a role to actually reject the system or just a rubber-stamping authority…" (SM3). Stakeholder buy-in does not come across as a strong theme, emerging in less than half of the participants, it is on par with the factor of clear vision, which is logical, as the two factors are closely related.

Communication. Out of all the critical success factors, communication is the most important and the enabling factor that binds all the other factors together [7]. Communication can be internal, amongst the organization employees themselves, or outside the firm [7, 13]. The communication of information in the department starts at the top level. However, each level may experience a communication break. For example, information is not relayed correctly from executive management to senior management. This pattern repeats itself at each level of the organization.

4.3 Technological Factors

The technological aspect of the TOE framework takes into consideration the technologies which are important to the department, this includes technologies that are already being used by the department internally and also those which exist externally in the market [15]. The technological subthemes which have been identified have been grouped under the technological theme as shown in Table 6.

Table 6. Technological factors.

	SM1	SM4	EM1	EM4	EM3	SM5	EM2	SM3	MM2	MM3	MM1	SM2	MM4	MM5
Infrastructure Integration	X	X	X	X	X	X	X	X	X	X	X	X		X
IT Business Alignment	X	X	X	X	X	X	X	X	X	X	X	X	X	X
Skills	X	X	X	X	X		X	X	X		X	X	X	X

Infrastructure Integration. EA should enable the organization to move in any direction without the need for major changes to the existing infrastructure [23]. Projects are often started without consultation of IT and the steering committee. This leads to a lack of integration of the system being developed or procured. A consequence is systems being developed in silos, creating duplication. "I think at this stage people run with the project before IT is consulted on the infrastructure needs." (EM1). "There are a lot of other systems that are running which are not properly consolidated." (EM3).

This implies that the lack of integration leads to silo systems being developed. Thus infrastructure integration plays a role in the adoption of the EA within the organization.

IT Business Alignment. IT business alignment is a beneficial factor of EA [23]. This is one of the factors or aspects of the study which each participant acknowledged. Participant MM1 stated that: "... EA is essential for every organization, it outlines how every organization needs to align itself, the various aspects of EA technology and people in terms of the various aspects to achieve business goals." (MM1). "... all the executives need to input on what the needs and priorities are and then obviously IT needs to execute that and advise on the best way in how to solve that specific issue." (SM1). A benefit is a reduction of silos which results in reducing of duplication of systems. Business and IT are reliant on each other to achieve the respective goals. It is confirmed here that IT-business alignment is one of the most important benefits of EA.

Skills. Numerous projects have failed due to a lack of resources and skills [23]. The skills posed a barrier, resulting in many services having to be outsourced to external contractors leading to higher costs. "...we are outsourcing now because we don't have the skills which are fine because at the same time it might be expensive to have in-house..." (EM2). The department had an external resource for six months assisting with the enterprise architectural plan, but it took time for them to understand the business, resulting in the organization not receiving value for money. It was also mentioned by many participants that it would be best if the resource could be appointed internally and already understands how the organization operates.

4.4 Perceived and Tangible Benefits

The most important reason for the implementation of an architectural plan is to derive perceived benefits into tangible benefits [3]. The benefits of EA can be grouped into the following themes, managerial benefit, operational benefit, organizational benefit, strategic benefit and ICT benefit [2] as shown in Table 7.

The **managerial benefits** derived from literature such as business continuity, optimization, establishing and attaining organizational goals, project success and reducing of risks are confirmed. **Operational** benefits such as efficiency, performance, and standardization are confirmed. Therefore it can be said that an enterprise architectural plan has perceived and tangible operational benefits. The **organizational** benefits as derived from literature have also been confirmed in this study, as well as the **strategic** benefits. EA is used as a governance tool, enabling better controls. Because EA assists with better governance, this leads to less duplication of systems proving a better return on investment. The organization will no longer work in silos creating organizational integration. EA will provide a roadmap to the intended future state. The **ICT** benefits which EA provides are also confirmed in this study. It assists in reducing costs; new systems can be integrated into the current infrastructure leading to the reliability of systems. EA assists with the standardization of ICT systems, hardware, and software.

Table 7. Perceived benefits of EA.

		SM1	SM4	EM1	EM4	EM3	SM5	EM2	SM3	MM2	MM3	MM1	SM2	MM4	MM5
Managerial benefits	Business Continuity	X		X		X	X			X			X	X	
	Optimization	X	X	X	X	X	X			X			X	X	
	Organization Goals	X	X	X	X	X	X	X	X	X	X	X	X	X	X
	Project Success	X		X				X		X		X	X		
	Reduce Risk	X			X		X			X			X		
Operational Benefits	Efficiency	X	X	X	X	X	X	X		X	X	X	X	X	X
	Performance	X	X	X	X	X	X	X		X	X	X	X		X
	Standardization	X		X	X		X	X		X	X		X	X	
Organization-al Benefits	Communication	X	X	X	X	X	X	X	X	X	X	X	X	X	X
	Holistic View	X	X	X	X	X	X	X	X	X	X	X	X	X	X
Strategic Benefits	IT Business Alignment	X	X	X	X	X	X	X	X	X	X	X	X	X	X
	Decision Making	X	X	X	X	X	X	X	X	X	X	X	X	X	X
	Governance	X	X	X	X	X	X	X	X	X	X	X	X	X	X
	Increase ROI		X	X			X		X	X		X			X
	Organizational Integration	X	X	X	X	X	X	X	X	X	X	X	X	X	X
	Road Map	X	X	X	X	X	X	X	X	X		X			X
ICT Benefits	Cost Reduction		X	X	X		X	X	X	X	X	X	X		X
	Duplication		X	X		X	X	X	X	X	X		X		X
	Reliability	X		X	X		X			X	X	X	X		
	Standardization	X	X	X		X		X		X	X	X	X	X	

5 Conclusion and Future Research

This research revealed that EA is a requirement for government departments. The need for EA is driven by government policy and also seen as a need amongst staff. The perceived benefits are confirmed by all participants. The framework used for this study was the TOE framework, based on which the main themes were divided into technological, organizational and environmental factors. Additional factors emerged such as funding/budget, delayed IT consultation, the enforcement of governance and the tick in the box exercise. The comparative analysis confirmed the uncertainty and resistance to change in the EA adoption. The critical success factors act as enabling factors but if the factors are not present they hinder the adoption of EA. The aim of this study was to explain the reasons for the adoption of EA in the public sector. The TOE framework was used to group the main themes, then adding subthemes while leaving room for emerging themes. The organizational theme had the most factors influencing adoption, government regulation from the environmental aspect and skills from the technological aspect. The finding that emerged from the study was the perception or understanding of what EA entails. The different views by the participants resulted in unclear objectives but the benefits still outweighed the challenges.

One **limitation** of this research is that it took place in one national government department only. Due to the fact that the department did not have a dedicated enterprise architect within the organization, all participants answered the questions based on when the resource was part of the organization for a limited time and also basing the answers on the CIO of the department seen as the enterprise architect.

Future studies into the comparison around EA from a national department, a provincial department and local government using the newly found insights could be interesting to see if any new findings surface. It will be interesting to see how all the various departments manage to plan and implement this concept in order to abide by the regulation, harness the benefits and overcome its challenges.

References

1. Lankhorst, M.: Enterprise Architecture at Work. The Enterprise Engineering Series. Springer, Heidelberg (2009)
2. Jusuf, M.B., Kurnia, S.: Understanding the benefits and success factors of enterprise architecture. In: Proceedings of the 50th Hawaii International Conference on System Sciences (2017)
3. Schmidt, R., Möhring, M., Härting, R.-C., Reichstein, C., Zimmermann, A., Luceri, S.: Benefits of enterprise architecture management–insights from European experts. In: IFIP Working Conference on the Practice of Enterprise Modelling, pp. 223–236. Springer, Cham (2017)
4. Wan, H., Luo, X., Johansson, B., Chen, H.: Enterprise architecture benefits: the divergence between its desirability and realizability. In: 14th International Conference on Informatics and Semiotics in Organizations (ICISO2013, IFIP WG 8, 1 Working Conference). SciTePress (2013)
5. Seppänen, V.: From problems to critical success factors of enterprise architecture adoption. Jyväskylä studies in computing (2014)
6. Depietro, R., Wiarda, E., Fleischer, M.: The context for change: organization, technology and environment. Process. Technol. Innovation **199**, 151–175 (1990)
7. Hope, T.: The critical success factors of enterprise architecture (2015)
8. Dang, D.D., Pekkola, S.: Root Causes of Enterprise Architecture Problems in the Public Sector. In: PACIS 287 (2016)
9. Hope, T., Chew, E., Sharma, R.: The failure of success factors: lessons from success and failure cases of enterprise architecture implementation [best paper nominee]. In: Proceedings of the 2017 ACM SIGMIS Conference on Computers and People Research, pp. 21–27 (2017)
10. Niemi, E., Pekkola, S.: Using enterprise architecture artefacts in an organisation. Enterp. Inf. Syst. **11**, 313–338 (2017)
11. Tamm, T., Seddon, P.B., Shanks, G., Reynolds, P.: How does enterprise architecture add value to organisations? Commun. Assoc. Inf. Syst. **28**, 10 (2011)
12. Vallerand, J., Lapalme, J., Moïse, A.: Analysing enterprise architecture maturity models: a learning perspective. Enterp. Inf. Syst. **11**, 859–883 (2017)
13. Chuang, C.-H., van Loggerenberg, J.: Challenges facing enterprise architects: a South African perspective. In: 2010 43rd Hawaii International Conference on System Sciences, p. 10. IEEE (2010)
14. Hoti, E.: The technological, organizational and environmental framework of IS innovation adaption in small and medium enterprises. Evidence from research over the last 10 years. Int. J. Bus. Manag. **3**, 1–14 (2015)
15. Baker, J.: The technology–organization–environment framework. In: Information Systems Theory, pp. 231–245. Springer, New York (2012)
16. MacLennan, E., Van Belle, J.-P.: Factors affecting the organizational adoption of service-oriented architecture (SOA). IseB **12**, 71–100 (2014)

17. Jia, Q., Guo, Y., Barnes, S.J.: E2.0 post-adoption: extending the IS continuance model based on the technology-organization-environment framework (2016)
18. Saunders, M., Lewis, P., Thornhill, A.: Research methods for Business Students. Pearson Education Limited, London (2012)
19. Wisker, G.: The postgraduate research handbook: succeed with your MA, MPhil, EdD and PhD. Macmillan International Higher Education (2007)
20. Cassell, C., Symon, G.: Essential Guide to Qualitative Methods in Organizational Research. Sage, Thousand Oaks (2004)
21. State Information Technology Act (SITA). In: Education, D.o.B. (ed.), vol. 583, pp. 1–4. Government Gazette (2014)
22. Maramura, T.C., Thakhathi, D.: Analyzing the electronic governance policies and legislative frameworks on public service delivery in South Africa. J. Commun. **7**, 241–245 (2016)
23. Velumani, M.: Adoption of Agile Enterprise Architecture in Large Organization: A Case Study (2017)

Evaluating Market Pricing Competition with Game Theory Model

João Paulo Pereira[1,2](✉) and Murillo Ferreira[3]

[1] Instituto Politécnico de Bragança, Campus de Santa Apolónia, 5300-302 Bragança, Portugal
jprp@ipb.pt
[2] UNIAG (Applied Management Research Unit), Bragança, Portugal
[3] Instituto Politécnico de Bragança, 5300-253 Bragança, Portugal
murillo.hpf@gmail.com

Abstract. Recently in the literature, there have been many attempts to expand classic models of market competition analysis. Considering firms are competing globally against many different sellers over different markets, recent works proposed a model where it is possible to represent competition among companies where they compete against each other directly and indirectly, using a hypergraph to represent the competition structure. This document presents an attempt to demonstrate how the young and maturing networked price competition model, which allows finding the best price for the companies from the competition structure and market sizes, can be used in any case of study.

Keywords: Nash equilibrium · Bertrand network · Market competition

1 Introduction

Game theory has been applied in the economic scenario since its beginning, becoming one of the main mathematic tools used to analyze the market [1]. Game Theory has the special property of showing the best set of options or probabilities of each player, used mainly to predict social and economic behaviors [2].

Many authors have been trying to build a mathematical model that allows to represent competitions where one seller compete against multiple sellers at the same time (directly and indirectly). While made sense to represent only direct competitions in the past, nowadays firms are reacting in real time to price changes with online tools [3], such strategy has been named as Fast-Changing Web Prices [4]. A more robust model is needed to represent a competition where firms are directly and indirectly competing against each other, being recently developed and improved both for games where sellers choose prices [3, 5–8] and for games where sellers supply an amount of a good [9, 10], leaving the price to be decided by an auctioneer [11].

Although there are many works for the Networked version of the Cournot competition (where sellers supply quantities of a certain good rather than specifically ask for prices, like in the Bertrand competition), the Networked version of the Bertrand model is still young and under development. A thorough search of the relevant literature has not shown any study that has used the Bertrand Network model in a real scenario.

Á. Rocha et al. (Eds.): ICITS 2021, AISC 1330, pp. 603–613, 2021.
https://doi.org/10.1007/978-3-030-68285-9_56

2 Related Work

2.1 Game Theory

Game theory presents a mathematical model for decision-making that can be applied in the real world, firstly introduced in 1994 [1]. A game is defined as something composed by a set of players, they can be persons, objects, companies, everything capable to make decisions, simultaneously or sequential to other players. Each movement (or strategy) may have different outcomes, either good or bad. If each player wants to get the best out of the game, each of them must consider the set of options of all the other competitors. It is said that, if a player is always choosing the best option, it is said that the player is being rational [12].

There are many types of games, specially focused in the economic field. Reference [1] presents a case of study from an economic region of Taiwan. Considering local governments with different interests it is proposed a mathematical model for a cooperative game (every player tries to optimize the outcome of both at the same time) and a non-cooperative game, where each of them tries to selfishly optimize their own outcome. The authors concluded that a cooperation between them is very beneficial, ending with a suggestion of how they could cooperate. The usage of Game Theory to show that a cooperation is possible also appeared in other studies [13, 14]. Another study has also shown that merge of companies can be profitable [10].

Games can be applied either for external process (competition among firms, for example) or for internal process of a company. A study has been presented by [14] where it has been constructed a payoff matrix, representing the gains and losses of each department for each third-party company that could be chosen. At the end, the Nash Equilibrium has been found, concluding which outsource company should be hired. Game Theory can also be used to assist banks in a decision-making game: accept or decline loan requests [15]. The most interesting part of this study is that they point out the possibility to use Game Theory to enhance data-mining (analysis of a large dataset with the objective of extracting implicit information).

2.2 Existing Model

This warm-up example will be useful to understand the structure of the Bertrand Network. Consider the following competition among three vendors: firm f_1 sells devices to users that are interested in high-performance and has a captive market $\alpha_1 = 20$ that represents the set of buyers that will buy from firm f_1 no matter the price, as long it is not higher than 1 (loyal buyers). Firm f_2, on the other hand, sells devices to buyers that are interested in both high and low performance devices (participating in two different markets). Finally, firm f_3 are only interested to sell to buyers that are interested in low-performance devices. After this, buyers were categorized into three different markets. Firm f_1 competes for the market $\beta_{1,2} = 150$ against f_2, while firm f_3 competes for the market $\beta_{2,3} = 185$ against f_2. Neither f_2 nor f_3 has a set of loyal buyers, therefore, $\alpha_2 = \alpha_3 = 0$.

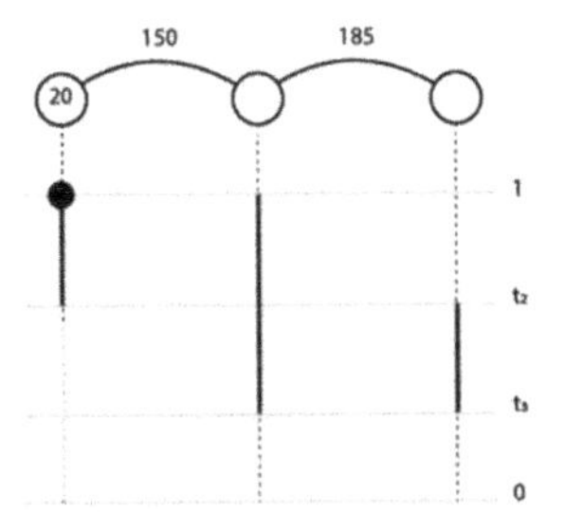

Fig. 1. Example of competition. Nodes represent firms, each hyperedge represents a market share. Captive markets are represented with a weight within the node, in which an empty node represents a captive market of size zero. Thick lines represent the pricing strategy of each firm that falls within their support, as a subset of [0, 1]. The black circle represents an atom at 1. It is said that when a firm has an atom at 1, firm chooses the price of 1 with a positive probability.

In the given example, f_1 faces a tradeoff between extracting the maximum surplus from its captive market by asking for the monopolized price of 1 and lowering the price to increase the likelihood of winning market share $\beta_{1,2}$. On the other hand, f_2 must decide for a price that, at the same time, will extract the maximum surplus from both shared markets. Figure 1 shows the competition network and the *equilibrium sketch* for the network. An *equilibrium sketch* is defined as a specification of the finitely bounded pricing range (support) for every seller i and the set of sellers that has an *atom* at 1. (Babaioff et al., 2013) have formally defined it in ***Definition 5.1*** and any support can be set, as well as any set of sellers with an atom at 1 can be chosen, but the results from the *equilibrium sketch* must satisfy the properties in ***Definition 5.2.***

From the results in Babaioff et al. (2013, pp. 7–8), the unique *Nash Equilibrium* (N. E.) for this topology is that each company must mix its pricing strategy by following its support (see Fig. 1) and randomize by following its Cumulative Distribution Function (CDF). To each company be indifferent in choosing any price that falls within its support, the utility level for each vendor must be the same by pricing at $x \in [t_m, t_n]$ (where $0 < t_m \le t_n \le 1$ are the boundary points of its support, for a network that has at least one firm with a nonzero captive market). Formally, the following must hold:

$$\begin{aligned} u_1(t_1) &= u_1(t_2) = u_1(x)\, \forall x \in [t_2, t_1] \\ u_2(t_1) &= u_2(t_2) = u_2(t_3) = u_2(x)\, \forall x \in [t_3, t_1] \\ u_3(t_2) &= u_3(t_3) = u_3(x)\, \forall x \in [t_3, t_2] \end{aligned} \tag{1}$$

By working with the equalities above, it is possible to find the CDF function that represents the probability of each firm to choose the price of at least x. The utility level of each firm is defined by Babaioff et al. (2013) as follows:

$$u_i(x, F_{-i}) = u_i(x) = x\left(\alpha_i + \sum_{j \in N(i)} \beta_{i,j}\overline{F}_j(x)\right) \tag{2}$$

Where N(i) is a set of neighbors of i, $\beta_{i,j}$ is the weight of the edge that connects i and j, α_i is the size of the captive market of i and $\overline{F}_j(x)$ represents the inverse CDF function (CDF for short), the mixed pricing strategy for firm j. The inverse CDF is a nonincreasing linear function in x^{-1} which can be explicitly found for every $x \in [0, 1]$ by working with equalities in (1). For a detailed description of how to find the explicit CDF function for every firm i (for the same graph), see Babaioff et al. (2013, pp. 7–8).

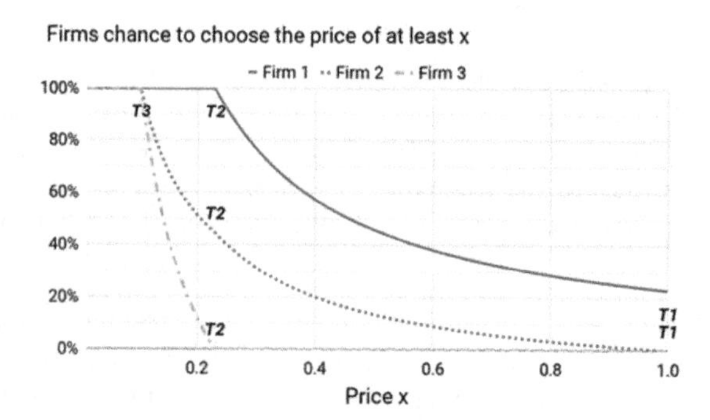

Fig. 2. Optimal Inverse Cumulative distribution function for each firm after working with the equalities defined in Eq. (1). This represents the expected behavior for each firm. For instance, Firm 1 has the chance of 100% of choosing the price of 0.2 for higher, which means that it is not expected to see Firm 1 asking for the price lower than 0.2.

For this example, by working with the equalities in (1), it has been found that $t_1 = 1$, $t_2 = 0.2294520548$, and $t_3 = 0.102739726$. Those values are the optimal pricing ranges for each firm i that allows mixing where none of firms can deviate to another CDF function (pricing strategy) or to another pricing ranges to get a better payoff. In Fig. 3, it is possible to see that f_2 is more likely to succeed in this scenario by having the highest utility level.

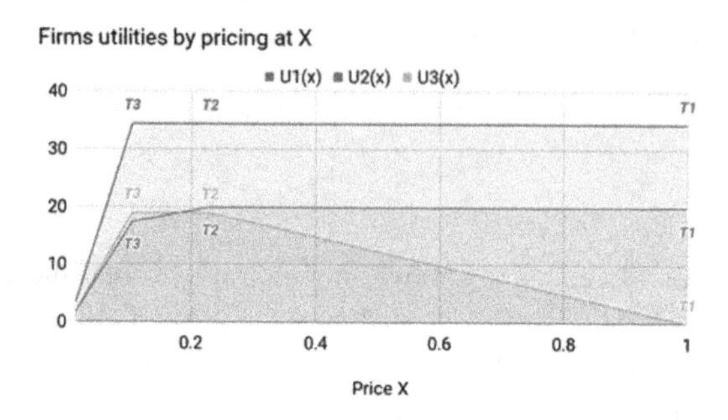

Fig. 3. Utility level for each company. Note that between the pricing range that fall within the support of each firm, the utility level is the same.

3 Applying the Model in a Reverse Manner

To verify if it is profitable for a firm to join an existing market, it is needed to apply this model to one existing network and then, simulate an entrance from some firm (see Fig. 3 for an illustrative example). Comparing the results will show information such as: Changes in pricing decision of each firm (reaction); Expected profit of each firm before and after the entrance of a firm in the network.

While the existing Bertrand Network model allows to go from the size of markets to the optimal ranging prices for each company, it is suggested to do the opposite in this study. If the distribution of buyers is not known, it will not be possible to repeat the calculations like the one briefly shown in the Sect. 2.

First, one must model the competition network by checking who is competing against whom and in which market. Then, figure out the *equilibrium sketch* of the built network because it is not known any algorithm that can give the equilibrium of any given network, although the *equilibrium sketch* is known for some graph structures (Babaioff et al. 2013). Also, a data set of prices is needed, so an information about prices over time is required to known how sellers are pricing. It is assumed that the companies are playing the price competition game rationally, i.e., are indeed following their best CDF to choose prices within the optimal pricing range. Recalling the Sect. 2, the same graph structure is modeled, but now the market sizes are assumed to be unknown and it will be shown that it is possible to get similar results from the original example by only knowing the prices that each firm was choosing.

3.1 Collecting Prices

Consider that prices were collected from day d_1 to d_n and let $P_{i,k,j}$ be the j-th price from the k-th day of the i-th firm. Let P_i be the non-decreasing sorted set of prices from the i-th firm. Finally, let P be the union of every P_i. As the existing model uses prices in $[0, 1]$, let the collected prices be mapped from $[0, M]$ to $[0, 1]$, where M is the highest price collected. Formally, let $f : P \rightarrow [0, 1]$. Recalling the graph presented in the Sect. 2, the *equilibrium sketch* defines that the seller f_1 is pricing in $[t_2, t_1]$, f_2 in $[t_3, t_1]$ and f_3 in $[t_3, t_2]$. Therefore, t_1 must be the greatest value in P, t_3 the minimum and let t_2 be the minimum value of f_1. Let $C_i(x)$ be the number of elements in P_i that are less than x. To calculate the probability of randomly choosing the value of x or higher from the data set, let $\overline{F}_i(x) = \frac{|P_i| - C_i(x)}{|P_i|}$.

3.2 Assumptions

Suppose that an analyst has collected a data set that shows:

- $t_1 = 1$ as it would be in any equilibrium (Babaioff et al. 2013);
- f_1 chooses the price of 1 with positive probability (has an *atom* at 1). Consider that after analyzing the data set, it has been found that the probability for f_1 to choose the price of t_1 or higher is $\overline{F}_1(t_1) = 0,23$, like in the original example (see Fig. 2);

- With similar reasoning, $\overline{F}_2(t_2) = 0,45$;
- $\overline{F}_1(x) = \overline{F}_2(x) = \overline{F}_3(x) = 1 \forall x \leq t_3$ since every firm is choosing prices higher than t_3;
- $\overline{F}_2(t_1) = \overline{F}_3(t_2) = 0$ by claiming the results from Babaioff et al. (2013, p. 9), ***Lemma 4.2*** that states: *"Fix any valid tie breaking rule. In any network and any equilibrium, no two sellers who share a market both have an atom at the same positive price"*;
- $\overline{F}_3(x) = 0 \forall x \geq t_2$ since t_2 is the higher price for f_3.

Note that the CDF for each company has been defined at the boundary points in their support, which would be possible by observing a data set.

3.3 Calculating the Market Share from Prices D CDFs

Now, the calculation must be started by defining the following equalities that in equilibrium must be satisfied (see Eq. (1)).

$$\begin{gathered} u_1(t_1) = u_1(t_2) \\ t_1 * \alpha_1 = t_2 * \left(\alpha_1 + \beta_{1,2} * \overline{F}_2(t_2)\right) \\ \frac{t_2}{t_1} = \frac{\alpha_1}{\alpha_1 + \beta_{1,2} * \overline{F}_2(t_2)} \end{gathered} \tag{3}$$

$$\begin{gathered} u_2(t_1) = u_2(t_2) \\ t_1\left(\alpha_2 + \beta_{1,2} * \overline{F}_1(t_1)\right) = t_2\left(\alpha_2 + \beta_{1,2}\right) \\ \frac{t_2}{t_1} = \frac{\alpha_2 + \beta_{1,2} * \overline{F}_1(t_1)}{\alpha_2 + \beta_{1,2}} \end{gathered} \tag{4}$$

$$\begin{gathered} u_2(t_2) = u_2(t_3) \\ t_2 * \left(\alpha_2 + \beta_{1,2}\right) = t_3 * \left(\alpha_2 + \beta_{1,2} + \beta_{2,3}\right) \\ \frac{t_3}{t_2} = \frac{\alpha_2 + \beta_{1,2}}{\alpha_2 + \beta_{1,2} + \beta_{2,3}} \end{gathered} \tag{5}$$

$$\begin{gathered} u_3(t_2) = u_3(t_3) \\ t_2 * \left(\alpha_3 + \beta_{2,3} * \overline{F}_2(t_2)\right) = t_3 * \left(\alpha_3 + \beta_{2,3}\right) \\ \frac{t_3}{t_2} = \frac{\alpha_3 + \beta_{2,3} * \overline{F}_2(t_2)}{\alpha_3 + \beta_{2,3}} \end{gathered} \tag{6}$$

To be able to calculate the proportion of each market, the following equalities will be used:

$$\alpha_1 = a * \beta_{1,2} \quad \alpha_2 = b * \beta_{1,2} \quad \alpha_3 = c * \beta_{1,2} \quad \beta_{2,3} = k * \beta_{1,2}$$

This will simplify the calculations since now it will be possible to find the ratio of each market over $\beta_{1,2}$ (chosen by convenience, note that every shared market will be greater than 0, otherwise that market share would not exist). From Eq. (3):

$$t_2 = \frac{a * \beta_{1,2}}{a * \beta_{1,2} + \beta_{1,2} * \overline{F}_2(t_2)} \qquad t_2 = \frac{a}{a + \overline{F}_2(t_2)}$$

$$a = t_2 * \left(a + \overline{F}_2(t_2)\right) \quad a - t_2 * a = t_2 * \overline{F}_2(t_2) \quad a = \frac{t_2 * \overline{F}_2(t_2)}{1 - t_2}$$

Since ‘a’ only uses constant values, from now on ‘a’ can be treated as a constant value. Proceeding with (4):

$$t_2 = \frac{b * \beta_{1,2} + \beta_{1,2} * \overline{F}_1(t_1)}{b * \beta_{1,2} + \beta_{1,2}} \qquad t_2 = \frac{b + \overline{F}_1(t_1)}{b + 1}$$

$$t_2 * b + t_2 - b = \overline{F}_1(t_1) \quad b * (t_2 - 1) + t_2 = \overline{F}_1(t_1) \quad b = \frac{\overline{F}_1(t_1) - t_2}{t_2 - 1}$$

From Eq. (5):

$$\frac{t_2}{t_3} = 1 + \frac{\beta_{2,3}}{\alpha_2 + \beta_{1,2}} \quad \frac{t_2}{t_3} = 1 + \frac{k}{b+1} \quad k = \left(\frac{t_2}{t_3} - 1\right) * (b + 1)$$

Finally, from Eq. (6):

$$\frac{t_3}{t_2} = \frac{c + k * \overline{F}_2(t_2)}{c + k} \qquad \frac{t_3}{t_2} * c + \frac{t_3}{t_2} * k - c = k * \overline{F}_2(t_2)$$

$$c * \left(\frac{t_3}{t_2} - 1\right) = k * \left(\overline{F}_2(t_2) - \frac{t_3}{t_2}\right) \qquad c = k * \frac{\overline{F}_2(t_2) - \frac{t_3}{t_2}}{\frac{t_3}{t_2} - 1}$$

This concludes the calculation and now it is possible to find the buyer distribution over the network that would lead to the same pricing range and CDF in the original example by just setting a value for $\beta_{1,2}$. Note that there are many possible values for $\beta_{1,2}$, and each of them will lead to different utility levels. By choosing $\beta_{1,2} = 150$, let see the value of the other markets that would be found by using the results that was calculated:

$$\alpha_1 = 20.16233766 \quad \alpha_2 = 0 \quad \alpha_3 = 0.0455 \quad \beta_{2,3} = 195$$

Note that those results are very close to the first example in Sect. 2. By calculating every market size as a ratio over one of them (in this case every market size was being calculated over $\beta_{1,2}$) it is possible verify, for T being the total of buyers that will be distributed over the structure:

$$\alpha_1 + \alpha_2 + \alpha_3 + \beta_{1,2} + \beta_{2,3} = T \quad a * \beta_{1,2} + b * \beta_{1,2} + c * \beta_{1,2} + k * \beta_{1,2} = T$$

$$\beta_{1,2} = \frac{T}{a + b + c + k}$$

Similar reasoning can be applied for other networks. It is worth noting that the price and CDF values have been rounded at the beginning of this section, thus the approximated results for each market share.

4 Expanded Bertrand Network

To show what is willing to achieve, consider the following Figure (Fig. 4).

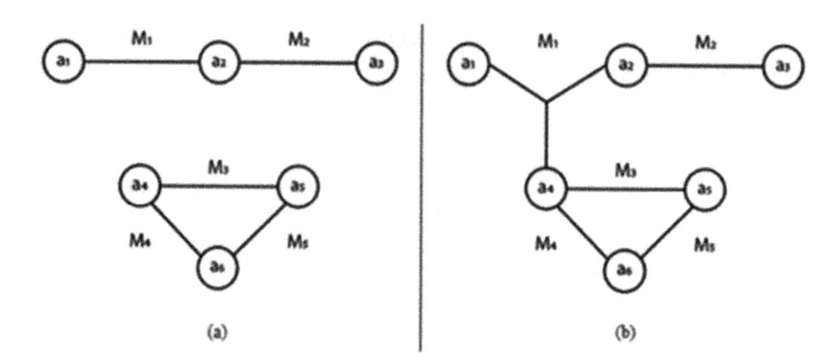

Fig. 4. Two connected components of the hypergraph in (a) representing two different market networks and (b) an expansion attempt of firm f_4 to join the market share M_1.

First, it is stated that any market which is being shared for more than two firms are called oligopoly shared market. The meaning of $N(\mathrm{i})$ has changed to be the set of shared markets that are being shared to exactly 2 vendors. Now the Eq. (2) is expanded to one form that considers the mix of duopolies and oligopolies:

$$u_i(x) = x * \left(\alpha_i + \sum_{k \in N^o(i)} \beta_k * \sigma_{-i,k}(x) + \sum_{j \in N(i)} \beta_{i,j} * \overline{F}_j(x) \right)$$

Where $\mathrm{N}^{\mathrm{o}}(\mathrm{i})$ is the set of the neighboring oligopoly shared markets of i, β_k is the size of the oligopoly shared market k and $\sigma_{-i,k}(x)$ represents some combination of the CDFs of all vendors that are participating in the oligopoly shared market k excluding firm i, as a subset of $[0, 1]$. Essentially $\sigma_{-i,k}(x)$ tries to represent the combination of the pricing strategy of all competitors of i in the oligopoly k.

That said, the attempt to define the aggregated function of CDFs for oligopoly markets is shown below. To let the utility function be linear (like in the original model), let the aggregated function be the average sum of all the CDFs. Consider any oligopoly k with n firms and let H(k, y) be a hash function that returns the index of the y-th firm in that oligopoly k, for $1 \leq \mathrm{y} \leq \mathrm{n}$:

$$\sigma_{-i,k}(x) = \frac{\sum_{y=1}^{n} \overline{F}_{H(y)}(x) - \overline{F}_{H(i)}(x)}{n-1}$$

For the sake of simplicity, k has been omitted from H(k, y). The belief that, in any oligopoly with at least 2 firms that has a captive market, ***Lemma 4.2*** from Babaioff et al. (2013, p. 9) does not apply is shown below.

Conjecture 1. Fix any network composed by one oligopoly shared market with at least 2 firms that has a positive captive market. There exist more than one firm that does choose 1 with positive probability.

Assume that ***Lemma 4.2*** does apply. Sort firms and label them from 1 to n such as $\alpha_1 \geq \alpha_2 > \alpha_3 \ldots \geq \alpha_n \geq 0$. Then, if f_1 has an atom at 1, the following holds:

$$\sigma_{-1}(1) = 0$$

Then:

$$\sigma_{-1}(1) = \frac{\sum_{j=2}^{n} \overline{F}_j(1)}{n-1} \quad \sum_{j=2}^{n} \overline{F}_j(1) = 0 \tag{8}$$

Which is true since none of the firms has an atom at 1. Therefore, for every $1 < i \leq n$:

$$\sigma_{-i}(1) = \frac{\overline{F}_1(1) + \sum_{j=2}^{n} \overline{F}_j(1) - \overline{F}_i(1)}{n-1} \quad \sigma_{-i}(1) = \frac{\overline{F}_1(1)}{n-1}$$

Thus, For every $1 < i \leq n$:

$$u_i(1) = \alpha_i + \beta * \sigma_{-i}(1) \Rightarrow u_i(1) = \alpha_i + \beta * \frac{\overline{F}_1(1)}{n-1}$$

For firm n there exist a minimum price t_2 that allows mixing, thus the following must hold:

$$u_n(1) = u_n(t_2) \qquad \alpha_n + \beta * \frac{\overline{F}_1(1)}{n-1} = t_2 * \left(\alpha_n + \beta * \frac{\overline{F}_1(t_2)}{n-1}\right)$$

$$\alpha_n + \beta * \frac{\overline{F}_1(1)}{n-1} = t_2 * \left(\alpha_n + \beta * \frac{1}{n-1}\right) \qquad t_2 = \frac{\alpha_n + \beta * \frac{\overline{F}_1(1)}{n-1}}{\alpha_n + \beta * \frac{1}{n-1}}$$

Every firm that will mix must be mixing at the same interval $[t_2, 1]$, so none of them can deviate and ask for a price that will allow him to win the shared market with probability one (Babaioff et al. 2013; Guzmán 2011).

$$t_2 = \frac{\alpha_i + \beta * \frac{\overline{F}_1(1)}{n-1}}{\alpha_i + \beta * \frac{1}{n-1}} \quad \frac{\alpha_n + \beta * \frac{\overline{F}_1(1)}{n-1}}{\alpha_n + * \frac{1}{n-1}} = t_2 = \frac{\alpha_i + \beta * \frac{\overline{F}_1(1)}{n-1}}{\alpha_i + \beta * \frac{1}{n-1}}$$

This is only possible if $a_n = a_i$ for every i, contradicting the assumption that each firm may have different sizes of captive markets. This suggests that another firm might be choosing the price of 1 (which would lead to $\sigma_{-i}(1) > 0$ for another $i > 1$), or some firm is choosing the price of 1 with negative probability, from the equality in Eq. (8). This concludes the belief that Lemma 4.2 must be reworked, in addition to another prof by Guzmán (2011) that asserted that the duopoly model is the solution of the oligopoly model, but with a different aggregated function.

5 Conclusion

This work has presented a way to analyze market competitions using the maturing Bertrand Networked model. In the case of study used in this work, finding the properties shown demanded a lot of work, mainly because there is not many works available using this competition model that could have helped, therefore many properties was being found during many attempts to find the size of the captive markets and market shares that could have motivated the sellers to use price on the way it was observed.

Although the results are not precisely, it provides bounds in which can be used to have more information about an existing competition, such as possible number of buyers in the market and how much buyers would be willing to pay after have observed the behavior of the companies.

References

1. Chen, L., Yu, N., Su, Y. : The application of game theory in the Hercynian economic development. In: 2014 International Conference on Management Science Engineering 21th Annual Conference Proceedings (2014)
2. Dixit, A.K., Nalebuff, B.J.: The Art of Strategy: A Game Theorist's Guide to Success in Business and Life, W. W. Norton & Company, New York (2010)
3. Babaioff, M., Nisan, N., Paes Leme, R.: Price competition in online combinatorial markets. In: Proceedings of the 23rd International Conference on World Wide Web, Seoul, Korea (2014)
4. Angwin, J., Mattioli, D. : Coming Soon: Toilet Paper Priced Like Airline Tickets. https://www.wsj.com/articles/SB10000872396390444914904577617333130724846. Acesso em 4 Aug 2020
5. C.L. a. o. Guzmán: Price competition on network, Banco de México (2011)
6. Babaioff, M., Blumrosen, L., Nisan, N.: Networks of complements. In: The 43rd International Colloquium on Automata, Languages and Programming (ICALP) (2016)
7. Babaioff, M., Lucier, B., Nisan, N.: Bertrand networks. In: Proceedings of the Fourteenth ACM Conference on Electronic Commerce, June 2013
8. Ferreira, M., Pereira, J. : A demonstration of an application of the Bertrand Network: guessing the distribution of buyers within the market. In: 18th Conference of the Portuguese Association of Information Systems, Santarém (2018)
9. Abolhassani, M., Bateni, M.H., Hajiaghayi, M., Mahini, H., Sawant, A.: Network cournot competition. In: Lecture Notes in Computer Science (2014)
10. Bimpikis, K., Ehsani, S., Ilkiliç, R. : Cournot competition in networked markets. In: Proceedings of the Fifteenth ACM Conference on Economics and Computation, Palo Alto, California, USA (2014)
11. Kreps, D.M.K., Scheinkman, J.A.: Quantity precommitment and bertrand competition yield Cournot outcomes. Bell J. Econ. **14**, 326–337 (1983)
12. Bruce, L.M.: Game theory applied to big data analytics in geosciences and remote sensing. In: 2013 IEEE International Geoscience and Remote Sensing Symposium – IGARSS (2013)
13. Budler, M., Trkman, P.: The role of game theory in the development of business models in supply chains. In: 2017 IEEE Technology Engineering Management Conference (TEMSCON), Santa Clara, USA (2017)

14. Xia, Y. : The research on internal game of enterprise based on game theory and analytic hierarchy process. In: 2016 International Conference on Logistics, Informatics and Service Sciences (LISS) (2016)
15. Alskheliwi, T., Jim, C., Lateef, K., Penn, S., Salem, A.: Applying game theory rules to enhance decision support systems in credit and financial applications. In: 2014 Computer Games: AI, Animation, Mobile, Multimedia, Educational and Serious Games (CGAMES), Louisville, USA (2014)

Author Index

Á. Rocha et al. (Eds.): ICITS 2021, AISC 1330, pp. 615–617, 2021.
https://doi.org/10.1007/978-3-030-68285-9

CPSIA information can be obtained
at www.ICGtesting.com
Printed in the USA
LVHW081918080221
678722LV00001B/8

9 783030 682842